T0261619

Life of the Past   James O. Farlow, editor

Indiana University Press   Bloomington & Indianapolis

CAMBRIAN
OCEAN WORLD

ANCIENT SEA LIFE OF NORTH AMERICA

JOHN FOSTER

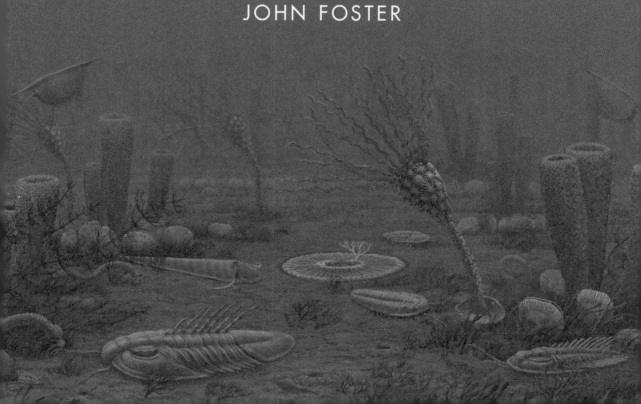

This book is a publication of

Indiana University Press
Office of Scholarly Publishing
Herman B Wells Library 350
1320 East 10th Street
Bloomington, Indiana 47405 USA

iupress.indiana.edu

Telephone orders   800-842-6796
Fax orders   812-855-7931

© 2014 by John Foster

All rights reserved

No part of this book may be reproduced
or utilized in any form or by any means,
electronic or mechanical, including
photocopying and recording, or by
any information storage and retrieval
system, without permission in writing
from the publisher. The Association of
American University Presses' Resolution on
Permissions constitutes the only exception
to this prohibition.

⊛ The paper used in this publication
meets the minimum requirements of
the American National Standard for
Information Sciences – Permanence of
Paper for Printed Library Materials, ANSI
Z39.48-1992.

*Manufactured in the
United States of America*

*Library of Congress
Cataloging-in-Publication Data*

Foster, John Russell, [date].
  Cambrian ocean world : ancient sea
life of North America / John Foster.
    pages cm. – (Life of the past)
    Includes bibliographical
references and index.
    ISBN 978-0-253-01182-4 (cl : alk.
paper) – ISBN 978-0-253-01188-6 (eb)
  1. Marine animals, Fossil – North America.
  2. Animals, Fossil – North America.
  3. Paleontology – North America.
  4. Paleontology – Cambrian. I. Title.
    QE766.F67 2014
    562.097 – dc23
                                    2013038554

3 4 5 6 7   26 25 24 23 22

For ReBecca and Ruby, who have endured my many mental and physical absences with good humor and went along on a number of the expeditions (snow, rain, heat, or decent weather); and for my father, Russ Foster, who instilled in me a love of the desert without even realizing it. See the influence a little landsailing can have?

Old Ocean, none knoweth thy story;
Man cannot thy secrets unfold.

MARTHA LAVINIA HOFFMAN

Just what happened on Earth about 542 million
years ago is still a bit of a mystery.

PATRICIA VICKERS-RICH

Contents

C

Preface

SOMEWHERE IN NEARLY EVERY STATE OR PROVINCE IN NORTH America are Cambrian rocks recording the history of life in one of the most important time periods in Earth history. In Vermont, Washington, Virginia, Wisconsin, California, Alberta, British Columbia, Sonora, New Mexico; in Nevada, Utah, Colorado, Pennsylvania, South Dakota, Oklahoma, Idaho, Texas, Alabama, Wyoming—the list goes on. Rocks of Cambrian age are almost everywhere. And in them we find some of the earliest complex animals to appear on Earth. The diversification of animals in the Cambrian is astounding.

By any human measure, the Cambrian period was an incredibly long time ago, but in terms of the story of our planet it only began after most of Earth history so far had already unfolded. It would be another 315 million years before mammals or dinosaurs would appear on the scene—or nearly five times as many years as have passed since the dinosaurs (other than birds) disappeared and left the world to the mammals, in our chauvinistic view. From the perspective of the most diverse major animal group of the Cambrian (the arthropods), however, the world of the Cambrian was theirs then, and still is now, as today the number of insect species alone is nearly one million. The only major loss to them since the Cambrian is that of the trilobites, the proverbial fossils of the Paleozoic era. Even trilobites outstripped other famous fossil groups in terms of diversity. We know of around a thousand species of dinosaurs from the fossil record; modern mammals number around six thousand species; birds are all the way up around ten thousand. The lowly trilobites? Twenty thousand species! Although they may have inflated numbers due to high preservation potential, as fossils trilobites own the Paleozoic.

The Cambrian period on Earth might as well have been another planet, compared with what we are used to today. This is the story of a different time and place. The time is incredibly distant; the place, not so.

The Cambrian period was obviously a very long time ago, but why was it important? It was nowhere near "early" in Earth history; 80 percent of that history occurred before the first years of the Cambrian. The world of the Cambrian represented a time quite different from most before it in terms of environmental conditions. This parallels differences in the biotas. But the Cambrian was probably most important because it was, quite simply, the birth of our modern biological world. Whereas the previous 3 billion years were occupied almost exclusively by microbes, and only shortly before had multicellular animals appeared, almost all the modern groups of animals that we know today trace their origins to the time interval between 542 and 488 million years ago. Crabs, lobsters, insects, and horseshoe crabs? Their ancestors were there in a myriad of arthropods. Lions and tigers and bears? The first members of our phylum of chordates and vertebrates appeared during the Cambrian. Corals and jellyfish? Their ancestors were there. Worms of all kinds? Those, too.

The Cambrian radiation, or explosion as it has been called, has been argued about for ages. No less in recent years. More, in fact. Was the speciation rate for animals actually higher during the Cambrian than at any point since? Was the explosion an artifact of preservation? Were there more phyla (body-plan groups) of animals than today? Were phyla "weeded down" or did the modern phyla only appear then and continue? What does all this mean for the mechanisms of evolution? We will review some of these debates, but mostly we will concentrate on the Cambrian and the subaqueous Garden of Eden of modern animal diversity.

Plenty has been written about the Cambrian previously, and much of it is deservedly about the Burgess Shale and its spectacular window to the Cambrian biota. We will see that here, too. But we will also visit a lot of other places with fossil records that contribute to that picture as well, each in its own way. I hope that the picture painted here leaves you with an impression of just how ubiquitous Cambrian rocks and fossils are and how important this period is to the history of life on Earth. I will not assume all readers have backgrounds in geology or biology, and so chapter 1 will introduce some aspects of the Cambrian period and biology in general, and chapter 2 will introduce some key concepts in geology through a trip into the Grand Canyon. The rest of the book is a journey. We will travel forward in time, first through the Precambrian in chapter 3, and then through the Cambrian itself in chapters 4–8. We will visit the Early Cambrian in chapter 4, the early Middle Cambrian in chapter 5, the Burgess Shale in chapter 6, and the late Middle Cambrian in chapter 7. Chapter 8 takes us through the Late Cambrian. In each chapter, we will see some localities that exemplify each of the time stages, and the animal groups preserved at them will be presented in an order dictated in part by their abundance or preservation at each site. The journey through time dictates the order of sites we visit, and those sites dictate the order in which we discuss individual animal groups. Chapter 9 takes a look at the data from sites throughout the Cambrian, focusing in large part on the Burgess Shale due simply to its almost unbeatable record among North American sites. Chapter 10 is a brief summary of where the animals of the Cambrian have gone in the millennia since.

Scattered through the chapters are several boxes containing profiles of current researchers working on Cambrian issues. This is a small sampling of the modest army of people worldwide who study this time period, and it is, of course, not close to being wholly representative. But I hope it will give readers a better view into how we came to know what we do about the Cambrian and its fossils.

All specimens illustrated in this book were collected by Museum of Western Colorado crews and photographed by the author, except where noted. A number of specimens were photographed by the author in the field. Specimens collected by other institutions or individuals, photographed by others, or in other institutions' collections are noted.

Acknowledgments

THIS PROJECT IS PART OF A RETURN TO MY PALEONTOLOGICAL FIRST love, a renewed interest in fieldwork and research on the topic that led me into paleontology in the first place. Although paleontology of all types was of interest to me as a young undergraduate, it was the Cambrian of the southern Great Basin that really got me hooked. But after that initial undergraduate work, graduate school led me down a long, winding path of Mesozoic vertebrate paleoecology that I continue to tread today. After more than 15 years of this, however, I knew I needed to expand my research horizons and work on some additional (and totally different) project so as not to start plodding the same ground repeatedly – for my sake and everyone else's. The fact that I had started out working in the Cambrian in the Mojave Desert, and that I had returned regularly over the years just for the fun of it, made the answer obvious. I needed to get back to where I started, to the rocks and fossils that had never really been left behind. The Cambrian called again. It was time to stop picking around and start trying to answer some of the questions that I had begun to ask as I worked in the Mojave. Thus began my return to the Cambrian. I had a lot of catching up to do because a lot had happened in the ensuing years, and the self-imposed crash course was intense. The idea for this book came a little later, on a cross-country drive a few years ago. But it has only intensified the learning curve.

Because of the journey begun as I just outlined, and because the Cambrian is 54 million years of appearances of a whole range of animal phyla and dozens of geologic units (just in North America), it is a massive task to attempt to bring this together in a manageable way. I have been pleasantly surprised by how much help I have been willingly given by a whole range of researchers who have assisted in ways small and large, but all important. As a "lost son" of the Cambrian who has only relatively recently returned to the fold of the most important period in Earth history, I have appreciated this help beyond my ability to fully express to those that have assisted. It is almost cliché to say, but it is true: I could not have done this without them.

First off, I have to thank my undergraduate advisors, Don Prothero and Jim Sadd, whose fault it is I ever got addicted to the Cambrian in the first place. In recent years, trilobite specialists Stew Hollingsworth and Fred Sundberg have answered many questions, shared specimens, and provided many references for a number of projects and have served as de facto postgraduate advisors. A number of others helped at various stages of this and related projects by answering one or a whole host of questions; generously providing reprints, figures, or data; introducing me to field areas; donating specimens; and offering advice on collecting and preparation of some more challenging specimens. These are, in no particular order, Dave Rudkin, Mark Webster, Eben Rose, John Taylor, John Sibbick, Dave Liddell, Simon Conway Morris, Bruce Lieberman, Steve Rowland, Derek Briggs, Fredrik Terfelt, Per Ahlberg, Rachel Wood, Brigitte Schoenemann, Paul Strother, James Hagadorn, Graham Budd, Malgo Moczydlowska, Xingliang Zhang, Jean-Bernard Caron, Talia Karim, Loren Babcock, Joachim Haug, Marilyn Kooser, Kevin Peterson, Paul Myrow, Krista Brundridge, Bob Gaines, Laura Wilson, Brian Pedder, Martin Smith, Andrew Milner, Josh Bonde, J. W. Schopf, Arvid Aase, Mark Fahrenbach, Norm Brown, Joe Collette, Melissa Hicks, Val and Glade Gunther, Mike Cuggy, Diego García-Bellido, Andrew Knoll, Harry Mutvei, and Dave Comfort. Becky Bernal of the Colorado Mesa University library was instrumental in obtaining many references, and never met an interlibrary loan she could not track down. Jill Hardesty (KU) and Tom Jorstad (Smithsonian) helped with some of the borrowed figures. The artists John Agnew, Terry McKee, Matt Celeskey, and Karen Foster-Wells really brought the Cambrian to life with their work. Joe Fandrich and Pete Bucknam provided equipment and expertise for the thin-section photo-micrographs. Thanks to the Museum of Western Colorado field crews (and associates from the Raymond Alf Museum) that collected many of the data and specimens seen here. These included ReBecca Hunt-Foster, Ray Bley, Zeb Miracle, Mike Perry, Darrell Bay, Tom Lawrence, Andy Farke, and Amy Jackson-Ayala. Thanks also to the Monterey Bay Aquarium and to the folks at Western Slope Aquatics in Grand Junction. Access to museum collections was generously provided by Colleen Hyde (GRCA); Dave and Janet Gillette (MNA); Bruce Lieberman (KU); Pat Holroyd and Mark Wilson (UCMP); Andy Farke (RAM); Marilyn Kooser (UCR); Oklahoma Museum of Natural History; Mary Stecheson (LACM); Mark Florence, Finnegan Marsh, Doug Erwin, and Conrad Labandeira (Smithsonian Institution, NMNH); and Logan Ivy (DMNS). Special thanks to those who agreed to be profiled for the boxes: Malgo Moczydlowska, Diego García-Bellido, Eben Rose, Xingliang Zhang, James Hagadorn, Andrew Knoll, Kevin Peterson, and Bob Gaines. Special thanks also to Bob Sloan, Jim Farlow, Jenna Whittaker, Nancy Lightfoot, Dawn Ollila, Jamison Cockerham, Tony Brewer, and everyone else at Indiana University Press for their hard work making this project happen. Finally, thanks to the Bureau of Land Management, National Park Service, USDA Forest Service, and Parks Canada. These land management agencies are charged with protecting the land on which most of the localities mentioned in this book are found, and without their work these Cambrian sites would not be preserved for future generations nor would the data about the Cambrian world be available to researchers and, ultimately, the general public.

Abbreviations

DMNS Denver Museum of Nature and Science, Denver, Colorado

GRCA Grand Canyon National Park Museum, South Rim, Arizona

KUMIP University of Kansas, Lawrence, Kansas

LACMIP Natural History Museum of Los Angeles
County, Los Angeles, California

MNA Museum of Northern Arizona, Flagstaff, Arizona

MWC Museum of Western Colorado, Fruita, Colorado

RAM Raymond Alf Museum, Claremont, California

UCMP University of California Museum of
Paleontology, Berkeley, California

UCR University of California, Riverside, Department of Geology
Paleontology Collections, Riverside, California

USNM National Museum of Natural History (NMNH),
Smithsonian Institution, Washington, DC

1.1. Adam Sedgwick, the man who named the Cambrian period, in an 1867 photograph by William Farren. Sedgwick was 82 at the time and had named the Cambrian 32 years earlier.

Courtesy of the Sedgwick Museum of Earth Sciences, University of Cambridge. Reproduced with permission.

Natural Mystic:
An Introduction to the Cambrian

IMAGINE A TROPICAL MORNING ON THE OCEAN. THE AIR IS comfortably warm and moist, but it is not muggy or hot; scattered clouds are slightly pink with the last colors of dawn as the sun glares orange, low on the eastern horizon. The ocean on which you are floating is deep blue, and the surface waves are only a few feet high. From all appearances, it could be offshore Hawaii. Pitching lightly on a large, inflatable dinghy, we prepare technical diving gear and extra tanks and notice that there is no land in sight; we are probably at least several tens of miles from the nearest land beyond the horizon to the south, but how far we can't tell. We are going deep, and this type of diving requires special training, equipment, and experience. As the sun climbs in the sky we notice that the sea surface color transitions to a lighter blue away to the south and west.

Ducking under the waves in our diving gear we see a world of medium blue all around and below us. There are no fish in sight. We aim toward the deep and begin kicking our way down into the azure world below. As the pressure increases to nearly 100 pounds per square inch on our bodies and puts pressure on our lungs and ribs, the water around us is turning darker and darker blue. The sea around us now is midnight blue and we can barely see; we have reached the edge of available light at nearly 91 m (300 ft.) down. Knowing that we are near the bottom of the ocean in this area we pull out our dive lamp and turn it on. The bottom soon comes into view as a flat, muddy plain, and suddenly we see movement as indistinct animals shrink into burrows in the bottom sediments. Hovering 2 m (6 ft.) above the seafloor and watching the scene below us, we notice an inch-long, segmented arthropod moving slowly like a sowbug across the bottom muds. There is a scattering of tube- and cup-shaped red and purple sponges; a few tufts of green algae; and one or two short, orange, stalked organisms that look a little like spindly flowers. Suddenly a few feet below us a silvery shape flutters into our lamplight, startling us and reminding us that this ocean is an exotically distant one. The animal is about 45 cm (1.5 ft.) long and is shaped like a segmented halibut. A rounded head contains stalked eyes, followed posteriorly by multiple flap-edged segments that ripple consecutively in waves along each side as on a hovering cuttlefish, and the tail consists of pairs of elongate blades reaching up and out from a central segment. The animal moves along smoothly, thanks to the waving of the lateral flaps. Most strangely, the animal has two spined appendages protruding down from under the head. This odd inhabitant of the deep disappears out of view of our light and we are left looking at each other is amazement. *What was that?* Moving slowly over the bottom we also notice in the water column, just below us, small shrimp-like arthropods with nearly translucent double-valved shells over their backs. Eventually our dive watches indicate our bottom time is almost over and it is time to begin the slow process of staged ascent to the surface.

During our dive, our ocean world consisted of animals familiar and foreign, but there was a very good reason that we saw no fish beneath the waves nor birds above them. As much as we can see strange things while diving the deep ocean even today, our bizarre destination on this trip was that of the Cambrian ocean world of offshore Utah about 505 million years ago. We saw an area that was probably 113 km (70 mi.) offshore at the time, in 91 m (300 ft.) of water, and whose bottom muds became a relatively thin unit of shale exposed now in the desert of the Great Basin, one that yields fossils of trilobites and other animals by the thousands. The world we visited in our imaginations is not only based on real evidence, but it is also important to remember that this ocean world was a very real place, one that really existed at a particular time and in a particular place. Although we can see that world only in mental imaginings now, its tactile reality was once every bit as solid as the fossils and rocks that record its existence today. It was the Earth's reality at a time that proved to be critical to our planet's evolutionary history. The Cambrian ocean world was one of the most fascinating in Earth history, and it has a tremendous story to tell. In many ways, it made the world of today.

Indeed, one might argue that the modern biological world began nearly 540 million years ago. The story we are beginning here paints a picture of the birth of that world in a journey through the 54 million years of the Cambrian period, illuminating the creatures and environments that populated such a wondrous time and set the stage for most of the lineages of modern biology. One could argue any number of beginnings of our current **biota,** from just a few tens of thousands of years ago back to the time before the Cambrian–it all depends on one's definition of "modern." Some animals around during the Cambrian resemble modern species, but many do not, and certainly by my claim in the first sentence of this paragraph I would have to argue not only that today's biological world began during the Cambrian but by extension so did that of the Carboniferous or Triassic periods, for example. I do not claim that species and ecosystems were the same–only that nearly all the major animal groups we know today (and which were present for much of relatively recent Earth history) originated or diversified into recognizable forms during the Cambrian period. The Cambrian set the cast of characters in place–act one, you might say. This appearance of animals, particularly its rapidity, is the phenomenon you have likely heard of as the Cambrian "explosion." It is the nature of that "explosion" that has intrigued, mystified, and lured paleontologists for generations.

Life was not new at this time. It had been around for a while, and so had the Earth. What was new was that suddenly, just before the Cambrian, and after untold millennia of life on the planet consisting of microorganisms, algae, and not much more, animals appeared and during the Cambrian diversified in form and function, setting the stage for the modern biological world and leaving the previous six-sevenths of Earth history in the metaphorical dust.

The Cambrian period (and also to some degree—as we have been discovering in recent decades—the period right before it, the Ediacaran) was a time of outright biological revolution, the likes of which had never been seen before and haven't been since (or at least it hadn't been seen before except perhaps since the origin of life itself). No extinction and recovery event since the Cambrian has resulted in the kind of biotic expansion and ecological and morphological invention that occurred during the Ediacaran–Cambrian times—not the recovery from the end-Cretaceous extinction, not that after the Permian–Triassic boundary. The origins of animal groups during the Ediacaran and Cambrian and the explosive diversification of these groups throughout the Cambrian turned the Precambrian world on its head and set the stage for everything we know today. But as an introduction, let's take a look now just at the Cambrian world itself—and what a world it was!

Nelson Horatio Darton started walking from the Santa Fe train tracks at Siam Siding in the hot and very dry Mojave Desert of California; he had nearly a mile of flat alluvial sand to cross, and wound his way through creosote bushes the entire time. What lay ahead of him was a ridge several hundred feet high known at the time as Iron Mountain. He found there rock types that indicated they had been laid down in shallow seas, but—better yet—he found fossils that could indicate the rocks' age.

Next Stop, Mojave

Darton was the son of a civil engineer who had helped construct the Civil War ironclad *Monitor*. Because Darton rather enthusiastically took to the math and science his father taught him beyond his class work, he quickly tired of school and by the age of 13 had become a chemical apprentice. Through collecting minerals he became interested in geology and eventually ended up as an essentially self-taught geologist working for the United States Geological Survey, along with the likes of John Wesley Powell and Charles D. Walcott. Darton was an excellent mapper of geologic formations and structures, and he spent years in the field mapping and naming rocks all over the United States—but particularly in the Rocky Mountains.

Darton was in his early 40s when he came through the Mojave Desert of southeastern California on the train, riding a line that cut across the heart of the desert between Needles and Barstow. He studied the rocks at Iron Mountain, measured their thickness, and collected some fossils—but had, as he said, limited opportunities for a more detailed study of the site. Darton showed the fossils to Walcott, the director of the U.S. Geological Survey, who identified them as probably Middle Cambrian in age. Darton had discovered a previously unrecognized outcrop of Cambrian rocks in the Mojave and published a short note on the site in 1907.[1]

What Darton probably could not have predicted was that he had first recognized Cambrian outcrops that would ultimately yield thousands and thousands of fossils of Cambrian animals of various kinds and that the locality, now known as the Marble Mountains, would eventually be known as one of a handful of sites found to preserve the less well known elements of Cambrian **faunas,** species without hard skeletons. Sites that preserve rare species, or soft tissues, or animals in great abundance are particularly prized in paleontology. Although he didn't realize it, Darton had found in these Cambrian rocks a special type of fossil locality known as a **lagerstätte,** and his site would be worked by paleontologists, geologists, amateur collectors, students, and folks on vacation for now a hundred years running.

Of course, Darton's Marble Mountains site was not the first identification of Cambrian rocks in the region, as he himself acknowledged in his paper. And Darton's site was neither the first nor nearly the best of the Cambrian lagerstätten that have been found. Walcott and others had worked in the area earlier and found Cambrian rocks to the north near Death Valley, near Las Vegas (one can only imagine what Las Vegas was like in 1907), and of course along the length of the Grand Canyon. But Darton had shown that Cambrian rocks and important fossils could be found even in a tiny range of hills in the middle of the forgotten center of the dry Mojave, where most of the surrounding ranges consist of mangled igneous rocks of much younger age and where the intervening valleys are nothing but sand and gravel. Darton's little piece of the Cambrian is not a Rosetta Stone of Cambrian paleontology, but it is yet another piece of the puzzle that many paleontologists have been working on putting together for a long time. And still we struggle.

First Steps: The Early Stratigraphers

So, how did we end up with the Cambrian as a period and a concept in the first place? The story begins more than 300 years ago with early "geologists" (most were hobbyists – it wasn't really a job in the eighteenth century) trying to make sense of the rock record they saw around them. No one could fail to notice the layering of the sedimentary rocks seen in various countries, but most believed in the permanence of rocks. They had always been there in the shapes and form we now see. It was a Danish physician and priest named Niels Stensen (sometimes spelled Steensen, also known by the Latinized "Steno") who realized that the presence of fossils in rocks suggested that the particles of such rocks had once been soft and had hardened around the fossil. The origin of sedimentary rocks included a loose, soft stage of deposition and a later hardening of the matrix. With this came the realization that tilted sedimentary rocks had not only been hardened but also had been bent upward out of their original orientation. Based on these observations Stensen elucidated three absolutely cornerstone concepts for the understanding of sedimentary rocks (and of extrusive igneous rocks): (1) such rocks were originally laid down as flat-lying, horizontal beds; (2) the beds were originally continuous up to the point that they were bound by a valley or basin wall, for example (i.e., discontinuities in a bed not caused by basin edges were likely due to subsequent erosion); and (3) unless overturned or otherwise deformed by faults or folding, overlying beds are younger than underlying beds (i.e., beds pile up one at a time starting at the bottom). Thanks to Stensen's observations, all who work in sedimentary rocks since his time instinctively think in order from bottom to top. With these three concepts in mind one can begin to understand the sedimentary rocks in the landscape.

One of the first of the early hobby geologists to recognize the significance of the structure of the layering and some of its complexities was James Hutton, an English farmer with lots of time for walking the hills and wondering about the rocks he saw. What he did see helped him realize that the only way to understand the past processes, including the formation of rocks, was to assume that processes we observe today also operated in ancient times and likely factored in to the origins of what he was seeing. This is the concept of **uniformitarianism,** and Hutton was among the first to use it to interpret rock formation. Previously, those interested in geology had mostly assumed that most rocks were in fact made by unusual processes such as large floods and sea level fluctuations. Among the phenomena Hutton observed that suggested to him the cyclicity of rock formation and thus the antiquity of the Earth was the **angular unconformity.** In such a situation one sees underlying beds standing on end and cut off on a horizontal surface, on which is lying another set of overlying beds. The angle is that between the sets of beds and the unconformity is the separating surface between them, which often represents a significant period of time. Using Stensen's three laws, Hutton was able to determine that at these angular unconformities he saw ancient sediments that had piled up in layers, been solidified into rock, uplifted to a high angle, and eroded off, at which point sediments piled up again in layer after layer and solidified into rock also. In order for Hutton to see the angular unconformity at Siccar Point on the Scottish coast, then, the entire package had to again be uplifted and eroded a second time. What Hutton eventually wrote about—and what it took a long time for his contemporaries to accept—was that the erosion of rock, formation into transported sediments, their deposition, their solidification into rock, their uplift, and their erosion, is a cycle that has repeated itself over and over throughout all of Earth history, and that understanding this is a key to both the processes of sedimentary geology and the antiquity of the Earth.

It is the fossils, as we will see, that are the key to defining all geologic periods, including the Cambrian, and the geologist who laid the foundation for the modern understanding of biostratigraphy was a man who actually made his living working on rocks. **Biostratigraphy** involves using the fossils in rocks to tie together in time, or correlate, layers of rock that are separated by some distance, and William Smith was the engineer/surveyor who first convinced the English geological community that each layer of rock had its own distinctive group of fossils that allowed him to do just that. Robert Hooke had suggested that this might be possible years earlier, but Hooke studied fossils in detail, often with a microscope, and didn't have Smith's depth of practical field experience, so it was not until Smith put the idea into practice that Hooke was proven correct. Smith worked on many construction projects across the countryside that involved digging into fresh rock, and he noticed that the groups of animals found in the lowest layers were quite different from those of the upper layers. Stensen's law told Smith that this was not random. The rocks were ordered oldest to youngest from bottom to top. Unlike the rock types, which often repeated themselves in a stack of rocks, the fossil faunas did not. Particular associations of fossils always occurred low, for example. This is the principle of **faunal succession,** and it is a key to the understanding of Earth history and to the definition of geologic periods. Smith had shown that the faunas of England had changed through time and that he could use this fact to help him determine the relative age of rocks in widely separated parts of the country. Meanwhile, in France, naturalist and zoologist Georges Cuvier was noticing the same phenomenon on the Continent. Of course, Smith and Cuvier were also among the first to show that faunas on Earth in general had changed through time – and it took decades before scientists figured out how such changes may have occurred.

So with Smith's and Cuvier's example other geologists took the faunal succession concept and ran with it, eventually determining that most of the "recent" rock record (as we will see) can be divided up into what geologists call **periods** of Earth history, based on distinctive faunas of each. The period is just one rank of several divisions of time. Above it, and inclusive of several periods, is the **era;** below it, and what a period is divided into, are the **epochs.** It is important to point out that the period is a unit of the time scale; the name for the layers of rock equivalent to a period is a **system.** Thus, the Cambrian period is represented by rocks of the Cambrian system. One is time and the other rocks. (It may be a little confusing at first, but we geologists don't invent seemingly redundant words for no reason – in the interest of clear communication we like to have very specific words with very specific meanings.) What are now well-known periods of Earth history began to be named as geologists described very distinctive faunas from around Europe. The Jurassic, age of giant reptiles, was named in 1799 based on faunas in the rocks of the Jura Mountains in Switzerland. The Cretaceous and Carboniferous were named in 1822 based on marine chalks in northern France and coal-bearing deposits in the British Isles, respectively.[2] Most of the periods were described based on fossils of sea creatures from rocks deposited at the bottoms of shallow oceans. The Jurassic, for example, although famous for gigantic dinosaurs like *Brachiosaurus* and others, was not designated on the faunal succession of these giant terrestrial reptiles but on that of the invertebrate animals in the marine rocks.[3]

The Cambrian period was born in 1835. In that year, Adam Sedgwick (fig. 1.1) named the period after the Roman name for Wales (*Cambria*) based on rocks he had studied in the area for several years. Sedgwick was the son of an Anglican vicar from Yorkshire. He was born in 1785 and attended Trinity College at Cambridge University, and by the time he was 33 he was a professor of geology at the university despite never having been formally trained in the science. He took to the field with enthusiasm, however, and was soon out studying England's rocks and making contributions to stratigraphy building on Smith's. He was quite popular as a lecturer in class and opened his classes to women—a rather progressive move for the time. During his career he held several important posts within the university and in geological societies, met the queen, and continued lecturing into his 80s. In 1831 Sedgwick was headed out to do some geology field work in Wales and chose a recent Cambridge graduate named Charles Darwin as his field assistant. The two would continue to correspond for years.

It was during the summer field seasons in Wales during the early 1830s that Sedgwick and another geologist named Roderick Murchison began noticing rocks that seemed to contain fossils rather different from the oldest ones they had seen farther east. They worked together for several years on the Welsh rocks, Murchison taking the slightly higher ones and Sedgwick those at the base. In 1835 they presented their findings jointly and named two new periods of the geologic time scale: the Silurian was named by Murchison and the older Cambrian was named by Sedgwick. Unfortunately, whereas Murchison defined the Silurian based on its fossils, Sedgwick for some reason described the Cambrian based on the nature of its rocks—in Wales. As we have discussed briefly, although fossils can be unique to a particular time, rocks of different ages can appear quite similar and rocks of the same age can look quite different depending on where you are—thus, Murchison's Silurian was recognized elsewhere in Europe rather quickly (being based on fossils as it was), but Sedgwick's Cambrian was initially difficult to distinguish outside Wales. It took further study of the fossils of the Cambrian and Silurian to fix this problem. Because of the differences in definition the boundary between the Cambrian and the Silurian bounced all over the place—Murchison thought stratigraphically low-occurring fossils looked a lot like his and must be Silurian, and Sedgwick thought the opposite based on rocks. Unfortunately, part of the process of straightening out this mess involved Sedgwick and Murchison, friends going in, getting into a bit of a dispute over the matter. Eventually fossils of the Silurian and the Cambrian were found to be different, and when the fossil occurrences were truly refined it turned out there was a third period, between the two, named the Ordovician by Charles Lapworth—but not until 1879.[4] Cambrian rocks were soon recognized elsewhere besides Wales—and now they seem to be everywhere.

The geologic time scale continued to develop over the years, and soon there were a dozen periods defined. After the Cambrian period, within the Paleozoic era, came Lapworth's Ordovician, then Murchison's Silurian, and then in order came the Devonian (which Sedgwick and Murchison had described together earlier), the Mississippian and Pennsylvanian (together the Carboniferous), and the Permian (fig. 1.2). The Mesozoic era (Age of Dinosaurs and other spectacular reptiles on land and in the sea) was split into the Triassic, Jurassic, and Cretaceous, and the Cenozoic Age of Mammals was split into the Tertiary and the Quaternary. Nowadays there are fourteen periods in the geologic time scale defined by fossils; the Ediacaran has been added before the Cambrian, and the Tertiary has been replaced by the Paleogene and Neogene. All others are the same. In addition there are now nine additional periods named that fall before the Ediacaran, each defined not by fossils, which are extremely rare and tiny in such old rocks, but by numeric ages. These nine go back to 2500 million years ago!

But once the Cambrian was an official period, back in 1835, and even after its rocks were beginning to be recognized officially all over the world, this was not the end of the controversy. In fact, it's never really ended. The trick now was to define exactly what the Cambrian period (time) and system (rocks) are. The beginning and ending of the Cambrian have been agreed upon; the beginning of the period is defined by the appearance of a particular type of **trace fossil,** and the end of the Cambrian (or more correctly the beginning of the next period, the Ordovician) is defined by the first appearance of a particular species of a type of fossil called a **conodont.**[5] It is the modern subdivision of the Cambrian that is still in progress – the project was officially begun by the International Commission on Stratigraphy in 1977. Let me begin with a little background.

As I mentioned earlier, the timescale subdivisions known as periods are divided into epochs which are in turn divided into **ages.** The rock sequence equivalents of these are systems, **series,** and **stages.** Generally periods are divided into Early, Middle, and Late epochs. (When referring to rocks we call these Lower, Middle, and Upper.) For years, the Cambrian was also divided into Early, Middle, and Late epochs. Recently, however, the Cambrian has been divided into four epochs/series and a total of ten ages/stages, and Lower, Middle, and Upper have been in declining use.[6] Some of these ten ages/stages have been named, others haven't. The International Subcommission on Cambrian Stratigraphy has also been working on designating Global Boundary Stratotype Sections and Points (**GSSPS**) for each of the stage boundaries. These are the so-called golden spikes (although in practice they are not necessarily either one) – permanent markers affixed to the outcrops marking first appearances of fossils that define the beginnings of a system (or stage) right on the rock at its stratotype section. The top and bottom of the Cambrian are defined, as well as three age/stage bases. The bases of the remaining six Cambrian

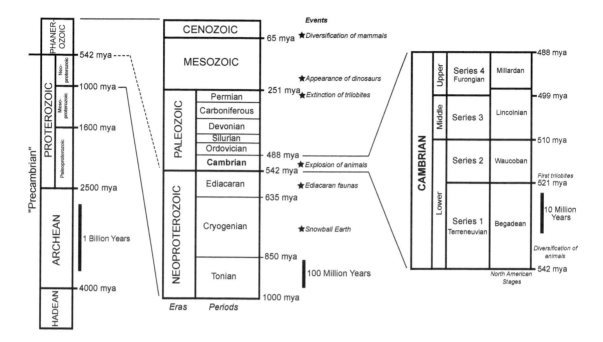

1.2. Geologic Time Scale showing position and subdivision of the Cambrian. On the left is the time scale of all of Earth history showing subdivisions of Precambrian time. Center shows subdivisions of Phanerozoic (542 mya to modern times) and Neoproterozoic with major events in geologic and biologic history marked with stars. On the right is the Cambrian Period and subdivisions. Also see the scale bar for each section. Mya = million years ago.

ages/stages are not yet agreed upon. Only two of the four Cambrian series have been named. These stratotype sections need to be accessible, not likely to erode or be built over or otherwise destroyed, and they must be representative. The commission is international and the potential sections numerous and globally dispersed. So coming to agreements takes a while. Part of the problem here is that unlike the other systems of rocks, that of the Cambrian suffers from a comparatively spotty fossil record, especially in its older levels. Deciding on what fossil to define a stage base is a problematic first step. Then there is the issue that many trilobites, on which a great number of the stage bases are defined, can be very localized rather than globally distributed, making internationally traceable base definitions difficult.

Despite the shift toward a four-part subdivision of the Cambrian, I will still use Lower, Middle, and Upper as well as Early, Middle, and Late in this book. The new system is still in progress and such terms as Series 3 or Stage 4 will likely have little meaning to most readers. Be aware, however, that my "Early Cambrian," for example, is really in modern use the informal "early Cambrian," which officially comprises the Terreneuvian and the yet-unnamed Series 2. We will use the more refined subdivisions, too, when it is helpful. But enough with the disclaimers.

The international Cambrian system subdivisions currently stand as shown in figure 1.2. The Cambrian is divided into four series: the oldest, which consists of Cambrian time before the trilobites appeared, is the aforementioned Terreneuvian; the youngest, roughly equivalent to the traditional Upper Cambrian, is the Furongian; and the middle two are still known as Series 2 and Series 3. These latter two are, respectively, the early days of trilobites in the later, traditional Lower Cambrian, and roughly the Middle Cambrian equivalent. They have not yet been named but are due to be soon—or at least eventually. Each series is divided into two or three stages. These have been numbered 1 through 10 from bottom to top. Sixty percent of the stages still need to be named.[7]

Again, the above international stratigraphy is the overall way to describe the age of various Cambrian units, and it is still a work in progress. Within North America, we can also use a system of series and stages, set up for that continent only, to discuss relative ages of localities. Because Cambrian faunas can tend to be endemic to continental areas, each continent has a system of series and stages unique to its region. The stages are defined by boundaries marking distinct trilobite faunal turnovers (extinctions followed by new faunas), and these are identifiable over large areas. The utility of these North American stage names is that they are more easily defined for workers in this region than are the equivalent global-scale units, and they are very useful within a region. The Cambrian in North America is divided as shown in figure 1.2.[8] The early part of the Cambrian that is devoid of trilobites is the Begadean (after a river in Canada), the early part of trilobite times is the Waucoban (a name first used by C. D. Walcott long ago), the approximate Middle Cambrian is the Lincolnian (after Lincoln County, Nevada), and the approximate Upper Cambrian is the Millardan (Millard County, Utah). The Begadean is undivided due to its lack of trilobites. The Waucoban is subdivided into the older Montezuman stage and the Dyeran stage (see chapter 4); the former is characterized by the oldest trilobites in North America, forms such as *Fallotaspis*, and is named for a mountain range in Nevada, and the latter is named after the town of Dyer in Nevada, and is characterized by younger but still relatively primitive trilobites like *Olenellus*. The Lincolnian is subdivided into the Delamaran and Marjuman stages. These are characterized by kochaspid ptychopariid, zacanthoidid, and dolichometopid trilobites for the Delamaran, which is typified by the Pioche Formation in Nevada, and by ehmaniellid, marjumid, asaphiscid, and menonomiid trilobites for the Marjuman. Finally, the Millardan contains the Steptoean and the last stage of the Cambrian, the Sunwaptan. These are defined by the first appearance of the trilobite species *Coosella perplexa* for the Steptoean and *Irvingella major* for the Sunwaptan.[9]

So what do we know about the age of the Cambrian? Estimates change now and then, but currently the best data indicate that the period began 542 million years ago (plus or minus 1.0 million years). The best estimates for the end of Cambrian time put it at 488.3 million years ago (plus or minus 1.7 million years). That error estimate, basically meaning give or take a period of time nearly as long as the Pleistocene, is in fact an incredibly small percentage of the total estimated time—the precision on these estimated numeric dates for geologic periods is impressive.

People often ask how geologists come up with the numeric age estimates of fossils, and it is frequently believed that fossils themselves are dated. This usually isn't the case, except in more recent material.

Generally, the age of older rocks and fossils is estimated by radiometric dating of minerals in rocks containing the fossils (or as near above or below as possible). Because the rates at which certain elements change from one isotope to another are constant and have been experimentally determined, we can use these isotopes to estimate the age of the elements in the minerals of the rocks. Some rock-dating techniques rely on rather different factors but isotope ratios are a common form of dating. The most important thing you need is a mineral that was formed at the time you are trying to date. For example, dating grains in a sedimentary rock doesn't help because the grains that make up that rock were eroded from rock that is often many millions of years older than the beds you are trying to date (sometimes multiple generations). You can date igneous rocks such as basalts because these were cooled from molten magma and thus formed their mineral crystals as the rock formed. What to do about sedimentary rocks then? This is very important since we usually are trying to tell what age fossils are, and fossils are almost always found only in sedimentary rocks. Luckily for us, there are volcanoes. The ash from a volcano is usually made of crystallized minerals formed contemporaneously with the sedimentary rocks in which it is often found. Alternatively, lava from a volcano can cool as a layer among sedimentary beds. As molten rock is blown (or flows) out of a volcano and cools and crystallizes, it settles in a layer within regular sedimentary units several miles up to several hundred miles away. Sometimes individual new grains will get mixed in with sedimentary rocks like shallow ocean mudstone or lake deposits. In either case, because these tiny volcanic minerals were formed at the same time as the deposits in which they get buried, we can use them to date the rocks immediately above and below them even though we can't date the sedimentary rocks or fossils themselves.

So how does the radiometric dating of the ash crystal work? Remember that as soon as the mineral crystal formed, as it was erupting from the volcano, for example, it formed comprising several elements, and some of these elements may have consisted of unstable isotopes. For example, radioactive decay of uranium's isotope U-238 causes atoms of this element to lose 32 protons and neutrons, changing the weight and atomic number of each atom such that it becomes lead-206 (Pb-206). The rate at which this decay takes place is constant, and by comparing the relative amounts of U-238 and Pb-206 we can tell how long it has been since that crystal formed. There are also other elemental isotopes (with other known decay rates) that can be used for such age estimates, including potassium, carbon, and U-235.[10] The rate is usually expressed as the number of years required to convert half the amount from one isotope to the other: the half-life. And the element will not run through its total mass in just two half-lives; i.e., the amount of an element will be down to 50% of the original amount in one half-life, to 25% after two, and to 12.5% after three . . . and on from there. Thus, if, for example, we know that the half-life of a particular element's isotope is one million years and if we have measured that the element in our rock crystal accounts for just 3.125% of the mass now, we know that it has been stabilizing for five half-lives and then is 5 million years old (assuming we know that it was 100% of the mass initially). If the percentage had been 50/50, we would know that it had been one half-life and the rock was 1 million years old. This is how radiometric dating of rocks works, and it has been used to date igneous material within the sedimentary units in many Cambrian and Precambrian sections around the world. This is how we know the ages of the rocks we work with.[11]

The geologic timescale as it stands today is divided into four eons: the Hadean (one of the great names in geologic timescale history, in my opinion, even though it is informal), the Archean, the Proterozoic, and the Phanerozoic (fig. 1.2). The Phanerozoic is the only one of these with which most people have any familiarity; it contains the Age of Mammals, the Age of Dinosaurs, and the Paleozoic, time of the trilobites. In fact, most geologists are more familiar with details of the Phanerozoic than any of the other three eons as well. The Hadean eon includes the time from the formation of the earth until about 4000 million years ago. The Archean consists of the time between then and 2500 million years ago, and the Proterozoic consists of then up to 542 million years ago. In some sections of this book we will refer to the Proterozoic and the eons before that collectively and informally as the Precambrian. The Phanerozoic then includes the last 540+ million years and contains the Paleozoic, Mesozoic, and Cenozoic eras. The Cambrian period is the first period of the Paleozoic, and thus accounts for the first few tens of millions of years of the Phanerozoic.

The period boundaries, as we have seen, were recognized starting hundreds of years ago based on turnover of one fossil assemblage to another at each of the boundaries; these have been refined over the years and the chronological dates have been added as the boundary rocks have been able to be radiometrically dated. It is important to keep in mind that the **formations** of rocks (mappable units that are assigned names) are independent of both the period and system boundary definitions (and the dates assigned to them) and may cross those boundaries. Thus the Wood Canyon Formation in Death Valley straddles the Proterozoic–Cambrian boundary, and the Bright Angel Shale in Nevada and Arizona crosses the Lower–Middle Cambrian boundary.

The Paleozoic era of the Phanerozoic contains seven periods in North America. From oldest to youngest these are the Cambrian, Ordovician, Silurian, Devonian, Mississippian, Pennsylvanian, and Permian (fig. 1.2). The Pennsylvanian and Mississippian together comprise the Carboniferous, which is named after all the coal found in those rocks. So the Cambrian period kicks off not only the Paleozoic era but also the Phanerozoic eon. Adam Sedgwick's rock system from Wales, what the Romans called *Cambria*, is the beginning of the "old life" era and the "large life" eon — and it would soon be recognized as containing evidence of the biggest diversification of life in Earth history.

Many analogies have been used to describe just how long geologic time is. One popular version is the 24-hour clock in which humans appear in the last few seconds; another is a calendar year in which the dinosaurs (other than birds) disappear around December 26. I've never quite been able to relate to these all that well, however. I suspect in part this is because in the calendar example, for instance, each day is about 12.6 million years. I can relate to a day, but not to that many years—other than as a geologist who, like all of our kind, can throw around tens of millions of years as if they were playing cards. Such periods of time just don't faze us in the context of our work, but we can no more relate to that amount of time personally than anyone else. In the clock example each second is still equivalent to more than 53,000 years—again, I lack an ability to connect with that. I prefer analogies that put both aspects of the comparison into units I can relate to.

I once compared traveling back along a geologic timeline to a road trip in which one drove back in time through the ages, passing one year with every 0.04 in (1 mm) traveled—at normal highway speed we would be pulling up to the very *end* of the Cambrian period after five long hours of driving at 60 mph. (To travel through the Cambrian back to its beginning would take another half hour.) Perhaps this time we should slow down. In order to get a more personal idea of how long ago the Cambrian period was, and indeed all of geological time, let's take a walk along a timeline back to the Cambrian. Our timeline will be set up along the equator, and we will be walking east from Quito, Ecuador. We will need to climb mountains and hack our way through a lot of jungles along this rather gigantic geologic timeline, but most importantly we will need to swim across many hundreds of miles of ocean! We will be walking and swimming a straight line along the equator and we won't stop for anything, not even rest, as we will be walking or swimming 24 hours a day, 7 days a week—sort of a Great Geologic Timeline Biathlon. At a pace of one 3-foot step (or stroke) every second equaling 10 years along the time scale, our journey will take us back a millennium approximately every minute and a half. With each step or stroke we will travel back in time a full decade—an entire set of fads, fashions, and bad clothes in the space of a second. Although we will be creeping along our timeline at about 2 mph (~3.2 kph) we will walk or swim past time representing 219,000 sunrises in every minute of our travels.

As we start off from Quito, we'd still be in town less than half a mile later when we passed the time of the building of the Fourth Dynasty pyramids at Giza in Egypt. Within 17 minutes of walking we'd be back to the end of the last ice age, but it would take us two days to reach the beginning of the Pleistocene 165 km (102 mi.) away on the east slope of the Andes. Two and one-half full months of round-the-clock walking and swimming through the Age of Mammals would bring us to the time of the extinction of non-avian dinosaurs at the end of the Cretaceous period—and we'd be in the middle of the Atlantic Ocean! Eventually we would hit the beach in Gabon and switch back to walking, still east, through the forests of Congo. As we topped out on Mount Kenya in eastern Africa, we would be approaching the Jurassic–Cretaceous boundary. Our swim across the Indian Ocean would be largely through the Jurassic period and its time of giant dinosaurs like *Apatosaurus*. Fully eight and one-half months into the journey we would reach western Borneo (12,000+ miles from our starting point) and would be back to the time of origin of both dinosaurs and mammals in the Triassic period. More than five months of (mostly) swimming later, we would reach the time of the first terrestrial vertebrates in the Devonian period, and we would be in the middle of the Pacific Ocean south of Hawaii (almost 23,000 miles into the trek). Finally, almost a year and a half after we started walking and swimming nonstop, we would approach Quito again from the west and would be nearing our goal of the Ordovician–Cambrian boundary, ten years passing with each splashing stroke. All the way around the world at the equator and our objective is finally in sight! We crawl back on land, keep heading east, and eventually pass through town, waving to those who'd seen us off 18 months ago (and who had probably nearly forgotten we'd set off). We would then continue to the east slope of the Andes where the Ordovician–Cambrian boundary finally awaits us. We have been traveling 565 days (13,560 hours) over 44,918 km (27,727 mi.) and have taken nearly 50 million steps or strokes, passing nearly 50 million decades along the way–all the way around the world and then some, just to get *back* to the Cambrian along an equatorial geologic time scale on which every 10 cm (4 in.) is one calendar year!

The Cambrian period was nearly 54 million years long; nearly as much time separates us today from the last dinosaurs. Walking our timeline through the Cambrian now, back to its beginning, will take us just over two months and we will travel more than 3000 miles through this first period of the Paleozoic era. Our first stop along the way is the Upper Cambrian deposits of the Deadwood Formation of South Dakota, the Lodore Formation of Utah and Colorado, and Colorado's Dotsero and Sawatch formations; this would take us roughly two days to get to once we had crossed into the Cambrian. Another 17 days later we would be in the middle of the northern Amazon basin (again) and would be passing the times of the Burgess Shale deposits of eastern British Columbia, the Marjum Formation and Wheeler Shale of western Utah, the Langston Formation of Utah and Idaho, and the Bright Angel Formation of Grand Canyon. Almost a week and a half later and still in the Amazon we would pass over the Early–Middle Cambrian boundary, the time of the Pioche Formation in Nevada. And on we would slog until eventually we would be swimming once again, just off the coast of South America, where we would pass the time of the Chengjiang fauna of China and the faunas of the Latham Shale and lower Carrara Formation in California; this would be approximately 625 miles of walking and swimming the timeline since we'd passed the Pioche Formation at the Early–Middle Cambrian boundary (i.e., about 11 million years before the boundary). Another week would go by before we would reach the time of the appearance of the first trilobites still well into the Early Cambrian. Finally, after another 682 miles (two weeks) of swimming we would reach the beginning of the Cambrian period (in the middle of the Atlantic) and would pass into the Precambrian. Our journey would be complete.

Exhausted, we would stop and float on the surface for a long time, relaxing. Perhaps then we would crawl into our chase boat and collapse. We would *not* want to continue along the timeline through the entire Precambrian. As appealing as traveling back through the Proterozoic, Archean, and Hadean sounds, to do so would take us nearly 13 more years—about 9.2 more laps around the globe! In fact, during our journey of 20 months of walking and swimming to the Cambrian we've only made it about 12% back through Earth history.

The Wild World of the Cambrian

Lost time travelers landing in the Cambrian world might be forgiven for thinking they had accidentally landed on another planet. The continents we know today were in what would be to us strange orientations, and they also were in strange places. Some were dismembered in a sense, with small parts attached instead to what are now other continents. Parts of Europe and Newfoundland were joined together as part of a mini-continent called Avalonia. Florida was attached to South America and Africa. Now tropical land was then polar (Africa); in North America west was north; there were no animals living on land (that we know of; some ventured onto it temporarily); and although there probably were microbes on terrestrial rocks, and, as we will see, possibly algae and mosses, there were few if any land plants. Either our time travelers would be very confused or they'd naturally assume they were nowhere near Earth. In addition to the strange geography, the atmosphere was of a slightly different composition, the Moon was somewhat closer to our planet, and the days were shorter.

Paleogeography

During the Cambrian, North America did not extend from the edge of the tropics up to the Arctic Circle; it was instead a tropical continent on the equator. And North America was rotated approximately 90 degrees clockwise relative to its current position, so that the modern western half of the continent faced north and the eastern half was to the south. During the Early and Middle Cambrian central North America was just south of the equator and it moved paleo-northward so that it was straddling the equator by the Late Cambrian. North America was joined together with small elements of the modern European continent (Scotland, some of Ireland, and a fragment of Russia) in the ancient continent called **Laurentia** (fig. 1.3). Most of the northern hemisphere was ocean during the Cambrian. The north shore of Laurentia (today's western North America) faced what is called the Panthalassic Ocean; waves lapping on the south shore of Laurentia (modern eastern North America) were generated by wind blowing over the Iapetus Ocean, which was open for thousands of

1.3. Global paleogeography during the Cambrian. (A) Hemisphere with Laurentia. (B) Opposite hemisphere.

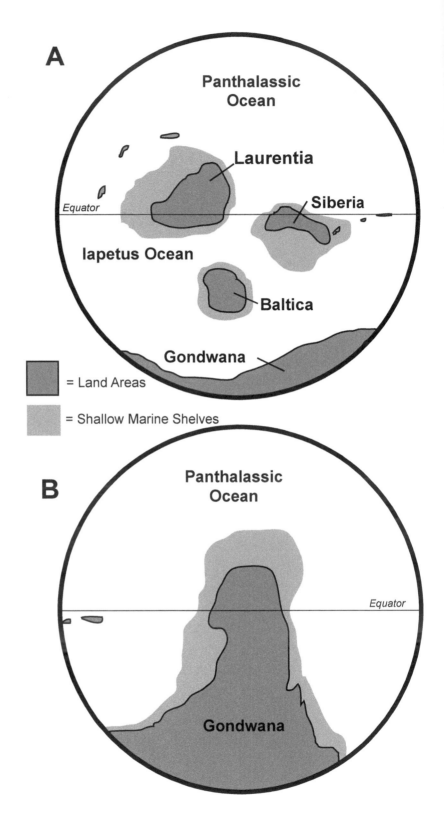

A

Panthalassic Ocean

Laurentia

Siberia

Equator

Iapetus Ocean

Baltica

Gondwana

☐ = Land Areas

☐ = Shallow Marine Shelves

B

Panthalassic Ocean

Equator

Gondwana

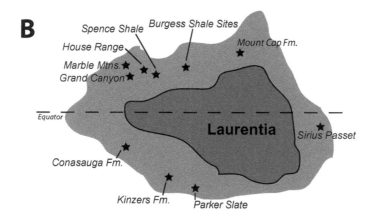

1.4. Geography of the Cambrian in North America. (A) Locations of some of the sites visited in this book. (B) Locations of sites relative to the Cambrian paleogeography of Laurentia. Dark gray is land area; light gray is continental shelves.

miles to the south before coming to the shores of the polar supercontinent of Gondwana. Gondwana contained four of the seven modern continents (Africa, South America, Australia, and Antarctica) plus smaller elements of other continents. Florida, parts of New England, and Nova Scotia, for example, were part of Gondwana. About midway between Laurentia and Gondwana, in the middle of the Iapetus Ocean, was the continent Baltica, which consisted of most of current northern Europe, including Scandinavia and much of European Russia.[12] Laurentian fossil sites found in North America today were scattered mostly on the continental shelves in shallow-marine water around the ancient continent during the Cambrian (fig. 1.4).

Although we describe the continents as having strange positions back in the Cambrian, the landmasses had in fact been engaging in a bit of plate tectonic square dancing for probably several billion years and had never stopped. And after the Cambrian they of course kept going. What we see in the Cambrian is simply where each continent was at the moment—mid-dance, so to speak.

Plate Tectonics and Paleomagnetics

In order to understand why Cambrian geography was so different, we must first understand a little about how the Earth works internally. The following is a short background on Earth structure and plate tectonics, the geologic theory that explains so much of the evidence—and allows us to determine that during Cambrian time the modern continents were scattered all over the globe like so many puzzle pieces. It started as a crazy idea in a world where continents were thought to be stationary pillars of rock, dutifully recording history, and they did record—only we eventually determined that the history they recorded was far more interesting than we'd suspected.

Although the Moon has moonquakes, and although other rocky planets are not entirely motionless, compared to the Earth all these other members of the solar system are asleep in orbit. The Earth pumps out heat, gas, and cubic miles of new rock in volcanoes. It rocks the oceans and continents with sometimes large earthquakes. Its molten core moves in fluid swirls. Each of these is important. The new rock forms new land, and often formerly new land is consumed in **subduction zones**—land sucked back down into the Earth by the movement that causes earthquakes. The gases released from volcanoes through Earth history have affected the chemistry of our oceans and atmosphere. Even life, as we will see, has had profound effects on these aspects of our planet. All of the geologic activity happens because much of the center of the Earth is relatively soft and even molten and fluid.

A quick review of Earth's structure will help explain the bizarre geography of the Cambrian and other times deep in Earth history. The Earth consists of three basic onion-like layers: the crust on the surface, the mantle in the middle, and the core in the center (fig. 1.5). The core is about 7030 km (4340 mi.) in total diameter and consists of an inner solid iron center about 2203 km (1360 mi.) across surrounded by an "onion" layer of liquid metal about 2414 km (1490 mi.) thick (likely iron as well). The mostly solid mantle (the next "onion" layer up) is about 2916 km (1800 mi.) thick and contains rocks much darker and heavier than in most of the Earth's surface, but it also has rocks with lots of silicon and oxygen, elements that make the mantle less dense than the core. The crust, the rocks we see day to day, is just 5–65 km (3–40 mi.) thick and consists of rocks even less dense than the mantle, due to their incorporating even more silicon, oxygen, and other light elements.

The crust consists of oceanic crust (about 3 miles thick) under ocean basins and continental crust (about 22–40 miles thick) where our continents and shallow seas are. The crust is divided up into tectonic plates that move around the surface of the Earth through geologic time. Some tectonic plates are mostly oceanic crust, others mostly continental crust. Oceanic crust is also denser than continental crust so that when two such plates collide the oceanic sinks below the continental. Also divided up into the tectonic plates along with the crust is the uppermost layer of the mantle, known as the **lithosphere.** It is this cooler and harder shell of the mantle that moves with the crust over the softer, more easily deformed underlying mantle (the **asthenosphere**).

We understand that tectonic plates move, but what causes them to move? We are not really sure. Slow motion flow of the rock in the asthenosphere (part of the mantle) may help facilitate the movement of the lithosphere and continental and oceanic crust, but how this may happen is unclear. This mantle convection is thought to result from the heat of the planet's core, similar to the convection currents in nearly boiling water. These mantle convection currents may pull the tectonic plates around from below, the eruption of new plate material may push the plates from their edges, and "pull" from the sinking of cold, oceanic plates may drive movement as well; we don't yet know the exact cause of the plate movement, just how they move.

As plates move around, they can do one of three things in contact with their neighbors: move away from each other, collide with each other, or move laterally past each other (fig. 1.6). A **spreading center** is where plates move apart, and the Atlantic Ocean's mid-ocean ridge is an example of one of these. Three types of collision are possible: continental crust with continental crust, continental with oceanic, or oceanic with oceanic. An example of the first is India crashing into Asia—the result: the Himalayas, the highest mountain range in the world. Africa has been crashing into Europe for some time, and there you have the Alps with a little chunk of Africa sitting atop the Matterhorn, thanks to some of the former's getting thrust up over Europe. When oceanic plates collide with continental plates the cooler denser oceanic plate generally sinks under the continent. Examples of oceanic-continental and oceanic-oceanic collisions (subduction zones) are all around the Pacific, from oceanic plate diving under Asia creating the Marianas Trench and Mount Fuji, to the other side of the same Pacific plate diving under South America creating the Andes and many volcanoes there. A major **strike-slip fault** is formed where plates move past each other; a textbook example of this is in California, where the Pacific Plate (oceanic, mostly) is grinding past the North American Plate (mostly continental) along the San Andreas Fault. Movement in the mantle causes tectonic plate motion and strain builds up several miles under the Pacific–North American plate boundary and (*Voilà!*) every so often you get jolted out of bed when you're in California. This is how plates move and have for hundreds of millions of years.

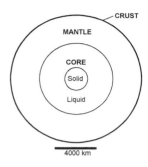

1.5. Structure of the interior of the Earth. Layers approximately to scale relative to each other.

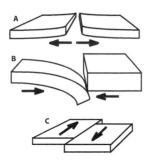

1.6. The three main types of plate tectonic boundaries. (A) A divergent margin (e.g., the Mid-Atlantic Ridge). (B) A convergent margin with subduction zone (e.g., the south shore of Alaska). (C) A transform margin with a major strike-slip fault (e.g., the San Andreas Fault in California).

Paleomagnetic work is how geologists determine the ancient positions of continents. Plate tectonics explains *how* continents have moved around through time, and geologists specializing in paleomagnetics can tell us *where* continents were through time. Within a rock, a needle-shaped piece of magnetite will align itself with the Earth's magnetic field, not only pointing at the magnetic north pole but also pitching up or down relative to its latitudinal position on the planet; it will be approximately horizontal at the equator and nearly vertical at the poles. When magnetic minerals such as this become locked in rocks of a particular age, geologists can use them to determine the direction of the pole and approximate latitude of the rocks at the time. Long ago when continents were thought to be stationary, the apparent shifting in the position of the pole recorded in the rocks was thought to reflect actual variation in the position of the magnetic pole in ancient times. It was referred to as "polar wander." It was noticed, however, that the paths taken by the magnetic pole didn't match up from one continent to another; it would appear to have one wandering route through time from one continent, while from Europe, for example, its path would seem to be totally different. With the plate tectonic revolution in our understanding of Earth dynamics, however, it became clear that the polar wander seen in the paleomagnetism of the rocks on any one continent recorded not the movement of the pole itself but the movement of the continent *relative* to the (mostly) constantly positioned magnetic pole.[13] This explained the incongruence between continental records—two continents that are close today often pointed completely opposite directions (and latitudes) to the magnetic pole for a particular ancient time simply because although the pole was relatively constant, the continents could be rotated opposite directions from their current orientations and may have been at wildly different latitudes back then. Eventually, geologists studying paleomagnetism were able to track the movements of the continents through time and learn of the existence of ancient supercontinents and where current continents ripped apart from others. As with plate tectonics, many more things that had been previously observed now made sense.[14]

Paleomagnetic studies show us that during most of Earth history continents have been moving all over the globe and, as we have seen, that during the Cambrian North America was an equatorial land with the current west coast facing north and today's east coast facing south. Several parts of modern North America were attached instead to other continents during the Cambrian. Plate tectonics tells us why such strange continental positions are possible.

Perhaps the most important lesson from the above story is the significance of such a high degree of geologic activity on the Earth. One of the key aspects of Earth's ability to sustain life is the fact that it is not dormant. The movement of the fluid core helps generate the Earth's magnetic field, which deflects the solar wind and helps keep our planet's oceans and atmosphere where we need them—tightly enveloping our home. The fact that the inside of the Earth is hot, fluid, and moving means a dynamic and life-sustaining surface for us and other inhabitants.

Ocean Levels and Tides

During the Cambrian, sea levels were generally quite a bit higher than they are today. Cambrian rocks record a general rise in sea level throughout the 54 million years of the period, but within that overall rise there appear to be approximately 28 minor rises and falls of sea level. The important part is that only a handful of times did the sea level fall to or below the present day level, but during nearly every rise it came up higher than it had last time (thus the overall increase throughout the Cambrian). How high? Sea level started the Cambrian a bit lower than today, but in its first rise it came up to approximately 99 ft (30 m) higher than sea level is today. After the continued rise throughout the period, sea level by the Late Cambrian was nearly 530 ft (160 m) higher than today (fig. 1.7). Such a level in modern times would flood many low-lying parts of our continents, but during the Cambrian Laurentia appears to have been much flatter than modern North America and was in fact nearly as much under the ocean as above it. The high Cambrian sea levels and relatively low, flat continents translated to a flooding of about 40% of continental areas as compared to barely more than 5% today. Areas such as the North Sea and English Channel, Hudson Bay, Baltic Sea, and the Gulf of Carpentaria and Torres Strait between Australia and New Guinea are today shallow because they in fact are not ocean basins but rather flooded continental regions. Such flooded continental areas were far more extensive during the Cambrian.[15]

The twice-daily high and low tides that we are used to today would have been similar, if slightly more pronounced, during the Cambrian. Because of the transfer of angular momentum from the Earth to the Moon over time, the Moon has slowly been getting farther and farther away from the Earth, and if we go back in time more than 500 million years to the Cambrian period the Moon would have been some 21,000 km (13,000 mi.) closer to our planet than it is today.[16] Because of this, the Moon's gravitational pull would have been slightly stronger, and thus the high and low tides seen on the beach every day would have been just a bit higher and lower. This transfer of angular momentum has a whole host of other effects. Earth days have been getting longer through geologic time; the Moon's orbit of the Earth has been taking longer. In fact, during the Cambrian period an Earth day (the time it takes our planet to make one rotation on its axis) may have been as short as 20 hours, and the Moon's orbit of the Earth may have been around 25 days instead of the 27.3 days it takes now. Far back in geologic time an Earth year would have consisted of more days than currently, during the Cambrian perhaps as many as 400 days. So we can imagine the Cambrian world with a slightly larger Moon in the night sky (perhaps about 5–10% larger), with somewhat shorter nights (and days), and with tides that were a little more extreme than what we know today.[17]

We still are not certain how the Moon originated. Was it captured by Earth and pulled into its orbit of our planet? This seems unlikely as the composition and age of the Moon suggest it formed in the same neighborhood of the solar system as Earth.[18] Was it formed of material blasted out of Earth by an early strike by a gigantic asteroid? Was it formed at the same time as Earth or soon after from some of the same local material that made up our planet? We don't know for sure. Some have suggested that the Moon may have begun orbiting our planet as recently as 1 to 2 billion years ago; this is based on extrapolating the rate the Moon moves away from Earth back in time to the point at which the Moon would have been so close it would have disintegrated (~2.9 Earth radii) – that is, the closest the Moon could ever have been.[19] The age of the Moon indicates it would already have been a couple billion years old by this point,[20] so this scenario requires remote capture as the mechanism of the Moon's origin as an Earth satellite. Many scientists see remote capture as less likely than formation nearly synchronous with Earth, mainly due to the difficulty of the mechanics of such an encounter resulting in the Moon becoming a satellite.[21] So the Moon more likely formed from the same swirling dust and gas as the Earth, or maybe from material blasted into orbit around Earth very soon after it formed. Although remote capture is not impossible, these scenarios are more likely. Either way, as different as the Cambrian may have seemed, long before the Cambrian, deep in the Precambrian, Earth days would have been very short, the Moon very close, and the Earth's tides incredibly pronounced. Such conditions would have made the Cambrian look familiar to us in comparison.[22]

Climate and Atmosphere

The Cambrian world appears to have been largely devoid of ice. There is little or no evidence for polar ice during the period – no continental ice sheets, no rivers of glaciers. There had been ice ages before the Cambrian, ones still unrivaled in Earth history, and there were a couple ice ages to come later in the Paleozoic, but the Cambrian period itself was significantly warmer than today. The Mesozoic appears also to have been mostly free of ice, and it is only in relatively recent ages of the Cenozoic that the world has gone into its current period of glacial cycles. During the Cambrian the average global surface temperature was approximately 22°C (72°F) (fig. 1.7). This is as compared to today's average of approximately 12.5°C (54.5°F). The global average accounts for daily, seasonal, and latitudinal variation; there are parts of our globe that today are almost always warmer or cooler than 54 degrees, but that is the average. So at more than 70°F, the Cambrian was quite a bit warmer than what we are used to. The temperatures of the oceans' waters would have been comparatively warm also. Part of the reason for this atmospheric warmth was

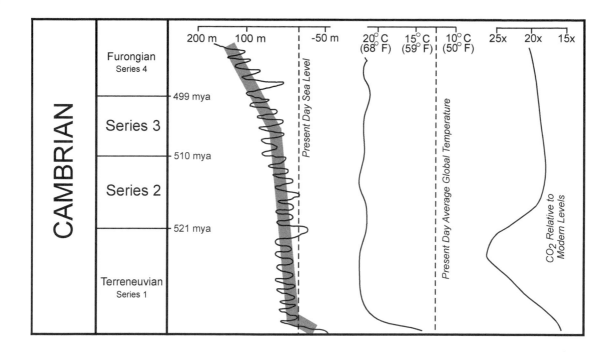

| | | 200 m | 100 m | -50 m | 20°C (68°F) | 15°C (59°F) | 10°C (50°F) | 25x | 20x | 15x |

CAMBRIAN

Furongian
Series 4

499 mya

Series 3

510 mya

Series 2

521 mya

Terreneuvian
Series 1

Present Day Sea Level

Present Day Average Global Temperature

CO₂ Relative to Modern Levels

that the CO_2 level of the atmosphere during the Cambrian was several times higher than it is today–possibly as many as 15 to 25 times higher (fig. 1.7).[23] With that much CO_2 trapping heat from the sun in Earth's atmosphere it is no wonder the average temperature was so much higher than today–and no wonder that ice was so rarely (if ever) seen during the Cambrian.

Overall higher precipitation during the Cambrian appears to have made for a wetter world. In terms of climate belts, Laurentia occupied parts of the tropical and arid zones of the Cambrian.[24] Remember that the ancient continent straddled the equator (or at least was very close to it), so being in a tropical climate is hardly surprising. The shallow seas probably looked much like the light blue carbonate banks seen today in parts of the Caribbean, while the land surfaces looked rather different from your standard tropical landscape of today. We need not picture familiar palms, cycads, and banana trees because none of these existed yet. In fact, there were no plants on land at all–possibly no greenery at all, save for some terrestrial algae and mosses if recent results are correct, but more on that later. The main idea is that Laurentia was in a warm and, in places, humid part of the globe 500+ million years ago.

Today, oxygen accounts for approximately 21% of the gases of the atmosphere. Nitrogen, of course, accounts for most of what we breathe, but oxygen, so critical to us, is still a significant amount of the air we take in. During the Cambrian it appears the oxygen level was a little bit lower, perhaps about 20–15% of the air in the atmosphere.[25] The organisms of the time would have been adapted to it, so as strange as having nearly 25% less oxygen sounds to us, it was not prohibitive to the life of the time.

1.7. Trends during the course of the Cambrian period. *Left,* a sea level curve showing the many sea level rises and falls and the overall trend toward sea level reaching nearly 200 m (656 ft.) higher than it is today by the end of the Cambrian. *Middle,* the average global surface temperature trend for the Cambrian, showing the generally warmer climate of the period compared to today. *Right,* the relative CO_2 level curve for the Cambrian (in number of times higher than modern preindustrial levels), showing much higher levels for all of the period.

Based on data in Holland (1984), Haq and Schutter (2008), and Riding (2009).

Also keep in mind that this relative amount of oxygen is comparable to what mountaineers experience when they climb very high peaks, so such conditions would be survivable for us, if a little disorienting.

The oxygen content of the atmosphere has varied throughout the Phanerozoic, sometimes rising to a high percentage, other times falling to a rather low one, but always maintaining a level to sustain life as we know it, always staying a little higher or lower than the amount we know and love. This was not always the case. And life didn't always love oxygen. As we will see in chapter 3, the world before the Cambrian was about as alien as one can get and still be on Earth, and one aspect of that alien world was that for almost half of Earth's history the atmosphere was comparatively devoid of oxygen—probably containing less than 1% of the amount we know today! The switch to an oxygenated world was a revolutionary one, and as comparatively different as the Cambrian atmosphere was relative to today's in terms of lower oxygen levels, it was definitely modern in the sense that it contained *far* more oxygen than had been in Earth's air for most of the planet's history up to that point.

Cambrian Animal Farm: Background Biology

Now that Cambrian rocks have been identified in so many parts of the world, and fossils found in so many of those outcrops, what do we know of the life of the time? Quite a lot, but not as much as we'd like. There are significant fossil records from the Cambrian in Australia, Africa, South America, and even Antarctica.[26] In Asia, sites such as Chengjiang, along with European localities in Poland and Sweden, are telling us more than we'd learned about Cambrian faunas in possibly the previous hundred years. In North America, Cambrian fossils have been found in many—probably most—states and provinces. Chances are your state or province or one very nearby has some Cambrian fossils in it. Some of these sites stand out, however, for their soft-body preservation. These include sites such as the House Range and Wellsville Mountains in Utah, the Kinzers Formation in Pennsylvania, Sirius Passet in northern Greenland, the Marble Mountains in California, and of course the Burgess Shale Formation sites in British Columbia (see fig. 1.4).

What life forms occupied the Cambrian oceans? Although the organisms can appear rather strange to us, many are in fact members of groups we still see in parts of the ocean today. There were algae, glass sponges, calcareous sponges, demosponges, sea pens, sea anemones, comb jellies, jelly fish, brachiopods, molluscs, priapulid worms, annelid worms, velvet worms, many types of arthropods, echinoderms, sea cucumbers, and chordates, just to name a few. We will see these in more detail later.

As much as we would like to believe in our so-called ages—the Age of Dinosaurs, the Age of Mammals, or alternatively, the Age of Insects, which might be said to have been going on since the middle of the Paleozoic—Earth has only truly ever been in one age. In terms of diversity and importance, the story of life on Earth has always been and continues to be the age of microorganisms. The vast majority of organisms at any given time, by far, are simple and single-celled. They form the basis of food chains, and the increased complexity of animals is only possible because of the symbiosis of various formerly lone bacteria-like organisms. Neither we nor any ecosystem would function without them.

It's a Trap

We must avoid the temptation to view the animals of the Cambrian as primitive. Although in a systematic sense there is nothing derogatory about the term – it refers only to characters that are shared by descendants as well as their ancestors – in everyday usage it has come to imply animals that somehow survive despite being outdated or even ill adapted. In truth, of course, the animals of the Cambrian were adapted just fine for their environments. Thanks to the popular notion of something "going the way of the dinosaur" there is an impression that the simple status of "extinct" automatically implies a poorly adapted animal that died out because it deserved it. Of course, we think this more of groups of animals that are large, diverse and well known; after all, no one says some species has gone the way of the pantodonts, even though these large, ancient mammals are every bit as extinct as non-avian dinosaurs. There is an irony in that, because in order to have been such quintessential failures as the dinosaurs you first have to have been spectacularly successful. Who would notice your extinction otherwise? And the trilobites outshined the dinosaurs in terms of species diversity (by quite a bit) and in terms of group longevity. And trilobites were arguably more morphologically and ecologically diverse than dinosaurs. Any extinct group of animals used as an example of poor adaptation might well tell us humans, "Get back to us when you have been around for 100+ million years and have diversified into thousands of species." Whatever extinct animals were ill adapted, the *last* ones we should be using as examples for our colloquial sayings are dinosaurs or trilobites.

These sayings and the impressions behind them all likely derive from a misunderstanding of extinction. The natural disappearance of a species is sometimes popularly viewed as resulting from some inherent deficiency.[27] This is a little bit of a misrepresentation. In fact, species may be out-competed by new forms or find themselves in a quickly changing environment, but whatever does them in, it is only necessary for their birth rate (or hatching rate) to drop below a certain level (relative to mortality rate) and not recover and eventually they will die out. All kinds of factors can influence this. Change in the species' food resource, loss of a food resource, a new predator, any interference with reproduction – anything that lowers rate of hatching or birth relative to mortality (or raises mortality) and does so at a greater rate than genetic mutations or sexual recombination can offer up morphological variations as potentially beneficial adaptation – any of these can, with time, drive species toward extinction. The important point is that it is not perceived intraspecific deficiencies that are the problem, but rather the rate of change in external factors to which the species needs to adapt. In most cases, and on average for approximately 1 million years, each species is able to adapt and evolve to accommodate its environment and those changes before conditions change too severely for the species to keep up. This in fact happens to all species. You may have heard the estimation that 99% of all species that have existed are now extinct. It is probably a lot more than that. It is unlikely that any modern animal species can trace its line of members back more than a few million years.[28] In a way, species follow a pattern similar to their member individuals: they appear at some point, sometimes give rise to new lineages, and eventually disappear. Extinction is more the rule than the exception. On a group level, many lineages have been around for hundreds of millions of years, but plenty of groups have disappeared as well – within a group it is all a function of extinction versus origination at the species level (just as within a species it is a function of mortality versus reproduction rate at an individual level). In order for groups to be large enough to leave an impression on us when they become extinct they must first be well adapted and diversify into many species. This is, of course, a sign of success and should impress us as such as much or more than the group's extinction indicates what we perceive as failure. What we must remember primarily, though, is that old and extinct species are and were *not* inferior. At their time and in their environment they were every bit as successful as species today.

Shaking the Tree

One of the traps we are in as large vertebrates is that we think of the world in terms of what we can see. And we can't see most of the life that exists in the world. Life isn't so much a tree with a big trunk and neatly, progressively diverging braches; it strikes me as a bush of wildly branching chaos with many different lines going off in many different directions. Vertebrates may form a tree but a full representation of life might be a tumbleweed. The Bush of Life is a branching mess of groups of which plants and animals are only two tiny sticks. Almost everything we know of 99.9% of Cambrian fossil localities is about the animal branch of this bush. So let's go back a bit and visit a little of the diversity of life that is on the other branches, many of which were probably branches back during the Cambrian but which we rarely see in the fossil record.

There are three main lines of life, called "domains": the Archaea, the Bacteria, and the Eukarya (fig. 1.8). We and all other animals are in the Eukarya, along with algae, plants, mushrooms and other fungi, and other interesting organisms like slime molds and ciliates (including *Paramecium*, the bacterium-eating single-celled organism of many a biology lab). Remember that among the animals are, of course, not just cats and lizards and insects and squid, but things as immobile and – for lack of a better word – simple as sponges and corals.[29] So most of what we see of the biological world around us today, and most of what we see as fossils in Cambrian rocks as we dig, is a tiny fraction of the Eukarya limb. The differences in organization among these three groups—Archaea, Bacteria, and Eukarya—are fairly simple, although the ecologies differ greatly within and between the three. Bacteria and Archaea are generally simple single-celled organisms lacking nuclei, but differing from each other in other aspects of their structure. Cells of these two types contain their genetic material (DNA) in an unseparated region within the cell called the nucleoid. Eukarya may be single-celled or multicelled, and of course vary greatly in structure, especially among plants and animals, but they are unique in having cells (however many) with a nucleus. The nucleus contains the cell's DNA and is a separately enclosed organelle within the cell.

The Archaea contains microorganisms that have cell wall and plasma membrane composition different from Bacteria. Although they are single-celled and lack nuclei like Bacteria, Archaea have other aspects of their composition (ribosomal protein and RNA polymerase) that are more similar to Eukarya, and suggest a closer relationship with this latter group. Archaea consists of a probably very ancient group of organisms that today (and probably in the past) live in extreme environments where little else can survive. There are several subgroups of Archaea. The methanogens are poisoned by oxygen and must live where it is nearly absent. Some live in the guts of animals, others in swamps. Methanogens are not heterotrophic; that is, they do not consume other organisms for energy as do, for example, herbivorous or predatory animals. Nor do they produce their energy through photosynthesis as do plants. Instead, methanogens use environmental hydrogen to reduce carbon dioxide to methane. Organisms such as these that metabolize environmental components rather than sunlight for energy are called **chemosynthetic**.[30] A second group of Archaea, the halophiles (also called halobacteria; fig. 1.8), thrive in salty environments, in some cases ten times more saline than seawater. The thermoacidophiles live in places both hot and acidic. How hot? Generally, 60–80°C (140–176°F) is a comfortable zone for these life forms, the kinds of temperatures you find in hot springs and other geothermal areas but in few other places on the surface of the planet. On the extreme end, one archaean can live around deep-sea hydrothermal vents at temperatures reaching 113°C (235°F)—water so hot that if it were up at the surface and not in the pressures of the deep ocean, it would boil! In terms of acidity, the thermoacidophiles prefer a pH of around 2 to 4. By contrast, seawater is around pH 8.3 generally. Some archaeans can tolerate pHs as low as 1.

On the Bacteria limb we find a number of groups that live in a variety of environments and with a range of ecologies. Some of them are familiar, others less so, but many are important. Like the Archaea, these organisms lack a nucleus and carry their DNA within the cytoplasm of the cell, but as we saw earlier, the composition of their cell wall and plasma differs from that of the Archaea. Although most are single-celled, some have **flagella** and are capable of movement. Among the groups of Bacteria

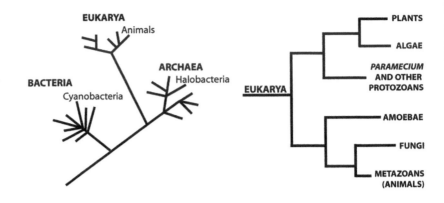

1.8. The major subdivisions of life. Cyanobacteria are just one branch of the Bacteria; halobacteria are just one branch of the Archaea; and plants, animals, and fungi are just three of the branches of the Eukarya.

Based on data in Knoll (2003a).

1.9. Relationships within the Eukarya.

are some interesting organisms. Chemoautotrophic bacteria are common in aerated soils and are a diverse group. Nitrogen-fixing aerobic bacteria are less diverse but are important in that some forms live in root nodules of some plants and help them grow. Enteric bacteria are anaerobic and live in the intestinal tracts of animals. Some are foreign and make the animal sick, but many are benign and live in the gut permanently. *Escherichia coli* (*E. coli*) is a notorious member of this latter group, but it is actually one of the good bacteria that many animals have in their intestines naturally. It only becomes a problem when you accidentally consume an additional, external source of it in reasonably large amounts ("Well done, please."). Another group of Bacteria, some 15 genera in number, consists of parasitic organisms that work their way into animals, most often arthropods, mammals, or birds. A large and important group within the bacteria is the **cyanobacteria,** a group we will encounter later as well. These bacteria are photosynthesizers and, like plants, they use chlorophyll. Some cyanobacteria are solitary, other species are multicellular, and still others live in large colonies, often forming macroscopic filaments. There are species that live in freshwater (most commonly) or soils, but some live in the ocean. Cyanobacterial colonies, as we will see, used to be quite abundant in the oceans.

Finally, the Eukarya consists of the organisms with a true nucleus in their cells. This group includes amoebae, slime molds, ciliates, algae, fungi, plants, and animals (fig. 1.9). Most of the rest of this book is dedicated to the animal fossils of the Cambrian, so let's take a look at some of the other groups on this limb of the bush. **Amoebae** (or rhizopods) are single-celled organisms that lack flagella but can move anyway through extension of parts of the cell outward and readjustment of the rest of the cell. Amoebae live in freshwater and ocean environments as well as on land (in soil), and most are free-living heterotrophs. One form is parasitic and is the cause of amoebic dysentery. A slime mold is convergently similar to a fungus in form and also consumes dead organic matter, but the cellular structure is different. Ciliates include *Paramecium* and other forms that move with the assistance of short hair-like **cilia** around the outside of the cell. They are generally solitary heterotrophs that live in freshwater.

Algae include eukaryotes that are mostly photoautotrophs. This simple demarcation, blurred as it is by exceptions, "defines" algae. Some algae may be green, but they are not plants. Many are solitary and single-celled, so the diversity within the group is impressive. The algae include a variety of forms. Euglenids consist of the textbook eukaryote genus *Euglena* and its relatives. *Euglena* is a single-celled alga that has a flagellum and can photosynthesize, but if it finds itself in the dark it also may ingest food particles—an organism that is capable of either **autotrophy** or **heterotrophy** as the need arises! Dinoflagellates are very abundant in the modern sea as tiny phytoplankton, forming the base of most of today's marine food chains. Most forms are photosynthetic but some are parasitic or carnivorous. Other algae that are important include members of the Chlorophyta, Phaeophyta, and Rhodophyta, respectively, the green, brown, and red algae, marine members of which today constitute seaweeds. These again are not plants, but colonial algae that can grow in some cases up to 100 m (330 ft.) long.

Fungi are familiar to us as things as unpleasant as molds or as tasty as shiitake mushrooms. Although the wide variety of mushrooms we know from field and supermarket are the most familiar fungi, there are many that are important for a wide variety of uses, from yeasts to lichens. Commonly growing in leaf litter, on lawns, or coming out of logs in the forest, fungi are heterotrophic, but rather than ingesting food, they secrete enzymes into the food on which they grow and absorb the nutrients. In the process they decompose the material. In this sense fungi are, along with bacteria, important environmental decomposers, returning otherwise unconsumed organic material to the nutrient cycle. While many fungi break down dead organic matter, some fungi are beneficially symbiotic with other organisms such as plants, and some are parasitic, infecting living hosts with detrimental effects.[31]

Plants are of course familiar to us today in the wide variety of trees, grasses, bushes, and flowers all around us. All plants are photosynthesizing, multicellular organisms with abundant chlorophyll and differentiation of tissues (e.g., leaves, wood, seeds, fruits, etc.). They are far more complex than algae or any of the other organisms we have discussed so far. They use sunlight to convert water and carbon dioxide into oxygen and the sugar glucose. The most primitive plants are the bryophytes, or mosses. After this come the vascular plants lacking seeds; these plants, such as ferns and horsetails, reproduce with spores rather than seeds. The gymnosperms (naked-seed plants) came next and include plants like conifers and cycads, and these were finally followed only some 125 million years ago by the angiosperms, or flowering plants. These include many of the plants we know today such as grasses and flowers and palms and fruit trees and others. Unfortunately, even by Cambrian time, more than 80% into Earth's history so far, almost none of these types of plants existed. Only mosses, it appears, may have been on the scene yet.

Animals (the **Metazoa**), consist of multicellular heterotrophic eukaryotes. They are differentiated from other organisms (in most cases) in having nervous tissue and muscle tissue and in having a unique set of reproductive/life cycle characters. Metazoan cells also lack the cell walls that characterize plant cells. Sponges (**Porifera**) and **cnidarians** (e.g., jellyfish and corals) are among the most simple animals, and vertebrates among the most complex. Metazoans are divided into several major groups (fig. 1.10). The first and most inclusive group is the **Bilateria**, which includes all animals with bilateral symmetry and a front and back orientation. In practice it includes all animals except the sponges and cnidarians. As we will see, it also includes the pentaradial **echinoderms** such as starfish, because the ancestors of these five-limbed forms were bilaterally symmetrical.

Bilateria is divided into the **protostomes** and the **deuterostomes.** The difference between these relates to the development of the respective embryos and whether the primary folded opening of the dividing mass of cells becomes the mouth or the anus. Echinoderms and chordates comprise the deuterostomes; all other bilaterians are protostomes.

The protostomes are further subdivided into the **Lophotrochozoa** and the **Ecdysozoa.** Lophotrochozoans include creatures like lamp shells (brachiopods), segmented worms, clams, snails, and squids; the ecdysozoans include such forms as roundworms and arthropods.

The metazoan branch of our bush of life contains about 34 **phyla,** major divisions based on body plans. The relationships of the major phyla are shown in figure 1.10. Each phylum has from 1 to more than 1 million species described from it. Most have a few dozen or a few hundred or several thousand; the diversities of different phyla vary greatly. The most diverse animal phylum is the **Arthropoda,** which includes more than 1,093,000 species of insects, spiders, shrimp, crabs, and their relatives; the next most diverse is the Mollusca, which includes only 93,000 or so snails, clams, squids and their relatives. That's a big drop, a million. In terms of numbers, arthropods rule the animal world. Our own phylum, the **Chordata,** those of us with backbones plus similar basal forms that have a notochord but no vertebrae, includes only a little less than 50,000 species. The **Annelida,** a group of worms? A healthy 16,500 species. Approximately 96% of living animal species are invertebrates, members of all the animal phyla other than Chordata plus all non-vertebrate (non–backbone bearing) members of Chordata (this would include, for example, sea squirts but not basal fish). Back in the Cambrian, when there appear to have been just a handful of chordates, and only a few known, very basal vertebrates, the percentage of invertebrates probably would have been even higher. It was then, and really still is, an invertebrate world. We just live in it.

The Cambrian Tumbleweed

One of the things that is just plain fun about Cambrian paleontological work is that you can find remains of representatives of many of these 34 animal phyla in rocks that are about 540 to 490 million years old. Among

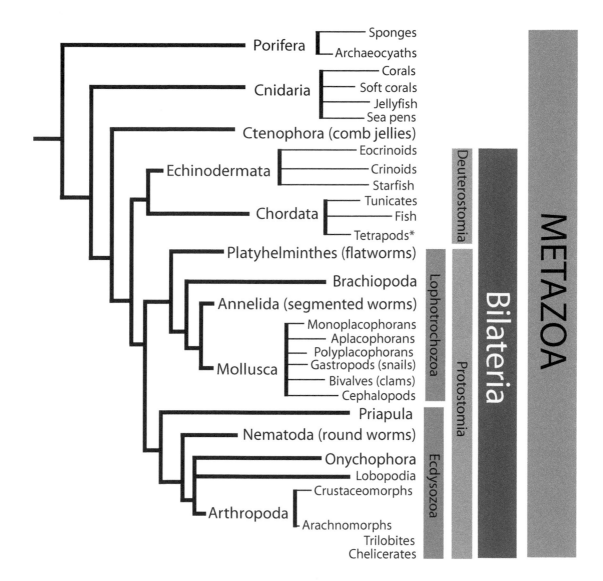

1.10. Relationships within the Metazoa (animals). The groups Bilateria, Protostomia, Deuterostomia, Lophotrochozoa, and Ecdysozoa are also shown.

the groups that we find are the Porifera (sponges); Cnidaria (sea anemones, sea pens, corals, jellyfish); **Ctenophora** (comb jellies); **Priapula** (priapulid worms); Annelida (polychaete worms, sister group to the earth worms); stem group relatives of **Onychophora** (velvet worms); **Tardigrada** ("water bears"); many Arthropoda (trilobites, bivalved arthropods, plenty of basal arthropods of chelicerate or crustaceomorph grade); **Mollusca** (monoplacophorans like *Scenella* plus others); Brachiopoda (brachiopods, lamp shells); Echinodermata (sea lilies, sea cucumbers); and Chordata (basal cephalochordates, basal vertebrates/jawless fish). In coming chapters, we will get to know each of these groups in more detail, but for now figure 1.10 shows us a general hypothesis of how they are related to each other. Two main things are in flux in these classifications, at least as far as Cambrian bushes of life go: the identification of the fossil material may jump from one group to another, and at any one time it may not even be agreed by all parties which one it should go in; and the relationships of the groups to each other switch around a lot, too. Some of the groups' relationships are based on molecular studies of modern representatives (but at least we *have* modern relatives of some of these fossils; that's not always the case) and these can be relatively stable, but it is very difficult to untangle the relationships of basal arthropods, for example. In the Cambrian there are many arthropod fossils, many overlapping characters, and few species that appear to be from the **crown groups**—the more recent radiations of arthropods with living descendants of whose relationships we can be more certain. We have found many forms that simply don't fit easily into the arthropod branch, and they appear to be **stem group** arthropods—extinct members that split off before the common ancestor of the modern groups. We have ancient representatives of some of the modern groups (malacostracans, for example), so the fact that stem group arthropods occur in the same beds simply indicates that the split occurred a little deeper in time. The real question is how these stem group species (of which there are many in Cambrian rocks) relate to each other and to crown group arthropods. As we will see in coming chapters, this is both a headache for working on Cambrian arthropods and a wonderfully exciting opportunity. But first we need to familiarize ourselves a little more with the Cambrian rocks in which our evidence is found.

2.1. Siliciclastic sedimentary rocks shown in thin-section photographs. (A) A shale, the Middle Cambrian Spence Shale from near Liberty, Idaho, composed mostly of clay and some silt. (B) A sandstone, the Upper Cambrian Sawatch Formation from near Minturn, Colorado, composed of angular to rounded quartz grains and some rounded feldspars. Both (A) and (B) are shown at the same scale to facilitate comparison of their respective grain sizes. Scale bars in (A) and (B) = 300 microns (about 1/3 of a millimeter).

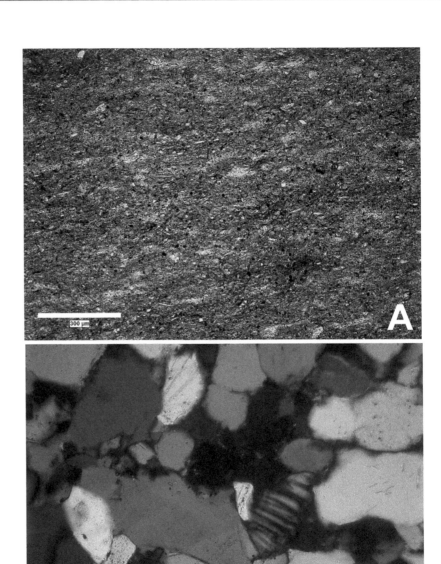

DRIVING ACROSS THE COLORADO PLATEAU TODAY IS A VERY SCENIC and relaxing trip. Such a journey is a destination in itself. The roads are usually not crowded, and the pavement and landscapes seemingly stretch on to infinity. Rivers, canyons, balanced rocks, beautifully exposed brick-red layers of rock bending up out of the Earth and scoured by erosion—it's a wonder to all, not least the sedimentologist. In Monument Valley, hulking rock monoliths seem to float slowly past along the drive, miles away but towering. Stone ships scattered across the plateau give the appearance, from a distance of miles, of a massive sandstone regatta being battled out over the millennia, the competitors seemingly frozen in action. All this rock is young, relatively speaking, and it is the uplift of all this rock—flat and intact, the whole plateau—that has allowed the Colorado River to cut its spectacular Grand Canyon. In the Grand Canyon are exposed rocks of the Paleozoic era, among them three formations of the Cambrian period. We are headed to visit these soon. Before we dive into the Cambrian we should dive into some more specific geology and prepare ourselves for some of the things we will see in our target rocks that will tell us more about what kinds of environments our favorite animals were living in.

Because rocks are the most basic subject in geology, we should start with them. As you may remember from high school geology (if you are lucky enough to have had it), the three basic rock types are igneous, metamorphic, and sedimentary. Igneous rocks are those that form by cooling of magma (below the Earth's surface—intrusive) or lava (erupted from a volcano—extrusive); metamorphic rocks are any rocks that have been buried deeply enough and put under enough pressure to partially melt and recrystallize. You can start with a number of different original rocks and get a particular type of metamorphic rock. Partially melt and recrystallize granite (a type of intrusive igneous rock) and you end up with (metamorphic) granitic gneiss. Metamorphose sandstone and you get a quartzite, metamorphose limestone and you get a marble, metamorphose shale and you get a slate, and so on. We don't really need to worry about igneous rocks in this case, because as a rule (with its wacky exceptions) fossils do not occur in them.[1] Fossils are also pretty rare in metamorphic rocks since the heat and pressure usually destroy them or mangle them beyond use. But some fossils do occur in them. It is usually enough to know the source rock of the metamorphics, as outlined above; but you

Sedimentary Rocks: The Wrapping

will find the occasional deformed trilobite in lightly metamorphosed slate. Metamorphose any rock enough and you won't see any fossil in it.

What we are really interested in in this book is the third type of rocks: sedimentary. Sedimentary rocks include two main types: siliciclastic,[2] mostly **sandstones** and **shales,** which are formed by the deposition of grains of sand or mud ultimately derived from the erosion of previously existing rock on land; and non-siliciclastic, which are either chemically precipitated from water or accumulate by the piling up of biogenic material that is also precipitated by organisms in the water. These rocks include **carbonates,** which we will see a lot of later on.

The siliciclastic cycle runs a little bit like this: (1) take any source rock (igneous, metamorphic, sedimentary) and erode it by beating it with water and wind and gravity;[3] (2) wash it down streams and rivers for many millennia until the fist-sized cobbles break down into pebbles and then into grains of sand or mud of any of a number of minerals or fragments such as quartz, clays, micas, or rock fragments; (3) wash these grains into an ocean or a lake or a lazy river, anything slow-moving enough that the grains will settle out of the water and accumulate in thick layers on the bottom (or blow them into a sand dune desert); and (4) do this until the sediments (the grains of sand or mud) pile up under the landscape so thick that they are miles deep in the Earth—there with heat and pressure the sand grains will be cemented together by minerals filling the pore spaces between grains and before you know it (geologically speaking) you will have sedimentary rock. It will then be available to be uplifted to the surface, where it can be eroded and start the cycle all over again. The non-siliciclastic cycle is a little simpler in the number of steps and includes simply piling up chemically or biogenically precipitated mineral grains, such as calcium carbonate crystals or microplankton shells, in a lake or ocean bottom until the layers are buried fairly deep.[4]

Sandstones and Shales

Siliciclastic rocks may be classified in part based on their grain size. Grain composition is another way to fine-tune our classifications, but we will mainly be concerned with general siliciclastic rock types, which we can discern on the approximate diameter of the grains of sand or mud composing the rocks. We are mostly interested in getting a general idea of the paleoenvironmental settings of the fossils we will be seeing in the rocks, and so a lot of detail is not really necessary—plus the fact that this way we avoid getting too bogged down in thrilling geology jargon like "subarkoses" and "feldspathic litharenites." Siliciclastic rocks are composed of grains ranging from "boulders" to "clay." Don't be fooled; those terms have very specific meanings in sedimentology. Those equate to grain sizes ranging from 25.6 cm (10 in.) and larger down to about a couple millionths of an inch, respectively. The categories for the rest of the scale between boulder and clay are, in descending order of size:

cobbles, pebbles, granules, sand, and silt. Each has its own defined range of grain diameters, and sand and silt are even subdivided themselves. The size of the grains in a siliciclastic rock, then, determine what you call it; and conversely if I tell you what kind of rock I have, you will know the approximate size of the grains in it. Rocks with mostly pebbles and cobbles would be **conglomerates;** because sand usually fills in the pore spaces between large clasts, conglomerates require a mixture of clast sizes. Sand-sized grains form sandstones, and silts and clays form **mudstones.** Some sandstones may have a few thin layers or lenses of granules or pebbles in them. Shales are a particular type of mudstone that splits in fine sheets of rock. You may also find mudstones that crumble rather than splitting in sheets; the grain size is the same, the key is in how it weathers. Generally there are few crumbling mudstones in Cambrian outcrops; you almost always find splitting shales. Sometimes the shales will be metamorphosed – lightly into an **argillite,** moderately into a slate, or heavily into a mica schist or phyllite. Either way, grain size and the specific name of the kind of siliciclastic rock are tied together. So, you say you have a very coarse sandstone? That means you have one with sand grains ranging from 1 to 2 mm (0.04–0.08 in.) in diameter. A fine sandstone? Grains range from 0.25 to 0.125 mm (0.0049–0.0098 in.) in diameter. These may sound incredibly small but they are distinguishable out in the field with just your eye and a hand-lens magnifier. Get down into the clays, of course, with the diameters in the ten-thousandths of a millimeter, and it is often difficult to determine grain size even with a standard binocular microscope. One needs to look through a scanning electron microscope for that. Fossils are really nothing more than big grains ("clasts" in geology-speak) in sedimentary rocks.[5]

The main siliciclastic rock types we will become familiar with in the Cambrian are sandstones and shales, rocks with grains ranging from various subcategories of sand down to silts and clays (fig. 2.1). There are a few conglomerates with pebbles and cobbles and even boulders, but they are relatively rare. In general, the grain size preserved in a siliciclastic rock is proportional to the energy in the environment in which it was deposited. Small equals low energy; big equals high energy. Shales are thus usually indicative of quiet-water settings where tiny grains have time to settle out of a nearly still water column, whereas a pebbly sandstone was more likely formed, for instance, on a seashore where the high-energy wave action was able to roll the pebbles back and forth, but the water movement was not enough to transport the pebbles any farther out to sea. As a rule, as water movement slows, siliciclastic grains drop out of the water and settle to the bottom from largest grains down to smallest. The largest rocks will be added to the pile of sediments at the bottom of a river or ocean in higher-energy settings than pebbles, which will settle to the bottom, before sand – a gradient of these grain sizes essentially equates to a gradient of slowing water movement. Therefore, the places that you see huge chunks of rock in big boulder conglomerates are only in deposits

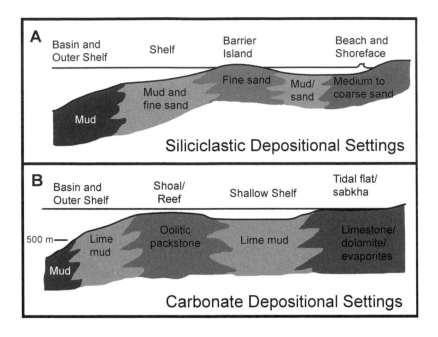

2.2. Relative depth and sediment types from proximal to distal settings on a continental shelf (not to scale). (A) Siliciclastic settings showing relationships of sand, mud, and silt to depth and distance from shore. (B) A similar schematic for carbonate settings.

representing the most plunging of rivers and the most pounding of coastlines,[6] and the shales represent quiet, almost unmoving water. Marine sandstones generally represent shallow, near-shore deposits, and shales are formed of mud from out beyond the sandstones in water slightly deeper than the sands. Shales also appear out on to the edge of the continental shelf where they can represent fairly deepwater deposits (fig. 2.2a).

Carbonate Rocks

Non-siliciclastic rocks are formed through the precipitation of minerals from water by chemical or biochemical processes, and they consist of several types. A few of these are carbonates (which will be very important to us in this book), evaporites, and cherts. Evaporites include rocks containing high abundances of minerals such as halite (sodium chloride, basically salt) or gypsum. These form in beds on the bottoms of shallow seas, usually in areas of restricted circulation where evaporation rates are high. Salt deposits, for example, may form in these areas as the mineral halite is precipitated out of the seawater as concentrations increase with the evaporative loss of water. Eventually such salt layers are buried like any other kind of rock and may be mined, but in some cases such layers—due to their low density and relative softness—may begin to move upward through the rock column creating "salt domes"—bulges in the layers of rock in a region caused by upward movement of low-density salt masses. Salt layers in rocks are mined in a number of places around the world including western Austria, for example, and salt domes rising out of Paleozoic rocks cause many of the uplifts and (along with erosion) much of the spectacular scenery in the canyon country around Moab, Utah.

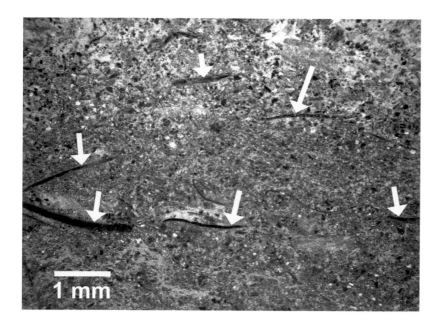

2.3. Limestone shown in thin-section photograph. Matrix is fine-grained calcium carbonate with silt grains mixed in. Arrows indicate abundant fossil fragments. Rock is Lower Cambrian Chambless Limestone from California.

Cherts are rocks made up almost entirely of silica, and they are usually biochemically formed. Although many sandstones may be mostly silica in detrital grain form, cherts are different in that they are almost solid crystalline silicon dioxide and they are often formed in bedded layers at the bottom of the ocean through the raining down of billions of siliceous skeletons of microorganisms. These usually planktonic organisms precipitate the silica from the seawater to form their outer shells, which eventually rain down on to the seafloor and make up whole layer upon layer sequences of almost solid silica. Bedded chert layers can be very hard in outcrop and may be formed from the skeletons or skeletal parts of diatoms, radiolarians, sponges or mixtures of these organisms.[7]

Carbonates are a group of non-siliciclastic rocks consisting of a variety of minerals, the most important of which in ancient rocks are calcite, which forms limestones, and dolomite, which forms dolostones (often simply called dolomites themselves). Calcite is calcium carbonate ($CaCO_3$) and is the most common mineral found in ancient limestones; modern oceanic lime muds are composed of calcium carbonate of a slightly different crystal structure (still $CaCO_3$) called aragonite. But for limestones of the Paleozoic, it is the calcite form that we most often find. Paleozoic limestones also often have fossil fragments preserved in them in abundance (fig. 2.3). Dolomites are very similar in mineral structure to the calcite in limestones but have an added magnesium atom in the calcium carbonate: $CaMg(CO_3)_2$. Most of what we see in the Cambrian are these two types of carbonate rocks: limestones and dolomites. Limestones are formed largely by the accumulation of broken microscopic skeletal supports of calcareous algae (and probably some other forms) that biogenically precipitated calcium carbonate from seawater. Some direct, inorganic precipitation of calcium carbonate may contribute to accumulations of lime mud that eventually becomes limestone, but it appears that this is a minor component. Most lime muds today occur in the shallow, warm waters of the tropics. They can occur down to 3500 m (11,480 ft.) depth but most are at less than 500 m (1640 ft.). Given this source of formation, limestones generally are thought to indicate deposition in relatively shallow-marine environments in relatively warm water. As with chert deposits, it is amazing that so many billions of microorganisms can contribute so much material to the seafloor as to form whole beds of rock, but compared to carbonates, chert beds are a tiny factor in the sedimentary rock record. The calcite production of marine (and some freshwater) microorganisms has, throughout Earth history, contributed so much material that carbonates constitute fully 10–20% of the sedimentary rock record. That is a lot of calcareous algae!

The formation of dolomites is not well understood. In fact, in laboratory settings geologists have had little luck replicating dolomite at normal temperatures and pressures. It appears some dolomite is formed through diagenetic replacement of limestone with magnesium–calcium carbonate, but finds of naturally occurring dolomite in modern environments suggests that somehow this mineral does form at normal atmospheric temperature and pressure. Although we don't know how it forms, we do know dolomite is found today in some areas where lime muds are exposed to air and evaporative conditions, including areas in the Persian Gulf and the Bahamas, suggesting that where dolomites are not clearly formed by diagenetic replacement of calcite in limestones, they may represent **sabkhas** and other arid **supratidal** flats. There may be other environments in which dolomites are formed, but such tidal flats could be a significant source of such rocks.[8]

Some carbonate rocks contain individual grains that are microscopic, and others contain larger crystals; the first is usually indicative of quiet depositional conditions and little alteration of the rock, whereas large crystal size shows that there was some diagenetic alteration of the material, either dolomitization or recrystallization of the calcite, for example. Often times one can find fossils or fragments of fossils in carbonate rocks; sometimes one may find algal structures forming olive-sized clumps in the samples; and high current activity such as tides or waves can cause small, rounded **ooids,** spherical snowballs of carbonate material formed by the rolling action of a single grain that accumulates multiple thin layers around it. Sometimes fossil fragments are the nucleus of an ooid; other times the nucleus may be a sand grain.

Whether limestone or dolomite, most carbonates form in shallow-marine environments away from major siliciclastic input from the continents. Some limestones can form in lakes, but in the Cambrian when you see a limestone you think of shallow-marine environments on the continental shelves but down to perhaps 500 m (1600 ft.) or so. Carbonates generally represent environments deeper than sandstones in the Cambrian, and they may be either deeper or shallower than the shales (fig.

2.2b). With our introduction to sedimentary rocks complete, it is time to pack our bags for our first journey into the Cambrian.

The Grand Canyon is known worldwide for its incredible views both from the rim and deep within its gorge, and the rafting on the Colorado River through several hundred miles of its journey through the canyon is some of the best on the planet. Although its layer upon layer of sedimentary rock is itself an obvious lesson in geology, few who journey to this well-visited destination actually appreciate the true significance of the classroom laid out before them.

The Cambrian rocks of the Grand Canyon are unremarkable from the rim and are probably not ones that most people naturally focus on – certainly not to the degree the casual visitor would tend to notice the cliffs of the Redwall Limestone or the Coconino Sandstone, for example. Even the deep, dark gorge formed by the Colorado's eroding through the Vishnu, Brahma, and Rama schists may catch more eyes than the dull grays, greens, and tans of the rather nondescript Cambrian section just above that gorge. Even the 250 m (820 ft.) of Cambrian rock in the eastern Grand Canyon is just a fraction of what is preserved in the walls of this natural feature.

Twenty-three formations of sedimentary rock representing about 700 million years of geologic history are spread out over a range of approximately 1400 million years in the story of our planet (fig. 2.4). To most, the canyon is scenery, and indeed it excels in that capacity, but visitors can be excused if they don't quite share the geologists' passion for hearing what the Grand Canyon is saying, not because of any inherent dullness in the message – it probably is as fascinating to many tourists as it is to those who study the Earth full time – but simply because the language is as hieroglyphic to the average citizen as is the writing on ancient temples. Few catch all the clues to past worlds that are at their fingertips (literally!) during their hours at the park. Indeed, visitors regularly pass Permian-age synapsid tracks in the Coconino Sandstone without noticing. These tracks of a predecessor of today's mammals are not an arm's length off the South Kaibab Trail and yet few ever see them. The same goes for corals and brachiopods along other trails. In the canyon are also the remains of metamorphosed mountain ranges eroded off and capped with beach sands, sand dunes, reefs, and carbonate shoals. To see some of these elements, however, requires either knowing what one is looking for or having the luck to look in the right spot at the right moment of a hike, and most folks are not primed for either in an average visit. Edward Abbey wanted to free his tourists from the confines of their cars and get them out on the remote trails of Arches to see what was behind the next sandstone fin (and he was willing to free them of roads as well, in order to force the issue!). Visitors to the Grand Canyon generally are away from their cars, but between the scenery and the squirrels there is so much else to see

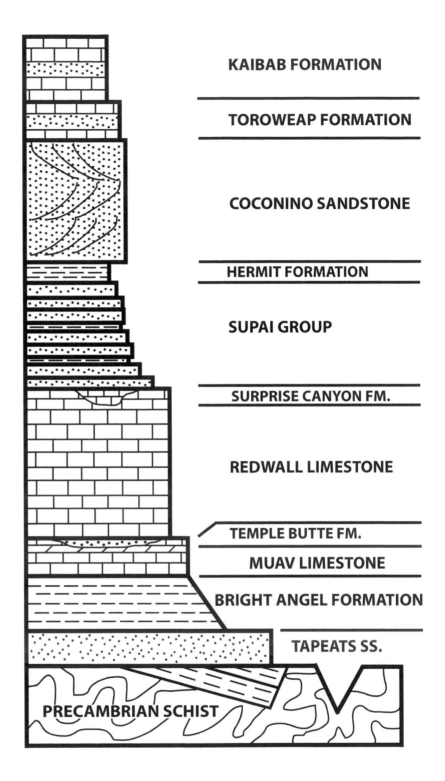

2.4. Stratigraphic section showing geologic formations of the Grand Canyon. Stipple pattern = sandstone; dashed pattern = shale; brick pattern = limestone.

KAIBAB FORMATION

TOROWEAP FORMATION

COCONINO SANDSTONE

HERMIT FORMATION

SUPAI GROUP

SURPRISE CANYON FM.

REDWALL LIMESTONE

TEMPLE BUTTE FM.

MUAV LIMESTONE

BRIGHT ANGEL FORMATION

TAPEATS SS.

PRECAMBRIAN SCHIST

that guides would need a figurative "raising of the voice" to tune people in to the saga in stone under their feet. It is difficult not to shout out to the crowds on the rim, "Slow down! Leave the squirrel food and cameras behind and let's take a slow walk down into the abyss of this canyon and see what there is to see."

Standing on the rim of the Grand Canyon we can look down at our route, but the cliff under our feet is so high and so steep that we can't see the first three miles of the switchback trail. Although we are just over a mile away in straight-line distance we don't visually pick up our planned course until it has finally hit the flats and come out northbound from the base of the walls of rock underneath us. Leaning over the edge we must strain to see thousands of feet nearly straight down to portions of the trail exposed here and there far under our boots. It's steep, and it's high, but somehow the Bright Angel Trail manages to be inviting all the same. What is astounding about the layers of the Grand Canyon is that from across the divide of the entire canyon south to north, units such as the Coconino and Redwall seem like long horizontal strips with little significant thickness. But when you hike down the trails and look across Bright Angel Canyon, for example, as you descend it, each of these formations is a towering wall hundreds of feet high that fills your entire field of vision. And there are hundreds of miles of such dizzying cliff outcrops all around the canyon. The rocks in the Grand Canyon are so flat lying and the canyons dug in the plateau so steep that looking at a geologic map of the area shows color-coded rock layers following close to the topographic lines indicating elevation. From the map we can tell we have about 914 vertical meters (3000 ft.) to drop on our hike.[9]

No time like the present—so off we go, descending the dusty trail and trying to get ahead of the dust-generating mule trains that will inevitably come down behind us. We drop over the edge quickly and start our way down in the shadow of Kolb Studio, passing soon through a tunnel drilled through the Kaibab Formation. This layer of rock is about 130 m (425 ft.) thick and is Middle Permian in age, meaning it was deposited approximately 269 million years ago (yes, we're at the rim of the canyon; this is the young rock around here). As we amble down the trail checking out the wall of Kaibab on one side of us, we note that it is composed of a lot of limestone with some sandstone here and there; beyond the layers of sandstone, there is also some sand mixed in with the limestone. If we were to head east the number of sandstone layers would increase, but here we are mostly seeing limestone of an off-white to yellowish-tan color. And fossils are abundant in the Kaibab—there are fossil sponges, brachiopods, corals, gastropods (marine snails), scaphopods (molluscs with horn-shaped shells), bryozoans (colonial animals related to brachiopods), sharks, and conodonts (chordates with strange teeth). Some of the fossils are nearly complete and identifiable, but many are just crushed fragments in the limestone. Again, limestone represents shallow-marine deposition and sandstone the near-shore environments of a coastline. The mixture of limestone, sandy limestone, and sandstone in the Kaibab indicates that during the middle of the Permian most of Arizona was covered by a shallow, tropical sea, and the shore of that sea was not far to the east in what is now the Four Corners region. On shore from the beach there likely were sand dunes, and all the marine fossils found in the Kaibab were the ocean inhabitants of the shallow sea, which stretched north through Utah and Nevada and southeast into New Mexico. In passing through this limestone we are essentially walking along the bottom of a 269-million-year-old ocean–we will be visiting a lot of ocean bottoms in this book.

As we continue down the trail from what I call the "mule train curve" (a spot on the Bright Angel Trail just below the rim where lines of mules and their passengers sometimes stop for a briefing from the guides before continuing into the heart of the canyon) we descend a long, steep switchback and find ourselves in the Toroweap Formation, a slope-forming and tree-covered series of benches that from a distance is similar in color to the overlying Kaibab Formation and is not at first easily distinguished from it, although the Kaibab is a little more cliff-forming. The Toroweap Formation is 175 m (246 ft.) thick and consists of a mix of sandstone, limestone, and dolomite deposited in and on the edge of a shallow seaway that advanced and retreated slightly from the northwest. In age, the formation is Early Permian, approximately 275 million years old. Size patterns of the component grains, sedimentary structures in the beds, and fossils in these rocks indicate that they were formed in a variety of environments. Sandstones were formed on tidal flats and by shoreline sand dunes. The limestones and dolomites were formed in the shallow seaway when sea level rose, bringing into the area environments friendly to animals such as brachiopods, corals, crinoids, molluscs, and bryozoans. In areas to the north and west, evaporite minerals in the rocks of the Toroweap show that at times the seaway was very shallow and restricted in its circulation, and to the east of here, where the formation consists entirely of sandstone, we can see that the shoreline of the sea was bordered by extensive sand dune fields.[10]

As we continue through the benches of limestone and sandy dolomite past yet another tight hairpin turn, we are now approaching the lower tunnel and the bottom of the Toroweap. At this point we can see the offset of the Bright Angel Fault, which has uplifted the west side of this side canyon 60 m (200 ft.) relative to the east side on which we stand. After we pass through this second tunnel and take a break at the next hairpin (being careful of the small, dive-bombing canyon birds that appear none too pleased at our presence), we notice that we are facing an imposing, vertical, and nearly vision filling wall of sandstone several hundred feet high. This is the Coconino Sandstone, and it is a light tan to whitish formation that in some parts of Grand Canyon can be nearly 185 m (607 ft.) thick. Standing at this hairpin and reflecting on the astounding mass of rock facing us, we notice that within it are huge diagonal striations in sets

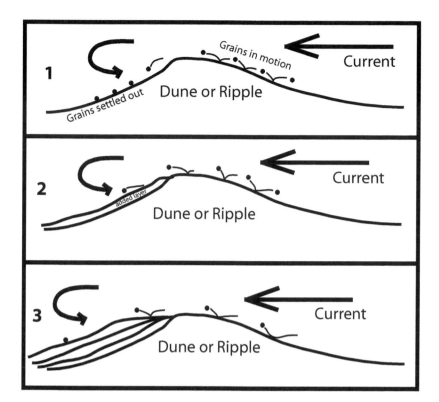

2.5. The formation of cross-beds. (1) Water or air current moves grains up face of dune or ripple; grains settle out in leeward-side current eddies, forming a layer on back side. (2) Process continues and back side layers accumulate. (3) With time, buildup of back side layers results in diagonal crossbedding at an angle to regular (horizonatal) bedding within the unit.

that vary in angle and sometimes direction. These are **crossbeds,** which are essentially the progressively stacked faces of laterally migrating sand dunes cut open by erosion of the cliff and now exposed in cross section. As sand grains are blown up the windward side of sand dunes they hop low across the surface; when they tumble down the opposite side, now in the lee of the wind, they often settle into sloping beds of sand, which are then buried by subsequent millions of sand grains. Sometimes the sands collapse and form beds of miniature sand slides on the leeward side of the dune. In one way or another, as the sand dune migrates down wind, grains of quartz sand pile up in beds that lie roughly diagonal to the ground surface (fig. 2.5). After several million years, these sets of sand dunes can pile up into thick layers of sandstone like the Coconino, which lines the upper levels all around the Grand Canyon as a nearly white band.

So the Coconino Sandstone represents a time in the Early Permian when the area was covered in a giant field of sand dunes before the Toroweap Sea came into the region. The Coconino was named in 1910 by N. H. Darton, whom you may remember from chapter 1 and his finding the Cambrian rocks in the Marble Mountains of California. The giant crossbeds in the Coconino give us an idea of the scale of the sand dunes that existed in this desert—some crossbed sets are up to 20 m (66 ft.) thick, indicating some respectable Permian dune faces to roll down or sandboard on if one had a time machine. Some other sedimentary features preserved in the Coconino Sandstone include windblown ripple marks and raindrop impressions.[11]

As we continue down from our stop we begin the "meat" of our descent of the Coconino. This formation is a sheer cliff and getting down it would not be easy, but thanks to the Bright Angel Fault our hike is actually simpler than we might have feared. Because of the break and movement along the fault, a cone-shaped pile of debris rock stands up against the cliff, and when we get to the top of this we begin a long series of short, tight switchbacks in the well-worn trail that ease us down the cliff along the debris pile. Each time we get to the right side hairpin of each switchback we find ourselves up against the sheer cliff of the Coconino Sandstone, and it is hard not to stare up for a few moments at the impressive pile of billions of sand-dune grains. At our feet here, however, are also some exposed dune face bedding planes, and it is also hard not to crawl around these surfaces to see if we can find what the Coconino is particularly famous for: footprints.

The Coconino Sandstone has footprints in abundance and in high diversity; this was not a desert barren of life in the Permian. Thanks to the footprints left behind by various animals, we believe that among the sand dunes lived scorpions, spiders, isopods (pill bugs), millipedes, and reptilian-like ancient relatives of mammals called synapsids. The tracks of synapsids are particularly abundant on the bedding planes of the Coconino Sandstone and are known by the ichnogenus name *Laoporus* (fig. 2.6).[12] These little oval tracks with up to five claw-like toe impressions are preserved often in long trackways on the bedding planes. The tracks are widely spaced across the mid-line of the trackway with no tail drag, and the steps are relatively short, so what we see is an animal that was usually about 1 foot (30 cm) long with four semi-sprawling legs and a relatively short tail. Crescent-shaped sand bulges on the heels of the tracks indicate that these synapsids were most often walking up the faces of the sand dunes and not uncommonly at an angle across it—we don't know exactly why. If we look on bedding planes in the Coconino it is not unusual to find *Laoporus* tracks; sometimes you can see them just hiking the trails of the canyon. And it is always exciting to see that synapsidian moment in time from so long ago, several seconds of the animal's morning walk among the dunes, preserved right under your nose with the chasm of the canyon spread out below you.

We reach the bottom of the tight cone switchbacks and begin a long, straight traverse across the slopes below the cliff. We notice that the nearly white cliff of the Coconino lies on a sharp, slightly overhanging contact with red mudstone. We have reached the Hermit Formation, which underlies the Coconino and consists of around 30 m (98 ft.) of nearly brick-red siltstone, claystone, and fine-grained sandstone. We can see what appear to be cracks in the top surface of the Hermit filled in with sandstone just like that of the Coconino. This suggests that there was a significant period of time missing between these two formations and that just before the sands of the Coconino were deposited, the top surface of the Hermit was exposed and cracked. The Hermit forms a tree-covered

slope here along our traverse down the trail, and the mudstones and sandstones represent deposition by low-gradient streams on a broad, semi-arid floodplain, as attested to by the terrestrial plant fossils found widely within the Hermit Formation. In age the Hermit is Early Permian at about 280 million years old.[13]

At the bottom of our traverse we come to a shade shelter and (if we're lucky) running water! It's not a mirage; it's about the only way you can get out of the canyon in the summer if you're on your own two feet. A pack animal to ride or carry water makes life easier, but most of us must hike, and hikers usually can't quite carry enough water to climb out of the canyon exposed in the sun in July, for example, and not get dehydrated or, much worse, heat stroked. The two water and shade stops between the rim and the Tonto Platform make exiting the canyon feasible—it's still hot, it's still steep, but it's a lot easier to avoid heat and hydration trouble if you make full use of the rest houses, as the water and shade stops are known. Since we are carrying packs and it is broad daylight in the summer, we will use the shelters even on the way down. Everything is more work in the heat.

Looking out from our vantage point, while sitting on our packs enjoying some water, we see the rock formation known as the Battleship and

2.6. Fossil footprints assigned to the ichnogenus *Laoporus,* from sand dune deposits of the Permian-age Coconino Sandstone near the Grand Canyon. Footprint length = ~1 cm (0.4 in.).

2.7. The upper stratigraphic section at the Grand Canyon, showing the Supai Group (from a point about halfway down to the Tonto Platform) up to the rim formed by the Kaibab Formation.

many thin cliffs of red sandstone stair-stepping down the next several hundred feet below us. We are sitting near the top of the Supai Group (fig. 2.7), and after our rest we begin down the trail again, passing the helicopter "pad" (more of a semi-flat, cleared spot just off the trail, perched precariously out on a rock peninsula just a minute or two's walk from the shelter), and paralleling deep red outcrops of sandstone and shale. These outcrops continue down long traverses and short switchbacks, and we pass a dry gully composed of many solid thin beds of sandstone stacked one upon the other.

The Supai Group is composed of four formations, each with a nearly tongue-twisting name. From the bottom of the Supai up: the Watahomigi, Manakacha, and Wescogame formations are Pennsylvanian in age (~318–299 million years old) and the Esplanade Sandstone is Early Permian (~282 million). These four formations combined (as the Supai Group) are approximately 235 m (771 ft.) thick, and they are composed of varying amounts of sandstone and mudstone with some beds of limestone and dolomite mixed in to some of the formations. Many of the sandstones have crossbeds of various sizes, some indicating sand dunes (of a smaller scale than those seen in the Coconino), while other crossbeds may have formed in water by the lateral migration of sandbars and large ripples in rivers. In these hydrologically formed crossbeds the mode of formation is essentially the same as that of crossbeds formed by sand dunes, except that the fluid moving the sand is simply water rather than wind. The Esplanade Sandstone appears to have been laid down largely by sand dunes. Many of the Supai sandstones have ripple marks indicating movement of sand grains across the dunes by wind. Fossils found in the Supai Group include vertebrate and invertebrate tracks and traces, burrows, brachiopods, shells of single-celled organisms called fusulinids, and plants. These fossils and the characteristics of the rocks that contain them indicate that the Supai Group formations were laid down in a shoreline setting where intermittent streams, bordered by dune fields, flowed into a shallow tropical sea to the northwest. As sea level rose and fell, these environments shifted around and resulted in the interbedded layers of sandstone, mudstone, limestone, and dolomite seen in the formations today.[14]

After much winding down through all the red rock of the Supai we find ourselves at another rest house (this one known as Three-Mile) perched out over a cliff. Looking back up the trail at where we started, we see we are now getting reasonably deep into the canyon, almost half way. The Supai, Hermit, Coconino, Toroweap, and Kaibab are all above us as we take another break. Below us we see another sheer cliff of a scale similar to that of the Coconino, but this one is of a red color just lighter than the Supai above us. This is the Redwall Limestone, and it is our next descent obstacle. As we did at the Coconino, we will navigate our way down the Redwall by way of many short, snaking switchbacks worked into another cone of debris rock near the Bright Angel Fault. After refilling water bottles and convincing a few of the local squirrels that no, they are not welcome to help themselves to our trail mix, we begin down the trail again. The first few switchbacks are cut deeply into the Redwall Limestone, and we soon notice that the rock is actually a light gray color; the red color is surficial and is just washed down from the deep red rocks above. With our noses nearly on the outcrops along the trail we see that the gray rock is a clean, fine-grained limestone, and after a few minutes we notice a coral fossil embedded in it. Down the trail a little further we are faced with yet another gigantic wall of this formation facing us from just across the way. We can see little else but a full vertical surface of red-stained limestone.

The cliff of Redwall Limestone near the Bright Angel Trail is around 150 m (492 ft.) high, but the formation can be up to 250 m (820 ft.) thick in other areas. The Redwall was deposited during the Mississippian period, ranging from ~345–328 million years old, and it consists mostly of limestone, although the lower third or so contains a lot of dolomite beds. In some beds of the limestone there are some crossbeds, and in others the small, rounded grains of calcium carbonate known as ooids. The crossbeds, as we have seen, are formed by currents within a fluid, and ooids are as well. Ooids, as we also saw earlier, form when calcium carbonate accumulates around a small nucleus grain as it rolls back and forth on the bottom of a shallow carbonate shoal (see fig. 2.2b), much as a snowball builds up snow. You will remember that limestones and dolomites accumulate on the bottoms of warm, shallow seas. The ooids and crossbeds seen in some parts of the Redwall Limestone indicate that sometimes there were currents on the bottom of that sea. Fossils from the Redwall also indicate shallow-marine conditions, as we saw up the trail with our coral. In addition to the corals, other types of fossils found in the Redwall include brachiopods, clams, cephalopods (squid relatives), bryozoans, crinoids (echinoderms), fish, algae, crustaceans, foraminiferans (shelled, single-celled organisms), and trilobites.[15] All of these types of organisms would have found the shallow-marine shelf setting of Redwall times inviting – although the sea level varied throughout this period, this part of northern Arizona seems always to have been covered by the sea during this interval. It was "Bahamas time" in Arizona – and the living was easy if you were a marine organism.

One rock unit occurs between the Supai and the Redwall, but only in isolated spots. This is the Surprise Canyon Formation, and it is later Mississippian in age. In the eastern part of the Grand Canyon where we are it occurs only as isolated, infilled scours into the Redwall Limestone. After our Redwall sea deposited all that calcium carbonate mud (millions of years later, but still during the Mississippian), the limestone was exposed and eroded into narrow gullies that were soon after filled with other sediment. Another formation with spotty distribution in this part of the canyon is the Temple Butte Formation, and it occurs here and there below the Redwall Limestone.

After many switchbacks down the Redwall we find ourselves nearing an area where the trail levels out and straightens somewhat. All of the Redwall is above us now. We are here! We are finally coming to the Cambrian rocks, and our first outcrop is just ahead. Passing the internally light-colored limestone we notice it is weathered dark on the surface and seems to consist of thick beds stair-stepped below the sheer cliffs of the Redwall. As the trail meanders north down the gentle slope nearing the Tonto Platform there is a dry streambed to our left. The trail now is particularly dusty and rocky, and we find it is significantly hotter than above. On the opposite side of the dry stream is a tall, crumbling slope of almost unnaturally green shale and sandstone. We continue hiking past a thick growth of cactus, and a few minutes further on we come to (what is this?) shade, in the form of large cottonwood trees. The growth of such large trees here is assisted by a spring which in places at Indian Garden makes the ground quite damp and makes possible the growth of abundant green grass and horsetails – a veritable oasis. The tree-mounted thermometer here reads 39°C (103°F) in the shade, so we decide to take refuge and camp in the shade of the trees. We will explore our Cambrian rocks in the morning.

Camped out in the canyon you are among the cottonwoods and cool groundcover, and the temperature is almost always noticeably warmer than on the rim. Often it is plain baking, even at night. The multiple layers of cliffs that comprise the canyon walls tower more than 914 m (3000 ft.) above you to the south. The shade of the cottonwoods is a relief, but the abundant mice and squirrels, all of them bold little demons not afraid to challenge you for your food, are not. At night in camp the Milky Way may come out and arch over the entire canyon in line with the Bright Angel Trail as if it were a span constructed for that purpose. The canyon walls may be lit up by moonlight, or a bright planet in the sky to the south may blend in with the lights of the lodges along the South Rim, leaving you unsure for a moment if you are seeing a celestial neighbor in our solar system or a seemingly more distant reminder that human civilization carries on just over half a mile above you. By dawn the orange-band light of sunrise begins creeping down the Redwall Limestone to the west of your camp and you pick up the pace of your preparations to get to work while the canyon still has this fleeting bit of shade. It is time to hit the trail again and explore the Cambrian rocks all around us.

Hiking the Tonto Platform is a joy. As beautiful as the Grand Canyon is from nearly any vantage point along the rim, and as stunning as it is to hike the disorientingly steep trails down from that rim, my favorite part of the Grand Canyon is being *in* it–down on the Tonto Platform hiking the flats with the canyon walls surrounding you, vertical barriers in every direction and stunning vistas no matter where you turn. Here you experience thunderstorms, rainbows, virga, and the singular smell of the wet desert–often in the course of a single hike. You are on flat, exposed terrain with no shade–a deep black canyon below you on one side, sheer orange walls rising above you on the other–but you are totally content where you are.

Although we have come down the canyon descending the rock formations top to bottom, going backward through time, now that we are in the Cambrian rocks, let us start at the base and work our way forward in time *through* the Cambrian. To do this we must hike the 2.4 km (1.5 mi.) out to Plateau Point, where are finally rewarded with a view of the Colorado River, still far below. Pipe Creek Rapid is visible down there some 396 m (1300 ft.) below our boots, and the Precambrian schist of the Inner Gorge is darkening the scene and making the depths of the canyon distinctly forbidding compared to the red sandstone beauty of the cliffs above the Tonto Platform.

Grand Canyon Cambrian: The Tonto Group

Standing on the point we find ourselves among outcrops and bedding planes of the base of the Tonto Group, a unit of three (now four) formations, all of Cambrian age, named by G. K. Gilbert back in 1874. The formations traditionally included are, from bottom to top, the Tapeats Sandstone, the Bright Angel Formation, and the Muav Limestone (fig. 2.8; plate 1); the recently designated Frenchman Mountain Dolomite unconformably overlies the Muav. The first three formations were named by Levi Noble in 1914. At Plateau Point we are walking around on the bedding planes of blocks of Tapeats Sandstone, and looking across to

2.8. The Cambrian section at the Grand Canyon showing the three main formations of the Tonto Group. Stipple pattern = sandstone; dashed pattern = shale; brick pattern = limestone.

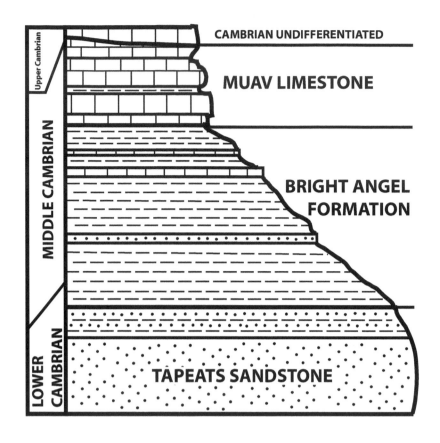

CAMBRIAN UNDIFFERENTIATED

Upper Cambrian

MUAV LIMESTONE

MIDDLE CAMBRIAN

BRIGHT ANGEL FORMATION

LOWER CAMBRIAN

TAPEATS SANDSTONE

2.9. Elements of the Cambrian Tapeats Sandstone in Grand Canyon. (A) The cliff of the main part of the Tapeats Sandstone resting on Precambrian metamorphic rock near Pipe Creek. (B) Sandstone of the Tapeats above and below a layer of green shale approximately 10 cm (4 in.) thick. (C) Crossbedding (about 30 cm [1 ft.] thick) in the Tapeats Sandstone. (D) An olenellid trilobite fossil (*Olenellus*?) from the upper transition beds of the Tapeats Sandstone in the western Grand Canyon, Museum of Northern Arizona specimen (MNA N.2202). Scale bar = 1 cm. (E) Sandstone bedding surface with numerous *Diplocraterion* burrow bottoms, near Plateau Point. (F) Trilobite or other arthropod trace fossil *Cruziana* from the Tapeats Sandstone, GRCA specimen. Scale bar = 5 cm.

the east we can see that this formation (at this point at least) rests on the schist of the Inner Gorge.

The Tapeats Sandstone in the Plateau Point area consists of a 164-foot (50-m) cliff of brown and tan sandstone (fig. 2.9a).[16] Some of the sandstone is pebbly, and in many places it contains crossbeds; there are even thin beds of shale within the formation (fig. 2.9b,c). As we saw earlier in this chapter, crossbeds and pebbles in sandstone suggest relatively high energy and shallow water; the fact that shale is mixed in with the sandstone indicates that the environment shifted to one of relatively quiet conditions, at least for short periods and at least in this local area. Many of the bedding planes in the Tapeats preserve ripple marks, indicating that currents influenced the sand, currents most likely caused by rivers, wave action, and tides. Trace fossils of Cambrian sea animals are abundant in some layers of the Tapeats, and trilobites are rare but have been found in it. Among the animal traces are those of trilobites (or similar arthropods) called *Cruziana* (fig. 2.9f) and possible worm burrows called *Diplocraterion* and *Arenicolites*. Among the fragmentary trilobite specimens from the upper layers of the Tapeats Sandstone (in the western part of Grand Canyon) is one named *Olenellus* (fig. 2.9d), which we will see more of later. At Plateau Point evidence of the marine life that lived in the sands at the bottom of the Cambrian ocean is all around us and beneath our toes as we walk around. Numerous short linear indentations in the rock

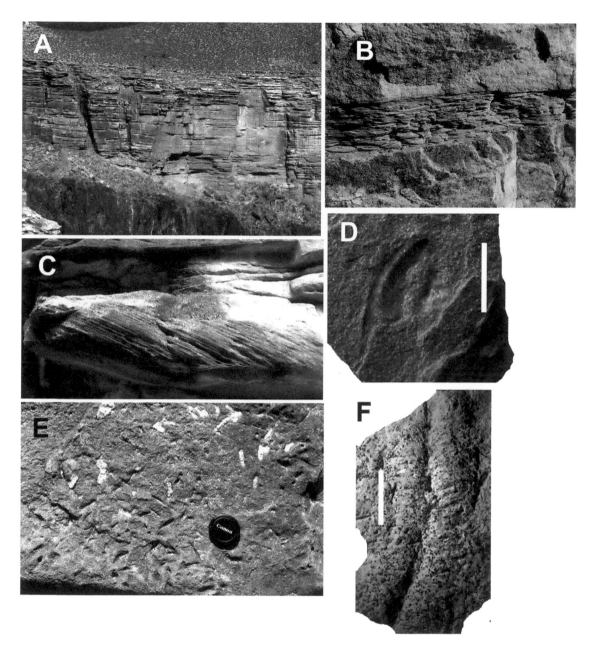

are actually the bottom ends of the U-shaped *Diplocraterion* burrows (fig. 2.9e). These were probably made by some type of wormlike organism living in the sand with a feeding apparatus sticking out of one end of the U shape. The abundance of these traces in the Tapeats Sandstone is nearly as exciting to a paleontologist as is the view of the Inner Gorge.

The characteristics of the Tapeats Sandstone indicate a mix of shallow, near-shore, and shoreline environments in a continental to marine setting: sandy sea bottoms up to 33 m (108 ft.) deep, subtidal channels, tidal flats, estuaries, beaches, beach dunes, and braided streams bordering tidal flats.[17] This was the shore, the shallow water, tidal zone, and beach playground (for worms) of North America's then-northern coast, and as we will see, the shallow-marine environment extended far off the coast. Mudcracks indicate periods of subaerial exposure in the tidal flats, and sand dune deposits near Phantom Ranch indicate that parts of the Tapeats were dry land coastal sediments.[18] In the area below us the Tapeats rests on metamorphic rocks of the Precambrian Rama or Brahma schists. There are places in which outcroppings of these metamorphic rocks stick up into the Tapeats, suggesting that the Tapeats and its beach and shallow-marine sands were deposited on top of, and in some cases around, outcrops or islands of this dark, much older rock.

In order to reach the next formation up, the Bright Angel Formation,[19] we will hike back south to the gradual slopes below the cliffs of Grand Canyon. Here we return to the green shale we saw on our way down the canyon. The Bright Angel Formation was once described by geologist John Strong Newberry as having sandstones that possessed "an indescribable look of antiquity,"[20] and indeed these rocks can appear so otherworldly as to punctuate their great age. The Bright Angel in this area consists of 86 m (282 ft.) of thin-bedded, green shale, siltstone, and fine-grained sandstone with some beds of brownish dolomite and brownish, coarse-grained to conglomeratic sandstone (fig. 2.10). The green color of the Bright Angel Formation is in places quite striking, and the color of the siltstones and sandstones is caused by **glauconite,** an iron silicate mineral that forms in shallow-marine continental shelf settings characterized by the presence of some organic matter and relatively low sedimentation rates. The green color of the shale beds is often caused by the clay mineral **chlorite.** Although it is dominated by shale, the Bright Angel contains a significant number of sandstone beds. The sandstones and shales often occur in beds that are more lenticular (on a large scale) than planar.[21]

Sedimentary structures seen in the formation include crossbedding in the sandstones, along with oscillation ripples and interference (or ladder-back ripples) in the shales. Oscillation ripples are what you see at the beach or when you are ankle deep in water at a lakeshore – parallel ridges and troughs of sand, symmetrical in cross section, caused by back-and-forth wave action over fine-grained sediments. These ripples can be somewhat larger than what you see at a lakeshore; some oscillation ripples are large (up to several centimeters high) and indicate wave energy affecting bottom sediments in reasonably deep water (still shallow marine, but deeper than you are likely to see just wading into the surf). **Interference ripples** also indicate current action on bottom sediments, but there are two sets of ridges and troughs so that the pattern resembles a grid with square depressions surrounded by ridges. Such a pattern results from current patterns switching approximately 90 degrees and superimposing a second set of ripples over a previously formed set. Interference ripples can often indicate that tidal currents influenced an area. As we saw earlier, crossbedding usually indicates current activity causing sand bar migration.

2.10. Characteristics of the Bright Angel Formation in Grand Canyon. (A) An exposed section of the Bright Angel Formation showing interbedded shale and sandstone, near Garden Creek. (B) Trilobite or arthropod trace fossil *Cruziana* from the Bright Angel Formation near Indian Garden. (C) Side view of U-shaped worm burrows (*Diplocraterion*) in sandstone within Bright Angel Formation, near Indian Garden. (D) Olenellid trilobite cephalon (*Mesonacis* cf. *fremonti*?) from the lower Bright Angel Formation of Bridge Canyon, scale bar = 3 cm.

(A)–(C), field shots; (D), Museum of Northern Arizona specimen (MNA 2279).

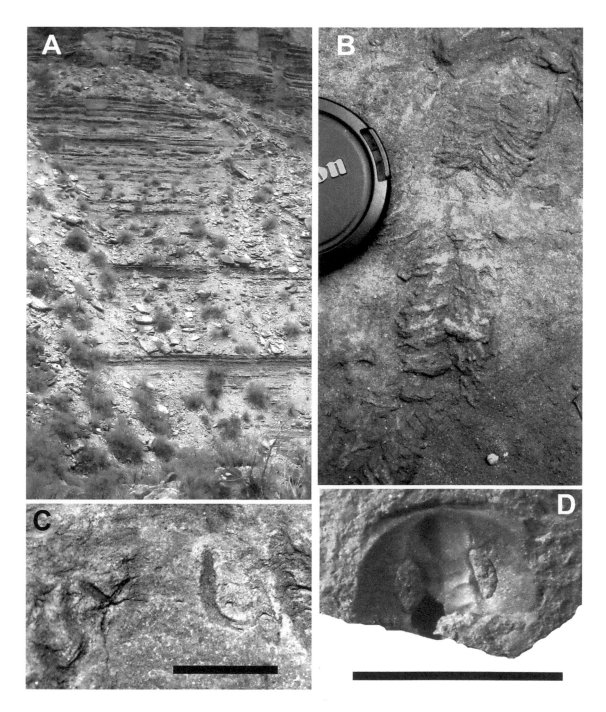

Of the three formations in the Tonto Group, the Bright Angel Formation appears to be the most fossiliferous. Fossils found in the formation (plates 2 and 3) include trilobites, brachiopods, hyoliths, stem group crustaceans, echinoderms, and trace fossils–many, many trace fossils. The trilobites include forms such as *Olenellus, Mesonacis*? (fig. 2.10d), *Glossopleura, Amecephalus,* and *Elrathina.* We will see these trilobites at many sites in coming chapters. The diversity of trace fossils is impressive. Some of the traces include the same types seen in the underlying Tapeats Sandstone – the trilobite trace *Cruziana* and the worm burrow *Diplocraterion* (fig. 2.10b,c). But there are others as well, including a different type of trilobite (or similar arthropod) trace called *Rusophycus*. The difference? *Cruziana* appears to be a trilobite plowing through sediment, whereas *Rusophycus* seem to be resting traces where an individual burrowed partway down and then hung out for a bit. Many other strange and wonderful traces are found in the Bright Angel Formation, and many of them make you ponder what animal was making such a mark and how was it doing it. The interesting value in trace fossils is not only that you can see behavior and know of the presence of some animals despite their lack of body fossils in a particular layer, but certain groups of traces are also known to be indicative of certain depths and environments in the ocean. This is based on comparisons with traces in modern environments. Some Cambrian traces aren't all that different from some we see today.[22]

The rocks, sedimentary structures, and fossils we have seen in the Bright Angel Formation indicate that it was deposited on an open marine shelf, in water probably deeper and farther offshore than that of the Tapeats Sandstone, yet not as far offshore as that of the overlying limestone formation above. The Bright Angel Formation consists of deposits representing a range of specific environments, including subtidal marine settings where the bottom sediments were influenced by tidal currents and storm action; coarse sediments representing storm deposits overlain by finer-grained sands, silts, and shales (some with ripplemarks); and shales and very fine-grained sandstones representing after- and between-storm periods when sediment settled out of calm waters. Recent studies have also suggested that much of the Bright Angel Formation was deposited closer to shore than we have traditionally thought – perhaps near an estuary. This is based in part on reinterpretation of some of the sedimentary structure and fossil-distribution patterns in the Bright Angel, along with the finding of apparent terrestrial moss spores in the formation – spores that would have washed or been blown into the shallows from land. Could the Bright Angel Formation have been deposited closer to shore than we thought? Could there have been mosses growing on what we thought for decades were essentially bare continental regions during the Cambrian? Perhaps. Recent studies have also suggested that continental areas in the Cambrian (and even earlier) might have been covered to some degree in terrestrial algae, as indicated indirectly by carbon isotope ratios. And it appears possible that Early Cambrian soils may have contained lichens and slime molds, too, at least in areas not far from the ocean.[23] In any case, one thing everyone seems to agree on is that the depth of the water in which the Bright Angel Formation was deposited was relatively shallow, and in some places very shallow.[24]

Either way, during Bright Angel times in the southwestern United States (early Middle Cambrian), most of New Mexico, Colorado, and eastern Wyoming was exposed, terrestrial continent – Laurentia. Rivers

flowed off this continent into the shallow sea that existed in what is now western Wyoming, almost all of Utah, Nevada, and most of western and southern Arizona. We don't know exactly where the shoreline was at the time, relative to this part of the Grand Canyon, but it seems to have been clearly less than 162 km (100 mi.) to the east–probably only a few tens of miles. Other parts of the shallow sea were quite far offshore. The rocks deposited in eastern Nevada at the time were also relatively shallow, still on the continental shelf, and they were probably close to 243 km (150 mi.) from the coast! This illustrates how extensive the area of flooded continent was during the Cambrian. The coastline probably ran north-south (modern directions) through central Wyoming, along the Colorado-Utah border, and then out into central Arizona before curving around to the southeast. The flooded continent encompassed most of the area west

2.11. The Muav Limestone in Grand Canyon. (A) Outcrops of the Muav Limestone. (B) Posterior thoracic section (with partial pygidium?) of the trilobite *Glyphaspis* sp. (GRCA 11985) from the Muav Limestone; specimen about 3 cm across.

of this line, and deep water was not reached until one hit what is now approximately the California-Nevada border.[25] That's a lot of shallow continental shelf and a lot of habitat for marine organisms.

The lower contact of the Bright Angel Formation with the Tapeats Sandstone occurs within a series of thin, interbedded sandstones and shales marking the transition from the predominantly sandstone deposition of the Tapeats to the Bright Angel's shale-dominated section. The contact of the Bright Angel with the overlying Muav Limestone is also transitional, but the contact occurs between an interlacing series of tongues of each formation. This is particularly apparent in exposure areas to the west, but at the Bright Angel Trail section there is at least one thick, brown, dolomitic tongue of Muav type that occurs in the upper Bright Angel Formation (with Bright Angel Formation both below and above the tongue).[26]

The Muav Limestone consists of thin- to thick-bedded, gray to white, brown-weathering, very fine grained, mottled limestone and dolomite with a mix of beds of greenish, calcareous shale, light-colored fine-grained sandstone, and silty limestone (fig. 2.11).[27] It is approximately 116 m (381 ft.) thick in the vicinity of the Bright Angel Trail and forms a stair-stepped cliff above the Bright Angel Formation and below the sheer cliff of the Redwall Limestone. There are four members of the Muav Limestone, all Middle Cambrian in age. From bottom to top these are the Peach Springs, Kanab Canyon, Gateway Canyon, and Havasu members. As mentioned above, similar limestones and dolomites occur in the underlying Bright Angel Formation but are beds or members of that formation. Beds of the Muav are generally thinner and comprise a higher percentage of sandstone and shale in the eastern Grand Canyon. In fact, in the Bright Angel Trail area there is a considerable amount of siltstone and sandstone in the formation. As with many Paleozoic formations in the Grand Canyon, the Muav thins to the south, suggesting the basin was centered to the west and north; in the eastern Grand Canyon we are closer to what would then have been the edge of the depositional basin. Most of the limestone and dolomite in the Muav is horizontally laminated or unlaminated, but in some places small crossbeds are apparent.[28]

Fossils in the Muav Limestone include brachiopods, molluscs, trilobites (fig. 2.11b), and sponges. There are also olive-sized balls of concentrically laminated carbonate in the Muav referred to as **oncoliths;** these structures may form by the rolling of ripped-up pieces of algal or cyanobacterial mats that accumulate on the bottom like a snowball or by algal or cyanobacterial growth in place, but they do appear to be related to photosynthesizing microbes somehow in their formation. The cyanobacterium *Girvanella* has been associated with these oncoliths in a number of other formations, and we will see oncoliths in limestones in coming chapters.

The Muav Limestone probably was deposited in warm, shallow-marine environments some distance from the shoreline, perhaps out near the shelf edge not far from deep water.[29] The water was probably deeper than that of the Bright Angel Formation in many areas. The abundance of sandstone in the eastern outcrops of the Muav suggests that sedimentary input to the ocean was not too far from these areas, however. The area of the Bright Angel Trail, with its sandy and silty Muav, was likely much closer to the coastline than was the far western end of the Grand Canyon.

A characteristic rock type in the Muav Limestone is the "intraformational flat-pebble conglomerates." Say it fast; it's fun. This phrase describes a number of beds that consist of dense accumulations of large flattened (rather than rounded) pebbles of carbonate material derived not from elsewhere but from within the Muav itself—pulled up, concentrated, and deposited all within the basin and during deposition of this one formation. Normally pebbles derived from older rocks are washed in from elsewhere, so what makes these IFPCs interesting is that the pebbles were not only derived from within the Muav's depositional basin, but also the pebbles are made of fine-grained carbonate laid down, lithified, transported a short distance, and redeposited within the formation's depositional timeframe.[30] They are Muav within the Muav. So the question is, since most rocks lithify after they are deposited and buried, where and how were these pebbles lithified in place as sediment before deep burial (to allow them to be ripped up and redeposited)? The best guesses are that they were lithified in either intertidal or possibly subtidal settings before being ripped from the sediment surface by storms or tidal channel currents. Modern carbonate tidal flats and their surrounding environments in some cases have been observed to contain carbonates that begin to lithify at the surface in humanly observable timeframes. After being pulled up by storm or tide currents the flat carbonate pebbles would have been transported a short distance, concentrated, and buried as these beds of IFPCs. Such beds are also abundant even in the western canyon. Subtidal channels and intertidal areas sound like very shallow water, however. In fact, rocks identified as being intertidal and even supratidal (exposed to the air) have been identified in the Muav Limestone even in its western outcrops. If this was offshore from the Bright Angel and in deeper water, how do we end up with such rocks? It appears that even among the vast miles of relatively deep carbonate sea bottom out on the shelf there were shoals that came so close to the surface in shallow water that during some low tides they were exposed as tidal flat islands. Such areas occur today on the shallow shoals surrounding some islands in the Bahamas.

Above the Muav and below the Temple Butte lie 8 m (26 ft.) of unnamed Upper Cambrian dolomite, the youngest Cambrian rock in the section. This set of rocks gets much, much thicker in the western canyon. It has been referred to as an equivalent of an Upper Cambrian unit known as the Nopah Formation,[31] which is found in the Virgin River gorge in far northwestern Arizona, south of St. George, Utah, along Interstate 15 as you drive toward Las Vegas. Little else is known about this "Cambrian undifferentiated" (or "Nopah equivalent") in the Grand Canyon. Its western equivalent has been named as a fourth formation in the Tonto Group, the Frenchman Mountain Dolomite.

The Tonto Group as Geology Text: Tradition and Developments

There are a couple ways to interpret the Tapeats Sandstone, Bright Angel Formation, and Muav Limestone. For years the Tonto Group has served as an example of the Cambrian transgressive Sauk Sequence. (Don't worry, I'll explain that.) Even when I was in school, not all that long ago, the Cambrian of the Grand Canyon was interpreted in that way. It is beginning to look as if the Tonto Group was deposited in a setting a little different from the traditional, transgressive, shallow-marine interpretation. But we will get to that. Now let us take a look at that traditional interpretation of the Tonto Group because there are other Cambrian sections in western North America where such a model is appropriate. Then we will take a look at the Tonto Group as it is beginning to look through a modern interpretation.

Tradition: The Tonto as Sauk Transgression

There are a few concepts we must hit first. The first of these is the **transgression,** an expansion of shallow-marine areas, often involving the flooding of continental areas, caused by a rise in sea level. As sea level goes up, shallow-marine deposits often encroach farther onto the continent until sometimes nearly an entire continent is covered in water.[32] Major transgressions such as these can pile up **sequences** of rocks that record the event, and in 1963 the six major sequences in North America were identified and named. The Cambrian transgression deposited what was named the **Sauk Sequence,** and most places you go in North America (or the world for that matter) record this transgressive pile of rocks in the outcrops.[33]

Next up is the **unconformity.** An unconformity is a surface between layers of rock that represents a significant amount of missing time. The surface may be there due to a lack of deposition during the time between the underlying and overlying layers, or it may be that the layers were deposited but later eroded before the overlying beds were laid down. Such surfaces are important, and there are four types of them. In an angular unconformity older rocks have been tilted up, eroded off, and had younger layers of rock deposited on top of them. A lot has to happen in order for there to be an angular unconformity–lithification of the older rocks, uplift to tilt the older rocks to an angle, erosion, more deposition, lithification of the younger rocks–and it takes a very long time for such things to happen, which is why John Playfair, colleague of the geologist James Hutton, way back in the 1780s, said of their study of a famous angular unconformity on the coast of Scotland: "The mind seemed to grow giddy by looking so far into the abyss of time." Indeed it does, and happily for the rest of us angular unconformities are not all that uncommon, so we all have the opportunity to experience the geological contact high that Playfair was describing if we can just find the right outcrop. A **disconformity** describes a type of unconformity in which parallel-lying beds above and below are separated by a clear erosional surface. Such an unconformity exists in the Grand Canyon where the Surprise Canyon Formation has been deposited over an erosional surface that cut gullies into the underlying Redwall Limestone. **Nonconformities** exist between sedimentary beds that are deposited on top of igneous or metamorphic rocks below. Such a contact exists in many places in the Grand Canyon between the schist of the Inner Gorge and the sandstone beds of the overlying Tapeats Sandstone. This is another situation in which you see into the abyss of time, as the metamorphic or igneous rocks also must be eroded off prior to deposition of the overlying sedimentary rocks. They aren't always eroded completely flat, however. In the Grand Canyon as well as other places the rocks underlying the basal Cambrian sedimentary formation often have some topography to them, and the Cambrian sands and the waters that carried them at one time surrounded islands of Precambrian rocks. A **paraconformity** is an unconformity between two sets of parallel beds with time clearly missing but without a clear erosional surface at the contact. This is a subtle type of unconformity. The contact between the Hermit Formation and the Coconino Sandstone in the Grand Canyon is sharp and fairly flat lying (aside from the cracks in the Hermit) without obvious angular differences to the beds and could be considered a paraconformity.

Where no unconformities exist between sedimentary rock layers the contact is considered conformable, and little or no time is missing, nor was there any erosion. Often, as one travels up through a section of conformable rocks, the rock types dominant below will begin to alternate in an interbedded fashion with those dominant above until soon the rock type has changed completely. This zone of interbedding will in many cases mark a formation boundary, but one that is sometimes difficult to define.

Why are unconformities and conformable contacts important to our discussion of Cambrian geology and the Grand Canyon? We'll see in a minute, but first there are just a few more concepts that we need to cover. Next up are **facies.** A facies in geology is simply a characteristic association of rock types (and often sedimentary structures, too) that characterizes a particular environment or depositional setting. It may consist of one or numerous kinds of sedimentary rock and specific (or no) types of sedimentary structures – but whatever the combination, it is indicative of a specific setting. Thus, the combination of fine-grained quartz sandstone with raindrop impressions and large-scale crossbeds (among other characteristics) that we saw in the Coconino Sandstone comprises an eolian sandstone facies that indicates large sand dunes in the area during the Permian. Similarly, if one found a mix of thin, ripple-marked, crossbedded sandstones interbedded with layers of coal, there would be reason to classify this facies as possibly representing a swampy deltaic setting. Facies

2.12. The vertical stacking of laterally adjacent facies. (1) Time 1 has the local section area occupied by beach sand with shallow marine silt and deep marine mud farther offshore. (2) By Time 2 rising sea level has moved the beach sand farther inland and now occupying the local section is shallow marine silt. (3) At Time 3 sea level has gone up more, the beach setting has moved far onto the continent and marine mud has moved into the local section. Note that the lateral relationships of beach/shallow/deep (sand, silt, mud) have remained the same during the entire sea level rise; only what has occupied the local section has changed. Note in Time 3 that the local section now contains mud above silt above sand, recording a sea level rise and the lateral movement of facies.

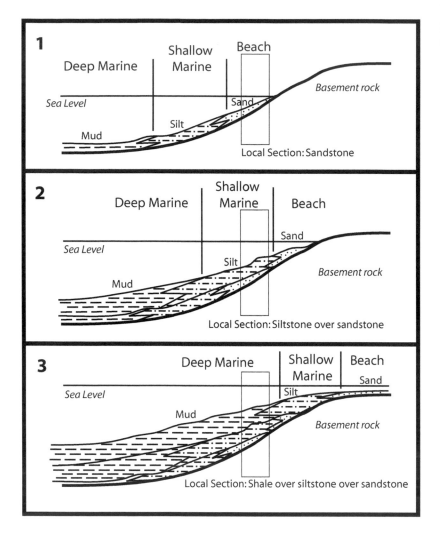

and their associations are how sedimentologists read the rocks for clues to the origins and depositional settings of the layers.

A fourth concept we need to cover is that of the **time transgressive** nature of many sedimentary formations. In order to be time trangressive, a formation must vary in age *laterally* across its extent. All sedimentary formations are, of course, oldest at the bottom and youngest at their tops, but from fossils we can tell that some formations also may be older on one end of their geographic extent and younger on the other end. Weird? Yes, but logical; this will all make sense in a minute. Trust me.

Finally, we have **Walther's Law,** named after one Johannes Walther, a German geologist who in 1894 made an observation key to our understanding of stratigraphy. What we call laws can be dangerous when dealing with nature. There are often exceptions that throw you off in your pursuit of understanding, but in general this guideline of Walther's states that within a continuous series of sedimentary rocks, the vertical stacking of facies should result from the lateral shifting of environments in the ancient depositional area.[34] So as long as there are no unconformities between them, a vertical transition of beds on any one point on a continent (say we call it Point X) that goes from a beach sand, for example, to a deepwater shale indicates not that the beach vanished but simply that the beach migrated laterally elsewhere and that by this later time our Point X found itself farther offshore (i.e., accumulating mud now rather than sand; fig. 2.12). Remember, again, that if there is an unconformity between the two facies, all bets are off, but Walther's Law tells us that within conformable sequences of rock, the vertical succession of facies we see can be read as a log of paleoenvironments that existed in the region and the order in which they occupied our particular Point X. One main reason that environments may shift is a sea level rise or fall, which would move the beach, for example, one way or the other. Alternatively, even in the absence of any change in sea level, there may be natural environmental shift within some environments. Out on a carbonate shoal, for example, currents and tides may randomly shift shallow intertidal shoals away and bring in to a spot subtidal carbonate muds, and this may be followed by a shallow oolitic shoal—no sea level change, just natural shifting of subenvironments.

Now, all of this is critical for assembling the big picture in the Cambrian. Everything we've just been discussing helps pull together the concepts and pattern we see in the Cambrian rocks of the Grand Canyon and will prove to be a familiar story as we explore the Cambrian through time and elsewhere in North America. Here we have it all coming together. First, remember that the main representative of the Cambrian in the Grand Canyon is the Tonto Group, comprised of the Tapeats Sandstone, the Bright Angel Formation, and the Muav Limestone, traditionally interpreted as representing the Sauk Sequence of the Cambrian transgression.[35] The bottom of this sequence is bounded by the nonconformity or the angular unconformity (depending on the spot) at the base of the Tapeats where it rests either on schist or uplifted Precambrian sedimentary rocks. The top of the sequence above the Muav is bounded by an unconformity also, but the contacts between the Tapeats and Bright Angel and the Bright Angel and Muav are gradational and conformable. This makes the Tapeats–Bright Angel–Muav sequence a single, related package of rocks consisting of a progression of facies. The progression of facies runs from sands in beach, shallow near-shore, tidal flat, and braided stream environments (Tapeats) to muds and sands in shallow shelf, lagoonal, and estuary environments (Bright Angel) to carbonate muds in offshore shelf and shallow carbonate shoal environments (Muav). This progression suggests a gradual deepening in the environmental setting of the continuous sequence.

The Cambrian formations of the Grand Canyon may be time transgressive, as exemplified by the Bright Angel Formation. The Bright Angel in the far western Grand Canyon contains Early Cambrian trilobites such as *Olenellus* in its lower levels. In the eastern Grand Canyon even the

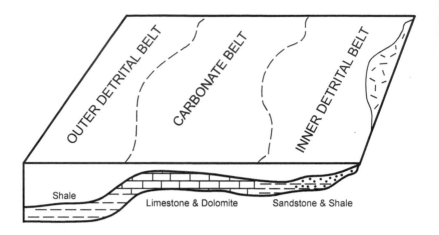

2.13. Relationships of inner and outer detrital and middle carbonate belts of Palmer (1960) to rock types and depths along a hypothetical continental margin.

Shale

Limestone & Dolomite

Sandstone & Shale

lowest levels of the Bright Angel contain Middle Cambrian trilobites such as *Glossopleura*. The Bright Angel (along with the Tapeats and Muav) seems to become, at any one stratigraphic level within it, progressively younger as you move from west to east. The Bright Angel was deposited along a wide belt, but this belt (which ran roughly north–south) moved slowly eastward through time as the sediments were laid down. Why? Put simply, sea level was rising and as it did it encroached farther and farther onto the continent, and the shoreline moved ever so slowly eastward.

This belt gives us an opportunity to discuss a concept introduced by Allison Palmer in 1960 to categorize different Cambrian formations of western North America depending on their paleogeographic location and environmental setting at the time.[36] What Palmer designated were the Inner and Outer Detrital belts separated by a Carbonate Belt (fig. 2.13). The Inner Detrital Belt consisted of sands close to shore and muds and some thin carbonate mud layers just offshore out on the shallow shelf. The Carbonate Belt was just out from this and consisted of thick accumulations of carbonate muds out on the edge of the shallow shelf, an area scattered also with shallow carbonate shoals. Beyond this was the Outer Detrital Belt, which consisted of dark muds on the edge of the shelf where the slope dropped quickly into deep water. If this sounds vaguely familiar, it should, because the Tapeats and Bright Angel would be part of Palmer's Inner Detrital Belt and the Muav would be within the Carbonate Belt. Here we have Walther's Law in action. At any given time during Tonto Group deposition, if we'd flown over the coast of western North America out to sea, the Inner Detrital Belt would have been represented by sands along the shore that were to become the Tapeats Sandstone, and shallow shelf muds, sands, and carbonates off shore that would become the Bright Angel Formation. Further out, the Carbonate Belt would consist of carbonate muds and shoals near the shelf edge that would become the Muav Limestone. Walther's Law tells us that these belts would have existed laterally even though in any one spot we may see them stacked vertically in the rock record. Through the Cambrian, as sea level rose and the shoreline moved eastward, these belts and their sediments (and the formations they were to become) moved east, too, and also became stacked on top of each other so that we now see Inner Detrital Belt Tapeats overlain by Bright Angel overlain by Carbonate Belt Muav. We see Outer Detrital Belt rocks a little less commonly because they are on the edge of deep water and thus encroach onto the continents less than do the other two belts, but we will see Outer Detrital Belt rocks later.

Once we know we have a conformable sequence isolated by unconformities above and below, the facies shifts, time transgression, Walther's Law, and now Palmer's belts help to demonstrate that the environments of our three Cambrian formations gradually moved across western North America and in the process ended up piled on top of each other in the rock record of the Grand Canyon. These concepts and rock successions are things we will encounter again in later chapters. Indeed, the succession seen in the Cambrian of the Grand Canyon will become a general pattern (although one not always followed to the letter) at many future stops on our journey, mainly in cratonic settings. So keep in mind for our coming travels what we have seen in the Grand Canyon; the beach sands of the Tapeats, the shallow-marine muds of the Bright Angel, and the offshore shoals of the Muav—we will see them again more than once.

A New View: The Tonto as Expansive Epicratonic Estuary

What I have just described works well for the traditional Sauk Sequence section, perhaps many cratonic sections, which the Tonto Group has for

---

The Cambrian Corps 1–Eben C. Rose

Eben Rose is a research associate at the University of Connecticut and has been researching the geology of the Tonto Group in the Grand Canyon since the early 1990s. Having spent his formative years in Bend, Oregon, Dr. Rose moved to Flagstaff, Arizona, after high school and by chance ended up working for a rafting company running trips down the Colorado River in Grand Canyon. It wasn't until a boatman for the baggage raft got injured on one trip that Eben took the oars and began working as a guide on these trips. After several years of river guiding, and the occasional, inevitable "epic bad run," as he calls them, through one rapid or another, he decided to continue his undergraduate degree at Northern Arizona University (NAU), while continuing to guide. It is the influence of the geology he was seeing firsthand along the Colorado River on his 40 or so raft trips through the Canyon that he credits with reinspiring his interest in geology. "I have been influenced by the canyon and mountain landscapes of the West all my life," he says, "but it was working in the Grand Canyon as a river guide that led me into taking the study of geologic history more seriously as a scientist." His formal work on the Tonto Group began when he started a master's degree at NAU, and he also credits fellow boatmen on the Colorado for helping inspire his work. "I became impressed with some of the stories that the more learned guides would tell about the geologic history of the canyon," he says, "and heeded the advice of one successful river guide who said to me, 'Know the canyon. Study it well, and you will have a story to tell of your own.'"

Dr. Rose's PhD dissertation at Yale University involved Archaean geology and the origin of life, and was thus a somewhat different project from his master's work, but he has continued to publish his work on Grand Canyon Cambrian geology, important studies that reinterpret the origins of the formations of the Tonto Group. In the course of several research river trips and hikes up many side canyons of the Colorado in Grand Canyon, Dr. Rose found evidence that the Tonto Group rocks were deposited in water shallower and closer to shore than traditionally thought—evidence such as mudcracked surfaces and petrified sand dunes in the Tapeats Sandstone, for example. "While these weren't entirely unexpected," he says, "based on other hints I was getting that these layers were deposited in shallower water than the classic model suggested, their extent and quality of preservation were an unexpected surprise." Formations such as the Tapeats Sandstone and Bright Angel Formation seem to have been deposited in a complex mix of near-shore terrestrial, tidal, estuary, and shallow-marine environments. This setting may be almost unique in that during the Cambrian the continents were largely flooded by high sea levels and were nearly unvegetated on their exposed surfaces; this created the "expansive epicratonic estuaries" that Eben envisions formations like the Bright Angel representing. "The Cambrian is a unique time in Earth history," he says, "because sea levels were higher than any other time that we know of . . . and sea level fluctuations are probably the most important driver of species diversification and extinction."

Comparing new interpretations of the Tonto Group to other cratonic formations of Cambrian age in the region should provide more insight into these continental margin settings in Laurentia—and the Grand Canyon certainly has more stories to tell.

---

years been thought to represent. Unfortunately, although such sections of rock do exist in North America, and after all these years as an example, the Tonto Group may not be one of them! Recent work since the early 1990s seems to show that the Tonto represents a mixture of environments along the flooded part of the craton close to the shoreline. These environments include sandy channel deposits of rivers flowing into the ocean, supratidal flats, intertidal zones, estuaries, and subtidal shallow-marine areas. There probably were restricted embayments in this area too, with fresh water from the continent flushing into shallow-marine waters and islands of Precambrian basement rock sticking up above sea level. Paleoenvironments of the Tonto Group are now suspected to have been nonmarine fluvial to perhaps shoreline for the Tapeats Sandstone (i.e., rivers flowing into the sea), an expansive estuary for the Bright Angel Formation (i.e., very shallow marine with inflowing fresh water and plant spores from nearby land areas), and shallow subtidal to sabkha-like intertidal expanses for the Muav Limestone. And islands of Precambrian rock seem to have been scattered in the area, too.

Some of the evidence for these environments includes (1) identification of isolated channel sandstone deposits, eolian dunes, and mudcracked surfaces in the Tapeats Sandstone; (2) a recognition that the time-transgressive nature of the formations of the Tonto Group is not as clear cut as it appeared earlier; (3) identification of possible moss-grade plant spores in shallow-marine rocks of the Bright Angel Formation; (4) carbon and strontium isotopic data indicating freshwater influence in the Bright Angel Formation; and (5) a recognition that glauconite is not exclusively formed in deep-marine settings. The abundance of glauconite in the Tonto Group, and its earlier interpretation as having to form in relatively deep-marine environments, may have been one of the factors influencing early workers' interpretations of the Tonto as a fully marine transgressive sequence, regardless of other factors, which then may have been shoehorned into a marine interpretation.[37] In this new interpretation, we need to imagine an incredibly flat, large, flooded area of the continent (the craton) with a mixed influence of rivers flowing into shallow-marine bays (estuarine settings), scattered barely vegetated islands, tidal flats, and perhaps slightly deeper (but still shallow) marine lagoons. The traces and body fossils show us that marine organisms were still abundant but that things may have been more shallow and closer to shore than we thought. It has been suggested that rather than serving as a textbook example of the transgressive marine sequence for the Cambrian, the Tonto Group might more appropriately be called an expansive epicratonic estuary, an environment not seen today and one that was different in existing in a very large, flooded continental region at a time before land plants had significantly affected the dynamics of either land or shallow sea. This may have been a land-sea-scape unique to the Cambrian, which makes the Tonto Group even more interesting than it was before.[38]

The Long Haul

But we've been admiring the Tonto Group long enough. We still need to get out of the canyon. This isn't necessarily the toughest hike we'll do (but it may be) and it is by far neither the steepest nor does it have the most elevation gain—but any climb out of the Grand Canyon is a bear nonetheless. I'm not sure why, but I suspect it's the sheer steepness of the series of cliffs you ascend one switchback at a time. There's no way around it, standing on the Cambrian and looking up your trail as it relentlessly zigzags up the Redwall, then the Supai, then the Coconino, then the Toroweap and Kaibab, each hundreds of feet thick and most almost completely vertical, is intimidating—and in a reversal of the mountain climbing mantra, in the canyon going down is optional, climbing back out is not. So we begin the long plod.

Standing on the rim once again hours later we look back down and realize it was slow and hot and not always that much fun, but overall the hike out wasn't all that bad. It's time to sit on our packs in the shade, enjoy an ice cream, and watch the other visitors trying to photograph the squirrels. It is a hive of activity on the rim compared to the relative isolation of the canyon, but we are cooling off now, sitting here in inactivity, and nothing at the moment is more welcome.

Gazing off across the canyon we enjoy the view as much as all those around us. But we narrow our eyes for a moment as we look down and notice something we'd not registered while down in the canyon. Across the way the Tapeats Sandstone rests not on Precambrian schist as we'd seen near Pipe Creek but on an angular unconformity. There are Precambrian sedimentary rocks below the Tonto Group! We've seen more than our fill of Precambrian igneous and metamorphic rocks; what lies in these sediments even more ancient than the Tapeats? Our legs are sore enough for one day, so rather than going right back down the trail to check out these rocks close up, we load into the geology van and drive north, headed for yet another outcrop of sedimentary rocks that predate the Cambrian radiation by millions of years.

3.1. Stromatolite in limestone of the Snowslip Formation (Proterozoic) of Glacier National Park, Montana. Laminated layers are calcium carbonate mud bound by cyanobacteria and built up in biscuit-like structures (seen here in cross section) over generations of growth of the colony. Specimen approximately 30 cm across.

A Long Strange Trip: The First 4000 Million Years of Earth History

3

OUR DESTINATION IS THE CAMBRIAN, BUT TOO MUCH OF EARTH history is sunk back in the depths of the Precambrian, previously the Dark Ages of the paleontological record, to ignore the prelude. We can't truly appreciate nor understand the Cambrian world without a little background on what led up to this circus of events 520 million years ago. Earth was, from our perspective, a weird place in the Precambrian, and seeing it will serve as a good lead-in to the Cambrian, but it may make even the Cambrian seem a little bit more like "home" in comparison. And this is why we are on the road from the Grand Canyon.

We drive around the Grand Canyon to the east and wind our way through Page and Kanab and through the south of Utah until we come out at Interstate 15. We then blaze north for 14 more hours through Salt Lake and Pocatello and Butte—almost to Canada. We approach from the east, rolling across the plains and coming up on grand blue mountains ahead. We are at the foot of the peaks of Glacier National Park. Established in 1910, this home to the most glaciers in the Lower 48 states contains U-shaped valleys, glacial lakes, moraines, and scenery of a caliber you just don't see in the mountains of the southern Rocky Mountains in Colorado or Wyoming. These areas and the Sierra Nevada around Yosemite were certainly glaciated heavily, but the sights in Montana's Glacier National Park are more like something you would see in Europe's Alps or the farther northern Rockies in Alberta and British Columbia.

In roughly the last 2 million years Glacier, along with Waterton, its sister park just over the border in Alberta, was ground up, scraped, and otherwise pulverized by ice that gouged deep valleys in the mountains on either side of the Continental Divide. During the multiple ice ages of the period from 1.8 million years ago until 10,000 years ago, snow fell in such great amounts and with such regularity for so much of the year that it piled up into ice that then flowed downhill in many of the valley drainages, grinding down much of the rock along the way. The ice in the valleys was hundreds of feet deep at the peaks of the ice ages, and the glaciers in many cases spilled out on to the plains, piling up debris rocks at their outer extents. As these glaciers melted off and retreated they left in the valleys more than 25 lakes in their place, some of them quite large. The glaciers also left behind moraines, piles of rock debris left over from the leading edge and surface of the glacier after melting, and valleys

There is nothing like the Cambrian until the Cambrian.

Andrew Knoll, 2003

73

gouged into smooth-sided trenches with distinct U-shaped cross sections. About 20 glaciers remain within the park, although they are tiny fractions of the rivers of ice they were 20,000 years ago. They make for spectacular scenery nonetheless.

We will drive up from the east along the north shore of Saint Mary Lake, headed west, and pass lonely Wild Goose Island and the Jackson Glacier overlook before we soon come to Logan Pass, where we will finally be making our next stop. Time to grab our packs and head off down another trail – this time the Highline Trail, which heads north from Logan Pass below the Garden Wall. This will be a longer but much flatter hike than our one in the Grand Canyon, and we head off down the trail through the alpine terrain with renewed energy. We work our way through a beautifully dizzying section of trail cut into solid rock just a few minutes down the path. It is high above Going-to-the-Sun Road in a small cliff of sorts, and we can't help noticing brecciated limestones in the outcrop next to us as we walk. Eventually we come out in more meadowy terrain and move along at a pretty good pace.

We've been seeing them for a while, passing a few here and there, but now we come upon one that is so well preserved and distinct we can't ignore them any longer. They aren't our imagination; they must be something in the rock. In large gray chunks of limestone we see bulbous structures of wavy lines like concentric rings in cross-sectioned cabbages (fig. 3.1). They are **stromatolites,** structures formed (in most cases) by the growth and buildup of colonies of cyanobacteria. We met these microorganisms very briefly in chapter 1. They are photosynthesizing bacteria that even today live in clear, shallow-marine or freshwater where there are few or no grazers, usually due to geographical barriers on the periphery of the environment or environmental extremes such as high salinity. They form branching or non-branching columns, bulbous mounds, or flat layers of microbial mats on the seafloor, and these may build up for years through repeated cycles of being covered by sediments and then growing another layer. As sediment gets incorporated into the mats and new layers grow, the structures are preserved as bulging layers of rock matrix even if no trace of the original bacteria remains. Computer models have been able to simulate different forms of stromatolites, including branching, columnar forms, through variations in layer growth of the bacteria versus layered sediment input.[1] We can hike mile after mile of limestones below the Garden Wall and see these stromatolites. The colonial cyanobacteria that made them apparently were very abundant at the time. And that time was about 1.1 billion years ago during the middle of the Proterozoic eon.[2] What we see along the Highline Trail demonstrates that life was abundant during the Proterozoic, and that life was in relatively simple forms that are still around today. The Precambrian was not a time of no life, just different ecosystems. But we'll come back to Glacier. Let's hop back to the beginning and take a journey through the Precambrian in chronological order.

Earth's first 600 million years might have been hell. At least that is implied in the informal timescale designation for this time: the Hadean (see fig. 1.2).[3] Soon after the planet formed and for millions of years afterward the surface of the planet was probably hot, sulfurous, and riddled with volcanic and meteoric input. The Moon was very close and perhaps appeared red in color, there was essentially no atmosphere at first, and even during the day the sky was black and starry.

The Earth formed from the swirling matter of the solar system around the same time as the other rocky planets 4.6 billion years ago; the Sun probably formed a couple hundred million years earlier. Accretion of the Earth may have taken 10–100 million years. Once the protoplanetary material had coalesced into the Earth, the early millions of years of the planet's history were a time of differentiation of the internal structure. The Hadean stretches from the time of the Earth's formation until about 4.0 billion years ago. Bombardment by debris from elsewhere within the new solar system was probably nearly constant until about 3.8 billion years ago, into the Archean eon. The end of the Hadean was as inhospitable as the early part: a heavily rifted Earth surface, with lots of impacts and volcanic eruptions, and an early atmosphere forming with the planet covered in water vapor, CO_2, and dust. During the late Archean even the level of radioactivity generated by the young Earth was probably close to three times higher, and thus the heat generated by the planet's core was likely greater. The atmosphere probably contained – in addition to CO_2 and nitrogen – carbon monoxide, methane, hydrogen sulfide, and ammonia. Not the most pleasant of combinations for us latecomers, so it's a good thing conditions eventually changed. At the time, the young Sun was possibly only 70% as bright as it is now, but the surface of the Earth was warm enough to contain liquid water, largely thanks to an intense greenhouse effect caused by very high CO_2 levels. The atmospheric surface temperature would have been a painful 85°C (185°F), despite that dim sun, and courtesy of the high CO_2 content and other factors, such as Earth's internal heat. In fact, there may have been (warm) surface water as early as 4.3 billion years ago. There was almost no free oxygen (O_2) in the atmosphere yet. Without free oxygen, and thus no filtering ozone layer, the surface of the planet was hit with the full brunt of ultraviolet (UV) radiation from the Sun. And UV radiation was more intense not just because of the lack of ozone, but also the radiated UV intensity was probably much greater at the time coming even from the dimmer Sun.[4]

Earth's dynamic system of crustal movement, plate tectonics, appears to have been functioning from very early on, perhaps as early as 4.4 billion years ago, and the formation of protocontinents would have begun around the same time. Continental crust of approximately that age has been identified in western Australia,[5] and almost all of the continental crust basement rocks that we see today appear to have been formed by 1.6 billion years ago. The amount of new continental crust added since the Archean eon seems to have been balanced out by destruction of old crust.[6] In North America, these basement rocks are exposed in a number of areas, but the most extensive is on the Canadian Shield, covering a significant percentage of the continent around Hudson Bay. Rocks older than 2.5 billion years are exposed in vast areas around the south and east of Hudson Bay, for example, and farther south in parts of Wyoming. The most important point is that these crustal rocks are different ages in different parts of North America because during the Precambrian the continents grew by accretion of smaller pieces of crust.

The Archean

The Archean eon began 4.0 billion years ago, and it lasted 1.5 billion years (see fig. 1.2). By 200 million years into the Archean (3.8 billion years ago), we see the first sediments deposited in water. The early oceans probably formed from H_2O brought to the surface of the Earth by volcanoes and by comets, although the latter's input was a bit less; this water in the form of steam vapor in the atmosphere rained out to form the oceans after the Earth cooled following its formation. This was not an ocean any modern beachgoer would have much interest in frequenting. The Archean sea appears to have been acidic, with a pH in the range of 2 to 4; that's about the same as lemon juice and vinegar.[7] The early ocean was also very low in oxygen and in places very hot–perhaps as hot as 57°C (135°F)! Not many would want to play in such water, not even the most dedicated surfer. No, the Archean ocean was one of science fiction films, and one that would have made for miserable swimming. But as we saw earlier, even some modern Bacteria and Archaea can tolerate higher temperatures and lower pHs that seem to have been present in the Archean ocean, so it is not surprising that were probably microorganisms even in these first waters.

By 3.5 billion years ago we see the first signs of life in possible fossil stromatolites and small structures that appear to be filamentous, microscopic fossils, evidence of tiny Precambrian bacteria that grew in elongate strands (fig. 3.2).[8] These are found in rocks of the Warrawoona Group near a place called North Pole, Australia–but, according to some researchers, these apparent bacterial fossils may instead be a form of graphite formed as a result of hydrothermal alteration of the host rock.[9] Few topics in paleontology are free of debate, least of all the status of the earliest fossils. Adding to the mix are signs of organisms that oxidize organic matter or hydrogen to produce sulfide, microbes that neither photosynthesize nor use free oxygen but rather are chemosynthesizing forms. There are few fossils of these, mostly isotopic signals in rocks that they probably were there,[10] but recently described fossils from 3.4-billion-year-old rocks in Australia appear to belong to microorganisms that metabolized sulfur for energy.[11] By 3.2 billion years ago there may have already been chemosynthesizing microbes living in hot environments below the seafloor in hydrothermal vents.[12] It appears that various types of organisms existed by quite early on in Earth history, but we need to be flexible in what we expect of their biologies–some will certainly prove to be strange to us,

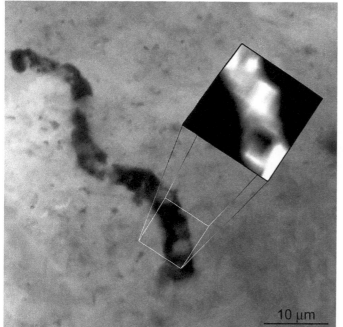

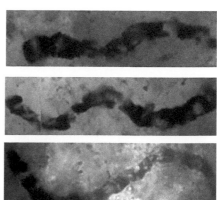

10 μm

3.2. Photomicrograph views of filimentous bacterial fossils *Primaevifilum,* from the 3.465-billion-year-old Apex Chert in Australia, some of the oldest evidence of life on the planet. Scale in largest view is 0.01 mm.

Courtesy of J. W. Schopf (UCLA).

and others may ultimately prove not to be life forms at all. The road to understanding the earliest organisms makes for a bumpy ride.

The apparent stromatolites in 3.5-billion-year-old rock are more than twice as old as the ones we just visited at Glacier National Park, but they indicate the presence of large masses of photosynthesizing cyanobacteria (or some type of bacterium) just 1 billion years into Earth's history. The apparent presence of photosynthesis on Earth so early in its history did not necessarily begin the widespread biogenic production of free oxygen. This is because many photosynthetic bacteria, even today, use other sources of electrons for photosynthesis, not water (as is used by plants and many other photosynthetic organisms). Without water as part of the process, these photosynthetic bacteria generated not oxygen as a byproduct of their work but sulfate and ferric iron. So for millions of years there would have been photosynthesis but no resulting oxygen. That would have to wait. "At some point," Harvard's Andrew Knoll says, "perhaps 2.7 billion years ago (but the date is uncertain), one group of bacteria evolved the capacity to obtain electrons by splitting water. This generated O_2, which was almost certainly quickly used by respiring microorganisms for respiration."[13] In any case, the evolution of photosynthesis probably increased the productivity of life by at least 100 times.[14] Productivity is the rate at which organisms grow and reproduce; such a huge increase suggests that the biosphere in general became more active with the advent of photosynthesis. The 3.5-billion-year-old stromatolite fossils are not gigantic, but they do appear as laminated domes and cones very similar in appearance to stromatolites known from younger rocks. They occur in rocks representing quiet shallow waters in partially restricted settings that probably had higher levels of salinity than the rest of the water body (exactly the conditions we see today in that living diorama example of stromatolite ecology, Shark Bay, Australia). Interestingly, it is not clear that the setting of the Warrawoona Group rocks was necessarily marine; it is possible it was a large saline lake.[15] The atmosphere at this time was still rather different from what we know today, consisting mostly of nitrogen, carbon dioxide, and water—with very little free oxygen. It appears that early Bacteria and Archaea existed by this time, and some of the Archaea might have had some strange biologies. Some, for example, probably produced methane gas as part of their main metabolism, just as plants produce oxygen and water and animals breathe out carbon dioxide. These biologies are not really so strange, of course, as most of them appear in some forms even today—it is just that we are mostly used to the workings of the familiar plants and animals. Although we often think of most early life forms as living in the ocean, the fact that some stromatolites may have lived in lakes indicates that life may have been almost everywhere by this time. In fact, it appears some microbes may have lived on the edges of temporary ponds on land as long as 2.7 billion years ago.[16]

The (Pre)Cambrian Corps 2— Andrew H. Knoll

Andrew Knoll is professor of natural history and Earth and planetary sciences at Harvard University and studies the interactions between evolving life and environment, particularly during the long interval of the Precambrian. Dr. Knoll grew up in Pennsylvania Dutch country in the town of Wernersville, close to the Appalachians and the coalfields, and collected fossils in the area as a teenager. Despite this early introduction to geology and fossils, paleontology did not seem like a career option at the time. He did major in geology as an undergraduate at Lehigh University, "after realizing that engineering and I weren't made for each other," he says. During his senior year he took a class on the diversity of photosynthetic life and became interested in the origin of the chloroplast. This was at the time "a hot topic in light of Lynn Margulis's proposal that the chloroplast originated as a free-living cyanobacterium that became engulfed by a protozoan and reduced through time to metabolic slavery." Through his term paper research on this topic, Dr. Knoll became aware of the pioneering research on Precambrian paleontology being led by Elso Barghoorn at Harvard, and it was here that Dr. Knoll studied for his PhD. "It was my good fortune," he says, "to work in Elso's lab as a graduate student."

Among Dr. Knoll's many research projects, one of the most enjoyable and surprising, he says, was the study of extraordinarily preserved animal eggs and embryos and early multicellular algae in late Proterozoic rocks in China. "It was a real treat to work with Chinese colleagues on these fossils," he says. In addition, he says, "the discovery that for most of the Proterozoic (from 2.5 to 0.54 billion years ago) the Earth's oceans had a bit of oxygen in surface waters, but commonly were oxygen-free below the surface, was a surprise." This surprise ended up being serendipitous, however. "It was a useful one in that understanding this has made other observations make sense," he says.

Studies of the Precambrian and other early time periods, such as the Cambrian, give us perspective on the interaction of our planet and its biosphere. "We've learned," he says, "that over the history of our planet, Earth and life have co-evolved. Changes in the environment affect life, and changes in life can transform the environment. This is a lesson to ponder as we think about our future as well as our past."

At the south end of the Wind River Mountains in western Wyoming is an outcrop of dark, laminated, and dramatically folded Archean rocks consisting of dark bands and orange layers that lie about on the slope, heavy but easily observed by hand. This is the Goldman Meadows Formation. These rocks consist largely of a mineral called hematite and are approximately 2.87 billion years old. They are of a type unique to Precambrian history—banded iron formations (often just BIF)—which formed on the bottom of the early oceans (fig. 3.3). The layering in the Wyoming rocks is wonderfully folded in tight rolls and chevrons, all a result of the high-pressure abuse the rocks have suffered deep in the Earth during their long history as part of North America—a lot can happen to rocks in 2.87 billion years. The banded iron formations in general were commonly formed in the Precambrian, but they do not form today and they are also uncommon in almost all rocks younger than 1.8 billion years. The rocks in what is now Wyoming formed nearly 2.9 billion years ago

3.3. Photo of polished surface of tight folding of layering of a banded iron formation from the Wind River Mountains of Wyoming, the 2.87-billion-year-old Goldman Meadows Formation. Scale bar = 5 cm.

at the height of Precambrian banded iron formation generation. Banded iron formations are also found in rocks exposed today in Australia, Africa, and Michigan, to name a few places; the Wyoming outcrops are some of the most southern in North America. Some banded iron formations are nearly as old as the oldest sedimentary rocks on Earth, and a few are just under 1 billion years old.

Banded iron formations often consist of layers of hematite interbedded with chert, but the lithologies can vary. The formations can even contain oolites, ripple marks, or crossbeds. The iron that piled up to form these rocks was in solution in the oceans, unlike today, because of the near lack of oxygen in the atmosphere and sea at the time. Geologists aren't entirely sure how the iron precipitated out to form the banded iron formation layers, whether it was a chemical precipitation or one initiated by biological activity, but the important point is that the iron was there in the seawater, a condition that is not possible today because of the high level of oxygen in today's oceans. Any iron that gets into seawater today immediately reacts with the oxygen. It appears that during the Precambrian, or at least the early parts of it, iron was dissolved in the anoxic seawater and only precipitated out and accumulated on shallow seafloors when it came into contact with free oxygen, probably produced near the surface of the ocean by very early photosynthesizing microorganisms.[17]

The Proterozoic

The banded iron formations almost disappear from the rock record after 1.8 billion years ago, around the same time that free oxygen appears to have begun building up in the atmosphere and oceans in significant quantities.[18] The Proterozoic eon began 2.5 billion years ago and lasted nearly 2 billion years, until the Precambrian–Cambrian boundary at 542 million years ago. The Proterozoic was a time of some interesting developments in Earth history, not the least of which was the oxygen revolution that I just mentioned. The buildup of the free oxygen level probably began early in the Proterozoic, around 2.2 billion years ago, and a large contribution to the buildup was the oxygen production by photosynthesizing organisms such as cyanobacteria. By 1.9 billion years ago stromatolite structures and tiny fossils of tube-shaped bacteria called *Gunflintia* were buried and preserved in chert now exposed on the shores of Lake Superior in Ontario.[19] Also in the rocks are other cyanobacterial microfossils called *Megalytrum* and bacterial fossils called *Eoastrion*. Similar fossils have been identified within carbonate rocks of similar age near Hudson Bay, and other carbonate rocks around the world of this age begin to show stromatolites in greater numbers.

It is not clear how much higher oxygen levels reached early on compared to what they had been, but it appears that Archean and earliest Proterozoic atmospheric oxygen levels were perhaps less than 1%. In comparison, the level today is about 21%, so prior to this oxygen revolution the Earth was, from our perspective, very choked for breathable air.[20] Of course, few, if any, of the organisms around before the change in the atmosphere would have had a metabolic need for oxygen like we do, because that is something that came later; oxygen was rather detrimental to most life of the time, and its buildup in the atmosphere was simply a product of the photosynthesis by cyanobacteria and other microbes. The anaerobic bacteria and other organisms that probably had lived freely during the Archean and earliest Proterozoic would simply have been slowly driven into marginal anoxic environments starting around 2.2 billion years ago as the amount of free oxygen in the atmosphere and oceans was increased by those cyanobacteria and other photosynthesizers. Although they had had Earth almost to themselves for close to 2 billion years, anaerobic bacteria and other organisms to which oxygen was in fact poisonous had to take refuge as the planet was overrun by oxygen, and they remain there today, safe in places too hostile for most life we are familiar with, in places like deep-sea hydrothermal vents and geothermal ponds and mud pots such as those at Yellowstone and Lassen national parks.

Interestingly, stromatolites appear well before the oxygen increase began, and the banded iron formations almost (but not entirely) disappear around 1.8 billion years ago, some 400 million years after the oxygen increase started. Why the lag time? Geologists who study these questions aren't entirely sure, but it is apparent that the buildup of free oxygen was a complicated process that involved more than just photosynthesizers cranking out O_2. Today, the oxygen produced by all the planet's plant matter is approximately matched in consumption by all of us that breathe that gas; similarly, the O_2 is not all consumed because what we produce in CO_2 is again converted to free oxygen and water by the photosynthesizing organisms. What of a world in which there are photosynthesizers producing the oxygen but few if any using it as we do? Why wouldn't the oxygen buildup not have been fast and in step with the appearance of cyanobacteria? It appears that the oxygen produced by photosynthesis may have been held in check by some process until about 2.2 billion years ago, after which its buildup took hold slowly. What this process was is not known for certain, but it must have been one that would have "consumed" oxygen prior to 2.2 billion years ago (when there were no oxygen-using organisms to take in what was produced by early cyanobacteria), stopped, and not reappeared since, as the oxygen producers and consumers of the world since the Proterozoic have been more or less in balance.

So, what could this process have been? One idea is that perhaps the early Earth produced, through volcanic eruptions, abundant gases that combined with the free oxygen coming from cyanobacteria to make the O_2 unavailable in the atmosphere. After that volcanism slowed, the oxygen was no longer being consumed at the same rate and began to build up. Other ideas are that oxygen buildup occurred during a period of decreased availability of elements that would otherwise bind with oxygen. This is sort of a chemical musical chairs idea. With cyanobacteria producing oxygen now, a net gain in oxygen levels would be achieved at any later point when elements that bind with oxygen were made less available. Two of these elements are hydrogen and carbon. If free hydrogen, light enough to escape the atmosphere, did indeed fly out into space at a rate greater than it was biogenically or chemically combined with other elements, a slow net gain in free oxygen would result as oxygen atoms found themselves with not quite enough hydrogen atoms to join up with. Similarly, if there were an increase in the rate at which carbon was taken out of the biosphere's cycle through burial in or as part of rocks, for example, less would be available for inclusion in reactions with oxygen and O_2 might slowly build up as well.[21] Eventually, of course, the oxygen level stabilized more or less; it did not keep increasing indefinitely.[22] This may have been in part because of the rise of oxygen-metabolizing organisms, which, of course, take in free oxygen and return CO_2 and water to the environment.[23]

Glacier's Record

By 1.4–1.1 billion years ago the major banded iron formations were gone, and it was the apparent time of origin of the eukaryotes, the cells that are so much more complex than those of the Bacteria and Archaea and that eventually gave rise to plants and animals. Oxygen has taken over the planet and cyanobacteria and the structures they built (stromatolites) are everywhere. Coming back to our break along the Highline Trail at Glacier National Park, we can sit and eat lunch on a block of carbonate rock the size of a car and be lounging on a solid, former shallow sea bottom completely covered with stromatolitic structures, the cabbage-shaped mounds of laminated limestone formed by continuous growth and covering (by sediment) of bulbous mats of cyanobacteria some 1.1 billion years ago. The sedimentary rocks of Glacier are mostly Proterozoic in age; and thanks to very light metamorphism, they are just as well preserved as rocks half their age. Because of this we can find in the rocks along the Highline Trail many stromatolites; we can follow whole beds of them and crawl over blocks of them right along the path. We also can see ripple marks and mudcracks, evidence of currents, tides, and the work of the sun all preserved in stone high on the mountainsides.

The rocks of Glacier National Park and surrounding areas are interesting because they include some 2896 m (9500 ft.) of Proterozoic sedimentary rocks. There are not many places in North America where the Precambrian rocks are sedimentary—as we will see later it is more common to find your Cambrian rocks resting on highly metamorphosed or igneous rocks. In fact, in the United States the main areas to see sedimentary Proterozoic (later Precambrian) rocks are at Glacier and surrounding areas, at the bottom of the Grand Canyon, in northeastern Utah, in central Arizona, around Lake Superior, and near Death Valley. Thick units are also known from the Yukon Territory. Glacier's Proterozoic rocks are very thick and are interesting for several reasons, so let's take a closer look at them.

The Precambrian sediments at Glacier are divided into seven formations within the Belt Supergroup, and are, in ascending order, the Altyn, Appekunny, Grinnell, Empire, Siyeh, Snowslip, and Shepard formations

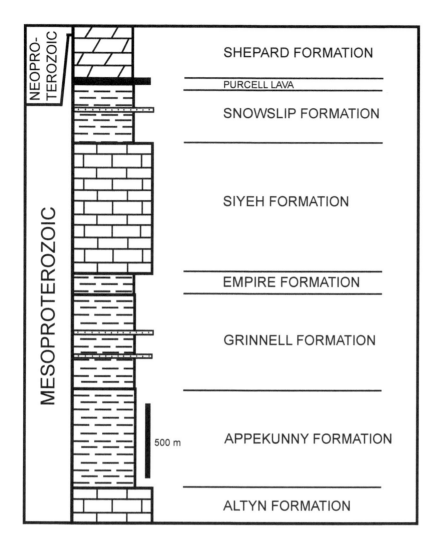

NEOPRO-
TEROZOIC

SHEPARD FORMATION

PURCELL LAVA

SNOWSLIP FORMATION

SIYEH FORMATION

EMPIRE FORMATION

GRINNELL FORMATION

500 m

APPEKUNNY FORMATION

ALTYN FORMATION

MESOPROTEROZOIC

3.4. Stratigraphic section showing the Proterozoic formations of Glacier National Park. Stipple pattern = sandstone; dashed pattern = shale; brick pattern = limestone; slanted brick pattern = dolomite.

(fig. 3.4). Our lunch rock along the Highline Trail, on which we sat on stromatolites, was in the Siyeh Formation. Each of these formations is between 122 m (400 ft.) and 1000 m (3300 ft.) thick, and the rock types preserved include limestones, dolomites, lightly metamorphosed shales, and sandstones, all representing shallow-marine to tidal environments. The Altyn Formation (and its western equivalent, the Prichard Formation) is about 1.3–1.4 billion years old, and consists of mostly limestone and dolomite containing several major beds of columnar stromatolites of at least five species. These stromatolites are of both straight and branching types that probably lived, respectively, in deeper, quiet waters and in subtidal zones. The Altyn also contains filamentous microfossils in some of its chert layers.[24] The Appekunny Formation is slightly younger than the Altyn and Prichard and consists of slightly metamorphosed shale and siltstone called argillite. The Appekunny Formation contains microscopic fossils of single-celled organisms as well as a larger fossil known as *Horodyskia moniliformis*, which looks like nothing less than a "string of beads," as it has been called. The beads are often about 2 mm (0.8 in.) in diameter

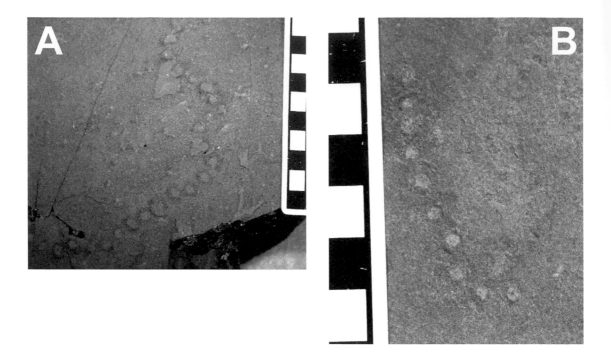

3.5. The Proterozoic fossil *Horodyskia* from the Siyeh Formation of Glacier National Park. (A) Larger specimen consisting of 20+ "beads." (B) Smaller specimen consisting of 10 smaller "beads." Scale bars in centimeters.

Photos by and courtesy of ReBecca Hunt.

and are strung out in straight or meandering lines along bedding planes, usually just leaving an impression in the sediment (fig. 3.5). Each of the beads appears to have been connected to those next to it in the string by a small tube. *Horodyskia* has also been identified from Bangemall Supergroup rocks of about the same age in Australia. Exactly what type of organism *Horodyskia* was (or even if it was an organism) has been debated, but it appears most likely that it was a colonial, bottom-dwelling form of sea life that may have had tissues and was possibly a metazoan—that is, an animal. This would nearly double the history of animals in the rock record. The matter of tissues is important because this indicates a level of sophistication not seen in organisms up to this time; tissues exist in an organism when the cells are differentiated into similar groupings, each with specific functions. If the interpretation of *Horodyskia* is correct, this fossil represents one of the earliest occurrences of relatively complex organisms yet found.

The Grinnell Formation is similar to the Appekunny in being an argillite and contains at least three species of mound-shaped stromatolites. The Grinnell also has some layers that represent deposition on tidal flats that got exposed to the Proterozoic sun, as evidenced by bedding planes in the formation preserving mudcracks (fig. 3.6). The Empire Formation is a relatively thin unit of argillite as well. The Siyeh Formation (also known as the Helena Formation) is the 1.1-billion-year-old unit of limestone beds exposed along much of our route along the Highline Trail between Logan Pass and Granite Park Chalet, among many other parts of the park. This is a highly fossiliferous formation, containing at least seven species of stromatolites and filamentous microfossils. The stromatolites are of columnar and mound shapes (fig. 3.7a), and they can be so extensive at

some levels that they are divided into particular beds and zones. In addition to the fossils, sedimentary structures like mudcracks and load-casts are found in the formation. Above the Siyeh is the Snowslip Formation, which consists of 1.0-billion-year-old argillites, siltstones, and sandstones with some stromatolites (fig. 3.7c). The Snowslip, like many of the formations we've seen, represents deposition in subtidal to intertidal environments near the shore of a Proterozoic ocean, and in places in the park you can see ripple marks of that ancient sea preserved in rock as if they were made yesterday – evidence of wave and current action from 1000 million years ago looking as good as new (fig. 3.7b). The overlying Shepard Formation is the uppermost formation of dolomite and siltstones in the park and contains several types of stromatolites.[25]

3.6. Mudcracks on a shale bedding plane of the Grinnell Formation near Glacier National Park. Surface may have been part of a sabkha or lakeshore during the Proterozoic. Rock hammer for scale.

By the time of deposition of Glacier's ridge- and peak-forming formations like the Snowslip and Shepard about 1.1–1.0 billion years ago, most of the continents were fused together into one supercontinent called Rodinia, an amalgamation that lasted from about this time until about 750 million years ago. Forget Pangaea, the relatively recent supercontinent of Permian to Triassic times in which continents such as Africa and South America were there in their approximately modern shapes and orientations, just joined together for a few tens of millions of years before they split apart and created the Atlantic Ocean. No, for Rodinia you have to imagine a random assortment of continental pieces mashed together into an earlier,

The Supercontinents

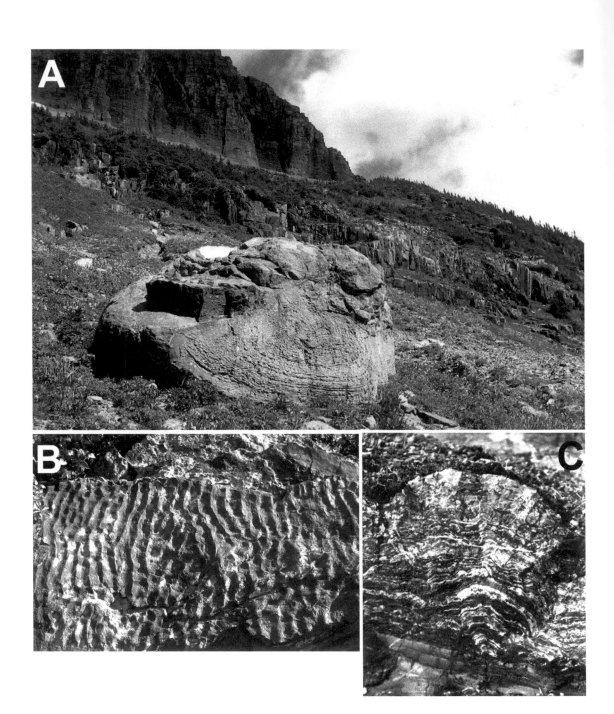

3.7. Structures in Proterozoic formations of Glacier National Park. (A) Large stromatolite about 2.5 m (8 ft.) across, loose and resting upside down near the Highline Trail; from the Siyeh Formation. (B) Well-preserved oscillation ripple marks from the Snowslip Formation. (C) Stromatolite from the Snowslip Formation, seen in cross section.

less recognizable supercontinent. North America's ancient counterpart, Laurentia, was nearly at the center of it at the beginning; Siberia was attached to the northern Yukon; Norway and the Baltic butted up against Greenland; Australia and Antarctica attached to the southwest with India and Tibet on the far side of them; and parts of South America clung to eastern (back then, southern) Laurentia. South of the South American pieces? Three blocks that would later become Africa. This arrangement eventually broke up, of course, starting about 900 million years ago, and there was a short-lived (only 50 million years) second supercontinent at the very end of the Proterozoic called Pannotia that lasted from 600 to 550 million years ago. Pannotia was formed by a continental collision known as the Pan-Africa Event, and when this supercontinent broke apart it formed the four main continental units known during the Cambrian: Laurentia (North America), Baltica (northern Europe), Siberia (Asian Russia), and Gondwana (the southern continental pieces that eventually became Africa, South America, Antarctica, India, China, Arabia, and Australia). As an indication of how mixed around the continental pieces were relative to what we know today: what is now northern Europe was near 60 degrees *south* latitude, and eastern Antarctica was near the equator! Even North America (Laurentia) was in the southern hemisphere. This was about 550 million years ago. As we saw in chapter 1, by the Cambrian, 542 million years ago, Laurentia was on still its own, surrounded by ocean, but was even nearer the equator.[26]

Another thing happened around the time that Glacier's sedimentary rocks were piling up and preserving stromatolite fossils. It appears that around 1.2 billion years ago, eukaryotic cells evolved. Until now, everything we've talked about in terms of fossils (other than possibly *Horodyskia*) has been Bacteria or Archaea. Our third major group, as outlined in chapter 1, has made its appearance. The Eukarya is the group to which plants and animals, among others, belong, and this represents a transition from simple cells without nuclei to more complex ones with not only a nucleus containing the organism's DNA but also **mitochondria** and, in the case of plants and algae, **chloroplasts.** It may seem logical that a more complicated cell structure would naturally take longer to evolve, but it appears that more than just time was involved. If our theories about the origins of mitochondria and chloroplasts are correct, it seems clear that certain types of non-eukaryotes needed to appear first. A particularly interesting idea regarding eukaryotic cell origins was put forth independently and six decades apart by a Russian botanist, Konstantin Merezhkovsky, and by cell biologist Lynn Margulis.[27] They argued that chloroplasts, the centers of photosynthesis in eukaryotic cells including algae and plants, were descended from none other than cyanobacteria and had simply become **endosymbionts** incorporated into the eukaryotic cells. In addition, Margulis suggested that the mitochondria, the centers of oxygen-using cellular respiration (and thus energy production) in eukaryotic cells, are endosymbiotic descendants of free-living bacteria that carried out cellular respiration first. Cellular respiration is the process of turning organic compounds (sugars derived either from food, in animals, or manufactured by the chloroplasts, in plants) and oxygen into energy, water, and carbon dioxide. The eukaryotic cell contains mitochondria and, in the case of plants and algae, chloroplasts. Mitochondria take the sugars and oxygen and convert them into water, CO_2, and energy in the form of ATP. It is the ATP that cells use to do their work. The oxygen mitochondria need is taken in from the environment; in animals, the sugars needed are

The Birth of
Cell Teams

ingested as food, or (in plants and algae) the sugars come from the chloroplasts, which produce sugars through photosynthesis (taking in water, sunlight, and CO_2 to produce sugars and O_2). The neat part is that mitochondria and chloroplasts reproduce independently and on their own within a eukaryotic cell. Their function and reproduction suggest that they represent the descendants of oxygen-metabolizing bacteria (mitochondria) and photosynthesizing cyanobacteria (chloroplasts) that became endosymbiotically incorporated into the newly evolved eukaryotic cells. So it makes sense that eukaryotic cells could form from more simple cells teaming up in an endosymbiotic relationship with formerly free-living cyanobacteria and with bacteria that had begun processing O_2 and had become oxygen-respiring organisms. Once these three elements had joined together, a whole range of eukaryotic forms could develop. In fact, some eukaryotes are even symbiotic conglomerations of host cells and other eukaryotes such as algae.[28] Each cell in a eukaryote is thus a tiny symbiotic team, more complex even in single-celled eukaryotes than in anything the Earth's biosphere had seen previously. When multicellularity and tissues arose (in part thanks to the versatility and possibilities introduced by the eukaryotic cell) complexity and diversity took off even more.

Eukaryotic cells could not have evolved as we know them until oxygen levels in the Earth's oceans and atmosphere reached significant, if not necessarily modern, levels. The oxygen-respiring bacteria that were "proto-mitochondria" had to evolve first as opportunists on the rising oxygen levels. Only then could such bacteria be incorporated into a partnership with other organisms to create the eukaryotic cell. As such, the Eukarya are a product of the Proterozoic oxygen revolution.[29]

There are molecular traces of possible eukaryotic biological processes in rocks as old as 2.7 billion years,[30] but actual fossils attributed to eukaryotes appear much later. The lag time could have resulted from a "slow assembly" of the various aspects now recognized as characteristic of eukaryotic cells. As we saw with the endosymbiosis theory of eukaryotic cell origins, you don't just go from a prokaryotic-grade cell to possessing a nucleus and acquiring mitochondria and chloroplasts overnight. Perhaps some of the aspects and biologic function of Eukarya had been developed by 2.7 billion years ago, but it took until sometime in the subsequent 800 million years to complete the transition.[31] Microfossils found by the hundreds in 1.9 billion year old rocks along those shores of Lake Superior (*Leptoteichos*) suggest a possible, nearly eukaryotic grade of evolution (including what at a cursory glance might appear to be a nucleus), yet they are not Eukarya. The diversification that led to the different fossil forms of eukaryotes we are more familiar with probably took place between 1.5 and 1.3 billion years ago.[32]

Among the first-known good eukaryotic fossils are elongate strings of microfossils of red algae named *Bangiomorpha* in rocks about 1.2 billion years old in Canada, found by Nick Butterfield. Interestingly, *Bangiomorpha* appears to represent the oldest-known occurrence of a sexually reproducing organism.[33] This allowed not only genetic diversity through recombination of genetic material but also more complex multicellular structure in eukaryotes. Recent research has suggested that multicellularity itself is not necessarily that difficult to evolve,[34] but more complex structure leading to tissues, organs, and animals probably is. As such, sex ended up being important to the subsequent success of eukaryotes. Unlike the asexual division of cells, which essentially clones the previous generation, sexual reproduction produced much greater genetic variety within each species and thus led to much greater potential for wider intraspecific morphological variation, more pronounced morphological disparity between species, and a potentially faster pace of evolution, all of which contributed to a much more dynamic and more quickly changing world. The biology of the planet picked up the pace and got a lot more interesting late in the Proterozoic.

The filamentous *Bangiomorpha* fossils compare well with a modern red alga known as *Bangia*. Red algae are mostly multicellular, mostly marine, photosynthesizing eukaryotes that can be loosely described as seaweeds. They are generally similar to the true seaweeds (or brown algae, of which kelp is the most recognizable), but are often smaller and have a different type of pigment. Most importantly, cells of red algae differ from all other modern algae in lacking **flagella,** the tail-like organs used for movement in so many microbes. Red algae are today more diverse than the brown algae (with reds having about 4000 species), and most of them live as leafy or filamentous macroscopic forms in warm, tropical waters. If you have snorkeled in the tropics you may well have seen red algae and not even realized it. So as the Proterozoic fossils from Canada and other places show, red algae have a long history. Fossil algae have also been found in 1-billion-year-old rocks in Australia.[35]

Deeper into the Canyon

Those Precambrian sedimentary rocks we saw deep in the Grand Canyon before our trip to Glacier contain a few important fossils as well. Despite their spotty occurrence in only certain parts of the inner part of the canyon, the Precambrian sedimentary formations of Grand Canyon are in fact quite thick; the Chuar Group alone is nearly 1550 m (5084 ft.) thick. The sedimentary rocks of Proterozoic age in the canyon are known as the Grand Canyon Supergroup and are divided into eight formations (and one formation of extrusive igneous rock, cooled lava). The oldest of the rocks in the Grand Canyon Supergroup are about 1.25 billion years old, and stromatolites have been found in these lower layers in the Bass Limestone in the Grand Canyon (fig. 3.8). The lower four sedimentary formations of the Unkar Group (including the Bass Limestone at the base) accumulated to a thickness of 1770 m (5805 ft.), recording mostly quiet, shallow-water marine deposition (good stromatolite habitat!), before volcanic activity piled on another 300 m (984 ft.) of basalt in the form of the Cardenas Lava. All this happened by about 1.07 billion years ago, so the amount of time represented by the Unkar Group is about 180 million

3.8. Stromatolite from the Proterozoic Bass Limestone (~0.9–1.0 billion years old), near Phantom Ranch in Grand Canyon. Scale bar = 5 cm.

Museum of Northern Arizona specimen (MNA P.755).

years. The top of the Unkar is an angular unconformity, caused when the group was uplifted, tilted slightly and part of the top of the Cardenas Lava eroded off before the next sedimentary rocks were deposited.[36]

Above the Unkar are four more sedimentary formations, among which are the Galeros and Kwagunt formations, which comprise the Chuar Group. These formations contain shallow-marine rocks featuring mudcracks and ripple marks, just like those in Glacier, but there are also significant stromatolitic structures called bioherms in the Chuar, also similar to what we saw in that more montane park. The Chuar Group rocks are approximately 800–750 million years old (around the time of the breakup of the supercontinent Rodinia), and they contain some interesting fossils. Among these are, of course, the stromatolites and possible algal filaments, but also the small, disc-shaped *Chuaria circularis*, a probable **acritarch** ranging in size from less than 0.1 mm up to 5 mm (0.003 in.–0.20 in.) in diameter. There are at least seven other genera of these tiny, presumably planktonic organisms preserved in the Chuar; they are thought to have been the ecological equivalents of today's oceanic **phytoplankton**. The Proterozoic forms probably were tiny, unicellular algae that floated in the water column metabolizing energy from sunlight.[37]

A unit approximately the same age as the Chuar Group is found in northeastern Utah. The red sandstones and shales of the 800–750-million-year-old Uinta Mountain Group are exposed in (believe it or not) the Uinta Mountains, a forested east–west trending range along the Utah-Wyoming border. These sediments are 4000–7000 m (13,120–22,960 ft.) thick! These often ripple-marked rocks represent braided rivers and marine delta settings and contain fossils such as *Bavlinella, Leiosphaeridia, Chuaria,* and acritarchs.[38]

Perhaps most interesting among fossils out of the Grand Canyon Supergroup are those of VSMs – vase-shaped microfossils. Not everything in geology is jargon. This is as appropriate a description as could be assigned; the fossils are small, generally about 0.1 mm (0.003 in.) long, were originally made of hard, secreted organic matter, and they are often shaped like many narrow-necked clay pots you might see in an archaeology collection at your local natural history museum (minus the handles, of course). But more interesting than what they look like is what probably made the shells: these VSMs probably encased testate (shelled) amoebae.[39] Amoebae, the blobby single-celled organisms that move under the microscope by the repeated extension of vagarious pseudopodia, are eukaryotes and are among the closest relatives of animals, after fungi (fig. 1.9).[40] And about 750 million years ago they probably did what no eukaryote had done up until that time: feed. All the eukaryotes we have encountered so far were autotrophic; that is, they fueled their cellular energy production through the conversion of sugars that they created themselves (generally through photosynthesis). (Many bacteria, such as the cyanobacteria of stromatolites, are autotrophic also.) Heterotrophic organisms, of which these Grand Canyon testate amoebae appear to be the first among eukaryotes, skip the photosynthetic step and consume directly the organic matter they need for cell energy production.[41] Eventually eukaryotes would evolve into the organisms of the terrestrial ecosystems we are familiar with as humans: autotrophic plants providing energy for heterotrophic herbivores, which in turn provide energy (not willingly, of course!) for heterotrophic carnivores. But at this stage in the Proterozoic, 800 million years ago, it appears that eukaryotic history had just turned another corner in diversification and developed the first heterotrophic Eukarya. The eukaryotic group had probably started nearly 2 billion years earlier, had developed a number of recognizable photosynthetic forms (red algae, acritarchs) by just 400 million years earlier, and now was poised for more, faster diversification. But there was a bottleneck ahead.

The Earth as Hoth

We now move into the Cryogenian period, the frozen interval of Earth history between approximately 850 million and 635 million years ago. But for the Cryogenian you might as well forget images of the recent Ice Ages, the times in the past 2.5 million years (the Pleistocene epoch) when even areas down to modern-day Iowa were in part or entirely covered in a thick ice sheet.[42] These were the times when northern North America (most of Canada) was covered by ice sheets and even the more southern mountainous west (including Glacier National Park) was scoured by montane glaciers. That was a geological amateur hour compared to what the Earth cooked up for the poor Earthlings of the late Proterozoic. Imagine nearly the entire planet covered in ice, miles thick over the continents, tens of meters thick over the oceans down to perhaps 1 m thick at the equator.

The equatorial sea surface looked like that of the modern day Arctic Ocean! This is the frightening time in Earth history known as "Snowball Earth." It is difficult to picture – even on our modern Earth with its frozen Antarctica co-existing with the intensely sunny and (wonderfully) warm equatorial regions – that the low latitudes, much less the equator itself, could even have been covered with ice. That in fact is what appears to have happened.

The Late Proterozoic ice ages came in several phases. The two main phases were the Sturtian from about 760–700 million years ago and the Marinoan (or Varanger) from about 620–580 million years ago. The transition from just cold to global deep freeze appears to have been pretty fast: in about 2000 years. In geological terms that's quick. Even more frightening was the temperature change: from a global average surface temperature of about 0°C down to -27°C (32°F down to -17°F). That is a cold day even for North Dakota in winter. Many of these estimates are based on computer models. How do we know that it happened at all? We have found glacially deposited rocks called **tillites** all around the world, but based on paleomagnetic studies we can tell that the rocks were at the time (during the Proterozoic) at a variety of latitudes, including near the equator, suggesting that the glacial periods involved ice all around the globe. Tillites of this age have been found in places such as India, Australia, China, Africa, Russia, Scandinavia, and North America. In some places the equatorial tillites are several thousand feet thick, indicating lots of glacial activity over many years. Among the equatorial glacial deposits from the Sturtian glaciation were those from North America.

How did this happen? The initiation of the Proterozoic ice ages may have resulted from mechanisms little different from those that probably started any other ice ages we know of, from those of the Paleozoic to those of the last 1.8 million years. The difference may have been that the cooling was severe enough to advance ice sheets to within 30–40 degrees of the equator (already worse than most glaciations we know of). This may have been a threshold after which total ice coverage was unavoidable. Once ice coverage extended from the poles to those low latitudes, the higher reflective capacity of the extensive ice cooled the Earth's surface even more so that ice build up continued. Because darker ocean absorbs heat and white ice reflects it, the more ocean froze (down toward 30 degrees latitude) the more the atmosphere cooled, and by the time the coverage got to the low latitudes there was no turning back. Things just kept on freezing until even the equatorial oceans would have needed an icebreaker.

Obviously, this ice coverage would have put a dent in the photosynthetic groove of many bacteria and eukaryotes dependent on light for their metabolism. Habitats for such organisms must have shrunken severely, but at least some of the microbes and algae made it through this nearly 200-million-year period of stress. Many species may have huddled around islands with internal heat sources such as geothermal vents and volcanoes, places where the ice would have been kept at bay and oases of relatively warm, open water would have provided habitats more suitable to some life forms.[43] Cherts about 700 million years old in northwestern Greenland contain evidence of communities of cyanobacteria that had lived in what were at the time tidal flats bordering a sabkha-like coastline,[44] presumably during an interglacial period.

With severely reduced interchange between the ocean and atmosphere (due to ice coverage), with continents sealed off from erosion by miles of ice, and with reduced numbers of organisms in the biosphere, there was little during the Proterozoic Snowball Earth glacial periods to remove from the air the CO_2 that was erupted from the Earth's volcanoes, which of course didn't stop just because of a little ice. So carbon dioxide levels rose slowly throughout each glacial period until another threshold was reached, this one at the point at which built-up greenhouse gases quickly warmed the atmosphere and melted off the ice. If you don't like the weather, wait 50 million years. It took a while, but the temperature swings into and out of the late Proterozoic glaciations were extreme. After the Sturtian glaciation there was about an 80-million-year break and then another round of nearly global ice sheets—the Marinoan, lasting about 40 million years. Immediately on top of the glacial deposits in many of these Proterozoic rocks are shallow, warm-water carbonate rocks, indicating a return to marine productivity soon after the bottleneck. And in many places these rocks between 580 million and 542 million years old contain some of the most interesting fossils we've seen yet.[45]

The Ediacaran period represents the end of the Proterozoic and of the Precambrian in general. It is the last of the ten named periods of the Proterozoic and represents the time from 635 million to 542 million years ago, the latter date being the beginning of the Cambrian. And the Ediacaran is a newly named period, having been formally approved in 2004.[46] This time interval includes some important milestones in the history of life and ones that are important for the story of the Cambrian as well. Included in these important events are the appearance of the oldest known animals (metazoans), the oldest large Ediacarans, the oldest animal burrows, and the oldest mineralized animals.[47] "Ediacarans" is a term for a group of fossils from this latest Proterozoic period that are truly enigmatic.[48] The first to be named was *Aspidella terranovica*, an odd, round marking in the rocks first noticed in St. John's, Newfoundland, and described by Elkanah Billings way back in 1872. It has since been found in other parts of Newfoundland as well as Australia, and in fact there are now close to 200 named species of Ediacarans from all over the world. Some are disc-shaped, some are leaf-shaped, and others are elongate or bulbous or apparently segmented. Here we finally have in the Precambrian macroscopic fossils, beyond the structures made by cyanobacteria (stromatolites) or rare elements like *Horodyskia*, and we have a lot of them. The diversity is impressive, and most intriguing is the fact that we are not even sure what most of them are. There are a couple hundred named taxa

Para-Animal Farm: The Ediacaran

and most bedding planes preserving these fossils in Australia contain 3–6 species and sometimes hundreds of individuals. Ediacarans occur in widely dispersed localities, in addition to Newfoundland and Australia, like the deserts of Namibia and Argentina, the mountains of western North America, and the forests of Russia and England.

The first animals may appear during the Ediacaran period. Among these are sponges, represented, purportedly, by spicules from Mongolia and, in one case in China, by tiny individual sponges less than 1 mm (~0.04 in.) across. The Chinese Ediacaran sponges were found in close, abundant association with seaweeds. The known possible sponges from this time appear to be related to modern hexactinellid and demospongiid sponges, showing that these modern animal lineages might have originated before the Cambrian. Other, larger sponge forms such as *Palaeophragmodictya* are known from Australia. Evidence for Ediacaran sponges is equivocal, however, and there is some debate as to the affinities of some fossils; at least some of the purported sponge spicules described from the Ediacaran appear to be diagenetic structures.[49] Recent studies on biomarkers even suggest that sponges may date from the Cryogenian, during the Snowball Earth glaciations.[50] Some analyses have argued, however, that *all* known Precambrian sponge "fossils" are either abiotic structures (such as arsenopyrite crystals) or are non-metazoan organisms (i.e., not sponges). The oldest sponges, it turns out, may be from the very base of the Cambrian, in Iran.[51]

One of the most abundant fossils in the Ediacaran period, at least in Australia and Russia, is *Dickinsonia costata*, a form shaped like a sand dollar but with a bilateral symmetry that divides the fossil into lateral halves which are ribbed with folds or ridges that radiate out to the edges in a bit of a sunburst pattern (fig. 3.9a). The end with the thicker ridges is assumed to be the head end, and the finer ridges would represent the tail. A much more elongate form, *Dickinsonia lissa*, has the appearance of a worm. These two *Dickinsonia* species have been proposed to be bilaterian metazoans, and they may have been grazers.

Among the more intriguing fossils from this period is *Spriggina*, a small, elongate, but oval-shaped form (fig. 3.9c) with many segments meeting at a midline and an apparently half moon–shaped head. This form appears to be some type of animal, perhaps an arthropod, and it is known from Australia and Russia. *Kimberella* is an oval-shaped impression with a relatively complex internal structure. Hundreds of specimens are known from Ediacaran rocks in Russia, and although it has been classified in some cases as a jellyfish, it may in fact be a very basal stem mollusc, possibly related to primitive modern forms such as monoplacophorans and chitons.[52] It has even been suggested that *Kimberella* may be related to the molluscs by way of the Ediacaran *Ausia* and the Cambrian-age, apparent stem mollusc *Halkieria*.[53] A small disc-shaped fossil from Australia, with five radially symmetrical ridges in the center, is known as *Arkarua*; it appears more common in deeper-water sediments. The newly named *Coronacollina* was a small bottom-dwelling organism

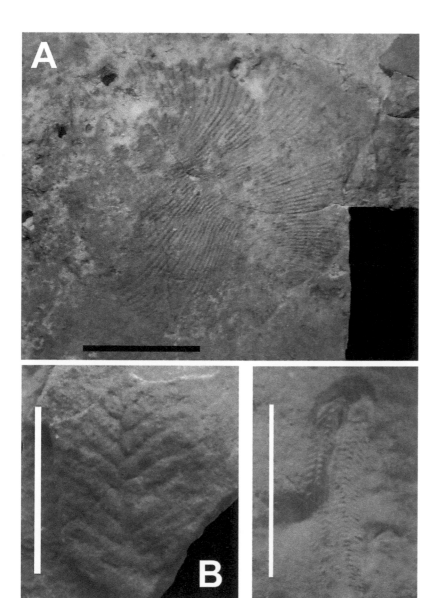

3.9. Ediacaran fossils from late Proterozoic sandstones of Australia. (A) Large *Dickinsonia*. (B) Poorly preserved rangeomorph. (C) *Spriggina*. Scale bars = 5 cm.

(A) and (C) courtesy of Smithsonian Institution; (B) Raymond M. Alf Museum of Paleontology specimen.

shaped like a short volcanic cone with four spine-like processes draped out from the "caldera" rim. It may have been sponge-like in its ecology, but more importantly it was constructed of a mineralized, multi-element structural support, one of the first recognized from the Ediacaran.[54]

Another type of fossil from the Ediacaran period is the leaf-shaped type known as rangeomorphs (fig. 3.9b), including forms such as *Rangea*, *Charnia*, *Charniodiscus*, and *Pectinifrons*. These leaf-shaped, branching fossils are preserved with simple to rather complex arrangements; some were shaped like single leaves with fractally branched fondlets within the piece, and others were branched in elaborate plumes. The fossils appear to have been held in the sediments on the bottom of the ocean by round holdfasts, which sometimes preserve as disc-shaped fossils by themselves. *Aspidella* was probably a sand-filled holdfast of one of these leaf-shaped organisms. The organisms may have been filter feeders, straining food from the water just above the seafloor. Rangeomorphs have been compared to modern sea pens, but it has recently been shown (at least for *Charnia*) that the two grow differently and probably are not related; *Charnia* adds leaf-like elements at its distal tip, whereas sea pens add them near the base and move older elements slowly upward.[55] An Ediacaran form called *Ivesheadia*, known informally as a "pizza-disc" because it rather resembles the top of a pie that had some of its cheese topping stuck to the lid of the delivery box, has recently been suggested to be the traces of heavily decomposed individuals of other genera such as *Charnia* and *Charniodiscus*. Several other genera in addition to *Ivesheadia* probably can be accounted for the same way; if so, the diversity of the Ediacaran was just a little less than we have come to think at this point.[56]

Although Ediacarans are found in significant abundance and good preservation in Africa, Australia, Russia, and Newfoundland, fossils of such quality have proven elusive in much of North America. Late in the Proterozoic Newfoundland was part of a tiny continental piece called Avalonia, which was on the other side of Rodinia from Laurentia (the bulk of modern North America). In order to find Ediacarans from Laurentia paleontologists have looked in the relatively limited areas in western North America with Precambrian sedimentary rocks of the right age. The rocks we saw in Glacier National Park—and most of the heart of the Grand Canyon, for example—were too old for Ediacarans.[57] On the western edge of the continent, however, there are rocks of the right age and sediment type, and these in fact seem to preserve Ediacaran fossils of several varieties. Among the fossils found in the Ediacaran-age lower Wood Canyon Formation in western Nevada is *Swartpuntia*, another stalked, roughly leaf-shaped organism with several petaloids, not just two in the same plane. This is a form that has also been found in Namibia. And in the Sonoran Desert of Mexico paleontologists have found the typical Neoproterozoic fossil *Cloudina*.[58]

Most Ediacarans appear to have been entirely soft bodied. Dolf Seilacher proposed that a number of the odd Ediacaran fossils that had eluded clear classification were in fact what he called **Vendobionts,** a new group of non-animal multicellular organisms that lived only in the Ediacaran period. Others believe a number of the Ediacarans are extinct, odd members of the Metazoa (animals) that have no modern descendants, evolutionary lines that did not make it past the Precambrian–Cambrian boundary, while still others believe that there were Ediacarans that despite their unfamiliar appearance were in fact basal members of some animal groups still around today. A few Ediacaran taxa have been suggested to have been possible fungus relatives, and it has even been proposed that many were giant unicellular organisms (amoeboid protozoans).[59] If some Ediacarans are in fact animals, this would not contradict molecular data that seem to suggest that a number of animal groups diversified during the Proterozoic. Based on fossils, some of the animal groups believed to be represented among the Ediacaran fossils (in addition to the sponges) include indeterminate Bilaterians, possibly worms (*Dickinsonia*); Arthropoda (*Spriggina*); stem molluscs (*Kimberella*); Echinodermata (*Arkarua*); and Cnidaria, including soft corals and/ or sea pens (Rangeomorpha). The vendobiont camp maintains that while some animals are present in the Ediacaran period, many of these forms listed above are mistakenly identified as animals and that most are from extinct lineages. Also keep in mind that there are literally dozens of other species of Ediacaran fossils for which paleontologists have nothing more than guesses as to how they lived, much less whether they might be related to any modern group. Recent finds suggest some members of the biota may have been capable of locomotion. Certainly, between the apparent sponge fossils and purported tiny (0.5 mm) fossilized metazoan embryos from the late Proterozoic of China[60]—not to mention divergence time estimates based on molecular studies—it is apparent that several animal groups at least had originated well before the Cambrian explosion. Perhaps the Ediacaran was a time of both some modern group origins and plenty of strange, extinct experiments. The roots of the Cambrian biota are beginning to look like they extend well back into the Precambrian.[61]

Not only did multicellular animals and possibly para-animals (vendobionts) appear during the Ediacaran, but in the last few million years before the beginning of the Cambrian there appeared on the scene an invention that years ago we thought didn't appear until the Cambrian itself—shells. In many parts of the world, including North America, we find in latest Proterozoic rocks forms such as *Cloudina* and *Namacalathus* that consist of tube-shaped shells of calcium carbonate, indicating that macroscopic organisms of some kind (possibly metazoans) began making exoskeletons a little before the Cambrian explosion. Was this development for support or protection? Was there already something to fear for the organisms in the oceans of the Ediacaran period? We have no fossils of predators, but perhaps these shells are indirect evidence of their presence. Such predators certainly were present by the Cambrian, and indeed they may have helped drive some of the evolutionary competition of the time, but the shelly fossils of the late Precambrian suggest that the origins of predation may extend back to before the great Cambrian diversification.

Be they weird dead ends or early animals, Ediacarans were unlike anything that had existed before in the Precambrian and they were unlike the world of the Cambrian. For billions of years Earth's biota had consisted of mostly microscopic organisms, but as we see in the Ediacarans, by the late Proterozoic life forms had gone large and multicellular and the ecosystem, although comparatively simple, was more complex than it had been. The increase in complexity and diversity would only continue and the rate of change only intensify in the next period, the Cambrian. So why now, after all this time, did the larger organisms finally appear in the Ediacaran period? What contributed to this beginning of the explosion? Several factors may have contributed. First of all, it may have taken a while for the genetic tools to be in place within the eukaryote genome to facilitate an increased rate of complexity acquisition.[62] Like many things, once a threshold in genetics was crossed, the complexity of organisms' systems could accelerate much faster than they had for hundreds of millions of years previously. Eukaryotes may have crossed this threshold late in the Proterozoic; after acquiring the genetic tool kit needed, animals may have diversified quickly. Among evidence that this tool kit was about all that was needed to produce the diversity of animals seen since that time is the fact that the most complex animals today do not have significantly more genes than very simple animals that have been around since the beginning of the Metazoa.[63] With what simple animals have it is possible to make most anything we know today.

Second, by the late Proterozoic, oxygen levels may have reached levels that finally allowed larger size and more multicellularity in organisms; it appears that the first simple animals would have required oxygen levels close to those of the modern atmosphere in order to function internally with simple diffusion systems of circulation.[64] This may have allowed increased complexity to begin to evolve. Most animals have tissues, and the cells of tissues need oxygen; single-cell organisms have no trouble accessing oxygen but getting oxygen to the cells of multiple tissues in a large but simple early metazoan would have been much easier in a more oxygenated world. From here, animals could have taken off into amazing levels of complexity, and it appears they did.

A few years ago, a hypothesis as to how late Proterozoic oxygen levels may have reached near-modern levels was presented by Martin Kennedy and others.[65] One of the things that can affect the levels of one element in the atmosphere or oceans is a change in the level of another element with which it commonly combines. In this hypothesis, oxygen may have undergone a net buildup due to a decrease in the amount of carbon available to the atmosphere. Because carbon and oxygen commonly combine, if some carbon were going somewhere else then a certain percentage of oxygen atoms would be left on their own, as in what I earlier termed a game of chemical musical chairs. And soon those lone oxygen molecules would build up, increasing the O_2 level of the atmosphere. So the question is this: What was removing the carbon from the equation? The Kennedy team noted that there was an increase in the amount of soil-formed clay in fine-grained rocks from about 800 million years ago through the Cambrian, indicating an increase in weathering and soil formation on the continents, probably as a result of the appearance of the earliest land vegetation.[66] Because organic carbon more or less "sticks" to the surface of clays formed in soils better than it does to those clay particles formed through the mechanical pounding of rock into dust, an increase in these soil-formed clays provided a way in which carbon was probably taken out of the atmospheric cycle by being buried in the fine-grained rocks. And this would leave more and more oxygen available to assist the development of our metazoan great-grandparents.[67]

Yet another possible cause of the very different biological world we see in the Ediacaran period as compared with the few billion years before it was the Snowball Earth bottleneck that life went through just before the Ediacaran began.[68] Bottlenecks, events that severely reduce the population size of a species, can accelerate the evolutionary rate of the survivors by changing the genetic makeup of the population. If a large population contains a wide genetic diversity, a bottleneck would reduce that genetic diversity (sometimes randomly) and potentially to a point that new mutations would actually increase the rate of morphological evolution in the recovering population.[69] The cause of this change in rate seems to be at the genetic level, but if such a process operated in the Late Proterozoic (and there seems no reason to presume it didn't) the early eukaryotes coming out of the Snowball glacial periods may have experienced an accelerated diversification that led to the appearance of multicellularity and a whole new host of ecologies that had not been seen before the Ediacaran—and that led to a continuing diversification in the Cambrian period.

Another contribution to this increasing complexity starting in the Ediacaran period was the rise of "macrophagous mobile metazoans." It may sound like a tongue twister from the current issue of *Alliteration Weekly*, but that phrase refers simply to animals that could move and ate relatively large organic material such as algal mats or other animals. It was the humble appearance of grazing and predation that may have help start a revolution, and we will revisit this issue later on as well.

Most of the Precambrian, the world of the stromatolites, seems to have been a time of primary producers (photosynthesizers mostly) that were limited more by available real estate (space to grow) and sunlight than anything else. Certainly there were heterotrophic microbes and single-celled photosynthesizers for much of the time before the Cambrian, but by the Late Proterozoic there were—in addition to the stromatolites—seaweeds and extensive mats of algae growing in the shallow seas, and these became a food resource for our macrophagous mobile metazoans by the Ediacaran. There seem to have been several types of grazers as well, from "mat scratchers" and "mat stickers" to "undermat miners." The evolution of this grazing feeding mode in early metazoans and their initial diversification inspired diversifications in other trophic levels of the community. As the food of the grazers now found their numbers cropped by grazing, they were no longer resource limited and could diversify more as they strove to find ways to avoid the grazers; meanwhile, one trophic level up from the grazers a host of new predators could evolve to specialize on the first-level consumers. Parallel to this, the Ediacaran saw a turnover of—and an evolutionary acceleration in—the groups of acritarchs in the planktonic realm.[70] This system of primary producers, primary consumers (herbivores), and secondary consumers (predators) had evolved in microbes fairly early in Earth history, but the Ediacaran appearance of such a system among the animals set the stage for the Cambrian radiation.

Finally, there is the issue of how much of the rise of life during the Precambrian is real. Darwin's view that evolution by natural selection should generate new species gradually but relentlessly through time hit a snag when it came to the Precambrian, one that he himself recognized and wrestled with. In fact, Darwin saw the complete lack of fossils (at the time) from Precambrian rocks as an argument against his idea, and he had no real answer for those that might raise the point. As it turns out, there are plenty of fossils from the Precambrian, but still far fewer than from the Cambrian on. Was the Precambrian–Cambrian transition perhaps a revolution of preservability? Did the ancestors of Cambrian animals run deep into the Precambrian and the fossils just not preserve well until 540 million years ago? That does not appear to be the case. If anything, it seems preservablility was *better* in the Precambrian and declined into the early Paleozoic.[71] So if chances of seeing fossils were better, say, in mid-Proterozoic rocks and yet all we see are microbes and stromatolitic structures (and maybe *Horodyskia*), and if chances of preservation were declining by the middle of the Cambrian and yet we find a gold mine of animal groups and fossils in these rocks, what does this mean for the rise of animals? Taken as a straight record, the lack of large organisms for most of the Proterozoic, transitioning into the dozens of Ediacarans of the end of the Proterozoic, transitioning into the incredible diversity of metazoans in the Cambrian indicates an explosion of animal life. Factor in the apparent inverse relationship in fossil preservability over the same period and we can conclude only that no matter how deep their roots, the rise of animals was even *more* explosive than it appears.

So the pattern is nothing like what Darwin might have expected. Whereas Darwin would have predicted the ancestors of trilobites to work their way deeper into the Precambrian in ever-simpler forms, we rather find trilobites fully formed and diversified well into the Cambrian and nothing before that save for the Ediacaran *Spriggina* as a possible ancestral arthropod. Even from Ediacaran to Cambrian periods there is the explosion of metazoan forms, and for all we have learned in the past 50 to 100 years, we still are unsure what happened during this interval. Happily for Darwin there were fossils and an incredible story in the Precambrian, but what he could not have predicted in 1859 was that the pattern of the rise of animals was as complex as it was. Certainly he never would have guessed the presence of organisms as different as the Ediacarans, especially if they were their own group of extinct multicellular para-animals, unrelated to metazoans. But even if we see among Ediacarans the ancestors of some modern animal groups, Darwin would likely have been surprised by their very late appearance in the Precambrian. What

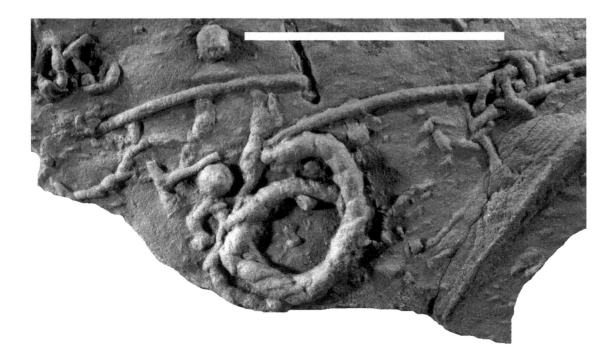

we understand now is that the *rate* of evolution is by no means constant. Earth affects life, life affects the environment, and the interactions of the two, combined with genetic changes and crossings of thresholds, can result in great revolutions and leaps forward now and then. Evolution can progress in fits and starts and not necessarily always slowly and gradually, and this makes for a more interesting story, one that Darwin himself might even prefer.[72]

The Cambrian period begins with the first appearance in the rock record of a trace fossil called *Treptichnus pedum*.[73] This is a curving, sectioned, shallow burrow that indicates an animal apparently adjusting its body position multiple times in the process of executing a circular turn (fig. 3.10). The first appearance datum of this particular trace was chosen as the base of the Cambrian because it is the first indication in the rock record of complex behavior. *Treptichnus pedum* occurs all over the world, so it has utility as an indicator of the beginning of the period. The official lower boundary of the Cambrian is marked at a particular bedding plane in a section of rock on the shoreline of Fortune Head, Newfoundland. Period boundaries have become very precise in recent years, and our definitions very specific. As J. W. Valentine and others wrote in a 1999 paper, "Whatever animal left that earliest trace in Newfoundland [*Treptichnus pedum*] began life in the Proterozoic and died in the Phanerozoic, achieving a special place in history."[74]

The Newfoundland *Treptichnus* level is a GSSP, a Global Boundary Stratotype Section and Point. The lowest bed of the Cambrian system

3.10. *Treptichnus pedum,* the trace fossil ichnospecies whose first occurrence indicates the base of the Cambrian. Specimen from the Bright Angel Formation. Scale bar = 5 cm.

Museum of Northern Arizona specimen (MNA 3864).

The Cambrian Explodes onto the Scene

here is really no different in appearance from the Precambrian rocks just below it, but above it occurs the critical trace fossil marking the beginning of the Cambrian. Unfortunately, the critical trace fossil is now known to occur in rocks about 4 m (13 ft.) *below* the marker too, and a different species of *Treptichnus* has also been found in rocks 3 million years back into the Proterozoic in Africa.[75] Now we know that the onset of complex behavior was a gradual process that crescendoed across the Precambrian–Cambrian boundary. In fact, as was pointed out by Frank Corsetti and J. W. Hagadorn a few years ago, a distinction needs to be made when talking about this interval in Earth history between the Precambrian–Cambrian transition and the Precambrian–Cambrian boundary. The latter is a specifically defined point in the rock, whereas the former encompasses a drawn-out series of changes in the Earth's biota—a dramatic and relatively sudden transformation in the history of the planet, yes, but one that does not present itself in the rock record as the single bedding plane we might prefer. The more we learn about this time, the more interesting and complex it appears to be.

We have finally stepped into the world of the Cambrian. It is time now to begin our journey through its 54 million years of history. For the next five chapters we will explore its menagerie as revealed to us by various fossil localities in what once, long ago, was a strange continent known as Laurentia.

4.1. Drawing showing the generalized structure of a solitary archaeocyathid sponge. Total height about 4 cm.

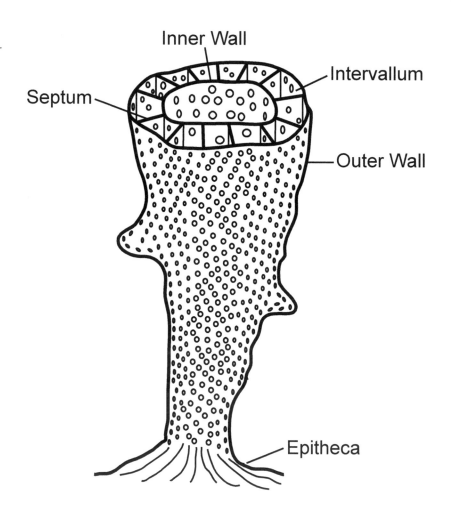

Inner Wall

Intervallum

Septum

Outer Wall

Epitheca

Welcome to the Boomtown: The Early Cambrian Seas

4

THE CAMBRIAN BEGINS AFTER THE FIRST ANIMALS HAVE APPEARED, after the first hard shells, but before the first appearance of those icons of the Paleozoic, the trilobites. In fact, trilobites do not appear until the Cambrian is already about 20 million years old. How, then, to define the beginning of this important period? As we saw at the end of chapter 3, the base of the Cambrian is marked by the first appearance of the trace fossil *Treptichnus pedum* (fig. 3.10), so that the Cambrian officially began not with the first animals, nor with the first shells, nor even with the first trilobites, but rather—as we discussed previously—with the first evidence of complex burrowing behavior. If this seems somewhat arbitrary, it is to a degree, but that does not mean it is entirely without justification or reason; simply, as geologists learned more about the Cambrian they realized that it began not with the appearance of all the fossils that characterized the period at around the same time but rather with a more drawn out (over 10+ million years), staggered appearance of the groups. Nevertheless, in geological terms it was still an explosive window of time. Something had to mark the beginning of the Cambrian, and it was decided the trace fossil was the best candidate. The first animals, after all, overlap with distinctly Ediacaran forms for a long time, and those faunas have mostly disappeared and been replaced by an abundance of shelled animals (which also appear early) in the earliest Cambrian, long before trilobites appear. So the trace fossil marker may be as distinct an indicator as we can ask for right now. Although shells appear in the Proterozoic, they do not become common until the Early Cambrian, about the same time as the appearance of *Treptichnus pedum*, so there is a coincident rise in abundance of these fossils in earliest Cambrian rocks also. One of the best places on Earth to see this Precambrian–Cambrian transition is in Russia, where the rise of these shelled animals is well preserved.

Small Shelly Fossils

Although *Cloudina* occurs in rocks of the Ediacaran and shows that hard protective shells had been used by macroscopic organisms well before the Cambrian, it was not until the Cambrian itself that these shells became widely employed and become abundant in the fossil record. "Small shelly fossils," as these mostly diminutive shells and skeletal parts are known, are in places common and very diverse in Lower Cambrian rocks, particularly before the appearance of trilobites. One of the early subdivisions of the Cambrian that was used in past years was known as the

Tommotian, and the rise of the first Cambrian shelly fossils, the first Cambrian diversification, has been sometimes (and very informally) referred to as the "Tommotian commotion." Many of the small shelly fossils that appeared at this time belong to unknown organisms but others, although only fragmentary, seem to indicate the presence of familiar animals in these earliest days of the Paleozoic.

Among the small shelly fossils found in rocks of the earliest Cambrian are sponge spicules, **archaeocyathids,** phosphatic and calcareous tubes of unknown organisms, phosphatic button-shaped plates of unknown organisms, possible plate fragments of the lobopod *Microdictyon*, phosphatic horn-shaped shells and plates, ossicles of echinoderms, shells of brachiopods, monoplacophoran mollusc shells, hyoliths, and miscellaneous molluscs of various shapes including probable gastropods (such as *Aldanella*). Some odd-shaped plates and rings, like those assigned to *Eccentrotheca*, probably were from conical, filter-feeding animals that lived attached to the bottom; these animals may have been covered in many of these small shelly sclerites as a protective layer. Although *Eccentrotheca* itself is known from Nova Scotia, a part of modern North America that was then part of the Avalonia subcontinent and not Laurentia, similar forms undoubtedly lived in Laurentian waters during the Early Cambrian.[1] There are also plenty of small fossils for which we have nothing but guesses as to what type of organism they represent. These faunas, which are known from various parts of the world but perhaps best from Siberia and Australia, suggest a diverse biota during the earliest millennia of the Cambrian and one in which animals had certainly diversified since the Proterozoic. Among the small shelly fossils in Russia were found what may be embryonic cnidarians,[2] the group that includes corals, jellyfish, and sea anemones. It appears that, contrary to some of our earlier predictions based on geologically later faunas, the small shelly fossil animals and organisms may have originated in relatively deeper water and later moved into shallow water; the opposite seems to be true of some younger clades.[3] Because these small shelly fossils were part of the initial diversification at the very beginning of the Paleozoic, however, their following a different pattern than later groups may not be as surprising as we might be tempted to think. Seemingly anything can happen in the first few million years of the Cambrian.

Dream Dive:
The Cambrian Reefs

Archaeocyathids

The archaeocyathids of this time are an interesting case. These are mostly small, conical fossils, pointed at the base (with small attachment "roots") and with an open circular top. The open top was also double walled concentrically, so that the middle of the cone was open (fig. 4.1). The space between the cone walls is called the intervallum, and the cones are supported by radial walls called septa. The archaeocyathids appear to have been composed of calcium carbonate, and in places their fossils are so abundant that it appears they may have formed archaeocyathid "reefs." Some were solitary individuals (although they often grew in groups close to each other), others grew in large, multi-cone masses (**bioherms**), and still others were thin and encrusting of other substrates. For years it was unknown what type of animal these fossils represented; similarities to sponges were noted, but it was not until fairly recently that it was decided that archaeocyathids were in fact an extinct form of sponge, although they lack the spicules that many sponges have. These fossils appear near the beginning of the Cambrian period but trail off in abundance and diversity at the end of the Early Cambrian, and their heyday of diversity and population numbers seems to have been during the time of the earliest trilobites and just before. Their peak diversity was nearly 170 genera, but only 2 genera have been found after the Early Cambrian; these forms indicated that archaeocyathids, severely reduced in numbers, hung on in some habitats into the Late Cambrian.

Archaeocyathids appear to have lived in shallow subtidal to intertidal marine areas, usually anchored in soft substrates. They seem to have preferred water depths of about 20–30 m (65–100 ft.), and generally lived in water shallower than 100 m (330 ft.). Optimal water temperatures for archaeocyathids may have been 25–30°C (77–85°F) although this is not well established. Individuals may have lived up to about 20 years, and what killed most archaeocyathids seems to have been storms that either uprooted the individuals or buried them in sediment. Archaeocyathid abundances seem to have been highest in the tropical to subtropical paleolatitudes between 30°N and S.[4]

Multiple low, domal bioherm buildups of archaeocyathids (and other organisms) in some cases coalesced into large patch reefs. Some of these reef buildups and patches of bioherms may have been as much as 32–40 km (20–25 mi.) across. These Cambrian reefs were a bit different from our modern reefs in that their main organisms in terms of construction were not corals, but several different species of archaeocyathids. Corals generally grow densely in carbonate environments and many contain symbiotic algae, whereas archaeocythids appear to have lived in carbonate and siliciclastic environments, and do not seem to have occupied more than 50% of the rock (and thus probably did not construct reefs as structurally strong as corals). According to some researchers, they probably did not live in symbiosis with algae.[5] So the ecosystems of these Cambrian archaeocyathid reefs were a bit different, but the forms of the large bioherms are similar to what we see in many modern reefs. The archaeocyathids in these reefs probably also only grew in areas with significant nutrient levels, and they may have been bound together at least to some degree by synsedimentary cementation—that is, hardening of the sediment while it was still being deposited in the environment rather than long after burial.[6] But archaeocyathids were not the only members of the reef-building community; these Cambrian reefs were also composed of calcified cyanobacteria, and corals and coralomorphs were also present. We met the cyanobacteria in chapter 3 as the photosynthesizing microbes responsible for the stromatolite structures see in many Precambrian rocks, but some species may form structural additions to reefs through the precipitation of calcium carbonate.[7]

Corals

Were there coral reefs in the Cambrian? If we could time travel, could we have snorkeled a coral reef 520 million years ago? To some degree, yes (but without the palm grove lining the beach, unfortunately). Corals were present in Cambrian reefs but were a more minor component, and they did not form the massive frameworks that we see in modern reefs. "Coralomorphs" such as *Cysticyathus* are fossils that seem to have the approximate form of corals but which we cannot be sure were true corals; in many cases, they may be ancestors of corals. Corals are cnidarians, a group that includes jellyfish and sea anemones, and the Cambrian coralomorphs are probably cnidarians or close relatives as well. Possible true corals have also been found in Lower Cambrian rocks in the Flinders Ranges of South Australia, and in Alaska, British Columbia, Montana, and Siberia.[8] Small (mm-scale) corals have been found in the Campito and Poleta formations in California, but the coral *Harklessia* was recently described from western Nevada and is a respectable size for a Cambrian reef former, about 12–18 cm (5–7 in.) across (fig. 4.2). The individual coral animals that precipitated the calcium carbonate that forms the coral structure appear to have been about 1–3 mm across.[9] The coral *Harklessia* occurs near the top of the Lower Cambrian Harkless Formation, thus the name. The second part of its species binomial (*H. yuenglingensis*) honors a popular East Coast beer.

Reef Structure

The cup-, platter-, and cone-shaped archaeocyathids and the masses of cyanobacteria, along with the coralomorphs, all seem to have grown on and around each other en masse, forming the "framework" of the reefs, with some individual sponges growing to 50 cm (20 in.) high. These types of reefs are particularly common in the Lower Cambrian of Siberia. In Nevada, the coral *Harklessia* grew among calcimicrobes and archaeocyaths to form reefs up to 7.5 m (24.5 ft.) wide. Although this coral looks generally like some corals one can see today while snorkeling the tropics, the Cambrian reefs were composed more of archaeocyaths and calcimicrobes, and the corals composed a respectable but minor 4.5% of the volume of the Harkless Formation reefs. Though perhaps not as strong as modern coral reef frameworks, some of these archaeocyathid-, sponge-, calcimicrobe-, and coral-formed reef structures from the Lower Cambrian may also have been wave resistant; they may not have been waveproof, but they were strong. Although *Harklessia* was large and would be recognizable to us as a coral, it was not particularly abundant, and so our Cambrian snorkeling adventure would cruise along a reef with plenty of

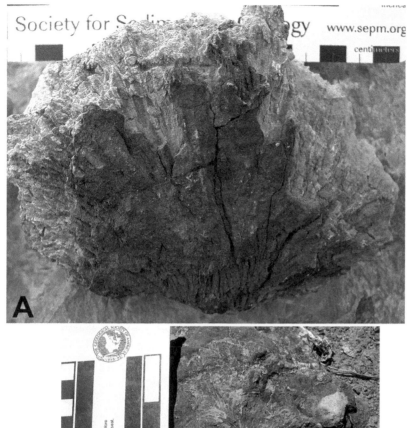

4.2. The Cambrian coral *Harklessia* from the Lower Cambrian Harkless Formation of Nevada. (A) Larger specimen in a gray rock unit, scale in centimeters. (B) Smaller specimen in red rock.

Photos by and courtesy of Melissa Hicks.

archaeocyathids, sponges, cyanobacteria, and small coralomorphs, but we would have to stay alert to spot the occasional large, true coral.

One of the most important aspects of the Cambrian reefs is the habitat they created for all other animals. If you imagine archaeocyathid-, cyanobacteria-, and coral-built bioherms growing over an extensive area on the otherwise flat plain of the Cambrian ocean bottom, you can see how this would provide shelter and habitat for organisms that might not live there otherwise, much as a sunken ship today becomes a haven for inhabitants of modern reefs. In among the archaeocyathids and other organisms may have lived a host of other Cambrian animals. On modern reefs, single coral heads provide habitat for hundreds of species and sometimes thousands of individuals of worms and arthropods, for example.

Animals that have been found in and among the Cambrian reefs include molluscs, brachiopods, trilobites, echinoderms, and hyoliths (fig. 4.3). Some of the habitat within the reef may even have been in crevices, pockets, and overhangs, as in modern coral reefs.[10]

Spot X: The Cambrian Begins in Laurentia

Although the global boundary marker for the Cambrian, the GSSP, is in North America now, in Newfoundland, during the Cambrian that particular part of our continent was not connected to Laurentia and was well down in the Southern Hemisphere. A good Precambrian–Cambrian contact in North America that was also solidly in Laurentia during the Cambrian is in the Death Valley area in eastern California. Here we see the beginning of the Cambrian in the first appearance of *Treptichnus pedum* in the upper part of the lower member of the Wood Canyon Formation. The Wood Canyon, you may remember, has Ediacaran fossils in its lower layers, and we will see more of this formation later on, but these top reaches of its lower member contain the top beds of the Proterozoic and the earliest beds of the Cambrian. Also tied in with the boundary here is a strong negative carbon isotope excursion,[11] which is often recorded in other parts of the world in carbonate rocks of the boundary age.[12] Above this Precambrian–Cambrian boundary in the western part of North America we also pick up small shelly fossils through many layers of rock before we eventually find the lowest occurring trilobites. Among the small shelly fossils are bits of molluscs and hyoliths and archaeocyathids. But through all these rocks representing millions of years, we still see no trilobites. In fact, to find these dinosaurs of the Cambrian we needn't bother looking too low in the section,[13] and then when we are closing in on the right horizon we have to look pretty hard. And the oldest trilobites in the world really occur only in three places: Morocco, Russia, and North America. To find Laurentia's oldest trilobites, the first trilobites to appear in what is now North America, we must travel to western Nevada.

Montezuma Range: The First Trilobites

Far out in western Nevada's Esmeralda County lies the Montezuma Range, a low and otherwise unremarkable range of mountains, or perhaps large hills, west of the mining town of Goldfield. This area has the distinction of yielding (along with Morocco and Russia) the oldest trilobites in the world. For all the places around the planet that produce trilobites in such well-preserved form or in such great numbers, there are only a few windows that show us the actual first appearance of these animals, and Esmeralda County, Nevada, is one of them. Here, on juniper-covered slopes of the range, trilobite paleontologists Bill Fritz and Stew Hollingsworth (along with Judy Fritz and Mary Hollingsworth) have put North America on the map by finding some of the oldest trilobites in the world and the oldest on the North American continent. The oldest trilobites in

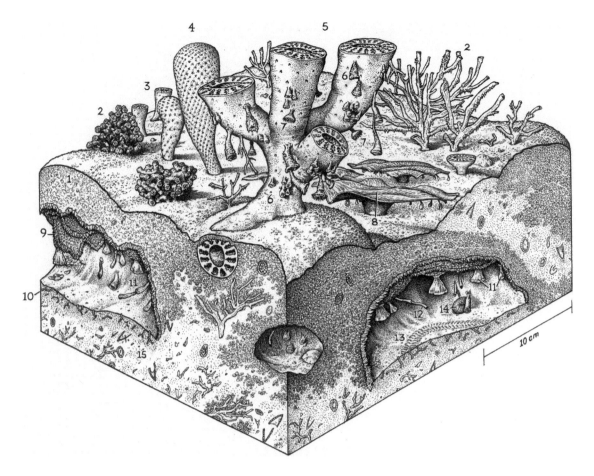

the world appear in Siberia, Morocco, and in Nevada nearly simultaneously in the late Begadean of the *Fritzaspis* zone.[14]

The rock units producing these first trilobites here include the Campito Formation and the overlying Poleta Formation. The Campito consists largely of fine-grained quartzite and siltstone and is a more-than-respectable 1065 m (3493 ft.) thick. The formation is divided into the lower Andrews Mountain Member, middle Gold Coin Member, and the upper Montenegro Member, and the formation is a lateral equivalent of the middle part of the Wood Canyon Formation farther south. The Poleta Formation is composed of limestone and siltstone (with some shale and sandstone) up to 580 m (1900 ft.) thick, and forms a carbonate-rich lateral equivalent of the upper part of the Wood Canyon Formation. Some of the limestones in the Poleta are composed of ooids, the small carbonate beads we first encountered in the Grand Canyon in chapter 2. Ooids are indicative of shallow carbonate shoals where currents and wave action roll the lime mud back and forth. Archaeocyathids are also relatively common in the Poleta, along with some trilobites.[15]

The Gold Coin Mine section in the Montezuma Range includes Laurentia's oldest trilobites as faint cephalon molds in sandstones of the Gold Coin Member of the Campito Formation. These trilobites include

4.3. Reconstruction of an Early Cambrian reef community. Components are (1) A calcified cyanobacterium; (2) Branching archaeocyathid sponges; (3) Solitary cup-shaped archaeocyathid sponges; (4) Chancellorid; (5) Radiocyath sponge; (6) Small, solitary archaeocyathid sponges; (7) Coralomorphs; (8) Flat archaeocyathid sponges; (9) Fibrous cement forming within crypts; (10) Microburrows; (11) Archaeocythids and coralomorphs; (12) Cryptic cribricyaths; (13) Trilobite trace; (14) Cement botryoid; and (15) Sediment with skeletal debris.

Drawing courtesy of John Sibbick. Key from Rachel Wood (1998).

4.4. Outcrops of the Campito Formation (foreground) in the Montezuma Range of Nevada.

Fritzaspis, Profallotaspis?, Amplifallotaspis, and *Repinaella* and are very rare. If we move upsection in the Campito into the lower Montenegro Member, more complete trilobites that are comparatively common can be found at the base of the *Fallotaspis* zone at a site on the west side of the Montezuma Range (fig. 4.4); these trilobites are from the base of the Montezuman stage and belong to a new species of *Archaeaspis* (fig. 4.5f,g). The *Archaeaspis* fossils are found at a dark, west-dipping outcrop of silty shale in the lower Campito and are typical of fallotaspids in having long eyes attached to the anterior part of the glabella. This site represents one of the oldest appearances of abundant trilobites in North America and indicates the beginning of the traditional trilobite-bearing Cambrian. A little higher in the Campito is a quarry that produces complete specimens of

4.5. Trilobites of the Lower Cambrian Campito Formation in Nevada. (A–B) Nearly complete specimens of *Nevadia weeksi*. (C) Articulated specimen of *Montezumaspis cometes*. (D) Cephalon of *Nevadia weeksi*. (E) Two cephala of *Esmeraldina rowei*. (F) Two articulated specimens of *Archaeaspis* sp. (G) Cephalon of small *Archaeaspis* sp. All scale bars = 1 cm.

Specimens (A)–(E) collected by Norm Brown; (F)–(G) by the author.

Fallotaspis itself.[16] In addition to *Archaeaspis* and *Fallotaspis*, the Campito Formation contains a number of well-preserved trilobite genera, including *Nevadia, Montezumaspis*, and *Esmeraldina* (fig. 4.5a–e).

The main question raised by the first trilobites is why they appear, spring, leap from the fossil record fully formed and diversified. *Fallotaspis, Nevadella*, and *Repinaella* do not appear to be partially formed "prototype" trilobites; as fallotaspids, they are a recognizable form of trilobite with differences from other families, to be sure, but nothing about them is half baked – they *are* trilobites and are every bit as developed as any later in the Cambrian. So why is our first view of trilobites so clearly of *true* trilobites? Why was there no slow-burn development period in which we see characteristics of trilobiteness appear one by one? Perhaps the origin and first diversification of trilobites really was so fast that we can't distinguish it in the fossil record, and what we can see consists of no trilobites at one geologic moment and suddenly fallotaspids the next. (The intervening geologic moment, of course, could have consisted of a couple million years.) Perhaps our fossil record (particularly that of the earliest Cambrian) is not as complete as we'd like to believe. Or, alternatively, there could have been a period of development of trilobites, during which they formed all their traits, but which we cannot see in the fossil record. It is possible, for example, that the calcified dorsal exoskeleton of trilobites only developed after the other characteristics of the animals, and because the new, mineralized exoskeleton preserves much more easily than the unmineralized one all trilobites would have had before, what we are seeing appearing in the fossil record of fallotaspids is full-formed trilobites, already in existence for some time, bursting onto the scene with the development of a hard dorsal exoskeleton. If this latter scenario is the case, the true lineage of trilobites probably extends somewhat farther back into the Cambrian or beyond, and the first appearance of trilobites in the fossil record is a preservational phenomenon.

This latter view has recently found itself with fewer supporters. It now appears that mineralized skeletons probably appeared soon after trilobites themselves did, but the long previous history of differentiation still seems to have occurred. In part this is based on phylogenetic studies of trilobites that factor in paleobiogeographic data. Taxonomic and morphological differentiation seem to follow paleogeographic patterns, suggesting separation of trilobite lineages occurring already in the latest Proterozoic![17] Why were there no trilobites until well into the Early Cambrian, then? Several factors may be hiding these first trilobites, and these hypotheses are small size, low abundance, and habitat preference for marginal environments. Paleontologists haven't been able to fully confirm or refute any of these hypotheses just yet, but they are the best explanations we have at the moment for why, like many other Cambrian animals, the group's origin seems to go back farther than its known fossil record, predating the Cambrian radiation itself.

The Poleta Formation has yielded a number of trilobites from the overlying *Nevadella* zone, including *Nevadella* and *Cirquella* (fig. 4.6).

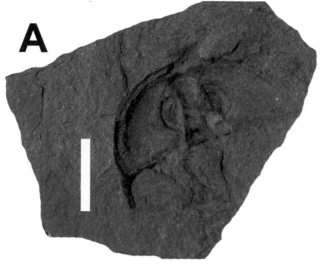

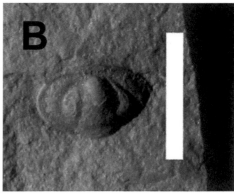

The Poleta Formation also preserves a number of taxa other than trilobites, particularly at a site known as Indian Springs. Sponges, hyoliths, brachiopods, molluscs, and chancellorids have been found here. One other important fossil type from the Poleta Formation is the group **Helicoplacoidea.** These spirally grooved organisms, shaped somewhat like an ice-cream cone, were probably the first true echinoderms, and they likely were bottom-dwelling suspension feeders. These fossils and others occur in the Poleta Formation in Indian Springs Canyon in western Nevada. Lower Cambrian rocks producing *Fallotaspis* and *Nevadella* zone trilobites (but not the oldest trilobites of the *Fritzaspis* zone) also occur in the Mackenzie Mountains in Canada.[18]

4.6. Trilobites of the Lower Cambrian Poleta Formation in Nevada. (A) Cephalon of *Nevadella eucharis*. (B) Cephalon of *Cirquella* sp. Both scale bars = 1 cm.

Specimens collected by Norm Brown.

Down South

In an area we will see more of in the next chapter, Sonora, Mexico, there are also Lower Cambrian rocks that have produced trilobites of the *Nevadella* and *Fallotaspis* zones. These fossils are in what is called the Puerto Blanco Formation. As in Esmeralda County, Nevada, these very early trilobites are not always easy to find and their skeletal parts have seen better days, but their presence is important and that is why researchers put in the (sometimes discouraging) effort to find them.[19]

Trilobites!

If there is an icon of the Cambrian period it is the **trilobite.** These often beautifully preserved marine arthropods appeared during the Early Cambrian and diversified into nearly 20,000 species over their 270-million-year tenure on our planet.[20] Trilobites have been contemplated by humans probably for millennia, as some fossil specimens appear from archaeological evidence to have been used as amulets in both North America and Europe going back perhaps as long as 15,000 years ago.[21] So the fascination with trilobites is nothing new; their modern study is a logical pursuit grown out of the same curiosity that affected our ancestors. The term *trilobite* was coined in 1771 by a German naturalist named Johann Ernst Immanuel Walch. Walch was a university professor who taught subjects as diverse as Latin and Greek, Roman history, biblical studies, and natural history. His 1771 publication, "The Natural History of Petrifactions," included a summary of the range of previous terms for various three lobed fossil specimens, and then he proposed the word "trilobite" to describe them as a group; the paper also included many illustrations of trilobite fossils.[22]

As discussed above, the trilobites burst onto the scene in the Early Cambrian with significant morphological disparity already in place, and their familial diversity increased quickly, peaking in the Late Cambrian. The trilobites' diversity remained high through the following Ordovician period and then slowly declined through the rest of the Paleozoic, but their sudden appearance and diversity peak in the Cambrian makes this a real age of trilobites. The trilobites became extinct at the end of the Permian period.[23] Throughout the Cambrian period there were at least 90 families of trilobites that existed during some part of the time interval.

Trilobites are probably the most common fossils in Cambrian rocks, and it would appear that they were incredibly successful and important in their ecosystems. The former, as indicated by their diversity, and the latter, as indicated by their ecomorphological disparity, certainly seem to be true, but interestingly their importance is probably exaggerated by their preservability. Go to any given Cambrian fossil locality and you are likely to turn up trilobites, but this is probably due more to the fact that these arthropods had a mineralized exoskeleton than to their original numeric dominance of the biota at that site. There were plenty of other animals around at the time, but these are rarely preserved because the species either were entirely soft bodied or because even their exoskeletons were not mineralized with harder material, as in the case of most arthropods. The result is that most Cambrian localities preserve plenty of trilobites, because of their calcite-hardened exoskeletons, along with the shelled brachiopods and hyoliths, but only rarely much else. In reality, these three groups could have been a minority in the original ecosystem, where a higher diversity of soft-bodied species may well have been numerically dominant. So were trilobites insignificant? Probably not. Their diversity of form and their sheer numbers at many sites, even considering their harder exoskeletons, indicate that they played an important role in many Cambrian ecosystems, but it was not a "trilobite world" to the degree suggested by the straight fossil record.[24]

That said, the trilobites of the Cambrian show an amazing diversity of form, and if this in any way reflects the complexity of their host ecosystem, hidden from us by the fact that so many members of it were soft bodied, then we can be fairly certain that the biological world of the Cambrian was not a primitive or simple one but rather that the oceans teemed with a beautiful diversity of interacting creatures of all sizes, shapes, and colors. Mechanisms of evolution and ecology may be difficult to study in some of these animals due to their comparatively restricted fossil record, but trilobites can serve as workhorse fossils for studying some aspects of the Cambrian world. So the trilobites, overrepresented as they may be, are important members of the Cambrian fossil record and in some cases may serve as proxies for the ancient things we can't see today.

Trilobites seem to represent a single, natural group, as indicated by at least seven characters common to nearly all of them. Among the characters are unique aspects of the eyes, suture patterns that assist with molting, the construction of the exoskeleton, and the form of the posterior end of the animal. The significance of trilobites forming a single, natural group is that they arose from a single common ancestor and not from several different lineages.[25]

Bug Anatomy: Structure

Let's take a look at the anatomy of a trilobite now, because we'll be seeing a lot of these on our journey. The first thing to mention is that the name "trilobite" of course means "three lobed." Although many trilobites break down easily into front, middle, and back sections, and are thus obviously three lobed, the three lobes in this case refer to those formed by breaking the trilobite exoskeleton into three zones along two anteroposteriorly oriented lines. The lines enclose a central **axial lobe,** which is bordered on each lateral side by a **pleural lobe;** the axial and two pleural sections form the three lobes of the trilobite.

The front-to-back zonation consists of the **cephalon,** the name for what is essentially the head of the animal; the **thorax,** a middle zone consisting of a couple, a few, or up to a hundred body segments; and the **pygidium,** the back end of the animal which often forms a shield-like cover nearly equivalent to the cephalon in size (fig. 4.7a). The cephalon is, in most species, composed of the **cranidium,** a central portion containing the eyes and other key features, and two lateral **free-cheeks** (or **librigenae**). Between the eyes along the midline of the cephalon is a raised area called the **glabella.** Some species have long or short projections off the posterolateral corners of the cephalon, and these are called the **genal spines.** Below the anterior part of the cephalon, on the ventral side of the animal, is a single, midline plate called the **hypostome** (fig. 4.7b). The hypostome was in most forms directly under the anterior part of the glabella and was either detached from the anterior edge of the cephalon (natant) or more or less firmly attached to it (conterminant). The attachment or detachment of the hypostome may have had a correlation with the feeding mode of particular species, as this plate formed a calcified ventral protection and support for the mouth and anterior gut of the animal. It is important to remember that even in species with detached hypostomes, the gap area between the anterior edge of this plate and the ventral edge of the cephalon was still covered with a cuticular exoskeletal material, it just was not mineralized. The form of the hypostome, both in shape and attachment style, plays an important role in many higher level classifications of trilobites.

4.7. Trilobite exoskeletal anatomy. (A) Features of the dorsal exoskeleton. (B) Features of the ventral part of the cephalon.

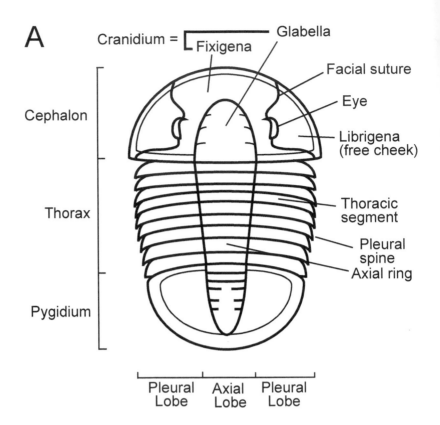

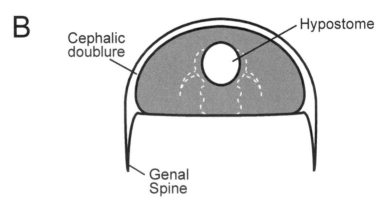

The thorax consists of a number of individual segments that articulate with the one in front and behind. These are called **thoracic segments,** and each is a single exoskeletal element from one side to the other, but like the trilobite in general, each segment can be divided up into several zones, including the central **axial ring** and the two lateral **pleurae.** Posteriorly directed projections off the sides of the thoracic segments are called the **pleural spines.**

All of the parts mentioned so far are part of the calcified exoskeleton of trilobites, a part that is mostly dorsal, or along the back. (Particularly well preserved trilobite specimens, from several species and formations of various ages around the world, suggest that the dorsal exoskeleton of

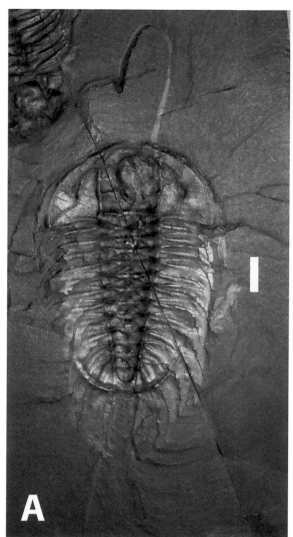

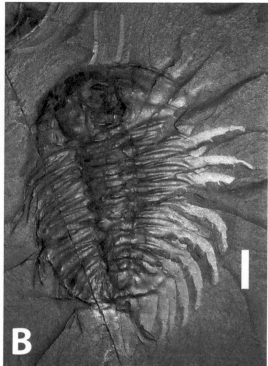

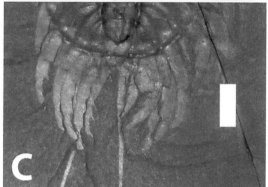

the living animals exhibited banded and spotted color patterns in some cases.) Below this calcified dorsal exoskeleton are the soft parts of the trilobite – the parts covered with an exoskeleton still but one that is not mineralized by calcite and, thus, is only very rarely preserved. In fact, trilobites had been known for years by just their calcified, mostly dorsal exoskeletons before anyone saw what the rest of the animals looked like.

Under the pleura on each side was a **ventral cuticle,** but there was little of the body between these; the pleurae probably served mainly as a protection for the appendages. Below the thoracic segments the trilobite body consisted of many elements, but the key structures were under the axial lobe. The gut, arteries, and nerve cords were contained between the axial rings and a ventral plate, while ventral and lateral to this central core of critical anatomy were the walking legs and feather-like branches.

4.8. Legs and soft-part anatomy of trilobites as exemplified by *Olenoides serratus* from the Burgess Shale. (A) Complete specimen (USNM 58588a) showing posterior legs and antennae. (B) Another specimen (USNM 58588b) showing structure of the legs, particularly on the right side. (C) Close up of posterior section of specimen from A (USNM 58588a), showing structure of legs and cerci. All scale bars = 1 cm.

All courtesy of Smithsonian Institution.

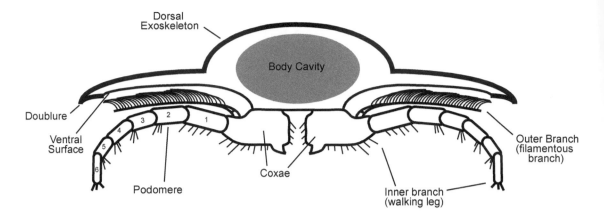

Dorsal Exoskeleton

Body Cavity

Doublure

Ventral Surface

Outer Branch (filamentous branch)

Podomere

Coxae

Inner branch (walking leg)

4.9. Cross-sectional view of a trilobite showing structure of the biramous appendages. Mineralized dorsal exoskeleton shown by thick line; thinner lines are unmineralized. Podomeres 1 through 6 numbered. Ventral surface is unmineralized and contains possible gill surfaces. Outer branch, once thought to be the gills themselves, probably waved to move water across gas-exchange surfaces on ventral surface.

Under the calcified exoskeleton, and only rarely preserved because they are unmineralized, are the trilobite's appendages. Trilobites generally have a single pair of antennae, the bases of which attach under the sides of the hypostome underneath the cephalon; from here the multi-jointed, elongate, and tapering sensory antennae would curve forward out from under the cephalon, where they could gather data about the world in front of the animal. In the trilobite *Olenoides serratus* (and potentially other species that we do not yet have evidence from), the last pair of appendages were also antennae-like organs called **cerci** which projected posteriorly from under the axial part of the pygidium.

Posterior to the front antennae in most trilobites there were three pairs of biramous (two-branched) appendages under the cephalon (sometimes four). Each segment of the thorax also had a pair of biramous appendages beneath it, and the pygidium had, anterior to the back antennae if they were present, several biramous appendages also. The leg structure of trilobites is sometimes well preserved in the Burgess Shale (fig. 4.8 and plate 17). Each biramous appendage consisted of an inner and outer branch (also called the **endopod** and **exopod**, respectively). The inner branch was the walking leg and was made up of a medial base called the **coxa** and distal elements called **podomeres,** of which most species had six on each leg (fig. 4.9). The podomeres often had ventral spines on each element, often extensive on the more medial ones, and there were usually several spines on the tip of the most distal one that contacted the substrate. The outer branch consisted usually of a single, stiff element branching off the dorsal surface of the coxa. The branch was divided into proximal and distal lobes, and in most species the proximal lobe had many filamentous rods that fanned up and back, overlapping the outer branch "fan" of the appendage behind. Fine, hairlike spines on the legs helped the trilobite sense the positions and movements of its legs and their relation to other parts of the thoracic skeleton.

Muscle scars on the insides of trilobite exoskeletons indicate how the multi-elemental body was moved. Each thoracic segment articulated with the segment in front of it by inserting its anterior edge below the posterior edge of that leading element. The axial ring of each segment

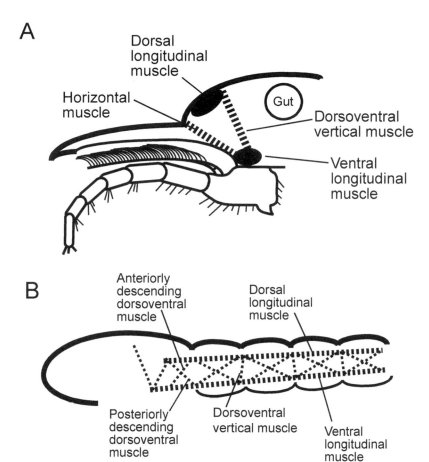

A

Dorsal longitudinal muscle

Horizontal muscle

Gut

Dorsoventral vertical muscle

Ventral longitudinal muscle

B

Anteriorly descending dorsoventral muscle

Dorsal longitudinal muscle

Posteriorly descending dorsoventral muscle

Dorsoventral vertical muscle

Ventral longitudinal muscle

4.10. Muscles of trilobites. (A) Cross-sectional view of one side of body showing muscles of body cavity in anterior view. (B) Schematic sagittal view of trilobite body showing muscles of thorax in left lateral view. These muscles moved body flexion and extension and allowed trilobite to enroll. Muscles in (A) and (B) shown as either solid ovals or thick dashed lines.

had an extended tongue (the **articulating half ring**) that projected forward to keep this articulation intact. Movement of the articulated series of thoracic segments was facilitated by a network of muscles that ran down the inside of the body like a girder bridge. This network consisted of ventral and dorsal longitudinal muscles running along the inside of each side (left and right) of the axial ring. The ventral and dorsal longitudinal muscles were attached to each other on each side by dorsoventral vertical muscles and anteriorly descending and posteriorly descending dorsoventral muscles (fig. 4.10). This network of muscles allowed the thorax to flex and extend, as needed, as an armored unit. For example, contraction of the ventral longitudinal muscle curled the trilobite up like a pill bug or armadillo and brought the pygidium up under the cephalon so that the entire, soft underbody and limbs were within the protection of the mineralized exoskeleton. Other muscle movements would have helped the trilobite wiggle out of sediment if temporarily trapped or out of its former exoskeleton when molting. Muscles also moved the legs during walking and digging and helped move food to the mouth. Walking legs probably had internal muscles between each podomere articulation and more stretching from the inside of the axial lobe down into the coxae.[26]

The nearly 20,000 trilobite species that have been named are classified into 10 orders, 9 of which are known from Cambrian rocks around the world. Unfortunately, the classification of trilobites is more clearly defined in groups for which we have good specimens of a number of growth stages from young to adult, and this is not the situation for many groups from the Cambrian.[27] We simply don't know enough about many Cambrian trilobite groups to be very confident about their relationships. But we do have a reasonably good view of the general picture.

All trilobites are in the class Trilobita, a group within the phylum Arthropoda (fig. 4.11a). Arthropods are defined by a number of characters, some of which are (1) a segmented body; (2) a well-developed exoskeleton; and (3) jointed, ventrally attached legs. The giant Cambrian carnivores of the Anomalocarididae are either just outside Arthropoda or possibly are basal arthropods, and there are several stem arthropods that we will encounter later. Arthropoda contains modern forms such as the crustaceans (all the tastiest seafood: crabs, shrimp, lobsters, and many others), the chelicerates (spiders, scorpions, horseshoe crabs), the hexapods (insects), and the myriapods (millipedes and centipedes), none of which we encounter in full form in the Cambrian, although several stem group taxa are present. Many Cambrian arthropods appear to be **stem taxa** to modern groups. Arthropods are sometimes split into the Crustaceomorpha and Arachnomorpha. The Crustaceomorpha consists of not only modern crustaceans but also Cambrian forms such as *Waptia* and *Branchiocaris* that are crustacean-like but may lack a few characters of the modern crown group. The Arachnomorpha may contain the Trilobita (sometimes the Trilobitomorpha; trilobites themselves plus related forms) and the Cheliceriformes, including horseshoe crabs, sea spiders, and arachnids (spiders, scorpions, and ticks). Alternatively, the trilobites may be closer to insects and crustaceans. Trilobites have also been included with naraoiids and aglaspidids within a clade called the Artiopoda. We will cover the classification of arthropods a bit more in chapter 6 after we have met a few of the non-trilobite arthropods of the Cambrian.

Of the nine orders of trilobites that have been identified in Cambrian rocks (fig. 4.11b), we will review the five that are among the most common in North America.

REDLICHIIDA

The **Redlichiida** is the oldest and most primitive order of trilobites, ranging from the Early to Middle Cambrian. Members of this order are characterized by having large, semi-circular cephala with large, crescentic eyes; spiny thoracic segments; tiny pygidia; and a long, segmented glabella (fig. 4.12). The two groups within the Redlichiida are the suborders Olenellina and Redlichiina. The Olenellina is characterized by having

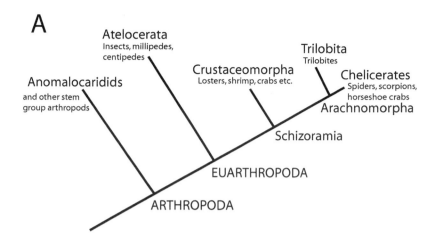

A

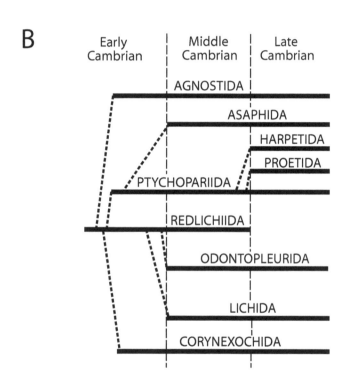

B

4.11. (A) One proposal for relationships among some modern and ancient arthropods. Note position of Trilobita. Some of the group names change under different classifications. For other proposed positions of Trilobita within the arthropods, see also figure 6.34. (B) Time distribution of the nine orders of trilobites present during the Cambrian, with very tentative relationships shown by dotted lines.

no facial sutures and includes the Olenelloidea and the Fallotaspidoidea, the latter of which contains the oldest trilobites in North America. Among the Olenelloidea are the families Olenellidae, Holmiidae, and Biceratopsidae, many members of which we will encounter later in this chapter, including the quintessential Lower Cambrian trilobite *Olenellus* and the comparatively odd looking biceratopsids *Bristolia* and *Peachella*. Within the Fallotaspidoidea are the forms *Fallotaspis*, *Judomia*, and *Nevadia*, and these and their closely related family members are characterized by, among other things, the eyes attaching to the anterior part of the glabella.[28] The Redlichiina include the few Australian species of the

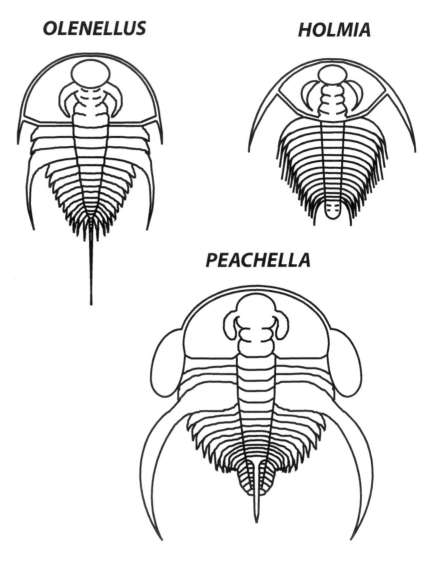

OLENELLUS

HOLMIA

PEACHELLA

Emuelloidea, a group whose members have a pair of enlarged pleural spines posterior to which the thorax tapers significantly and the thoracic segments become shorter anteroposteriorly. This results in there being often more than 50 thoracic segments in a species, as compared with the 13 or so in most *Olenellus*, for example. The Redlichioidea is a superfamily with many members, mostly from Asia, Africa, Europe, and Australia. Redlichioids have also been found in Antarctica; they were among the first Cambrian trilobites discovered on that continent in 1964.[29] There is a report of a redlichioid (?*Churkinia*) from Alaska, but that is about all from North America. The diversity of redlichioids in China is impressive. The Paradoxidoidea contains three families and some of the largest species of trilobites known (40–50 cm [16–20 in.] long). *Paradoxides* and *Acadoparadoxides* have both been found in Canada as well as South Carolina and Massachusetts, respectively, so this superfamily is the best represented in ancient Laurentia among the Redlichiina. Laurentia, however, was mostly a continent of Olenellina.[30]

PERONOPSIS

PAGETIA

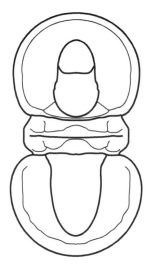

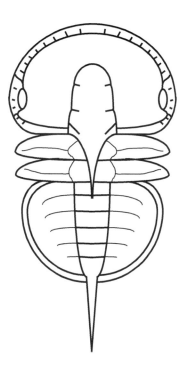

4.13. Two genera of trilobites of the order Agnostida. Note eyes on *Pagetia* and lack of eyes in *Peronopsis*.

AGNOSTIDA

The next important order of trilobites in the Cambrian is the **Agnostida.** These are very small trilobites with only two (or three) thoracic segments and pygidia that are approximately the same size and shape as the cephala (fig. 4.13). The glabella may have faint or nonexistent furrows or it may have very deep, complex ones, depending on the species. Many species have no eyes. There are two suborders: the Agnostina, which lack facial sutures and eyes and have two thoracic segments; and the Eodiscina, which have facial sutures and, in some cases, eyes and/or three thoracic segments. Eodiscines also may have a long, narrow pygidial axial ridge with four or more clear ring segments. Typical Cambrian genera of the Agnostina include *Peronopsis* and *Ptychagnostus*, and a common Cambrian genus of the Eodiscina is *Pagetia*. Agnostids were abundant and wide-ranging during the Cambrian, and we find the fossils in many parts of the globe.[31]

CORYNEXOCHIDA

The order **Corynexochida** includes trilobites with long, large glabellas that often reach the anterior edge of the cephalon and are mildly expanded at that end; they also typically have 7 to 8 thoracic segments (ranging 5–11) and large pygidia (fig. 4.14). The posteriormost furrows on the glabella may also be posteriorly directed in these forms. Among the Corynexochida are some of the iconic trilobites of the Cambrian,

4.14. Two genera of trilobites of the order Corynexochida.

Each image from Harrington et al., Treatise on Invertebrate Paleontology, *courtesy of and © 1959, The Geological Society of America and The University of Kansas.*

KOOTENIA

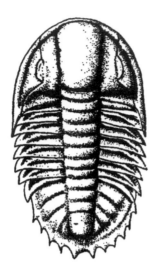

OGYGOPSIS

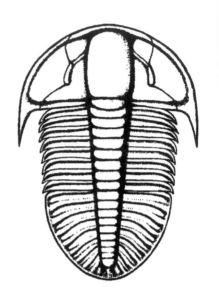

including *Bonnia*, typical of deepwater deposits of the late Early Cambrian; *Olenoides* and *Ogygopsis*, forms common in the Burgess Shale at the Mount Stephen Trilobite Beds locality (and at the Walcott Quarry, in the case of *Olenoides*); *Glossopleura*, abundant at some sites in the Bright Angel Formation in the Grand Canyon and also lending its name to a particular level (biozone) in the Middle Cambrian; and the spiny *Zacanthoides*, from many of the same formations as *Glossopleura*. *Kootenia* is also found in a number of formations of the Middle Cambrian, including the Bright Angel, and it has a distinctively serrated edge to the pygidium.[32]

PTYCHOPARIIDA

The **Ptychopariida** consists of many forms with a simple, unexpanded glabella that tapers anteriorly and rarely reaches the anterior edge of the cephalon (fig. 4.15). The cranidium often possesses eye ridges.

TRICREPICEPHALUS

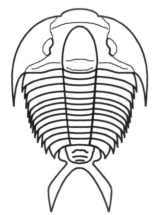

MARJUMIA

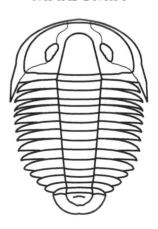

4.15. Two genera of trilobites of the order Ptychopariida.

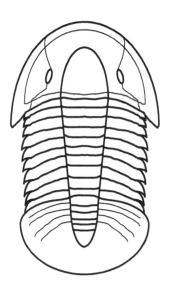

GLYPHASPIS

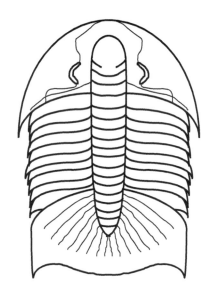

DIKELOCEPHALUS

Ptychopariids also generally have a clear, but relatively reduced pygidium and a moderate number of thoracic segments (usually 12–17). Among Ptychopariida the superfamily Ptychopariodea is particularly diverse, with at least 35 families included. The all-time iconic Cambrian trilobite, *Elrathia*, is a ptychopariid and is probably tied with *Phacops* for all-time iconic trilobite of any period, thanks in part to its presence in seemingly every rock shop ever opened. Among the other important ptychopariids from the Cambrian that we will encounter later are *Modocia, Tricrepicephalus, Mexicella, Marjumia, Cedaria, Elrathina,* and *Amecephalus.*[33]

ASAPHIDA

The order **Asaphida** becomes very important in Upper Cambrian rocks in North America (fig. 4.16). The range of species is morphologically diverse as adults but the order is united based on a shared condition of the protaspid cephalon; it is oval and smooth, without noticeable features. Among the Asaphida are the Cambrian genera *Glyphaspis, Dikelocephalus, Saukia,* and *Idahoia.*

The higher-level classification schemes for orders of trilobites remain unresolved, but ontogenetic factors and the style of attachment or detachment of the hypostome have played large roles in the debate. In one classification, trilobite specialist Richard Fortey argued that the detached hypostome condition, common in most ptychopariids, was a derived condition and that redlichiids were closer to corynexochidans and ptychopariids than to olenellids. As more early trilobites have become known, many with large preglabellar fields on the dorsal surface of the cephalon (suggesting relatively posteriorly placed hypostomes) and with short rostral plates (on the anteroventral margin of the cephalon), it appears likely that the fallotaspidoids and perhaps other early Redlichiida—the ancestors of later trilobite groups—had detached hypostomes, making this the primitive condition in trilobites. This would suggest a different classification, outlined in 2003 by Peter Jell, in which the ptychopariids and agnostids retain the primitive detached hypostome of early Redlichiida and olenellids, later redlichines, and Corynexochida quickly develop the derived attached hypostome condition.[34]

Bug Anatomy: Biological Systems

Trilobites had soft-bodied anatomy that generally does not preserve in the fossil record, but we can see indirect evidence of some of this anatomy in the fossils we find, and by comparing trilobites with their modern arthropod relatives we can draw a few conclusions about how these animals were built. Like their arthropod cousins (and like us for that matter), trilobites have internal body structure that can be divided into several systems: nervous and sensory systems take in information about the internal and external environments and communicate it to the brain and back out to muscles and organs; circulatory and respiratory systems oxygenate blood and transport it to tissues throughout the body; the digestive system processes and absorbs nutrients from food; and the reproductive system produces eggs or sperm cells and facilitates fertilization of eggs and development of embryos.

NERVOUS

The nervous system of trilobites probably consisted of paired nerve cords running down the body, with ganglia along the length at each segment of the thorax (and probably several within the cephalon and pygidium as well). The brain, or cluster of **cerebral ganglia,** was anteriorly positioned along the nerve cord, although its position within the cephalon is not certain (fig. 4.17). Comparisons with other arthropods suggest that the cerebral ganglia comprised two (possibly three) significant enlargements, the anterior of which (the **protocerebrum**) connected with the **optic nerve.** The ganglion (or ganglia) behind this would have innervated the antennae and cephalic appendages and enclosed a section of the gut, and a smaller ganglion in this region in some arthropods is associated with the mouth.

A key difference between the nervous systems of trilobites (and arthropods generally) and our own is that in trilobites the nerve cord is ventrally located along the body, whereas ours is dorsally located (along our backs). In fact, the circulatory systems of arthropods (and probably trilobites) are also different in not only being open but in having a dorsally located heart. This, too, is the reverse of the condition in vertebrates, wherein the heart and main artery are ventral to the spine. Dorsal circulatory organs and ventral nerves are far more common in the (bilaterian) animal world than is our strange setup of a dorsal nerve and ventral circulation organs.[35]

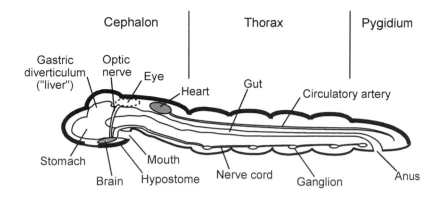

Cephalon | Thorax | Pygidium

Gastric diverticulum ("liver") | Optic nerve | Eye | Heart | Gut | Circulatory artery

Stomach | Mouth | Hypostome | Nerve cord | Ganglion | Anus

Brain

4.17. Schematic cross-sectional view of trilobite body showing various systems in left lateral view. Mineralized skeleton shown by thickest lines. Note ventral brain and nerve cord and dorsal heart and artery. Note also ventral mouth behind hypostome and location of key features in cephalon.

SENSORY

The eyes of trilobites are interesting instruments, and they are among the first sophisticated sight organs in Earth history. Most Cambrian trilobites had a single pair of what are known as **holochroal compound eyes,** meaning that they had many small lenses, often hundreds, packed together in each eye. Each of the hundreds of lenses in each eye may be only 0.1 mm in diameter, often less. Each eye is crescent- or bean-shaped in dorsal view, and although most specimens have been crushed, the eyes originally stuck up from the cephalon to some degree. Each lens within each compound trilobite eye was biconvex or elongate and was hexagonal in cross section (good for close packing), with a convex outer surface. Below each lens was likely an **ommatidium** with photoreceptor cells and an axon running the gathered light information to the brain.[36] The lenses were composed of calcite and were oriented perpendicular to the visual surface of the eye; thus, we can approximate the lateral sweep of the field of view of a trilobite from the shape of its eyes. Many olenellids may have had a total field of view of nearly 300 degrees, with most of the missing view directly behind the animal. Unfortunately, because of the common crushing of the eyes (and the rest) of most specimens, it is difficult to determine what the vertical range of view would have been in many species.

Most Cambrian trilobites probably formed an image by taking individual eye lens image data and compiling the "pixel" elements in the brain to form real-time composite pictures of the outside world, a mosaic of sorts. Unfortunately, we don't know as much as we would like about the eyes of Cambrian trilobites because the structure of trilobite cephala was not conducive to good preservation of the eyes until the Ordovician.[37] Holochroal eyes were employed by Cambrian trilobites of the orders Redlichiida, Corynexochida, and Ptychopariida, of which we will see many members in upcoming chapters. Interestingly, it appears the thickness of the eye lenses and mineralized exoskeleton of trilobites was correlated with environmental conditions; higher-energy environments tended to be occupied by thicker-shelled trilobite species, whereas thin-shelled and thin-lensed trilobites are found in relatively quiet waters.[38]

Cambrian eodiscoids (such as *Pagetia*) have what is called an **abatho-chroal eye**. This eye has relatively few lenses (fewer than 100 per eye), which are less closely packed so that they are separated from each other on the visual surface. And, of course, some trilobites are entirely secondarily blind; agnostids and some eodiscoids lack eyes, and even some ptychopariids (such as *Ctenopyge*) have very reduced eyes that may have sensed only general light levels. This may be due to their possibly **pelagic/planktonic** lifestyle in some species; floating in the water column filtering food from the water presents no need for sight. As neither predators nor individual prey (they at most would be caught only by chance by larger filter feeders), agnostids and certain eodiscoids would have no need for vision. Similarly, those trilobites living so deep in the water that there was no light at all (possibly including agnostids), or those living entirely within the sediment on the ocean floor, might be blind as well.

However, living in deep water where light is present but at very low levels may result in very large eyes, as is seen in some later forms. In fact, there can be a correlation between trilobite eye size and the depth at which the species lived. Some trilobites did strange things with their eyes, like *Symphysops* from the Ordovician of the United Kingdom, which had a single wraparound eye; likely derived from two-eyed ancestors, this trilobite must have fused the individual eyes into one horseshoe-shaped visual surface. Then there are the stalk-eyed trilobites, of which the oldest known is the Cambrian-age *Parablackwelderia* from China, a form with bulbous, compound eyes set on the ends of rigid stalks—like those of a cartoon alien or slug. What these eyes were for or how they arose is unclear, but it is possible that they may have evolved through sexual selection or have been employed by a predatory *Parablackwelderia* that hid in the sediment waiting for prey, with only its eyes sticking up to survey the world above.[39] The complete loss of eyes may indicate something about the habits (and habitats) of a trilobite, and great enlargement of the eyes may indicate the depth at which the trilobite lived; so medium to relatively small eyes may suggest habitats in good light, including shallow depths and clear water.

Trilobites also would have sensed their surroundings through their antennae and cerci along with the hairlike structures on the legs.

DIGESTIVE

Under the mineralized exoskeleton of the cephalon, under the front part of the glabella and tucked up inside the ventral protection of the hypostome is the digestive system of the trilobite. The mouth in fact faces down and back in most species, just behind the back edge of the hypostome; from here, the esophagus travels up and anteriorly to the stomach, where food is processed before heading down the intestine, which then stretches posteriorly down the length of the trilobite under the glabella and then through the center of the axial lobe of the thorax to the pygidial area (fig. 4.17). On either side of the stomach, under the anterior part of the glabella, there may have been lateral diverticulae, perhaps the equivalent of the digestive ceca of modern arthropods, which probably served some accessory digestive function. Some species seem to have had paired diverticulae down much of the gut; and others simple constrictions and expansions of the gut along its length; and still others smooth, tubular guts.[40] All of this suggests variety in diets among species of trilobites but what these were we cannot be certain. The anus was most likely near the ventral margin of the posterior end of the pygidium.

In most predatory species, prey would have been subdued with the legs, crushed partially, and then passed forward to the mouth by the coxae. The coxae in some species could be very stout and heavily spined.

CIRCULATORY

Trilobites had hearts, but their circulatory systems were probably open like most modern arthropods, so that the internal organs within the body cavity are bathed in blood. It is unclear where the trilobite heart was located, but it may have been dorsally positioned under the mineralized exoskeleton near the back end of the cephalon (fig. 4.17). If trilobites were like modern arthropods, the heart probably pumped blood through arteries to different regions of the internal body cavity, but after soaking the organs the now deoxygenated blood would work its way back to the heart freely, as there were not likely any veins.

RESPIRATORY

One big question regarding the open circulatory system in trilobites is where the gills were and how they oxygenated the blood. Trilobites needed to supply their blood and tissues with oxygen, and like most marine animals they would have gotten the oxygen from the seawater in which they lived. They did this by gas exchange through gills that were flushed with blood and also had seawater flowing over them almost constantly. The filamentous proximal lobe of the outer branch of each appendage may have served as an individual gill, so that the trilobite had a long series of gills underneath the exoskeleton of the thorax and above each walking leg. But it may be that those branches simply served as current generators for gas-exchange surfaces that in fact were located on the ventral cuticle of the pleural areas under the thoracic segments (fig. 4.15). The underside of the pleural exoskeleton in some trilobites has recently been found to contain an anastomosing pattern, likely resulting from the presence of this vascular system for gas exchange. A recent suggestion for circulation in trilobites is that blood was pumped by the heart laterally down the dorsal part of the cavity between the pleural exoskeleton and the ventral cuticle, and then it returned along the ventral gill surface toward the axial region and then to its target areas by arteries. Along the ventral return trip, through the anastomosing network of circulatory canals, blood would have been oxygenated as water was flushed over the gills by the waving of the filamentous outer branches of the appendages. From there, oxygenated blood would have been carried to the muscles and organs of the body cavity in the cephalon and axial lobe, as well as the limbs.[41]

REPRODUCTIVE

Nearly all modern arthropods have male and female sexes and most reproduce through mating, so we are probably safe in assuming trilobites were similar to this majority of modern forms. In most species of trilobites, embryos developed in eggs and were hatched at an early stage. In several cases paleontologists have found clusters of a few up to 1000 specimens of the same species of trilobite, possibly suggesting gregarious behavior and synchronous reproduction and molting, a behavioral characteristic that may have been typical of a significant number of species of trilobites. Very little is known of the reproductive system of trilobites except that some trilobite species that lack an anteriorly attached hypostome had some presumably female forms that appear to have had **brood pouches** meant for housing fertilized eggs. These brood pouches, which appear only in adults, were dome-shaped bulges at the front end of the cephalon, under which a batch of eggs was protected for some time before the embryos hatched.[42]

Ecology

With nearly 20,000 named species, the trilobites as a group must have demonstrated a wide range of life habits. Here we will take a look at some of these, starting off with their mode of locomotion. A lot of trilobites were probably **benthic,** meaning they lived on the bottom of the sea, and were mobile, moving across the seafloor by walking. We know what their legs looked like and approximately how they functioned, and we have the trails they left (*Cruziana*), but some probably burrowed, some swam, and some may have floated. And most trilobites seem to have been made to enroll for protection.

LOCOMOTION

The majority of trilobites probably moved along the bottom by walking, whereas others likely swam or floated in the water column, and still others plowed through the sediment itself. It all related to other aspects of the species mode of life, such as preferred habitat and diet. In some cases, walking species may also have pushed off the bottom to float or swim for a brief time within the water column before settling down to the bottom again. Muscles inside the exoskeleton moved each leg during walking, with the muscles inside the coxa and each podomere moving each one relative to the other so that each leg moved forward and back during each step. The legs moved in wavelike sequences so that the animal moved forward somewhat like a terrestrial isopod or millipede we might see today. In trilobites, it appears that the walking cycle (as exemplified by *Olenoides*) included two waves of four to five legs in contact with the substrate at a given time, with several more off the substrate swinging forward. Walking forms with more or fewer thoracic segments than *Olenoides* would have different numbers of legs involved but might well have had similar walking techniques. It is also possible that some forms (including *Olenoides*) may have progressed across the substrate occasionally by a push-and-glide technique, pushing off the sediment with many limbs at once and gliding for a distance with all limbs suspended before settling down and pushing off again.

Olenoides and other walking forms may have also, at various times, swum short distances through leg movements (probably during push-and-glide, described above), and plowed through the sediment at the sea bottom by walking with the cephalon edge tipped downward in front. Some forms may have moved through the surface of the sediment like this in order to feed. Movement through the sediment by trilobites, and the feeding and walking movements associated with it, are suspected to have formed the long, sometimes wandering, double-grooved trace fossils known as *Cruziana*. Shorter traces with similar overall features and proportions are usually no longer than the length of a single animal; these are thought to be single feeding or resting traces and are known as *Rusophycus*. Other trilobite forms, such as agnostids, may have floated or swam (again through leg movements) in the water column, and others may have made plowing through sediment a full-time habit and actually lived in the mud of the seafloor.[43]

Most trilobites seem to have been able to enroll themselves, somewhat like an armadillo, so that their mineralized exoskeleton was all that was exposed and so that their soft ventral side and legs were protected. This was especially developed in the phacopids, which are the order of trilobites that does not occur in the Cambrian. In well-protected forms, the pygidium wraps up underneath the cephalon and the hard, interlocking thoracic segments guard the rest of the animal so that all soft areas are safe. Agnostids, with their reduced number of thoracic segments and size- and shape-matching cephala and pygidia, are particularly good at wrapping up into what almost looks like a clamshell. Even olenellids, with their tiny pygidia and relatively huge cephala and spines all over, seem to have been able to enroll themselves for protection from predators.[44] Although not common in Cambrian rocks, in some formations it is possible to find trilobites preserved enrolled and in three dimensions.

Agnostids have been suggested to have been free-floating, often blind, planktonic species that may have lived in the water column rolled up like some modern, small crustaceans. But they are often found in large numbers in deepwater deposits, and at least some may have been benthic walkers rather than floaters or swimmers. In all likelihood, there was a range of ecologies among adult agnostids, from pelagic to benthic; most larval agnostids were pelagic.

Trilobites seem to have ranged across a wide spectrum of feeding ecologies. As we saw with their locomotion and traces, they lived in a variety of subenvironments in, on, and above the seafloor, and their impressive diversity would also dictate that they needed to feed on a number of different meal sources. Some probably were predators (and opportunistic scavengers), attacking prey living on the bottom of the ocean; some predators may also have targeted prey in the sediment or up in the water over the seafloor. Other trilobites may have lived in the water column and filter fed or preyed on the plankton. Some trilobites seem to have lived on the bottom and filtered food out of the sediment.

Predator-scavengers often had stout, spiny coxae on the legs for crushing prey items and passing them forward to the mouth. Recall that the mouth was to some degree protected by the hypostome; in suspected predatory species the hypostome is often anchored to the anterior edge of the cephalon (presumably for strength and stability). This is less often true in non-predatory species. Another apparent characteristic of predatory trilobites may be a relatively large glabella, which might be required for the larger "stomach" possibly necessary for processing larger food items than might be swallowed by, say, a filter feeder.

Free-floating or swimming trilobites were generally small and probably fed on plankton, either as predators of zooplankton or as primary consumers of phytoplankton. Certainly both ecologies were present among the small pelagic trilobites. Whatever their feeding mode, these small, plankton-eating trilobites were probably present in the water column in great numbers. Agnostid trilobites are generally thought to have been planktonic filter feeders because they are small, wide-ranging, and similar in very general form to the planktonic crustaceans so common in today's oceans. Recent work has suggested, however, that many of these often blind agnostids may in fact have been deepwater benthic predatory-scavenging species. Perhaps both ecologies were represented among agnostids by different species; some species have been found attached to pelagic, free-floating algae, and others have been preserved feeding inside hyolith shells or underneath benthic trilobite exoskeletons.[45]

Some trilobite forms appear to have had enlarged cephala that functioned like plows to push through the sediment and filter food out of the mud, which was then collected by the appendages. Some even had perforations in the rim of the cephalon for filtering. These filter feeders are not common but are interesting, and show how many different ways trilobites came up with to obtain food.

One of the most common feeding modes for trilobites, along with predatory-scavenging habits, was a generalized grazer and detritus feeder. These forms often lived on the ocean floor and crawled around feeding on particles of food on and in the sediment. Organic detritus from organisms living on the seafloor, or that rained down from the water column, would be picked up and eaten by these forms, along with pieces from

Malgo Moczydlowska-Vidal is professor of micropaleontology in the Department of Earth Sciences at Uppsala University in Sweden and specializes in the evolution and biostratigraphy of Precambrian and Early Cambrian acritarchs. Originally from Warsaw, Poland, Dr. Moczydlowska-Vidal graduated with a master's degree in paleontology from Warsaw University. Her master's thesis was on dinosaur-age sponge spicules from the Late Cretaceous, the last epoch of the age of reptiles. After this, she worked for about ten years at the Geological Institute in Warsaw, and it was there that she began working on Cambrian acritarchs and biostratigraphy. She left Poland and did some work at University of California, Los Angeles, with J. W. Schopf, then completed a PhD at Lund University in Sweden. After a postdoctoral position at Harvard working with Andrew Knoll, Dr. Moczydlowska-Vidal moved to Uppsala University back in Sweden, where she has been since 1992.

Her path into paleontology, she says, was a result of a lifelong interest in animals and paleontology, helped along by the finds of colleagues in Warsaw who were returning from the Gobi Desert in Mongolia with spectacular specimens of dinosaurs and other vertebrate fossils. These paleontological discoveries impressed her very much, but, she says, "I very soon realized that microorganisms are ruling the biosphere." After the early influence of dinosaurs and Cretaceous sponges, how did she end up working on Cambrian phytoplankton? "Lucky coincidence," she says. "There was a need to establish a biostratigraphy of the Cambrian successions in Poland below the stratigraphic occurrence of trilobites and small shelly fossils, and the only means to do this was to try microfossils, acritarchs," she says. Despite their small size and relative simplicity, acritarchs and other microfossils can evolve some attractive and interesting shapes. And their importance makes up for any lack of flash due to diminutive size. "These microorganisms," Malgo says, "are the most fascinating objects to study . . . because they generated organic matter through photosynthesis and thus provided the food for the evolving animals, and they produced free oxygen as a byproduct of photosynthesis for oxygenic respiration." So not only did acritarchs in a sense feed the Cambrian explosion, but the oxygen they contributed is still important. "We use the oxygen for our breathing that they produced," she points out, "and that their descendants produce continuously up to today." So if you like a good aerobic workout at the neighborhood gym or while working your way up a favorite local peak, thank a few hundred billion Proterozoic and Cambrian acritarchs.

The Cambrian Corps 3 – Malgorzata Moczydlowska-Vidal

green algae beds, seaweeds, or other primary producers. These were the trilobite omnivores of the time, and they appear to have been fairly diverse and numerous. Many ptychopariids seem to have been grazers or detritus feeders, and one of the characteristics of these forms would have been a relatively loosely suspended hypostome, in contrast to the strongly anchored one of predators. Benthic agnostids may have been detritivores also, feeding on smaller particles than the ptychopariids did.

It is also possible that some trilobites were suspension or filter feeders, gathering food particles from the water column either by passive raking of the water or by stirring up bottom material and filtering.[46]

GROWTH AND MOLTING

Trilobites were like many other arthropods and had to occasionally shed their exoskeletons as they grew. This process of molting is carried out today by many species of arthropods, particularly crustaceans, and by some other forms.[47] It is interesting to watch hermit crabs today in salt-water tanks as they shed their protective gastropod shell temporarily to retreat among some rocks, hide in isolation as they first shed their old exoskeleton (which soon comes floating out as a soft, crumpled mass to be eventually picked at by the fish) then wait while their new, previously underlying layer hardens into their new protective and supporting layer. Only then do they plant themselves back in their gastropod shell and begin crawling around the tank again. Trilobites would have also, every now and then, pushed out from their exoskeleton and hardened a new, larger one. Because of this, many of the fossils that we find today are probably trilobite molts.

In arthropods and a few related forms, molting is regulated by a hormone called ecdysone, and thus molting in these groups is called **ecdysis.** The period between molts is known as an **instar,** or intermolt stage, and although the animal appears at this time not to be growing, it is in fact the time during which tissue growth occurs. The growth of muscles and other tissues of the soft parts of the body eventually fill up the inside of the exoskeleton, and then it is time for another molt. The first step in the process is when enzymes begin to work to separate the inside of the exoskeleton from the epidermis, and calcium is absorbed back into the body from the exoskeleton. Once the exoskeleton is fully separated, the animal works its way out from inside the molt. Most trilobites accomplished this by popping loose a plate composed of the free-cheeks (librigenae) and rostrum-hypostome and then wiggling out from between that and the cranidium. Separation of the lower plate containing the librigenae from the upper cranidial plate would be somewhat like pulling open a clamshell-shaped to-go box; one plate pops up, the other drops down—and the trilobite crawls out the opening. A related but modified technique would have been employed by the olenelline trilobites, which did not have librigenae. These trilobites common in the Early Cambrian may have separated their hypostome and then crawled out from under the entire cephalon.

The next step in the molting process is to enlarge the new exoskeleton (which of course emerges from the old one no larger than it was,

PROTASPID

MERASPID

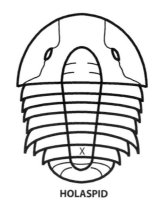

HOLASPID

4.18. The three main stages of trilobite development. Note appearance of first thoracic segment anterior to pygidium (x) in meraspid stage and subsequent addition of segments anterior to first (x) up to holaspid stage.

but growth is one of the goals here). Most marine arthropods do this by taking up water temporarily into the body to swell both it and the new exoskeleton. The animal will then go into a reclusive period during which the exoskeleton hardens and the calcium taken up earlier is redeposited in the new exoskeleton. After the new exoskeleton is hardened the animal gets rid of the excess water and returns to its previous, unbloated size. It then can emerge from hiding and continue about its business, with a brand new, larger exoskeleton to grow into.

It is important to remember that although externally growth appears to occur during molts, most growth of actual body tissues occurs during the intermolt stages. In some crustaceans molting (and thus growth) occurs throughout life; in other species of crustaceans, growth and molting stop at some point and the animal lives the rest of its life at that determined size. We don't know for sure if trilobites had this determinate or indeterminate growth—perhaps one or the other strategy was employed by different species. But growth allowed some species of trilobites from the Ordovician period to reach 70–90 cm (28–35 in.) in length![48] On the other hand, agnostids, for example, are tiny as adults, often not exceeding 1 cm (0.4 in.) in length.

Molting is common in crustaceans. Many of the modern arthropod species are insects, and insects molt, but often only once and not as adults. In many species of insects the molt during their life cycle is in the metamorphosis stage; for example, the transition from caterpillar to butterfly. So crustaceans probably serve as a better model for trilobite molting than do most other arthropods.

Growth in trilobites involves not only changes in size but also in exoskeletal morphology, as we see preserved in the fossils. There are four stages of development in trilobites: (1) the **phaselus** stage, during which there is only the cephalon and one pygidial segment, and there is development of the glabellar furrows and facial suture, a stage that has been confirmed as present only for a few groups due to spotty preservation of these tiny specimens; (2) the **protaspid** stage, during which the facial suture appears and pygidial segments are added; (3) the **meraspid** stage, during which thoracic segments are added; and (4) the **holaspid** stage, in which growth in size occurs after all segments have been added;

this is essentially an adult stage (fig. 4.18). Our friends the olenellines do not quite fit into this as neatly as we might prefer, as they lack facial sutures; in fact, we have specimens of these trilobites only in meraspid or holaspid stage. At some point in their development before or during the protaspid stage, most trilobites probably underwent a metamorphosis that included the beginning of calcification of the exoskeleton in the individual.

One of the important aspects of these developmental stages in trilobites is that growth by addition of segments and associated ontogenetic morphological change in the pre-holaspid stages procedes from back to front. Although the cephalon appears first, a trilobite did not just tack segments on behind the head one after the other; rather, the pygidium appeared behind the cephalon and then one segment at a time was added at a position just anterior to the pygidium.

Ecologically, these different stages may have varied. For example, many species may have been planktonic (or otherwise living in the water column) during the protaspid stage and been benthic from the meraspid stage onward. Other benthic species may have become so at some point during the protaspid stage after some time as planktonic animals. Conversely, some pelagic trilobites may have been benthic as larvae. Interestingly, it appears that (at least during a particular extinction in the Ordovician) trilobite species with both planktonic and benthic protaspid stages followed by a benthic adult period survived best; being entirely planktonic or benthic was a liability at this time. If such life-cycle diversity was present in Cambrian trilobites (and it probably was), perhaps extinctions during that time also had such variability in effects on species. Maybe there was a time when being entirely benthic or planktonic was a lucky key to survivability. Or perhaps the planktonic-benthic protaspid stage followed by benthic adult time was always an advantage. In either case, it is exciting to be able to see such patterns in the fossil record – life-cycle biology happening to play an unwitting role in the history of trilobite evolution.[49]

PALEOENVIRONMENTS

In what environments did trilobites live? In nearly every marine environment, it appears, from intertidal zones and sandy shallow surf zones to the (relatively) shallow shelf at or above storm wave base (~50 m [164 ft.] in depth), to carbonate shoals, to bioherms, to deepwater slopes. Trilobites seem to have been everywhere. It is possible they originated in relatively shallow water and spread out from there, invading deepwater environments later on. We find their fossils in shales, sandstones, and limestones of all variety of marine origins from near-shore shallows to complex reefs to the dark quiet of deep basins. Determining whether their fossils are preserved in the actual environment in which they lived can sometimes require some work, as some trilobite skeletons were transported out of their living environments and into another before they were buried. But there seem to have been few environments in the ocean that trilobites did not inhabit.

It appears that the shallows of the continental shelves had the highest percentages of **endemic** species, those that were peculiar to a particular region. Benthic (bottom-dwelling) trilobites of the deep, open ocean seem to have been more widespread geographically—sometimes on a global scale. If depth-variable oceanographic characteristics of the Cambrian seas were at all similar to those of today (and to some degree they may well have been), some of these trilobites may have lived below the **thermocline,** a thin layer within the ocean's water column at which the temperature drops significantly, separating a warmer upper layer from a cooler lower one; this suggests that these widespread species of trilobites were adapted to cold. Tiny pelagic species, those that lived in the water column out in the open ocean, may also have been widespread but do not seem to have overlapped extensively with shallow-shelf species. Some of these pelagic species may have been free swimming or free floating, consumers of phytoplankton or predators of small zooplankton.

A majority of trilobite species appear to have been benthic scavengers and grazers living on muddy terrigenous or muddy carbonate bottoms, with some in sandy areas or bioherms and reefs.

One interesting aspect of trilobite distribution is that different marine settings around Laurentia tended to have different associations of species. Some species were more characteristic of relatively shallow water, whereas others, found farther out to sea, were in deeper settings. These varied faunas have been called **biofacies realms** and contained characteristic sets of species associated with each environmental setting. The *cratonic biofacies realm* in North America includes species typically found in shallow-shelf settings around the continent (flooded continental shelf); the *intermediate biofacies realm* includes those farther out to sea in deeper-shelf settings; and the *extracratonic biofacies realm* includes those taxa typical of the deepwater settings (slopes and basins) near the shelf edge.[50]

GEOGRAPHY

As with the environmentally widespread occurrence of trilobites, there are few regions of the planet that don't have at least some Cambrian trilobites known. Every continent has them. I emphasize here, however, the extensive fossil record for trilobites of North America. Some of the best Cambrian fossils come from China, and trilobites are also known from other parts of Asia, including Korea, Mongolia, Pakistan, and Iran, to name a few. Some of the oldest trilobites in the world come from Morocco, and others from Egypt and Algeria in Africa. Trilobites and other Cambrian fossils have been found in many parts of Europe and Australia. Fossils of Cambrian age have been found in Argentina in South America, and Cambrian trilobites have been found in the Argentina and Neptune ranges in Antarctica. More North American states and provinces than not have produced Cambrian trilobites, including (by rough count) at least 31 states in the United States,[51] six Canadian provinces and all three territories, and at least the states of Sonora and Oaxaca in Mexico.

Some trilobite genera are widespread, but others are localized. For example, during the Early Cambrian redlichiines occurred mostly outside North America, in Asia and Australia. Perhaps not surprisingly, species that lived in deeper water offshore (i.e., more open ocean settings) tend to be more cosmopolitan, while species that lived near shore in shallow water of the inner shelf tend to be more endemic.[52]

Fossils Trapped in a High Cliff Wall: Trilobite Biostratigraphy

Biostratigraphy is the characterization and correlation of rocks based on the fossils in them; it also involves the study of the fossils and their distribution in formations of rock. Ideally, a good biostratigraphic chart of trilobite taxa would allow us to compare the ages of distant, previously unstudied faunas. In fact, this is what we are able to do thanks to the work of generations of paleontologists and their studies of trilobites.

The best species for biostratigraphy are abundant, distinctive and clearly identifiable, regionally widespread, and restricted to a relatively narrow stratigraphic interval. Rare, locally endemic, or long-lived species would of course be of limited utility. Thankfully, the Trilobita includes many good candidate species, and the Cambrian in North America has been subdivided into a number of trilobite-based **biozones.** Each biozone has its base defined by the first appearance of its namesake trilobite taxon; the upper boundary is indicated by the base of the next biozone up, the first appearance of another trilobite. Although many species are widespread, trilobite biostratigraphy usually involves separate biozone sequences for each Cambrian continent.[53]

It has also been recognized that the distinctive faunas for each zone may vary depending where the site is relative to the ancient Cambrian coast. Sites located in what are called restricted-shelf biofacies (close to shore, shallow) tend to contain lower-diversity faunas of trilobite species of restricted distribution, with few agnostids in the group; on the other hand, open-shelf biofacies (farther offshore, deeper) tend to contain a high diversity fauna of both endemic and widespread species with many agnostids. This has led to several biozonations for each time interval: restricted-shelf zonations that are often different for each Cambrian paleocontinent, and open-shelf zonations with which we can sometimes correlate between paleocontinental outcrops using cosmopolitan species.[54]

In Laurentia (most of modern North America) the Lower Cambrian is divided into four biozones, which follow a basal zone representing more than 20 million years that was trilobite free. So the first trilobite biozone occurs well into the Early Cambrian (fig. 4.19). That first biozone is the *Fritzaspis* zone, followed by the *Fallotaspis*, *Nevadella*, and *Bonnia-Olenellus* zones. The first three are named after the distinctive genera among the earliest trilobites to appear in Laurentia, ones we met briefly in the Esmeralda County, Nevada, section early in this chapter. The

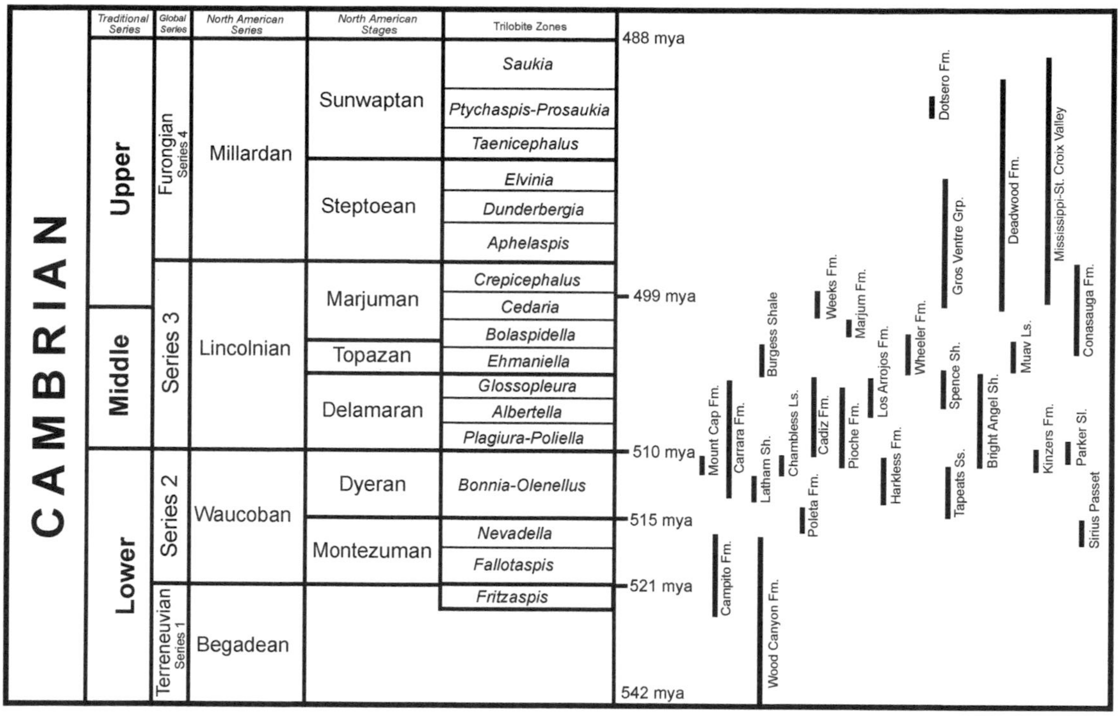

4.19. Timescale of the Cambrian period showing North American series and stage names and trilobite zones, with many of the formations mentioned in this book marked on right. *Fallotaspis, Nevadella, Bonnia-Olenellus, Plagiura-Poliella,* and *Albertella* trilobite zones shown here have recently been further subdivided; new zonations are so refined that all the new zones cannot fit on this figure (each would be too small to read at this scale). See Sundberg (2011), Hollingsworth (2011c), and Webster (2011a) for these more precise zonations.

Bonnia-Olenellus zone is named after the characteristic genera found in such abundance all around Laurentia, with *Bonnia* being a more restricted deeper-water taxon than the ubiquitous *Olenellus*.

The Middle Cambrian is subdivided into five trilobite biozones (fig. 4.19), each distinctive and characteristic of a particular level in the Laurentian succession. For example, collect trilobites from particular levels in certain formations in the Grand Canyon and northern Utah, the Mojave Desert of California, and the Sonoran Desert of Mexico and you will find many of the same genera or even species: *Glossopleura, Amecephalus, Zacanthoides,* and *Mexicella.* These occurrences indentify each fauna as being within the *Glossopleura* zone of the Middle Cambrian–four faunas stretched out over 1215 km (750 mi.) that can be identified as being the same age. Ages of new faunas could be determined similarly. This is the utility of biostratigraphy. The Upper Cambrian is divided into eight trilobite biozones (fig. 4.19), based mostly on genera first found in abundance in the central part of the North American continent. As we discuss various taxa and localities of different ages, we may occasionally refer back to our trilobite biozones to put each fauna in context. Note also that several of the formations we will encounter later in the book are on this chart also. Keep in mind that the biozones used in figure 4.19 are generally based on the restricted shelf species.

Arctic World: Sirius Passet

Far up on the northern tip of Greenland, east of Ellesmere Island and just seven and one-half degrees latitude from the North Pole, lies a rocky

Cambrian locality called Sirius Passet (on fig. 4.19). This is a site where almost no plants cover the outcrop—but plenty of snow does for most of the year, and not far away the Arctic Ocean is covered in sea ice. The site actually consists of at least five individual localities in the Buen Formation spread out over approximately 1 km (0.62 mi.) of Peary Land. Found on July 2, 1984, by two British geologists working for the Geological Survey of Greenland, Tony Higgins and Jack Soper, the Sirius Passet fauna is similar to the Burgess Shale in containing an abundance of soft-bodied forms. Sirius Passet, however, is a bit older; it is from the *Nevadella* zone, just younger than the first trilobites of the *Fallotaspis* zone. This makes the Sirius Passet site roughly equivalent in age to the upper part of the Esmeralda County section in Nevada. It is one of the oldest major Burgess Shale–type deposits in the world, with nearly 10,000 specimens collected from it so far. It is also one of the most currently active research areas, with many new forms named in just the last few years.

During the Cambrian, Sirius Passet was situated at the base of a steep submarine slope. Long before the sediments of this deposit were laid down, there existed a shallow carbonate mud shelf; this carbonate mud turned to limestone, and when the sea level dropped and the limestone was exposed, dunes formed on top of the highly eroded surface and blocks of limestone fell down the slope into the sea. When the sea level rose again, sand and shale were deposited above the eroded limestone in the shallows, while in the deep water at the base of the older limestone submarine cliff, fine muds were deposited and buried the animals of Sirius Passet.

Among the 25+ species of animals found at this site are the sponge *Choia* and archaeocyathids; hyoliths, trilobites such as *Buenellus*; bivalved arthropods such as *Isoxys*; brachiopods; three or four species of lobopodians; palaeoscolecid worms; the spiny halkieriid *Halkieria*, with two brachiopod-like shells on its body; the wormlike *Sirilorica*, a possible relative of nematomorphs; and a probable anomalocaridid (or lobopodian), *Tamisiocaris*. Many of the specimens at this site are unmineralized arthropods, such as *Siriocaris* (fig. 4.20). Annelid worms are found in the form of *Phragmochaeta*, the oldest known polychaete, and an arthropod with so-called great appendages (long feeding limbs) is known in *Kiisortoqia*. A form named *Ooedigera* was also named recently from the Sirius Passet deposit; this is a new member of Vetulicolia, a group that we will look at in a little more detail in chapter 7.

Kerygmachela from the Sirius Passet fauna possesses a combination of arthropod characters along with flexible, but not jointed, limbs that are more similar to the lobopods. This fossil illustrates the probable close relationship between lobopods and arthropods and suggests that arthropod origins may well lie within the former group.[55]

Darton's Marbles

If you drive out of the Los Angeles basin on Interstate 15 headed north, you climb out of the eastern end of the metropolis not far from Redlands

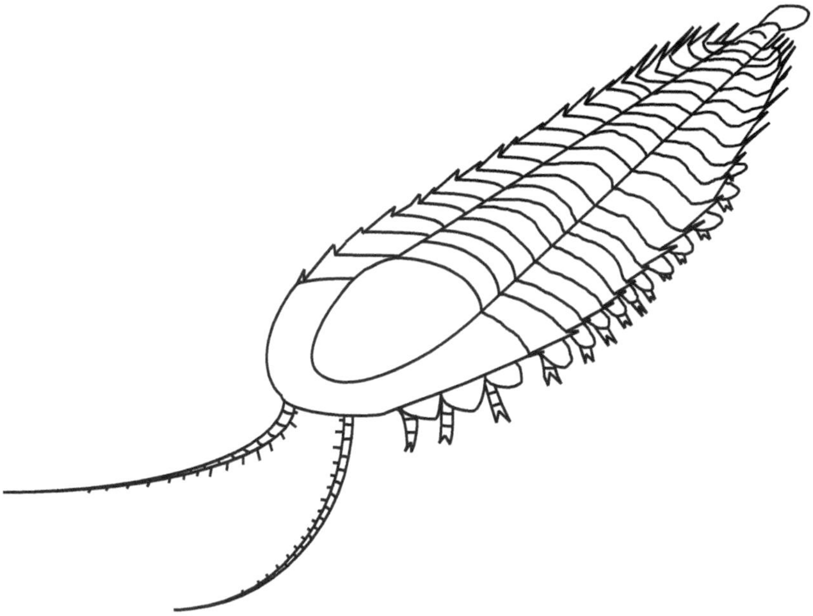

Figure 4.20. The arthropod *Siriocaris* from the Sirius Passet biota of northern Greenland.

and cross the San Andreas Fault. In a matter of minutes you have passed not only from the Pacific Plate to the North American Plate (i.e., from one major unit of the Earth's crust to another) but you also leave a major population center and start into the much less densely populated Mojave Desert. Home to desert tortoise, joshua trees, and creosote bushes growing in sandy soil, the Mojave is seen by most humans from air-conditioned vehicles streaming back and forth between Las Vegas and Los Angeles at 80+ miles per hour. If you take Interstate 15 to 40 and head east you pass through successively smaller cities from Victorville to Barstow to Ludlow, and at this point turn off onto what was once famed Route 66. At this point you are venturing onto a remote road whose surface has seen better days, and often you will fly down this two-lane highway through clear, open country and never see another vehicle in either direction for dozens of miles. Here you will find it hard to believe that you are barely an hour from a metro area of several million people. Just after the cone and lava flows of Amboy Crater you cross some railroad tracks and pass through the tiny town of Amboy. Ten miles farther down the road to the east is Chambless, whose general store closed several years ago. Turning off Route 66 to the southeast and proceeding a couple more miles you come to Cadiz, which is a service stop for the freight trains that pass through. At each step of the journey the signs of civilization diminish and you feel more intensely the forbidding isolation of the Mojave. North of Cadiz is the southern end of the Marble Mountains, and here are beautifully exposed and upturned, faulted beds of Lower and Middle Cambrian sandstones, shales, and limestones. When camped among these rocks you almost forget anyone else is around most nights except for the freight trains rumbling by in the darkness seemingly every ten minutes. This is terrain where even on moonless nights the surrounding landscape can be lit up just by the starlight. And occasionally while out doing fieldwork you will be unintentionally buzzed by a low-flying military jet from one of the bases in the region. In these otherwise quiet surroundings it is often easy to envision the Cambrian world.

Sometimes while doing fieldwork you will wake before dawn to the rhythmic off-and-on pulses of flapping of your tent in the ebbing breeze. The night before may have featured constant beating of the tent fabric by strong winds that did not let up – winds that bent the sides of the tent inward nearly to your face as you hunkered down in a sleeping bag, trying to sleep despite the racket. Your mind finally gave up on trying to block out the cacophony of the beating and bending fabric around you and found itself lulled to sleep by the simple constancy of it all. In contrast, the quiet of morning, punctuated now and then by several moments of slapping tent fly, as the breeze fights back in its waning hours, actually is more distracting than was the wind storm.

As the pre-dawn minutes become brighter you can start to see more details of the tent around you, and you sit up in the sleeping bag, stretching out a few kinks in the neck and back. You are on rock, after all. You pull on thick socks and sandals and a fleece hat, unzip the tent, and crawl out into the twilight. The eastern horizon is a lighter midnight blue now and the brightest handful of stars still shines to the west. Under your feet are thousands of fist-sized, angular cobbles, heavily desert-varnished and lacking grains smaller than gravel between them. We are camped on a small alluvial fan, and these component rocks form a surprisingly flat, although sloping, surface from which most sand has been blown away. Several yards away from camp on one side are the sandy flats containing scattered creosote bushes, and on the other, the rocky, exposed slopes rising several hundred feet above. As you start breakfast the sky lightens more and soon you see the orange glow of sunrise working its way down the slopes to the northwest. By the time the meal is finished the light is minutes away, shining off an outcrop just yards from camp.

Sunrise has come, and it's time to head up the trail for another day's work. All the surrounding rocky slopes are composed of rocks of Cambrian age, and trilobites and other fossils are abundant at sites just minutes from camp. It can be 49°C (120°F) here in the summer, but even in cooler seasons you get up before dawn, not because it's necessary to beat the heat but simply because this deceptively austere land is incredibly inviting when it comes to geologic and paleontological treasures. This is a land where you rarely see even small mammals, or plants other than creosote and a few small cactus, but in the spring you can see loads of wildflowers, beetles, and butterflies, and you will occasionally encounter large scorpions and chuckwallas. The small mammals are there, however; you may not see them, but in the middle of the night you may hear coyotes even when the moon isn't full. If you like sedimentary rocks or Cambrian fossils, this is the kind of strange, stark terrain you love.

Rocks of Cambrian age were first recognized in the Marble Mountains, as we saw in chapter 1, in 1907 by N. H. Darton. Building on

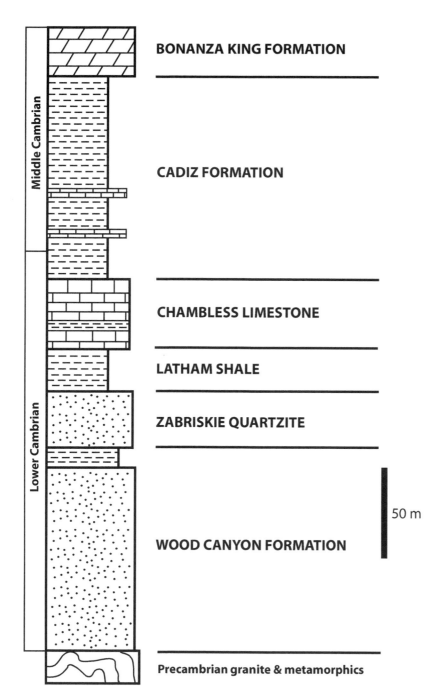

4.21. Stratigraphic section of Cambrian rocks in the Marble Mountains, San Bernardino County, California. Stipple pattern = sandstone; dashed pattern = shale; brick pattern = limestone; slanted brick pattern = dolomite.

BONANZA KING FORMATION

CADIZ FORMATION

CHAMBLESS LIMESTONE

LATHAM SHALE

ZABRISKIE QUARTZITE

50 m

WOOD CANYON FORMATION

Precambrian granite & metamorphics

Middle Cambrian

Lower Cambrian

Darton's first work, description and naming of the Cambrian rocks at this locality, and in the Providence Mountains about 30 miles to the north, was mostly the work of geologist and University of California graduate student John Hazzard.[56] The formations include, in ascending order, the Wood Canyon Formation, Zabriskie Quartzite, Latham Shale, Chambless Limestone, Cadiz Formation, and Bonanza King Formation (fig. 4.21). The Precambrian–Cambrian boundary occurs low in the Wood Canyon Formation farther northwest near Death Valley, but here in the

Marbles the Wood Canyon is entirely Early Cambrian in age, equivalent to parts of the Campito and Poleta formations in Esmeralda County, Nevada. The Lower–Middle Cambrian boundary is in the lower part of the Cadiz Formation so that the *Bonnia-Olenellus* zone in the Marble Mountains is represented mostly by the Latham Shale, Chambless Limestone, and the base of the Cadiz (fig. 4.19). The *Nevadella* and perhaps the *Fallotaspis* zones may be present in parts of the Zabriskie and Wood Canyon, but we cannot be sure because so few trilobite fossils occur in those units here.

Wood Canyon Formation

The Wood Canyon Formation rests in a nonconformity on Precambrian igneous and metamorphic basement rocks in the Marble Mountains. It is 110 m (361 ft.) thick in the Marbles, which is thin as things go in the Wood Canyon, as it is much thicker in the Death Valley region. The formation consists of gray quartzitic sandstone with abundant crossbeds (plate 4b,c) and some rip-up clasts, both of which indicate that the unit represents a tidal sand flat influenced both by braided streams flowing into the sea off the Laurentian continent and waves of the shallow sea itself. One good, nearly complete trilobite specimen has been found in the Wood Canyon in the Marble Mountains, but it has not yet been identified.[57]

Zabriskie Quartzite

The Zabriskie Quartzite is orange-brown to dark brown in color on weathered surfaces and is about 36 m (119 ft.) of medium-grained quartz sandstone that has been lightly metamorphosed to a very hard quartzite. The Zabriskie has faint traces of crossbeds and some clear *Skolithos* trace fossils, indicating a sandy shallow-marine environment with some braided river and tidal settings. In fact, the Zabriskie appears to represent a shallowing of the sea relative to the top of the underlying Wood Canyon, and also the beginning of a long transgression, or rise in sea level, that continued through the overlying two formations. The marine parts of the Wood Canyon and Zabriskie probably represent shallow and relatively high energy deposits, with plenty of wave action and currents influencing the sandy sediments that formed the bottom of the sea here. The trace fossils indicate that some types of burrowing organisms lived in the sand at the time.[58]

Latham Shale

The Latham Shale consists of 19 m (62 ft.) of green shale and orange-tan silty to sandy limestone and dolomite interbeds (fig. 4.22; plate 4d). The finer-grained sediments of the Latham indicate quieter and deeper water existed in the area at this time, compared to what we see in most of the Wood Canyon and Zabriskie. But the water was not tremendously deep.

A

B

4.22. Digging in the Latham Shale (Lower Cambrian) of the Marble Mountains, Mojave Desert, California. (A) Outcrops of Latham Shale in left and center foreground with middle distance outcrops in older Cambrian sandstones (across a fault). Ship Mountains in distance. (B) Digging for trilobites in shale outcrops of middle part of Latham Shale.

In fact, it appears that the water depth during much of Latham time was approximately 50 m (164 ft.)—shallow enough to allow plenty of light to still reach the bottom. This relatively shallow water existed over many miles of the northern shelf of ancient Laurentia (now the west of North America), and the bottom muds were probably only occasionally stirred up by stronger storms. And if the fossils we find in the excavated pits scattered around the Marble Mountains are only partially indicative, the bottom muds contained or were crossed by plenty of Cambrian organisms, from brachiopods and hyoliths to algae, possible conulariid cnidarians, eocrinoids, anomalocaridids, worms, and trilobites (plate 5). The brachiopods include the forms *Nisusia*, *Mickwitzia*, and *Paterina*, and a palaeoscolecidan worm that has not been named. There is a very nice eocrinoid specimen from the Latham Shale that was named *Gogia ojenai*.[59]

Palaeoscolecidan worms are relatives of the genus *Palaeoscolex*. These benthic, predatory worms are probably related to priapulids (see

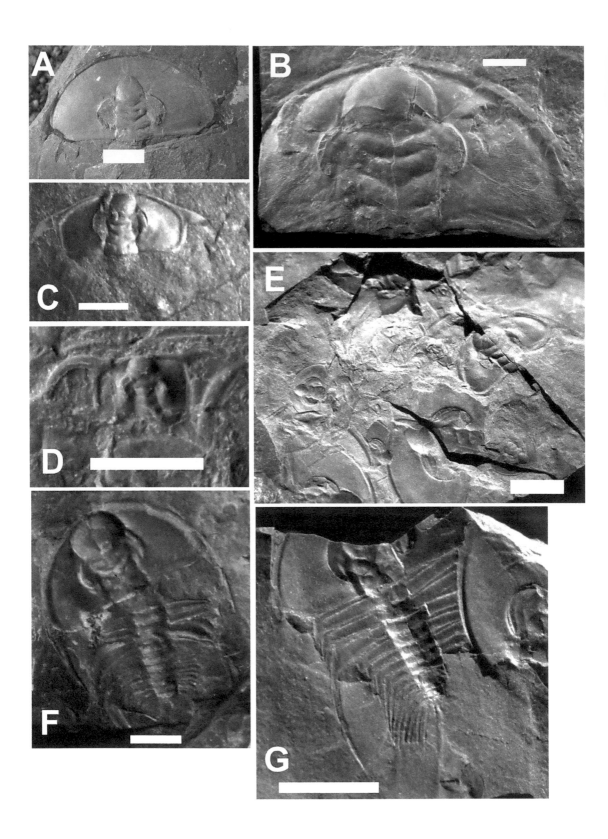

chapter 7) and several other **vermiform** ecdysozoans, but they have also been identified as annelids or nematomorphs. These burrowing worms are long and thin and appear segmented, although this seems to be just surface **annulation** and not true **segmentation**. The head contains a type of proboscis and a spiked collar, and each annulus or "segment" is covered by a ring of small plates. Although the specimen from the Latham Shale cannot be identified to genus (plate 27), its overall appearance is somewhat similar to that of *Palaeoscolex*.[60]

The Latham Shale is the most fossiliferous unit in the Marble Mountains, and it is dominated by trilobites, including: *Mesonacis fremonti, Olenellus clarki, Olenellus nevadensis, Bristolia bristolensis, Bristolia harringtoni, Bristolia mohavensis, Bristolia insolens, Bristolia anteros,* and *Peachella iddingsi* (fig. 4.23; plates 4 and 27). *Olenellus clarki* is a very abundant trilobite in the Latham Shale, and it is generally similar to the type species of *Olenellus, O. thompsoni,* which was described from outcrops on the other end of North America in the 1800s. *Olenellus* is a pretty prototypical trilobite for the Early Cambrian, with its large, half moon–shaped cephalon with moderately sized genal spines, its large, crescentic eyes, its enlarged pleural lobes and spines on the third thoracic segment, and its quickly tapering thoracic region with an elongate posterior spine and tiny pygidium (fig. 4.12). Any number of Lower Cambrian units all over the western United States, Canada, and the East Coast will produce trilobites of this kind. The related genus *Bristolia* (fig. 4.24) is interesting in that its genal spines are often very long and originate far forward on the cephalon–none more so than in *B. insolens,* in which the genal spines are in a position that they might initially appear like antennae. The genal spines are also far forward in little *Bristolia anteros* (fig. 4.23d). What these enlarged, anteriorly translated genal spines were for is not clear.[61] *Peachella* had large bumps in place of genal spines (fig. 4.12).

Olenellids such as *Bristolia, Mesonacis,* and *Olenellus* seem to have been generalized predator-scavengers, but with such variety of form and with sometimes as many as six species co-occuring in the same layer, they may have been feeding in different ways that we have not yet been able to determine. Such potential differences in feeding styles or targeted prey would have helped alleviate unduly intense competition, which is

4.23. Trilobites of the Latham Shale (Lower Cambrian), Marble Mountains, California. (A) Cephalon of *Olenellus nevadensis*. (B) Large cephalon of *Mesonacis fremonti*. (C) Cephalon of *Bristolia bristolensis*. (D) Cephalon of *Bristolia anteros*. (E) Small shale slab with at least ten cephala of olenellids. (F) Nearly complete exoskeleton of *Mesonacis fremonti*. (G) Nearly complete exoskeleton of olenellid. All scale bars = 1 cm.

(A)–(E), (G) from Museum of Western Colorado collections; (F) from Raymond M. Alf Museum of Paleontology collections.

4.24. Reconstruction of the olenellid trilobite *Bristolia bristolensis.* Note long, anteriorly placed genal spines and enlarged pleural spines of third thoracic segment. Cephalon width = ~3 cm.

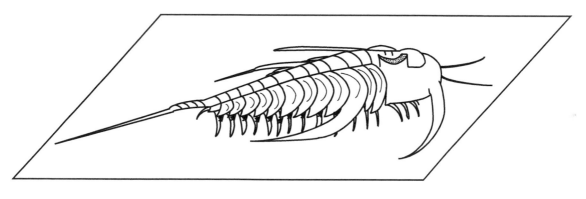

something any species would try to avoid. Although such trace fossils have not yet been found in the Latham Shale, traces such as *Cruziana* and *Rusophycus* in contemporaneous deposits like the lower Bright Angel Formation suggest some trilobites of this time may have plowed through the upper layers of sediment on the bottom, looking for live prey items or carcass debris to scavenge. Olenellids could not enclose themselves between similar-size pygidia and cephala like some other trilobite forms, but they could enroll to protect themselves from predators. Remember that the ventral sides of trilobites were covered with an exoskeletal cuticle that was unmineralized and thus not as hard and protective as the dorsal side that we see fossilized so often. So even with their tiny pygidia and many spines, olenellids would have protected themselves by tucking the interlocking armor of dorsal thoracic segments up under the hardened cephalon. But to protect themselves from what? As mentioned briefly above, there have been at least three feeding appendages of *Anomalocaris* found in the Latham Shale,[62] so we know that this animal, though not often preserved, was present in some numbers in the Latham Sea (plates 5 and 27). *Anomalocaris* has been often depicted as the terror of the Cambrian seas. But was it? There will be more on that in chapter 6.

Chambless Limestone

The Chambless Limestone overlies the Latham and consists of 48 m (157 ft.) of gray limestone with many dark oncoliths (which we first saw in the Grand Canyon in chapter 2), indicating abundant algal growth in shallow carbonate shoals of the time (fig. 4.25a,c). The Chambless probably represents a carbonate bank within the shallow shelf represented by the Latham Shale, because the Chambless is limited to areas near the Marble and Providence mountains; north of this area and elsewhere equivalent units are shale with olenellids like the Latham. So if the Chambless represents a local area of carbonate deposition, what might it have looked like? Most likely it was slightly shallower than the water represented by the Latham. The vision is of a shallow, clear-water carbonate shoal with a lime-mud bottom on which grew mats and small, round lumps of algae (plate 6). These algae occasionally got ripped up and rolled around by storms, and shells and skeletons of brachiopods and trilobites and possibly archaeocyathid sponges also got broken up and mixed in with the mud (fig. 4.25d).

The fossils found in the Chambless Limestone include (in addition to the brachiopods and archaeocyathids) hyoliths (*Novitatus*), eocrinoids (*Gogia*), and the trilobites *Olenellus terminatus, Olenellus puertoblancensis, Bolbolenellus euryparia,* and a species of *Bristolia. Bolbolenellus euryparia* is relatively abundant among the trilobites. Just identified recently from the Chambless Limestone is the corynexochid trilobite *Bonnima* (fig. 4.25b),[63] a form found in northwest Canada and previously unreported from the western United States. It was the first described dorypygid Corynexochida from the Chambless or a cratonic section in

4.25. Aspects of the Lower Cambrian Chambless Limestone of the Marble Mountains, California. (A) Outcrop of interbedded shale and thin carbonate layers between thick units of gray, oncolitic limestone. (B) Pygidium of the dorypygid *Bonnima* sp. (MWC 6961), internal view, in thin orange carbonate layer; scale bar = 1 cm. (C) Outcrop of typical gray, oncolitic limestone of the Chambless; note abundant oncoliths. (D) Thin-section micrograph showing abundant fossils in micritic matrix of calcium carbonate. (E) Cephalon of the olenellid trilobite *Bolbolenellus euryparia* (UCR 10186.1) in gray limestone, found by Pete Sadler; scale bar = 1 cm.

(B) from Museum of Western Colorado collection; (E) from the University of California–Riverside, Invertebrate Fossil Collection.

the western United States, and a second Chambless dorypygid specimen has been located recently as well.

The lower part of the Cadiz Formation is Early Cambrian in age, and the Lower–Middle Cambrian contact occurs somewhere probably in the range of 25 m (82 ft.) up from the base of the Cadiz. This means that most of the Cadiz is Middle Cambrian in age but we see the last days of the Early Cambrian in its rocks. The lowest part of the Cadiz includes shale layers interbedded with thin, gray limestone beds similar to the

Chambless Limestone, and these indicate a gradational contact with that underlying formation. The rest of the Lower Cambrian part of the Cadiz consists of interbedded shales and siltstones and contains a fauna including mostly the trilobites *Mesonacis fremonti*, *Olenellus fowleri*, *Olenellus gilberti*, *Olenellus terminatus*, and various species of *Nephrolenellus* and *Bolbolenellus*.

The olenellid trilobites of the Latham, Chambless, and Cadiz in the Marble Mountains appear to represent almost all of Dyeran time and the *Bonnia-Olenellus* biozone (fig. 4.19). The upper contact occurs within the Cadiz, and the underlying Wood Canyon and Zabriskie range into the *Nevadella* and possibly the *Fallotaspis* zones, but the record from these two formations is too poor to be sure where any discernible biozone contacts are. Most important is that a thin but fairly complete and fossiliferous record of Dyeran time on the craton close to the shoreline is preserved here in the Marble Mountains, and it shows a reasonably diverse group of Cambrian organisms living in the region. Darton would be proud.

Coastal Laurentia: The Early Cambrian Shelf

The Cambrian rocks of the Marble Mountains are very thin compared to their lateral equivalents to the northwest. For example, as we just saw, the earliest Early Cambrian formations in the Marbles and Esmeralda County (Wood Canyon and Campito formations) are, respectively, about 100 m (328 ft.) and 1000 m (3280 ft.) thick. In the Death Valley area the Wood Canyon Formation is a bit thinner than equivalent rocks in Esmeralda County, but is still much thicker than in the Marble Mountains. And unlike in the Marble Mountains, the areas to the northwest contain Precambrian sedimentary rocks below the Cambrian formations, so that the sedimentary sequences there are even more dramatically thick. Part of the reason for this difference is that the Marble Mountains formations, during the Cambrian, were being deposited on the **craton,** a low, tectonically stable part of the continent that was flooded by the ocean at the time. Such flooded areas today occur in the North Sea, Hudson Bay, and the Arafura Sea north of Australia. Beyond the craton, the deposits of the Death Valley and Esmeralda County areas overlay thinner continental and oceanic crust. The formations of the Marble Mountains were thin, were in relatively shallow water (probably less than 45 m [148 ft.] deep), and were close to the shoreline, although they still may have been tens or hundreds of miles out from land. The Death Valley and Esmeralda County formations, however, were thick and were located more than 204 km (126 mi.) farther out toward open ocean than the deposits of the Marble Mountains during the Cambrian.[64] But these outer deposits seem to have still been on the **continental shelf** in relatively shallow water (fig. 4.26). Deepwater deposits occur just northwest of the Esmeralda County sites, suggesting this was the edge of the continental shelf at the time. So the organisms of the Early Cambrian had hundreds of miles of relatively shallow water continental shelf and flooded craton to roam. And that is

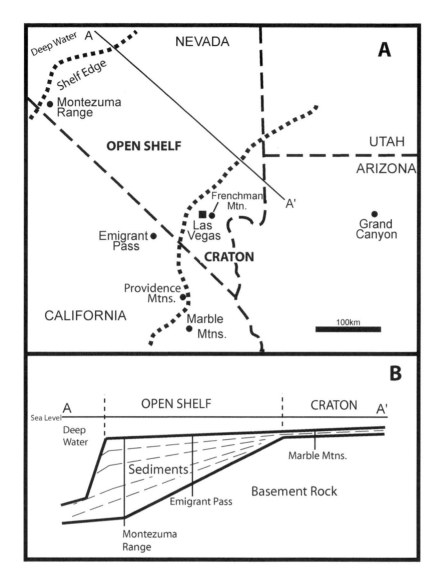

4.26. The late Dyeran shelf of Cambrian Laurentia. (A) Localities (solid circles) relative to craton and open shelf boundaries (thick dashed lines). (B) Cross section showing continental shelf and cratonic and open shelf settings and relative thickness of accumulated sediments in each area.

just from the shoreline out to the edge of the continental shelf where truly deep water began; parallel to shore there were thousands of miles of such shallow shelf circumscribing the continent of Laurentia.

On this continental shelf out beyond the cratonic areas the sea was likely not more than 100 m (328 ft.) deep in most regions, but it may have gotten as deep as about 200–300 m (650–1,000 ft.) in some areas. Continental shelves usually slope gently out from the shoreline to depths of about 200 m (656 ft.) at their outer limit. We have geologic evidence of very shallow water (< 50 m) too, however, and in areas out close to the shelf edge. Some of the deeper areas during the Cambrian may have been in tectonic basins within the shelf, and there may have been reefs and escarpments beyond which the depth increased. The edge of the shelf, however, marked the point at which the ocean floor really dropped off – down to depths of more than 500 m (1640 ft.) on the continental slope and 5000 m (16,400 ft.) in the open ocean basins. Cambrian organisms lived in the seas out here as well, and we find their fossils now and then, but most of what we know of the life of this time is from the various environments, deep and shallow, open shelf and craton, in the continental shelf waters that formed a wide belt around Laurentia.[65]

South Shore Laurentia I: New England Outcrops

We will journey to the other side of the continent now. In Vermont, the Parker Slate has yielded a large number of Early Cambrian species of trilobites, hyoliths, brachiopods, crustaceans, and other organisms. Trilobites first started to show up here in the fall of 1855, when Noah Parker collected a few and showed them to W. C. Watson, a high school principal in Georgia, Vermont, who contacted a geologist about the find. Soon early workers—including James Hall, G. M. Hall, E. Billings, and J. B. Perry—were collecting and naming trilobites from this unit.[66]

These were some of the first Early Cambrian animals studied in North America, and the fossils include the type specimens of several genera and species, including that of the classic trilobite *Olenellus* in the form of *Olenellus thompsoni*. The outcrops in this area, near and around the towns of St. Albans and Swanton in the far northwestern part of Vermont, contain several Cambrian formations, all from the later Early Cambrian. The formations are the Gilman Quartzite at the base, which is overlain by the Dunham Dolomite; there is an unconformity between the Dunham and the overlying Parker Slate, but the amount of time missing is not large; most of the Parker is also Early Cambrian and it ranges a little into the Middle Cambrian. Despite the relatively minor amount of time missing between the two, the Dunham appears to have lithified and been exposed before deposition of the Parker because the contact between the two shows erosion and excavated channels cut into the Dunham. The presence of *Olenellus* and associated fossils in these formations indicates their Early Cambrian age.

The Parker Slate consists of a gray to black micaceous slate or shale with minor amounts of interbedded dolomite and quartzite, and it can be up to approximately 330 m (1082 ft.) thick, although in the St. Albans area, for example, it is about 75 m (246 ft.) thick. The thickness of the formation is difficult to measure, however, thanks to tectonic faulting and folding of the rocks. The formation was named in 1932 with a type section at the Parker Farm near Georgia Center, Vermont.

The fauna that has been found in the Parker Slate is impressive, with numerous trilobites, including those typical of the late Early Cambrian, *Olenellus* (*O. thompsoni* and the elongate *O. vermontanus*; fig. 4.27) and *Bonnia*, numerous brachiopods including *Nisusia* and *Paterina*, the bivalved crustacean *Tuzoia*, the sponges *Leptomitus* and ?*Protospongia*, the possible molluscs *Salterella* and *Hyolithes*, and possibly, the large arthropod *Anomalocaris*.[67] These are many of the same taxa that are found commonly in rocks of the same age on the other side of the continent. The presence of rare unmineralized taxa such as *Tuzoia* and

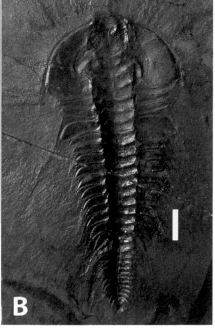

4.27. Trilobites of the Lower Cambrian Parker Slate from near Georgia Center, Vermont. (A) *Olenellus thompsoni* (USNM 15418). (B) The elongate *Olenellus vermontanus* (USNM 15399). Both scale bars = 1 cm.

Courtesy of Smithsonian Institution.

Anomalocaris make the Parker Slate yet another locality with Burgess Shale–type preservation.

Unfortunately, at least a few of these classic old localities in the Parker Slate have suffered a rather modern demise – demolition by construction. Little work has been done in recent years thanks to less availability of outcrops. In one case, an old fossil quarry was mined for its limestone and used in highway construction, leading paleontologist Alan Shaw to lament in 1955, "[I]t, thus, seems quite likely that the old quarry now lies strewn along some Vermont roads."[68] A boulevard of broken trilobite dreams.

West of Philadelphia, in the hills of southeast Pennsylvania around small cities such as York and Lancaster, lie outcrops of the Kinzers Formation, which has produced a paleofauna of Early Cambrian forms, including the trilobite *Olenellus* (fig. 4.28a), comparable in diversity to the Parker Slate and other lagerstätten of the Lower Cambrian. The Kinzers Formation lies above the Lower Cambrian Vintage Formation and Antietam Formation and below the Lower–Middle Cambrian Ledger Formation. The Kinzers is 45–155 m (148–508 ft.) thick and consists of dark shales and carbonates deposited seaward from a carbonate shelf that was situated off the south shore of Laurentia (now the east coast of North America). One specimen of *Olenellus getzi* from the Kinzers (USNM 90809) seems to have been bitten by a predator on the left side of its cephalon – a major injury that apparently healed and was thus survived by the trilobite. This is one of those rare glimpses into events in the day to day life of Cambrian animals. The paleofauna of the Kinzers Formation includes trilobites

South Shore Laurentia II: The Keystone State

4.28. Faunal elements of the Lower Cambrian Kinzers Formation of Pennsylvania. (A) A tectonically sheared specimen of the trilobite *Olenellus transitans*. Scale bar = 1 cm. University of Oklahoma specimen. (B) Reconstruction of the sponge *Hazelia*.

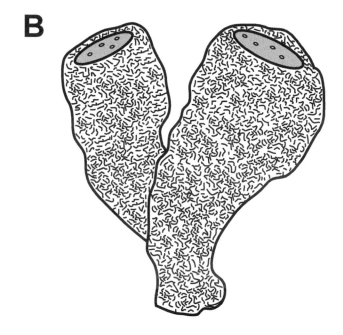

such as *Wanneria*, *Olenellus*, *Bonnia*, and *Lancastria*. Soft-bodied taxa preserved here include *Anomalocaris pennsylvanica*; a new, tentacled and rather enigmatic metazoan named *Kinzeria*; the bivalved crustacean *Tuzoia*; the arthropod *Serracaris*; at least four types of algae and cyanobacteria; and worms such as *Atalotaenia*. Also present are hyoliths, the primitive gastropod *Pelagiella*, brachiopods such as *Paterina*, the sponge *Hazelia* (fig. 4.28b), indeterminate small molluscs, chancellorids, the enigmatic *Salterella*, and several types of echinoderms.[69]

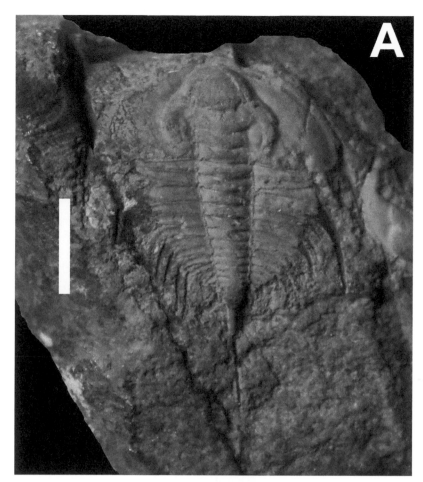

A

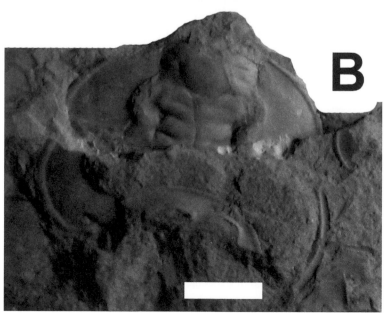

B

4.29. Trilobites from the Lower Cambrian Rome Formation of the southeastern United States. (A) *Olenellus buttsi* from near Montevallo, Alabama (USNM 94777). (B) *Olenellus romensis* from near Blue Ridge Springs, Virginia.

(A) Courtesy of Smithsonian Institution; (B) University of Oklahoma specimen.

South Shore Laurentia III: When in Rome

The Lower Cambrian Rome Formation runs through the southern Appalachians and is exposed in Virginia, Georgia, Alabama, and Tennessee. Among the specimens preserved in this unit are the olenellids *Olenellus buttsi* and *O. romensis* (fig. 4.29). These fossils show that in form and preservation, the material from the East Coast and Laurentia's southern continental shelf is very similar to that known from the Great Basin, Mojave, and other parts of the Cambrian north shore.

Suburban Paleontology: Frenchman Mountain

Nestled in the barren foothills on the northeastern edge of Las Vegas, Nevada, almost within sight of subdivisions and convenience stores, is a Lower Cambrian locality known as Frenchman Mountain (fig. 4.30a). It is actually the western, lower reaches of Frenchman Mountain that you are digging on, within earshot of Nellis Air Force Base. Dig here on the right day by pure luck and you may catch a performance by the USAF Thunderbirds, who may just fly right over your head. The sedimentary layers here dip steeply to the east (into the hill at 51 degrees), with the Tapeats Sandstone to the west on top of Precambrian basement. Above this the green and red shales and siltstones of the Bright Angel Formation are exposed in what appears to have once been a dump, complete with fragments of old tile and discarded mining truck tires for good measure.

The Bright Angel Formation here (along with overlying and underlying formations) has been translated west along the Las Vegas Valley Shear Zone from an area close to the Grand Wash Cliffs in the far western Grand Canyon. This is a glimpse of Dyeran rocks from far to the east of the Marble Mountains, for example, much farther onto the Laurentian craton. It was once close to shore. The green shales of the Bright Angel Formation here are thin, platy and fossiliferous, much like the Latham Shale. The red siltstones contain many trace fossils but few body fossils. The Lower Cambrian part of the Bright Angel Formation here may be equivalent in age to the Chambless Limestone or possibly the lowest Cadiz Formation (see fig. 5.1), but it is quite different from most of the Bright Angel Formation in Grand Canyon, which is Middle Cambrian in age.

Fossils found at one level at this locality include hyoliths, the trilobites *Olenellus terminatus* and *O. gilberti*, and the unusual trilobite *Biceratops nevadensis* (fig. 4.30b–d). Brachiopods are strangely rare here, and the trilobite fauna, although abundant, is not notably diverse. *Biceratops* is an unusual olenellid that is related to *Peachella*. Its cephalon lacks genal spines entirely and its glabella has only very indistinct furrows; also, the eyes are tucked close up against the glabella, positioned much closer to the cephalon midline than in, say, a typical *Olenellus*. The thoracic exoskeleton, however, is pretty typically olenellid, with enlarged third thoracic pleural lobes and spines and a long posterior spine. *Biceratops* has been found at few sites, but it is not uncommon at Frenchman Mountain in the right layers. Still, *Olenellus terminatus* often outnumbers *Biceratops* in the same layers at Frenchman Mountain, and elsewhere

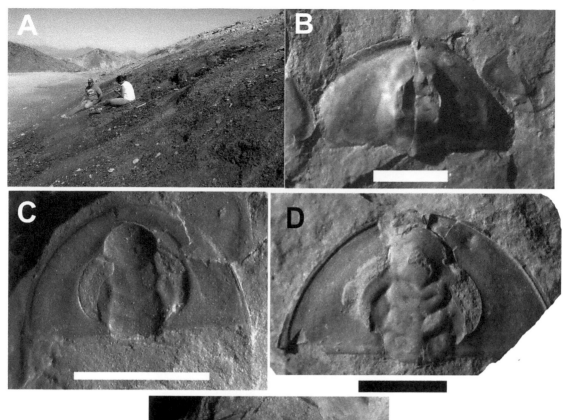

4.30. Trilobites from the Lower Cambrian Bright Angel Formation at Frenchman Mountain, Nevada. (A) The site outside Las Vegas, Nevada. (B) Cephalon of the trilobite *Biceratops nevadensis*. Note lack of genal spines and narrowly placed eyes of this species. (C)–(D) Two specimens of the common *Olenellus terminatus*. Note smaller size of specimen in (C). (E) Cephalon of the rare (at this site) olenellid *Nephrolenellus multinodus*. All scale bars = 1 cm.

(C)–(E), Museum of Western Colorado collection.

one specimen from the transition zone between the Tapeats Sandstone and Bright Angel Formation in the western Grand Canyon, figured by Edwin McKee and Charles Resser in 1945, appears to be *Biceratops*.[70]

The trilobite sclerites in the Bright Angel at Frenchman Mountain are smaller and more fragmentary than those out of the Latham Shale, and there are a number of broken cephala suggesting scavenging on the elements. Although high-energy agitation and breakage of trilobite molts on the bottom cannot be ruled out, the sedimentology of the site and experiments with agitation of modern eggshell suggest that this is not the case and that most breakage of trilobite pieces at these sites is possibly due to the feeding of scavengers.[71]

Thick sections of Cambrian rocks also occur in the rugged, bear-ruled mountains of the Yukon Territory. Here outcrops of the Illtyd Formation

Far Northwest

4.31. Two specimens of *Olenellus* sp. from the Lower Cambrian near Museum Peak in Alberta, Canada.

University of Oklahoma specimens.

and Sekwi Formations contain typical Lower Cambrian trilobites like *Bonnia, Olenellus,* and *Bonnima.* The Illtyd Formation in the Wernecke Mountains of the Yukon Territory of Canada contains thick sections of Lower Cambrian rocks, and many of the trilobites found here are the same genera as found farther south. Some were identified in Canada later. *Bonnima* was first named from this area, however, and was only identified elsewhere later.

The Sekwi Formation in the Mackenzie Mountains of the Northwest Territories has produced numerous trilobites including widely known forms such as *Olenellus, Nevadella, Bonnia,* and *Wanneria,* along with forms identified first (or only) from Canada such as *Sekwiaspis, Nehanniaspis, Yukonides,* and *Bradyfallotaspis.* Also found in this formation at the level of the lower *Bonnia-Olenellus* zone are specimens of the enigmatic sponge-like animal *Chancelloria* (along with the related form *Archiasterella*).[72] Specimens of *Olenellus* are also found in Alberta's mountains in a number of areas (fig. 4.31).

Inland Environments of the Early Cambrian

Among the environments preserved in the rocks of Early Cambrian age are some that we might not have expected but that are not at all surprising when we consider the world was not that different from today. The same physical forces operated, so why shouldn't there be such settings as the Early Cambrian coastal sand dunes along an ancient beach, found in what is now Sweden? There also appear to have been inland sand dunes on parts of some continental areas.[73]

A Look Back

So the Cambrian is rolling along. We have traveled forward in time nearly to the end of the Early Cambrian, about 30 million years' worth of travel already! But we are just now getting into some of the "meat" of the Cambrian period. It's not that an incredible number of important events have not already taken place in this part of the journey. It is just that we are reaching a point now where the evidence in many cases gets even better, and we can look at things in even greater detail. But we are reaching the end of the Early Cambrian. Our next stops look at these closing days of the epoch as we approach the Middle Cambrian.

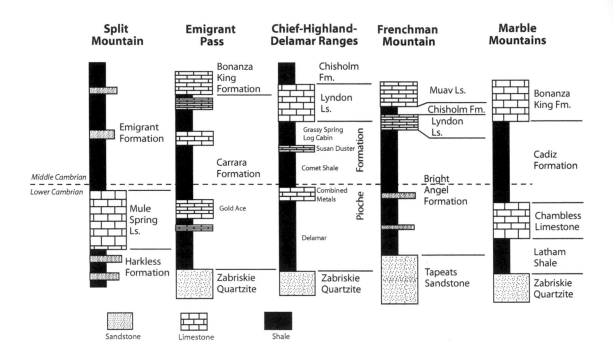

5.1. Stratigraphic section of the formations in the Chief, Highland, and Delamar ranges in eastern Nevada correlated with comparable sections in several other areas discussed in the text. Tie point for connecting each section is the Lower–Middle Cambrian boundary.

On Top of the World: The Middle Cambrian Begins

5

WE HAVE MOVED UP THROUGH EARLY CAMBRIAN TIME ACROSS MORE than two-dozen million years. Our next stop is the Lower–Middle Cambrian boundary. Driving north from Las Vegas on Highway 93 we travel up several of the basins of the Basin and Range Province and after about three hours, just over half-way to Great Basin National Park, we come to the town of Pioche, Nevada. This small collection of houses and old hotels, shops, winding streets, a "boot hill," and an opera house is perched on a hillside dotted above town with numerous old silver mines. Named after investor François Pioche from San Francisco, the town of Pioche sprang up between 1864 and 1868 as workers flooded the area in search of the precious metal that built the state of Nevada. The silver rush here peaked in 1872, a time when the town's population was about ten times what it is now and when mercenaries were hired to guard mine entrances, such was the wildness of the rush. Not that that kept things under control in town all the time, either; Pioche saw its share of shootouts – just like Tombstone, Dodge City, or Deadwood. Legend has it that the Pioche cemetery had seen nearly six dozen burials before the town had even been in existence long enough for any resident to die of natural causes.

At the heart of all the trouble, of course, was money, and the money came from silver. The silver, incidentally, formed when veins of granite intruded sandstones, shales, and limestones of some of the most important and fossiliferous Cambrian formations in western North America. The terrain around Pioche consists of mile after mile of sage-covered high desert (4500–5000 ft. elevation), and the valley south of Pioche is dotted with red outcrops of the Pliocene-age Panaca Formation, which records valley-fill, lake, and sand dune deposition from about 4 million years ago.[1] The mountain ranges are covered with low, open forests of junipers with a few piñon pines thrown in as well. This is a place of colorful geographic names such as Dead Deer Canyon and Slaughterhouse Gulch, along with the always-creative names of the mines in this part of the country. West and south of Pioche are the Highland Range and the Chief Range, relatively low,[2] tree-covered mountains composed of gently east-dipping Cambrian rocks that have been chopped up by faults and intruded by younger igneous rocks that have formed the silver deposits that gave the towns of the area their start.

There are numerous known fossil localities in these Cambrian rocks, which consist of, in ascending order, the Prospect Mountain (or Zabriskie) Quartzite, the Pioche Formation, the Lyndon Limestone, and

163

5.2. The Ruin Wash locality near the top of the Combined Metals Member of the Pioche Formation in the Chief Range of eastern Nevada. Foreground area was excavated years ago and still turns up specimens, even in loose shale.

the Chisholm Formation (fig. 5.1). The sites in the Pioche Formation in particular are important for the story of the beginning of the Middle Cambrian. The rocks of this area contain one of the best records of the early Middle Cambrian fossils anywhere in North America, and this is also one of the best places to see the Lower–Middle Cambrian boundary. Pioche is in Lincoln County, Nevada, and it is that county that lends its name to the Lincolnian stage in North America (fig. 4.19); this is an exemplary area for Cambrian faunas of this age.

Curtain Call for the Olenellids

Excellent Preservation: Ruin Wash

Heading west into the Chief Range southwest of Pioche, you eventually work your way up a dry creek bed past a crumbled lime kiln to a locality called Ruin Wash. Here among surprisingly barren ground and a smattering of juniper trees are several pits and trenches dug into the Pioche Formation (fig. 5.2; plate 7). Preserved here are eight species of the last of the olenellid trilobites, the most diverse assemblage in the western United States. This is the very end of the Early Cambrian. Among the species preserved here are the olenellids *Olenellus gilberti*, *O. chiefensis*, *O. fowleri*, and *Nephrolenellus geniculatus* (figs. 5.3 and 5.4; plate 28). Also found are *Olenellus terminatus*, a species we also saw preserved in the Bright Angel Formation at Frenchman Mountain outside Las Vegas, and the

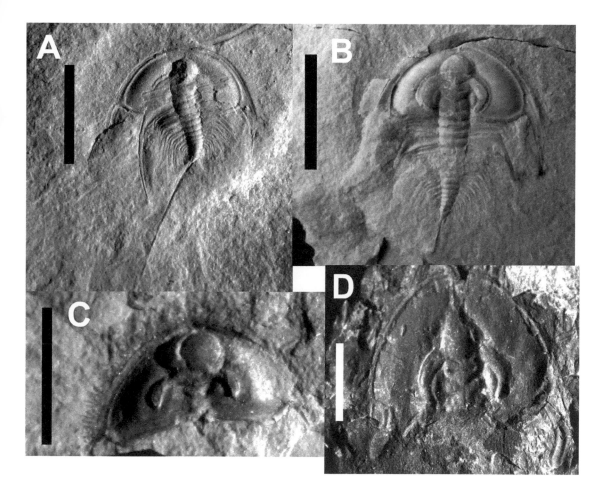

rare, very spiny corynexochid *Oryctocephalites palmeri* (fig. 5.4d; plate 28). An unusual number of the olenellid trilobites are preserved articulated and oriented convex up. Also preserved at Ruin Wash is another corynexochid, *Zacanthopsis*, but the olenellids dominate. But go just a meter higher in the Pioche Formation here and the olenellids disappear.[3] Soon one begins to pick up new species of trilobites, and the ptychoparids and corynexochids–relatively rare in Lower Cambrian rocks–become diverse and abundant. This Lower–Middle Cambrian faunal turnover of trilobites probably represents an extinction of olenellids and a diversification of the other orders of trilobites, but this is not entirely clear. If time is missing at this contact in many places, more may have been going on that we simply cannot see.

There are a number of other types of animals preserved at Ruin Wash, including the priapulid worm *Ottoia*, the bivalved arthropods *Tuzoia polleni* and *Tuzoia nitida*, the predatory arthropod *Anomalocaris pennsylvanica*, and conulariid cnidarians (fig. 5.5). The soft-body fauna at Ruin Wash constitutes another example of Burgess Shale–type preservation, and similar species of several of the same genera preserved just above the Lower–Middle Cambrian boundary in the Pioche Formation

5.3. Lower Cambrian trilobites from the Pioche Formation at Ruin Wash, Nevada. (A) Slightly taphonomically deformed specimen of *Olenellus gilberti.* Note that cephalon has shifted posteriorly below several thoracic segments. (B) Another nearly complete specimen of *Olenellus gilberti.* (C) Cephalon of *Nephrolenellus geniculatus.* (D) Cephalon of *Olenellus chiefensis.* All scale bars = 1 cm.

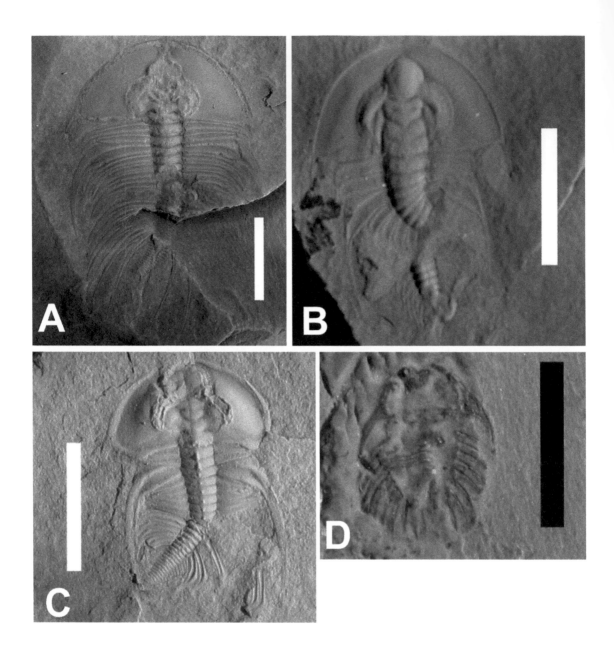

5.4. Lower Cambrian trilobites from the Pioche Formation at Ruin Wash, Nevada. (A) A nearly complete *Olenellus fowleri*, with elongate pleural spines. (B) A taphonomically deformed *Olenellus terminatus*. (C) A nearly complete *Nephrolenellus geniculatus*. (D) Tiny, spiny, and rare *Oryctocephalites palmeri*. All scale bars = 1 cm.

(A), (C), and (D), Museum of Western Colorado specimens collected and donated by Andrew R. C. Milner.

indicate that the extinction event associated with that boundary had less effect on these taxa than it did on the trilobite fauna across the same interval. Not all animals appear to have been under the same pressures at this time, and whatever was causing the turnover then for some reason affected trilobites to a greater degree than other taxa.[4]

BIVALVED ARTHROPODA

Splitting shale at Ruin Wash, you may reveal to light, for the first time in 510 million years or so, the unmineralized, bivalved arthropod *Tuzoia* (fig. 5.5a), which is one of a number of such animals known from various Cambrian localities. *Tuzoia* itself is known from the Burgess Shale,

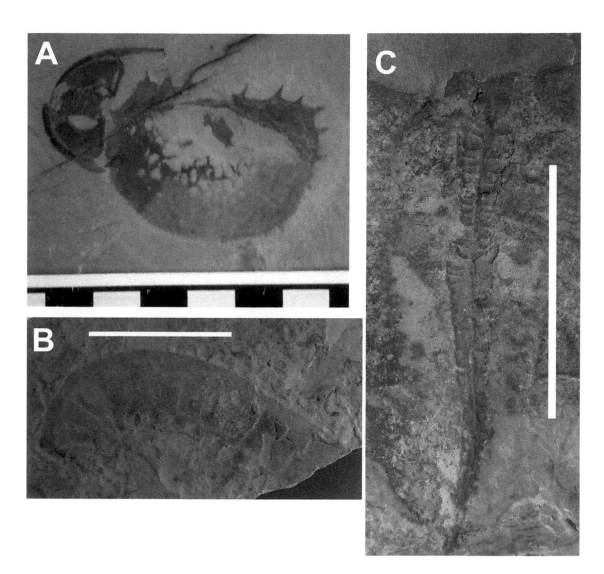

the Emu Bay Shale (Australia), and from China; other bivalved arthropods such as *Canadaspis* and *Isoxys* are also common at more than one site, too—more on those later. Although some bivalved arthropods from Cambrian rocks appear to be primitive stem forms, a few appear to be phyllocarid crustaceans, relatives of modern shrimp, for example. But the systematic position of some of these forms is quite unsettled. *Canadaspis*, for example, has been placed seemingly all over within Arthropoda (see chapter 6).

What we do know about *Tuzoia* is that its shells are characteristically spiky along the edges and along a midline, with a reticulate pattern comprising the bulk of an internal strengthening structure. The reticulate pattern was an elegant biologic design probably developed to maximize strength of the shell while minimizing shell thickness. This and the spines along the edges of the shells suggest defense. The shells reached about 18 cm (7.2 in.) in length, and the body of the animal (head,

5.5. Soft-bodied taxa from the Lower Cambrian part of the Pioche Formation at Ruin Wash, Nevada. (A) Valve of the arthropod *Tuzoia* (with olenellid cephalon next to it). Note reticulate pattern and marginal spines (KUMIP 293630). Scale is in centimeters. (B) Feeding appendage of a giant specimen of *Anomalocaris* (KUMIP 298500). Scale bar = 10 cm. (C) The conulariid *Cambrorhytium*. Scale bar = 10 cm.

All specimens from University of Kansas collections.

trunk, and limbs) may have been enclosed within the two shells, with the stalked eyes facing forward. Eyes preserved isolated in the Emu Bay Shale of Australia may belong to *Tuzoia*; what is interesting about these eyes is that they are nearly as advanced as those of modern insects, with both the number and size of lenses far outstripping the most advanced Cambrian trilobites. Whatever arthropod owned these eyes, *Tuzoia* or not, was apparently capable of predation in low-light conditions.

There are at least seven species of *Tuzoia*, with a global distribution that ranged from approximately 5°N to 40°S in subtropical marine settings. They probably were free swimming and possibly fed on plankton or detritus low in the water column in shallow-shelf settings and water depths of 100–150 m (328–492 ft.) or less.[5]

The Boundary: Oak Spring Summit and Beyond

The Lower–Middle Cambrian boundary in North America officially occurs in a nondescript gully in the Delamar Mountains, near Caliente, Nevada, not far from Pioche, at a site known as Oak Spring Summit (plate 7).[6] This continental boundary point, marking the beginning of the Middle Cambrian, along with the last of the olenellid trilobites and the beginning of the *Eokochaspis nodosa* zone, occurs at the contact of the underlying Combined Metals Member and the overlying Comet Shale Member of the Pioche Formation (fig. 5.1). The boundary is formed by the base of a 70-centimeter-thick limestone at the base of the Comet Shale. Just below the boundary is one of the densest accumulations of olenellid trilobite remains I've ever seen, representing the end of the Early Cambrian, and just above the boundary there are no olenellids anymore, only the occasional *Eokochaspis*.

The equivalent of this Lower–Middle Cambrian boundary in deposits representing deep water occurs in the lower Emigrant Formation in western Nevada at a site called Split Mountain (fig. 5.1). This is a site southwest of the town of Tonopah that lies a number of miles down a dirt road that crosses a beautiful, Joshua tree–dotted high desert plain within view of the snow-capped peaks of the White-Inyo and Sierra Nevada mountains to the west. The beauty of the approach, however, obscures the potential brutality of the rocky road toward vehicle tires.[7] At Split Mountain the section includes the Harkless Formation, overlain by the Mule Spring Limestone, which is in turn overlain by the Emigrant Formation. The Lower–Middle Cambrian contact occurs at about 1.5 m (5 ft.) up into the Emigrant Formation, just above the top surface of the Mule Spring Limestone. The last of the olenellid trilobites are rather rare in the lower 1.5 m of the Emigrant Formation; it is easier to find brachiopods. But at the base of this interval and at its top there are, along with the brachiopods, a few fragments of olenellid trilobites in thin, hard, nodular limestones. These fossils can be found with some persistent pounding with a rock hammer to crack open the limestones, but overall the fossils here are not abundant. Although the stratigraphic level is the same, the

preservation here is quite different from that at Oak Spring Summit or Ruin Wash.

5.6. Outcrops of the Lower Cambrian part of the Pyramid Shale Member of the Carrara Formation in the Nopah Range of California (Emigrant Pass). Dark limestone cliffs in the distance are the Bonanza King Formation.

Emigrant Pass

Rocks recording the Lower–Middle Cambrian transition also occur on the other side of the Great Basin from the Pioche area in eastern California. Here, at a site called Emigrant Pass in the Nopah Range (fig. 5.6; plate 7) and other areas, the Pyramid Shale Member of the Carrara Formation contains rocks of this age. Unfortunately, the rocks are not quite as fossiliferous as they are in the Chief Range or Delamar Range, for example, but they do at least record very late Early Cambrian olenellid faunas in abundance. The Carrara Formation consists of a number of members, the lower of which are of Early Cambrian age. The Carrara Formation rests on the Zabriskie Quartzite below and underlies the Bonanza King Formation (fig. 5.1), two formations you may remember from the Marble Mountains. In fact, the Eagle Mountain, Thimble Limestone, and Echo Shale Members of the Carrara Formation are approximately equivalent to the Latham Shale that we saw in the Marble Mountains in the last chapter. The Gold Ace Limestone Member is approximately equivalent to the Chambless Limestone in the Marble Mountains. The Pyramid Shale Member (like the Cadiz Formation in the Marble Mountains) contains a basal section of uppermost Lower Cambrian rocks, with an abundant

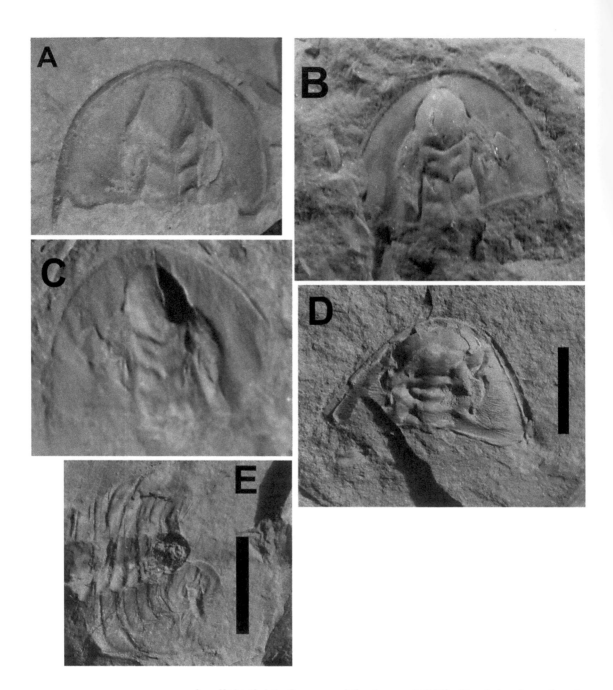

olenellid trilobite fauna, and the Lower–Middle Cambrian boundary in its lower levels. Middle Cambrian fossils above the boundary are present but rare compared to the olenellids lower down.

The uppermost Lower Cambrian fauna of the Carrara Formation, more or less comparable in age to those at Ruin Wash and Oak Spring Summit, includes the brachiopods *Paterina* and *Nisusia*; the olenellid trilobites *Bristolia*, various species of *Olenellus* (fig. 5.7), *Nephrolenellus*, and *Peachella*; the ptychopariid trilobites *Crassifimbra* and *Periomma*; and the mollusc *Novitatus*.[8]

Klondike Gap

The town of Pioche, Nevada, got its name from a man but it gave its name to a rock unit; the Pioche Formation was named by C. D. Walcott in 1908 for the town of Pioche. The Pioche Formation is usually about 240 m (787 ft.) thick, but depending on the area and local faulting, this thickness varies. The formation consists of tan-brown, gray, and light red to pink shale mixed with plenty of siltstone, sandstone, and limestone. It consists of six members, which are, in ascending order, the Delamar, Combined Metals, Comet Shale, Susan Duster Limestone, Log Cabin, and Grassy Spring members (fig. 5.1). The lagerstätte at Ruin Wash is at the very top of the Combined Metals Member. The Middle Cambrian begins just a meter higher than this and it is also a change in trilobite biozones: from the *Bonnia-Olenellus* (uppermost Lower Cambrian) zone to the *Eokochaspis nodosa* (basal Middle Cambrian) zone. Just above the level of Ruin Wash you lose all olenellid trilobites and start to pick up forms such as *Eokochaspis nodosa*, *Eokochaspis piochensis*, and *Oryctocephalites rasettii*. This zone continues through to near the top of the Comet Shale Member. The lower Comet Shale Member also contains some soft-bodied taxa, including the bivalved crustaceans *Canadaspis* (fig. 5.9), *Perspicaris*, and *Tuzoia*, and the predatory arthropod *Anomalocaris* (plate 28).

On the other side of the Chief Range from Ruin Wash lies another dry creek bed, this one with well-exposed outcroppings of east-dipping beds of shale, siltstone, and limestone. Surrounded as always by low junipers, this site lies not far from a narrow notch in a thick gray limestone known as Klondike Gap, where one can maneuver a high-clearance four-wheel drive vehicle from one side of the range to the other without having to drive the hour around on more passable dirt or gravel roads. Large, gray plates of siltstone near the wash east of Klondike Gap represent the top of the Comet Shale Member of the Pioche Formation and contain hyoliths and disarticulated and articulated trilobites such as *Amecephalus arrojosensis* and *Mexicella robusta* (fig. 5.8). The trilobite fauna has changed a bit from the base of the Comet Shale Member, picking up forms like *Nyella rara* and *Kochiella brevaspis*. This is an interval known as the *Amecephalus arrojosensis* zone within the *Plagiura-Polliella* zone (fig. 4.19). Trilobites at this level are also found back at Split Mountain, Nevada, in the Emigrant Formation; this interval and locality, higher than the Lower-Middle Cambrian boundary we just saw, yields an abundance of *Onchocephalites* and *Tonopahella*.[9] The Pioche Formation also yields a diversity of brachiopods from several of its members.[10]

Working up into the overlying Susan Duster Limestone Member and the Log Cabin Member, the Pioche Formation contains trilobite species of the *Plagiura-Polliella* zone, and in the Grassy Spring Member there occur species indicative of the *Albertella* zone (fig. 4.19). By the time we get up into the *Albertella* zone we begin picking up the genera that will become very familiar in the Burgess Shale and other units later in the

5.7. Olenellid trilobites from the Pyramid Shale Member of the Carrara Formation at Emigrant Pass, California. (A) Cephalon of *Olenellus gilberti*?, with a glabella less crushed than average specimens. (B)–(C) Typical, slightly tectonically deformed specimens of *Olenellus terminatus*. (D) Cephalon of *Nephrolenellus multinodus*. (E) Articulated cephalon (4 mm long) and anterior seven thoracic segments of *Olenellus* sp. next to an articulated series of much larger olenellid thoracic segments (MWC 7196). (A)–(C), cephalon widths ~3 cm. (D) and (E), scale bars = 1 cm.

All Museum of Western Colorado specimens.

5.8. Ptychopariid trilobites from high in the Comet Shale Member of the Pioche Formation (upper centimeters below the Susan Duster Limestone Member). (A) Locality (right half of photo) in wash east of Klondike Gap, Chief Range, Nevada. (B) Cranidium of *Amecephalus arrojosensis*. This species gives its name to this trilobite zone. (C) Cranidium of *Mexicella robusta*. Scale bars = 1 cm.

Middle Cambrian: *Pagetia, Olenoides,* and *Mexicaspis.* Each zone contains distinct collections of faunas but with some overlap and nowhere near the drastic change that marked the turnover at the Lower–Middle Cambrian boundary.[11]

The first agnostid trilobites appear in North America at the beginning of the Middle Cambrian. Eodiscoids appear during the Early Cambrian in other parts of the world but not North America, where the agnostoid and eodiscoid groups do not appear until later. And although ptychopariids and corynexochidans have been around during the Early Cambrian in North America they have been very rare. They now, during the Middle Cambrian, explode in diversity.

CRUSTACEA

Among the soft-bodied (unmineralized) arthropod fossils found low in the Comet Shale Member of the Pioche Formation are the bivalved

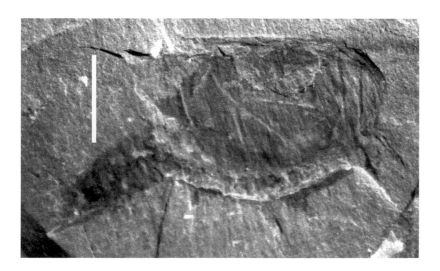

5.9. The bivalved arthropod *Canadaspis perfecta* from the Comet Shale Member of the Pioche Formation. Note preservation of abdomen. Scale bar = 1 cm.

Collected by Linda McCollum at One Wheel Canyon in the Highland Range, Nevada. University of Kansas collection (KUMIP 307021).

crustaceans (or stem-group crustaceans) *Canadaspis* (fig. 5.9) and *Perspicaris*, forms with paired shells that meet dorsally and enclose at least part of the body on each side, being open somewhat ventrally. These forms preserve much of the body in addition to the shells and we can see from their morphologies that they were at least stem group crustaceans, most likely ancient relatives of modern **phyllocarids,** which are in the group Malacostraca, the class of crustaceans including crabs and shrimps.

First, remember that as arthropods, *Canadaspis* and *Perspicaris* share with trilobites, insects, millipedes, spiders, and scorpions a segmented body with an exoskeleton, jointed legs, compound eyes (in most forms), an open circulatory system, and growth accommodated by molting. Modern crustaceans such as crabs and lobsters were not around yet during the Cambrian, but the preserved morphology of *Canadaspis* and *Perspicaris* show that these latter, Cambrian forms were nearly crustaceans nonetheless. Characters that distinguish the Crustacea, some of which these two forms from the Cambrian share, are (1) a five-segmented head with a long trunk divided into a thorax and abdomen,[12] (2) a cephalic shield or carapace, (3) jointed limbs comprised of either one or two branches, (4) segmented jaws that are in fact modified appendages, (5) gas exchange through gill-like structures, (6) eyes often on stalks (think lobster!), (7) often more than one pair of antennae, and (8) excretory organs within each segment near the base of each limb.

Modern phyllocarids, the group to which *Canadaspis* and *Perspicaris* appear to be related, have a "tail" consisting of seven abdominal segments (**pleomeres**) plus a **telson,** or modified last abdominal segment. They also have a "shell" (carapace) covering the anterior part of the body that is bivalved (similar to that of a clam) but is not hinged. It does, however, have an adductor muscle. The abdominal segments protrude behind the carapaces; phyllocarids also have stalked eyes and a movable, articulated rostrum. Modern phyllocarids, of which there are more than 30 species, are marine animals that are free swimming and live either deep in the open ocean or just above the sea floor; most live at water levels from the surface down to about 400 m (1312 ft.). Some live in areas with poor water oxygenation. Many are either suspension feeders that stir up bottom sediments and filter food particles out of the clouds of debris or are scavengers of bottom detritus.

Canadaspis (fig. 5.9) had seven abdominal segments and a telson and had gill-like structures associated with several limbs under the carapace; these were probably used for swimming and gas exchange. Its ecology was probably much like that of modern phyllocarids outlined above. *Perspicaris* was generally similar but had larger eyes and appears to have been more adapted for swimming.[13]

Another group of modern crustaceans with carapaces enclosing the body are the Branchiopoda, which are mostly small, freshwater forms today but existed in the marine plankton during the Cambrian.[14] *Canadaspis* and *Perspicaris* can be distinguished as being closer to phyllocarids than branchiopods, however, because of the lack of limbs on the abdominal segments, as seen in the Cambrian forms and in modern phyllocarids; modern branchiopods often possess limbs on nearly all the abdominal segments. And in fact, true branchiopods have been found in Cambrian rocks in China.[15]

Roll On: The Middle Cambrian Continues

We now need to move up from the *Albertella* zone into the *Glossopleura* zone (see fig. 4.19). In order to do so we head out from Pioche and go many hours south, through another desert, and eventually to a third.

Big Strike in the Sonora

In 1941, two Mexican geologists working for the oil company Petróleos Mexicanos, Isauro Gomez and L. Torres, discovered Cambrian trilobites in the Arrojos Hills near the town of Caborca in Sonora, Mexico. These appear to have been the first Cambrian fossils ever found in Mexico. In the ensuing years fossiliferous Cambrian and Precambrian rocks have been found in outcrops all around the Caborca area, and farther east and south as well. In fact, this is one of the most prolific areas in all of North America.

Caborca is in northern Mexico, about 113 km (70 mi.) inland from the northeastern shore of the Gulf of California. The dry Sonoran Desert here is much like the Mojave in that it has sand and gravel flats with scattered, low mountain ranges, but it has larger cactuses. The Cambrian units here consist of the Lower Cambrian Puerto Blanco, Proveedora Quartzite, Buelna, and Cerro Prieto formations and the Middle Cambrian Arrojos and Tren formations. These formations have produced previously known trilobites typical of some other parts of the Southwest such as the Lower Cambrian *Bonnia* and *Olenellus* and the Middle Cambrian *Amecephalus*, but they have also produced forms first found here and later found to be widely distributed in outcrops in the Mojave and

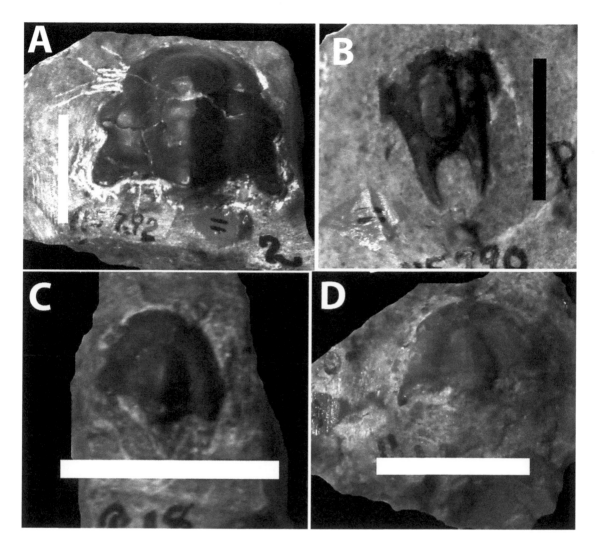

Great Basin. The Middle Cambrian *Mexicella, Mexicaspis,* and *Caborcella* (fig. 5.10) and the Lower Cambrian *Olenellus puertoblancoensis* are found at many localities to the north. And many of the accompanying non-trilobite fossils are found in these "rocas del norte" also: brachiopods, hyoliths, archaeocyathids, molluscs, and the enigmatic *Chancelloria*. The Caborca area is particularly rich in Middle Cambrian trilobites, such as *Amecephalus* and *Glossopleura*.[16]

Gulches, Hollows, and Narrows: The Spence Shale

Up north to southeastern Idaho we go. West of the town of Montpelier is the Bear River Range, low forested hills in the southeastern corner of the state. Now at higher elevation than we were in the deserts (now over 2134 m [7000 ft.]), we find ourselves in forests of tall pines and aspens with plenty of ground cover and forest litter. There is not a lot of natural outcrop here, but at a cut bank of a dry creek bed we find ourselves at an

5.10. Middle Cambrian trilobites from the Arrojos Formation in the Proveedora Hills near Caborca, Sonora, Mexico. (A) Cranidium of *Mexicaspis stenopyge* (USNM 115792). (B) Pygidium of *Mexicaspis stenopyge* (USNM 115790). (C) Cranidium of tiny *Mexicella mexicana* (USNM 115807). (D) Cranidium of *Caborcella arrojosensis* (USNM 115950). Scale bars = 1 cm.

Courtesy of Smithsonian Institution. Collected by G. A. Cooper.

5.11. Natural stream-cut outcrops of the Spence Shale (Langston Formation) at Spence Gulch, Idaho. This Middle Cambrian locality produces abundant trilobites, brachiopods, and hyoliths of the *Glossopleura* zone.

exposure of some 21 m (70 ft.) of gray to tan-green shale of a unit known as the Spence Shale. This layer is often described as a part of the Lead Bell Shale Member of the Langston Formation and is one of the most fossiliferous Cambrian rock units in the western United States. The Spence Shale is low in the Langston Formation here and lies a little above the top of the underlying formation, the Prospect Mountain Quartzite. The Langston Formation as a whole is about 150 m (492 ft.) thick in this area. The Spence Shale in the outcrop at this site known as Spence Gulch (fig. 5.11) is black or dark gray below near the dry creek bed and a color that has been described as "light tea green" higher up; trilobites and other fossils almost fall out of the outcrop at most levels at this site.

Fossils from Spence Gulch were first brought to the attention of paleontologists in 1896 when a man from Utah named R. S. Spence sent some to the leading authority on the Cambrian at the time, Charles Walcott. By 1906 Walcott had managed to visit the site himself. Digging at the cut bank today one still finds a number of types of trilobites typical of the *Glossopleura* zone: *Amecephalus*, *Zacanthoides*, *Pagetia*, *Peronopsis*, *Achlysopsis*, *Athabaskia*, *Glossopleura*, and *Oryctocara* (fig. 5.12; plate 28). Abundant hyoliths and brachiopods are also easily found, with the occasional eocrinoid (*Gogia*) turning up, too. In numbers of specimens, the trilobite samples are dominated by the Agnostida, mostly *Pagetia* and *Peronopsis*. Recent findings have suggested that some agnostids and some eodiscoids (such as *Pagetia*) may have been benthic animals, rather than

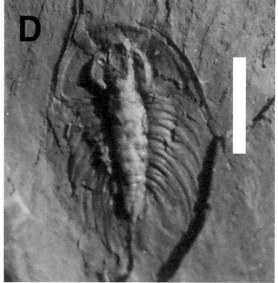

pelagic. Most of the trilobites are disarticulated, but a significant number are fully articulated.

The Langston Formation was named by Walcott in 1908 and above the Spence Shale interval it includes a significant amount of limestone and dolomite. The Spence Shale interval represents a fairly thick section of pure shale within the formation in this area. The Spence Shale was probably deposited in a relatively deep, low-energy open-shelf setting of the outer detrital belt, but one in which oxygen at the bottom was sometimes in short supply and burial was occasionally relatively fast. This allowed intermittent Burgess Shale–type preservation. Evidence of the relatively deep, open setting is also provided by the abundance

5.12. Trilobites of the Spence Shale from Spence Gulch, Idaho. (A)–(B) Nearly complete specimens of the ptychopariid *Amecephalus idahoense*. (C) Nearly complete specimen of *Achlysopsis* sp. (D) Complete specimen of the spiny corynexochid *Zacanthoides idahoensis*. All scale bars = 1 cm.

Museum of Western Colorado specimens.

5.13. The Spence Shale of Miners Hollow, Wellsville Mountains, Utah. (A) Dark, steeply dipping outcrops of the upper Spence Shale (Langston Formation) at Miners Hollow. Arvid Aase for scale. (B) The trilobite *Bythicheilus typicum* from the upper Spence Shale at the site (MWC 7894). (C) Calyx and brachioles of eocrinoid *Gogia* as found in the field, with thin layer of sediment covering the specimen. MWC 7898. Scales for (B) and (C) = 1 cm.

(B) and (C), Museum of Western Colorado specimens.

of agnostids, which are often associated with distal, deepwater settings farther from the coast.

The Spence Shale is also exposed in dark gray outcrops up on the steep western slopes of the Wellsville Mountains north of Brigham City, Utah, at a site known as Miners Hollow. This is a short, steep lactic acid fest of a hike up from the farmland east of Interstate 15. Hiking up to this site you stroll ever upward across the gravels of the ancient shoreline of Ice Age Lake Bonneville and eventually into less forgiving and much older outcrops of the Cambrian rocks. The Spence Shale beds are a foreboding dark gray and form craggy, steeply dipping outcrops on a slope that falls away more than 39 degrees to the west (fig. 5.13), and high above the forested mountains the outcrops only get more gnarly. On clear days when fog and drizzle aren't blowing up the canyon, giving the surroundings an appearance not unlike Tolkien's Mordor, you can see the snow-capped line of the Wasatch Mountains and the town of Brigham City below to the south.

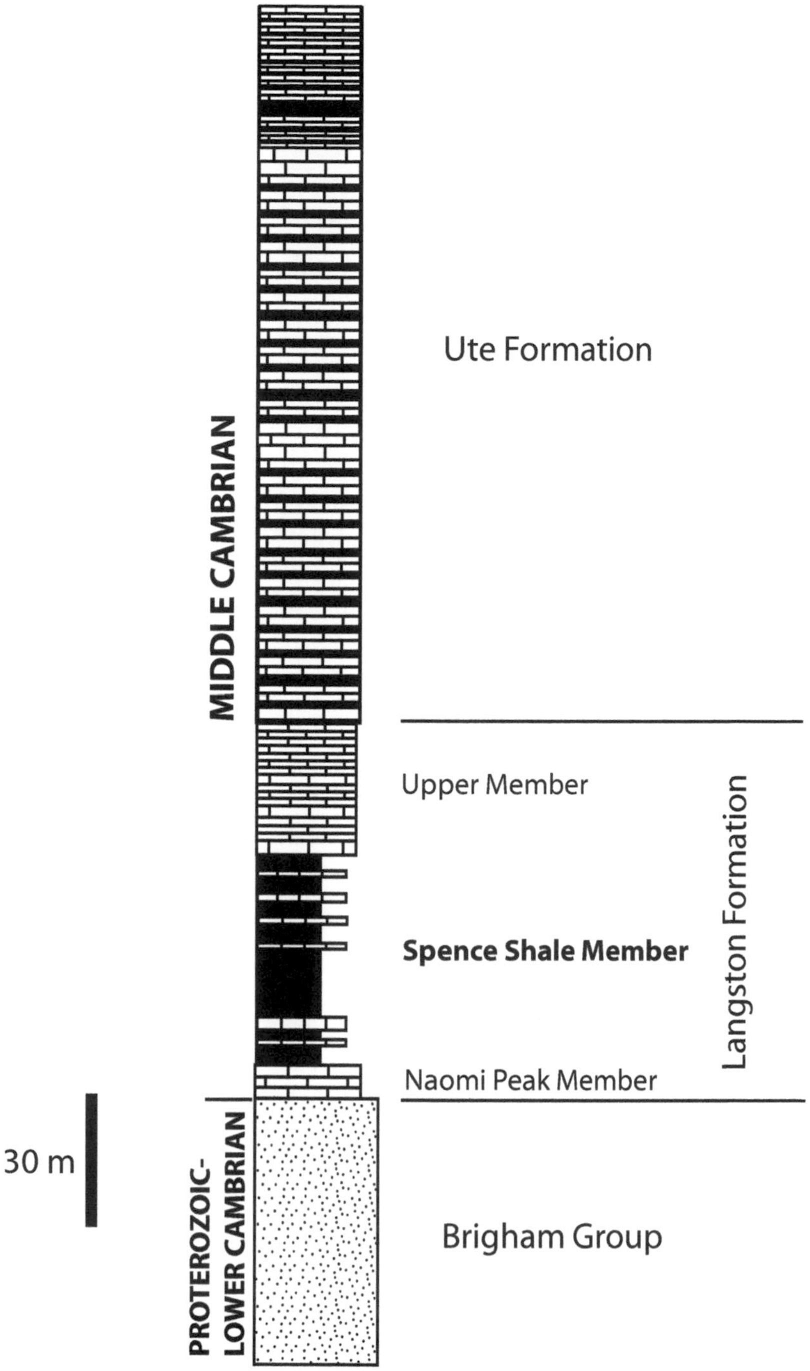

MIDDLE CAMBRIAN

Ute Formation

Upper Member

Langston Formation

Spence Shale Member

Naomi Peak Member

PROTEROZOIC-LOWER CAMBRIAN

30 m

Brigham Group

5.14. Stratigraphic section of the formations exposed in the Wellsville Mountains, northern Utah. Stipple pattern = sandstone; solid black pattern = shale; brick pattern = limestone.

In addition to Miners Hollow, there are a number of other sites in the Spence Shale in the Wellsville Mountains, most found or worked since the 1960s by three generations of the Gunther family of Brigham City. Trilobites had been found in the Wellsvilles going back a number of years, but it was Lloyd and Val (and now Glade) Gunther who did a little detective work and tracked down the localities, found new ones, and really started finding the important specimens that the Spence Shale had to offer in this area. Some of the specimens are beautifully preserved, and many are of species unknown from even the Burgess Shale.

The Spence Shale in the Wellsville Mountains is about 30 m (100 ft.) thick and, as in Idaho, it is part of the Langston Formation (fig. 5.14). It lies above quartzites of the Brigham Group, which are Middle Cambrian in age, and is below the thick, interbedded shales and limestones of the Ute Formation. The Spence Shale may represent a slightly shallower setting here than it does in Idaho, and the formation is composed of a series of shallowing-upward sedimentary cycles. The shale here is gray to dark gray, almost black, and has produced a number of trilobites and soft-bodied arthropods such as *Mollisonia*, *Waptia*, *Anomalocaris*, *Meristosoma*, *Utahcaris*, *Leanchoilia*, *Sidneyia*, *Isoxys*, and *Branchiocaris* (plates 11 and 12); we will see a few of these arthropods in more detail in the next chapter.[17] The trilobites preserved at Miners Hollow are generally the same genera as what we see at Spence Gulch, but agnostids are much less abundant at the former site. At Miners Hollow, the agnostids that are found, and trilobites generally, appear to be concentrated near the bottoms of the shallowing-upward cycles.

Recently described specimens of Spence Shale trilobites from the Wellsville Mountains include particularly well-preserved new species, a number of which have defensive spines of considerable length along the axial lobe and other areas of the exoskeleton. *Glossopleura yatesi* possesses spines of increasing length running down the axial lobe from thoracic segments 2 through 8, raising the question whether the axial nodes of trilobites such as *Anoria* and other species of *Glossopleura* may be the bases of now-missing spines in these animals. *Kootenia youngorum* and *Zacanthoides liddelli* are also quite spiny creatures, with not only axial spines but long genal and pygidial spines as well.[18]

In addition to the trilobites and arthropods, the diverse biota that the Spence Shale has produced from the Wellsville Mountains includes the cyanobacterium *Marpolia*, the algae *Acinocricus* and *Yuknessia*, sponges such as *Vauxia* and *Brooksella*, the worms *Canadia* and *Palaeoscolex*, the eocrinoid *Gogia*, the worms *Ottoia* and *Selkirkia*, the stylophoran echinoderm *Ponticulocarpus*, hyoliths, the possibly distant mollusc relative *Wiwaxia*, numerous types of brachiopods, and the enigmatic *Eldonia* (plates 11 and 12).

Back up in Idaho there is another outcrop of the Spence Shale on the other side of the Bear River Range. Just south of the verdant farmland of the Gem Valley the Bear River enters a steep-sided gorge, and just upslope from the river banks is an outcropping of dark shale, a site known as Oneida Narrows. This site is rich in deepwater trilobites such as *Oryctocephalus* and *Oryctocara*, along with agnostoids. Most of the forms we see at Spence Gulch (other than agnostoids) are rare or absent here. There is also the unusual trilobite *Thoracocare minuta*, a relative of the corynexochid *Zacanthoides* that superficially resembles agnostoids in having a pygidium as large as the cephalon, a thorax reduced to two segments, and in being up to only 3.6 mm (0.14 in.) long. *Thoracocare* may have lived in the water column out in deeper water. There are also a number of sponge spicules in the slabs at Oneida Narrows.

The Spence Shale sites Miners Hollow, Spence Gulch, and Oneida Narrows seem to form a gradient of sites progressing from moderately shallow to moderately deep settings, possibly in that respective order. Although we do not know exactly how deep the water was, and although some of our evidence is simply faunal differences between shallower and deeper settings, it is possible that the shallower Miners Hollow site was in water approximately 50 m (164 ft.) deep, while the Oneida Narrows site may have been in water approximately 100 m (328 ft.) deep. Some parts of the Spence Shale may represent water considerably shallower than this. In any case, it is interesting the range of relative depths and environments that can be represented within a relatively small geographical area.[19]

CYANOBACTERIA

The macrofilamentous cyanobacterium *Marpolia* is found in the Burgess Shale as well as the Spence Shale at sites such as Miners Hollow. Cyanobacteria are responsible for at least some (probably most) of the structures known as stromatolites, which we saw plenty of in chapter 3. But these organisms also grew in other forms—in the case of *Marpolia* as small, colonial groups forming filamentous strands that look like clumps of long, fine cat hair. They can be preserved in some abundance on some bedding planes in the Burgess Shale.

As we saw earlier, modern cyanobacteria are prokaryotic, often single-celled photosynthesizers—in a sense, they are the bacterial equivalent of algae. Some cyanobacteria are colonial and some are multicellular, and as a group cyanobacteria range across diverse environments from freshwater and marine settings to soils and lichens.

ALGAE

Both *Acinocricus* (plate 12d) and *Yuknessia* from the Spence Shale likely are members of the Chlorophyta, the green algae. Recall from chapter 1 that algae are photosynthesizing eukaryotes, but they are not plants. Chlorophyta today constitutes about 7000 species; most are freshwater, but many are marine forms. Single-celled green algae may be planktonic in the ocean, but other species live in snow, soil, or symbiotically with fungi as lichens—the green, yellow, or black encrustations you often see on rocks while hiking in the mountains. There are colonial green algae that form macrofilamentous groupings of cells, and there are truly multicellular green algae such as the modern seaweed *Ulva*. It is these modern, multicellular forms that may be the best analogy for the green algae found in the Spence Shale, large, apparently multicellular forms that might qualify as Cambrian seaweeds.

Acinocricus is preserved as several pieces up to 9 cm (3.6 in.) long and consists of a central stem with numerous whorls of spine-like structures.

Smaller branches among the spines possess smaller, leafy spines. This gives *Acinocricus*, whose name means "thorn ring," a rather spiky appearance overall. *Acinocricus* was named relatively recently based on type and referred specimens from the Spence Shale.[20] *Yuknessia* is named for small 4-cm (1.6-in.), tumbleweed-shaped tufts (known as **thalli**) composed of dozens of possibly tubular structures (**stipes**) growing up from the base. *Yuknessia* is also found in the Burgess Shale (see chapter 6) and in the Wheeler and Marjum formations in the House Range of Utah (see chapter 7).

POLYCHAETE ANNELID WORMS

Among the worms from the Spence Shale is the annelid *Canadia*, also found in the Burgess Shale (see chapter 6). Many segmented annelid worms today are freshwater or terrestrial, but during the Cambrian period these invertebrates inhabited only the shallow seas of the ocean world. The primitive **polychaete** annelids are known as bristle worms, and several kinds are known from the Cambrian. *Canadia* from the Spence Shale is a polychaete annelid and, of course, is far from alone among Cambrian annelids. We will see more on polychaetes later.

HAIR WORMS: THE NEMATOMORPHA

Palaeoscolex, another worm from the Spence Shale, is possibly in the group known as the **Nematomorpha**.[21] We have encountered palaeoscolecidan worms before, in the Lower Cambrian of several localities (chapter 4), but now we have *Palaeoscolex* itself in the Spence Shale. *Palaeoscolex* is reminiscent of an earthworm in appearance (they are not related) but has an anterior proboscis, something modern nematomorphs have as larvae. The body is covered in tiny (~0.04 mm) protective plates. Some fossil nematomorphs seem to have been infaunal deposit feeders, but the lack of sediment in the gut of *Palaeoscolex* suggests it was different, possibly an epifaunal predator. Modern nematomorphs are parasitic on arthropods as larvae, but grow up to have a reduced gut and a reduced or absent mouth.[22] So how do they feed? They don't—not in the typical sense at least. Instead, adult nematomorph worms absorb organic molecules across the body wall, molecules small enough to do so. This is obviously a specialized ecology, and if *Palaeoscolex* and other related Cambrian worms are in fact ancestral nematomorphs, it is an ecology that strays far from the apparent epifaunal and infaunal, deposit-feeding and predatory habits of the early species in the group.

STRANGE ARTHROPODS

One of the more unusual species found in the Spence Shale is *Meristosoma paradoxum*, a large (~17-cm [6.8-in.]), almost millipede-like arthropod with a tube-shaped body consisting of thirty-six anteroposteriorly short, ring-like segments, each containing a (probably) short leg (fig. 5.15; plate 12c). The head shield was large, rounded, and simple; the posterior

MERISTOSOMA

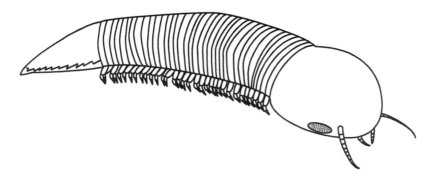

5.15. Reconstruction of the odd Spence Shale arthropod *Meristosoma.* Total length = ~15 cm.

shield tapered to a point and appears to have been somewhat serrated along the edges. *Meristosoma* may have been a scavenger or predator, but we are not sure, and it likely lived on the muddy bottom of the ocean, rather than in the sediment or in the water above the sea floor. It appears to be part of a clade quite separate from the trilobites and to be closer to *Waptia* and the crustaceans than its trilobite cousins.[23]

Grand Canyon

We visited the Lower Cambrian part of the Bright Angel Formation outside Las Vegas in the preceding chapter. In chapter 2 we visited the Bright Angel Formation deep in Grand Canyon. Most of the Bright Angel Formation, vertically through its section and along the majority of its exposure in the Grand Canyon, is Middle Cambrian in age, and the trilobite fossils found in it suggest that much of it is the same age as the outcrops of the Arrojos Formation in the Caborca area and of the Spence Shale in Idaho and Utah.

The Bright Angel Formation is exposed in two sections along 194 straight-line kilometers (120 mi.) of the canyon of the Colorado River, from the confluence with the Little Colorado in the east to the Grand Wash Cliffs in the west. It consists of a thick section of 182 m (597 ft.) of green shale, siltstone, and tan to green sandstone. Trace fossils are abundant, but body fossils are a little more rare. Still, plenty of fossil sites have been found in the Bright Angel Formation in the Grand Canyon, first by (of course) Charles Walcott and later, and perhaps even more so, by Edwin McKee, park naturalist for the Grand Canyon in the 1930s. Of the dozens of sites found by these two, few were quite as productive as one first worked by a man named Niles Cameron in 1911. This site was down below the South Rim and was also worked for several days by Walcott in 1915, while he was camping in the Grand Canyon, almost as a tourist, on his way back to Washington, DC, from Los Angeles. McKee worked the site and produced the largest collection in 1930. Trilobites were still in the ground. An in-place exhibit was installed in 1936, but it quickly got buried by natural erosion. I started trying to find the site again several years ago.

5.16. Specimens of the trilobite *Glossopleura* from around the western part of North America. (A) Type specimen *Glossopleura boccar* (USNM 62703) from the Stephen Formation at Mount Bosworth, British Columbia. (B) *Glossopleura stephenensis* (USNM 62702) from the Stephen Formation at Mount Stephen, British Columbia. (C) *Glossopleura mckeei* (USNM 62714) from the Bright Angel Formation of the Grand Canyon, Arizona. (D) *Glossopleura producta* (USNM 123356) from the Ophir Shale of the Oquirrh Range, Utah. (E) *Glossopleura similaris* (KUMIP 314054) from the Spence Shale at Antimony Canyon, Utah. All scale bars = 1 cm.

All courtesy of Smithsonian Institution, except (E), from University of Kansas collections.

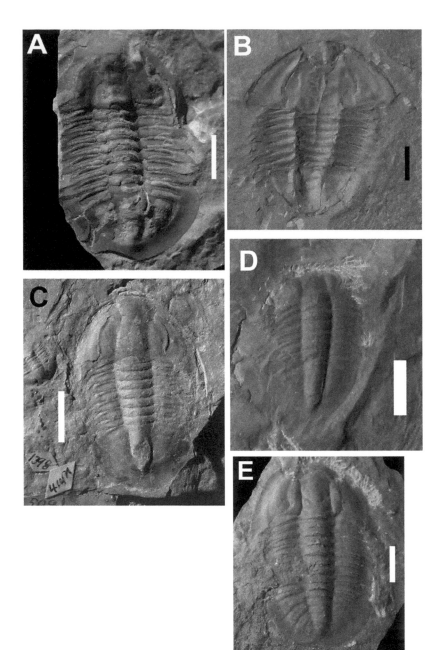

Thanks to photos McKee took, one of which happened to show a canyon skyline (and which his wife later donated to a library), I was able to relocate this site. It is one that has produced several hundred trilobite specimens preserved at the Grand Canyon and in the collections at the Smithsonian Institution. Among the fossils found at this site in the lower half of the Bright Angel Formation are hyoliths; eocrinoids; and the trilobites *Amecephalus piochensis*, *Amecephalus althea*, *Glossopleura mckeei*, and *Anoria tontoensis* (plate 3). The trilobites are in a green shale and many are well preserved; the ratio of articulated specimens is high. The *Glossopleura* and *Anoria* here are also fairly large trilobites, compared with some species. Also found in this deposit are brachiopods and traces left by walking and resting trilobites. The trilobite genera are many of the same ones we get at other sites of this age, but *Glossopleura* and *Anoria* seem to be particularly abundant here. *Glossopleura* is a common element of faunas from most formations of this age throughout Laurentia (fig. 5.16); thus, we call this the *Glossopleura* zone.

The Bright Angel Formation as a whole has yielded from the Grand Canyon at least 5 types of brachiopods, 2 types of eocrinoid echinoderms, 15 types of trilobites, and more than a dozen types of bivalved arthropods known as bradoriids.[24] Bradoriids were once thought to be members of the modern crustacean subclass Ostracoda, small, bivalved species that live in marine and freshwater worldwide. However, although we are not sure where bradoriids fit within the Arthropoda, it now is apparent that they were not actually ostracodes; it is possible they were stem group crustaceans. Most bradoriids were small—some very small—and their fossil record from Cambrian formations in China suggests that they were very abundant and served as a food source for larger animals of the time. Bradoriid arthropods of the Bright Angel and other formations probably lived near and on the bottom and possibly swam just over the sediment, stirring up sand or mud to access organic matter to eat. The bradoriids are just one group of bivalved arthropods that lived during the Cambrian; as we will see, this was a particularly common body design for arthropods at the time, although it appears to be a design arrived at independently by several groups. Well-preserved bradoriid specimens from China also indicate that, unlike many other bivalved arthropod species, which seem to have much of the body enclosed within clamshell-like valves, the two shells of bradoriids may have articulated but stayed open, serving as a two-part dorsal carapace over the animal. Specimens of the bradoriid *Kunmingella* from China also demonstrate that some species brooded their eggs, attaching up to 80 tiny eggs to their posterior appendages.[25]

The Cambrian Corps 4—Xingliang Zhang

Xingliang Zhang is a professor in the Department of Geology at Northwest University in one of the oldest cities in China, the ancient capital of Xi'an (which is also famous as the home of Qin Shi Huang's terracotta army and as the eastern end of the Silk Road). Dr. Zhang specializes in the Cambrian radiation and the origin of animals and has published on a wide variety of Cambrian animals from the Chengjiang biota and other deposits, animals including lobopods, trilobites, brachiopods, *Opabinia*, and a variety of non-trilobite arthropods such as bradoriids and *Sidneyia*. Originally from the northern part of Shaanxi Province in central China, Dr. Zhang was introduced to the Cambrian world by one of his academic tutors. "It was wonderful," he says. He points out that part of the appeal of the Cambrian is that the explosive appearance of animals "puzzles the public, and specialists too." Paleontologists and hobbyists share a fascination for the unanswered questions raised by the Cambrian biota. "There are more questions than answers in doing science," Dr. Zhang says. One of the important aspects of the Chengjiang biota in China is that it predates the Burgess Shale. "Chengjiang tells us that major animal body plans surely were established during the early Cambrian and that the divergence was abrupt," he says. And the discovery of vertebrates, of jawless fish, in the Early Cambrian Chengjiang fauna "really surprised me," he says. "It was unexpected." Dr. Zhang believes that one of the most important next steps in paleontology is to determine the timing and geographic origins of the ancestors of the Early Cambrian fauna.

As we saw in chapter 2, the Bright Angel Formation represents a very shallow marine setting in which influx of freshwater, spores of possible land plants, and tidal influences were common. The trilobites, hyoliths, and brachiopods that we find at sites like McKee's were apparently living in shallow water (plate 2), perhaps in large embayments on the shelf, not far from shore. Small islands may have been present in the area as well.

The Bright Angel Formation (along with the Tapeats Sandstone and Muav Limestone) is also exposed in the western Grand Canyon among the vegetative hostility of the cholla, ocotilla, agave, barrel cactus, and prickly pear of Peach Springs Canyon on the Hualapai Indian Reservation. This canyon is north of U.S. Highway 66 and runs down to the Colorado River. The heat at this site can be as unwelcoming as the tenacious plant life that calls the canyon home, yet the exposures of the Tonto Group here are worth the discomfort. As is true of much of the Bright Angel Formation, body fossils can be tough to find, but trace fossils are difficult to avoid. Traces such as *Rusophycus*, *Cruziana*, and *Diplichnites* are common, and trilobites such as *Glossopleura* have been found.[26]

HYOLITHIDA

Among the fossils from McKee's Bright Angel Formation site in the Grand Canyon is a hyolith (plate 3d; see also plate 12a). We have encountered sites with hyoliths all through chapter 4, and we will continue to see them in this and later chapters. Two genera are particularly common in formations we have visited, *Hyolithes* and *Haplophrentis*. The fossils are very distinctive in shape and are of a size easily noticeable while digging. They are a common but interesting element of many faunas.

The **Hyolithida** is an enigmatic group of fossils that, along with brachiopods and the ubiquitous trilobites, occur at many fossil sites in the Cambrian. Part of the reason for their abundance as fossils is that the animals possessed a nearly conical shell of calcium carbonate that preserved easily. This shell, the **conch**, is almost always flattened in fossil specimens, so that hyoliths are preserved as what appear to be very elongate triangles. Sometimes the fan-shaped **operculum** is preserved also; this is a sort of front hatch to the shell that covered the access to the living chamber of the animals' conch. The shape of the operculum suggests that in life the conch was convex ventrally and had two sloping sides dorsally; the conch tapered toward the back and probably curved very slightly upward. Most interesting in hyolith anatomy, and most debated, is the function of the thin, curved structures sometimes preserved as part of these fossils, the **helens**.[27] These structures, one on each side, extend from the anterior end of the conch near the operculum, and they look a little like antennae, although they are not. One reason that their function is so poorly understood is that they rarely preserve. Most often hyolith conches or opercula are preserved, sometimes the conch and operculum articulated, and only occasionally do you find conch, operculum, and helens together. These structures have been hypothesized to be stabilizers to keep the shell from rolling, supports for the operculum, and more recently, independent supports for respiratory or feeding structures for the animal. They seem to have been retractable and to have been able to be oriented by the animal when extended.

But exactly what kind of animal occupied hyolith shells is not clear. Most researchers classify the hyoliths as either primitive relatives of the molluscs (such as slugs, snails, and clams today) or as primitive molluscs themselves. Either way, a soft, perhaps clam-like animal is believed to have occupied the conch and to have extended and retracted the helens from behind the operculum. When the animal chose to venture partly outside the shell, it would flip the operculum up and move out as far as was necessary for the task at hand.

Hyoliths most likely were **sessile** (non-mobile) animals that lived on the sediment at the bottom of the ocean, and they seem to be common in shallow-shelf to distal-shelf settings. As far as we can tell, they do not appear to have moved around a lot or to have lived within the sediment. They may have been filter feeders, straining food such as microplankton and **detritus** from the water just above the seafloor.[28]

Cadiz Siding

As we saw previously, the Lower–Middle Cambrian contact is also preserved in the lower part of the Cadiz Formation in the Marble Mountains in southeast California, as olenellids have been found in the lower meters of the unit. This formation exposed in the area of Darton's casual exploration from the rail tracks contains rocks and trace fossils indicating shallow marine deposition in tidally influenced areas with carbonate shoals appearing now and then. The lower part of the Cadiz Formation is not always very fossiliferous, but it has produced earliest Middle Cambrian fossils. More productive, however, is a site high in the Cadiz Formation, just below the Bonanza King Formation. Here, on a barren slope above the sand-gravel fan valleys of the Mojave Desert are found numerous incomplete and complete trilobite specimens such as *Sonoraspis* (fig. 5.17), *Glossopleura*, and *Amecephalus*, along with forms also known from the

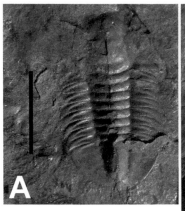

5.17. Dolichometopid trilobites from the Middle Cambrian Cadiz Formation of the Marble Mountains, California. (A) Paratype specimen of *Sonoraspis californica* (LACMIP 10786). (B) Holotype specimen of *Sonoraspis californica* (LACMIP 10785). (C) Referred specimen of *Sonoraspis californica*. All scale bars = 1 cm.

(A) and (B), Natural History Museum of LosAngeles County; (C), Museum of Western Colorado.

5.18. *Glossopleura* zone trilobites from the Middle Cambrian Chisholm Formation at Half Moon Mine near Pioche, Nevada. (A) Complete specimen of the ptychopariid *Amecephalus piochensis*. (B) Nearly complete specimen of the corynexochid *Zacanthoides typicalis*. (C) Nearly complete ptychopariid *Amecephalus althea*. (D) Articulated dolichometopid corynexochid *Glossopleura boccar*, missing the cephalon. (E) Complete specimen of the ptychopariid *Amecephalus packi* (USNM 90171). All scale bars = 1 cm.

All Museum of Western Colorado specimens, except (E) courtesy of Smithsonian Institution; (A) and (D) collected and donated by Andrew R. C. Milner.

Caborca area, *Mexicella* and *Mexicaspis*. All are found in dark brown to gray shale. Also found in the Middle Cambrian part of the Cadiz Formation, but somewhat lower than this site, have been chancellorids and free-floating, colonial relatives of jellyfish and corals called chondrophorine hydrozoans. The abundance of trilobites at this site in the Cadiz Formation is very similar to the collections seen in Mexico and at Spence Gulch and in the Grand Canyon.[29]

Half Moon Mine

Stratigraphically above the Pioche Formation, near Pioche, Nevada, is the Chisholm Formation, another unit that has yielded *Glossopleura*, *Amecephalus*, *Zacanthoides*, and other trilobites of the same age as those sites mentioned above (fig. 5.18). Many of these trilobites are found at the Half Moon Mine just outside Pioche, and there are at least three species of *Amecephalus* at this site. Plates and partial specimens of edrioasteroid echinoderms (see chapter 7) are also relatively common in the Chisholm Formation in this area. In fact, more than 900 specimens of the edrioasteroid *Totiglobus* have been found in the formation here.[30]

Abduction Territory

In a remote part of south-central Nevada lies the Groom Range, just off the Extraterrestrial Highway (U.S. 375) and east of the town of Rachel, home to the Little A'le'inn and epicenter of UFO sightings in the western United States. The Groom Range lies north of Area 51, the United States Air Force base and experimental aircraft test center that is so high security the rumor is they are cleared to shoot first and worry who you are later. Area 51 is also rumored in some circles to hide alien spacecraft and bodies—think *Independence Day*. Thanks to Area 51, at least two mountain ranges of Cambrian outcrops are off limits, so the Groom Range is all we have for Cambrian fossil data between the Pioche area and western Nevada in Esmeralda County.

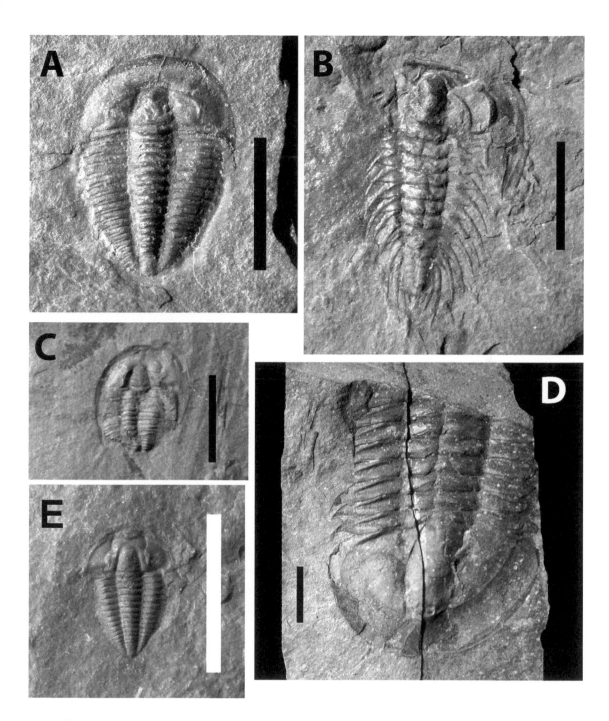

What the Groom Range holds on some of its upper, northwest-facing slopes is trilobite deposits in the Rachel Limestone that are the distal, deepwater equivalents of the Bright Angel Formation, Cadiz Formation, Chisholm Formation, and Spence Shale–*Glossopleura* zone faunas that were some of the farthest offshore of any we have encountered so far.[31] And the faunas are a bit different. Among the most abundant trilobites here are *Ogygopsis* and *Elrathina* (fig. 5.19), the former a genus believed to favor deep, **dysoxic** waters of the distal shelf. Out here, in the deep water

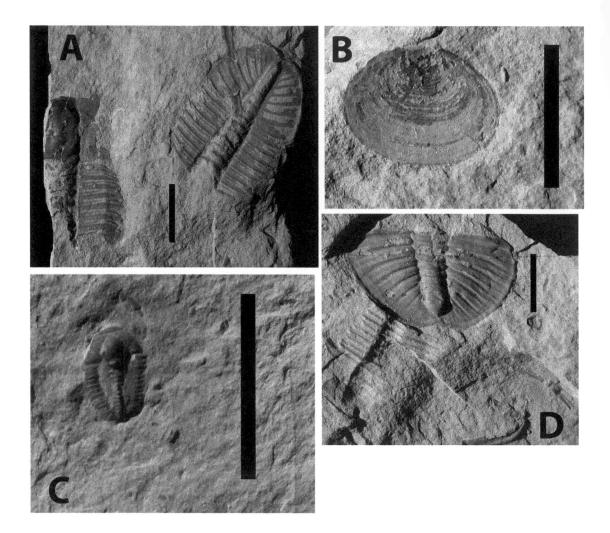

5.19. *Glossopleura* zone fossils from the Middle Cambrian Rachel Limestone in the Groom Range, Nevada. (A) Typical partial, articulated specimens of the corynexochid trilobite *Ogygopsis* sp. (B) Brachiopod shell. (C) Articulated specimen of the ptychopariid trilobite *Elrathina*?. (D) Pygidium of *Ogygopsis* with articulated and disarticulated thoracic segments of other individuals nearby. All scale bars = 1 cm.

All Museum of Western Colorado specimens.

on the edge of the shelf, the low-oxygen environment seems to have been literally crawling with these forms, in a setting and with a fauna both very different from what was occurring closer to shore at the same time. Brachiopods are also found in this fauna.

Farther Afield

There are numerous other field areas with trilobites and other fossils from the *Glossopleura* zone. The lower reaches of the Burgess Shale in British Columbia fall within the *Glossopleura* zone, and forms such as *Glossopleura* and *Amecephalus* have also been found in Washington state, the Ophir Shale of Utah (fig. 5.16d), and in the Wolsey Shale in Montana (fig. 5.20).[32]

Trace Fossils

Among the most important fossils found in marine rocks are the tracks and trails that animals unwittingly left in the sediment to be fossilized and preserved as fossils for 500+ million years. These traces are often

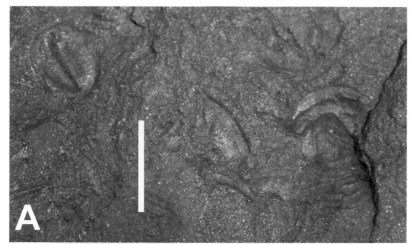

5.20. Trilobites from the Middle Cambrian Wolsey Shale of Montana. (A) Fragments, free-cheeks, and cranidia of ptychopariids and dolichometopids. (B) Nearly complete dolichometopid, possibly *Glossopleura*. All scale bars = 1 cm.

University of California Museum of Paleontology specimens.

preserved in stone if they are filled in quickly enough. One of the most significant aspects of trace fossils is that although we can sometimes identify what type of animal likely made the track, we can almost always determine a little bit about how it lived. As the saying goes, body fossils preserve the morphology of an animal, and trace fossils record behavior. A logical implication of this is that the same species of animal can leave potentially many kinds of traces, and because of that the trace fossil record of the Cambrian records a whole separate set of data about the animals that lived in the oceans of the time.

Some of the types of traces common in Cambrian rocks include *Skolithos*, a vertical tube trace fossil that is so common in some sandstone units that they are called "piperock." These traces probably represent the dwellings of small wormlike animals that lived in shallow marine, intertidal, and beach settings. These animals may have used tentacles or other structures to filter food from the seawater.[33] The trilobite (or similar arthropod) trails *Isopodichnus* and *Cruziana*, the former of which is made

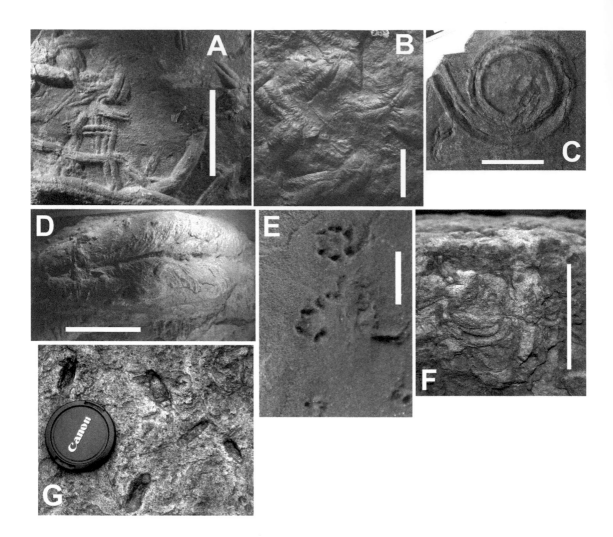

5.21. Trace fossils of the Bright Angel Formation of the Grand Canyon. (A) *Isopodichnus* (MNA N3726). (B) *Cruziana* and *Rusophycus* (GRCA 2126). (C) Indeterminate spiral trace (GRCA 2074). (D) Large *Rusophycus* (MNA N3757). (E) *Pholetichnus* (MNA 3863). (F) *Diplocraterion* in lateral view, field specimen. (G) *Diplocraterion,* bottom end of trace exposed on a bedding plane and viewed from above, field specimen. (A)–(D) and (F), Scale bars = 5 cm. (E) scale bar = 1 cm.

by young or small individuals, are also common. *Cruziana* is made by larger, adult trilobite-like arthropods that were crawling along the surface; both *Cruziana* and *Isopodichnus* were made by animals plowing through the upper layers of sediment, scratching the sediment and making long trails with their legs. Trilobite resting traces are common in Cambrian rocks, too, and are represented by the ichnogenus *Rusophycus*. Tubes lying parallel to bedding (usually on the bottom surface) are particularly abundant in many formations; these were made probably by wormlike organisms and are generally identified as *Planolites* and *Palaeophycus*. These traces were made by organisms burrowing through the sediment and feeding on the organics in the sand or mud (deposit feeding).[34] *Diplocraterion* consists of vertical, U-shaped burrow tubes in sandy sediment; these can be particularly abundant in some rocks of the Cambrian. They occur quite commonly in the Bright Angel Formation and Tapeats Sandstone of the Grand Canyon. Where it occurs in abundance in Cambrian rocks, *Climactichnites* looks a little like dune-buggy tire tracks crisscrossing the sandy bedding planes. These trails are believed to have been made by large molluscs.[35] A beautiful and massive slab covered with these traces is displayed at the Smithsonian Institution's National Museum of Natural History in Washington, DC. A possible indeterminate arthropod trace is known in *Pholetichnus*, a peculiar ichnofossil consisting of circular sets of six small round impressions, as if someone had stuck a pencil tip in the sediment. Another indeterminate arthropod trace is *Angulichnus*, which seems to show a body or tail drag down the center.[36] For illustrations of some of these trace fossils, see figure 5.21.

The trace fossil record of the Cambrian is interesting in that it shows a pattern of increasing density and complexity through the period, a pattern that probably resulted from the infilling of marine niches as the overall biota of the oceans expanded (we will discuss this more in chapter 9). Animals went from living on the bottom only during the Proterozoic to taking off swimming into the water column in the Cambrian; at the same time and perhaps partly as a result of both increasing predation and discovery of available food sources in the organics in the sediment, animals also took off burrowing down into the sand and mud for protection and an easy meal. Burrowing activity seems to have greatly increased at the beginning of the Cambrian, so it is appropriate that the start of the period is defined by the trace fossil *Treptichnus pedum*. Although the total amount of burrowing and trace fossil activity increases throughout the Cambrian, it is interesting that the depth to which animals burrowed still stayed within a few centimeters or inches of the surface. Cambrian burrowing was generally shallow. Deep burrowing would come later.[37]

6.1. Members and relationships of the Burgess Shale Formation, Cathedral Formation, and Stephen Formation. *Glossopleura* Zone range for Burgess Shale Formation shown on left; same range for Cathedral Formation on right.

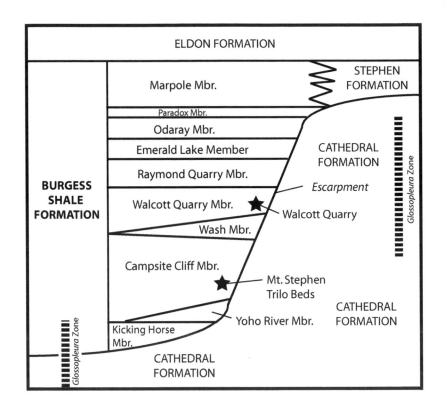

Magical Mystery Tour: The Biological Psychedelia of the Burgess Shale

OUR NEXT DESTINATION NEEDS LITTLE INTRODUCTION. QUITE THE opposite of the desert localities of the Great Basin, the localities we must visit next are mostly high on mountainsides. The mountains and glaciers of Banff, Jasper, and Yoho National Parks in Alberta and British Columbia, Canada, are some of the most beautiful on the continent, and their fossil treasures are as impressive and important. In 1885 the Canadian Pacific Railway (CPR) completed a line across the Rockies, a route that ran over a mountain pass from Lake Louise and off to the west into British Columbia. Along this route, just on the downhill side west of the pass along the Kicking Horse River, was a small encampment of rail workers that developed into the small town of Field, British Columbia. Named in honor of one Cyrus Field, an investor who – despite his namesake town – never did invest in the CPR project, the town of Field has remained small and incredibly scenic. In September 1886, soon after the railway was completed, a Geological Survey of Canada geologist named Richard McConnell climbed to the slopes of Mount Stephen above town and collected fossils from what would eventually become known as the Burgess Shale. These fossils proved to be particularly important because they were the first to be found in what has turned out to be one of the most important Cambrian formations in the world; McConnell's finds set off a series of small events that led to one of the most important discoveries in the history of paleontology, and within little more than a century whole books (and not a small number of them) have been dedicated just to this one formation's fossil record.[1]

Burgess Shale Formation Geology

The Burgess Shale Formation was named in 1998, although the name "Burgess Shale" had first been applied in 1911 by Charles Walcott. For years the Burgess Shale was considered part of the Stephen Formation, but it has now been separated out. The Burgess Shale Formation consists of dark gray to almost black calcareous, laminated mudstone that is largely argillaceous, meaning that it is slightly metamorphosed but not so far as to become pure slate. The formation is about 270 m (886 ft.) thick, and the shale splits in very hard, slate-like plates. There are silty interbeds in the mudstone that are millimeters to rarely centimeters thick. The formation is divided into 10 members (fig. 6.1); 2 key quarries we will visit are in the Campsite Cliff Shale Member and the Walcott Quarry Shale Member. These are just 2 among more than 35 Burgess Shale localities in the region with fossils. The Campsite Cliff Shale Member is low in the formation, and the Walcott Quarry Shale Member about midway up; the latter is about 23 m (74 ft.) thick and rests directly on the brecciated Wash Limestone Member.

More so than most areas we have encountered, the formations in the area around Field have complex spatial relationships with each other (fig. 6.1). As we see from figure 6.1, the Burgess Shale Formation represents the deeper, outer detrital belt deposit distal to a carbonate escarpment formed by the Cathedral Formation. The limestone of the Cathedral Formation is generally older, indicating that it was there as a submarine topographic feature during the time the Burgess Shale was being deposited; this can be seen in the figure by comparing the levels at which the *Glossopleura* trilobite zone occurs, respectively, within the Cathedral to the east of the escarpment and to the west of it in the Cathedral and lower Burgess. The Stephen Formation, which was largely contemporaneous with the Burgess, forms the thin shale deposits on top of the Cathedral Escarpment. The Stephen represents the shallow-water, carbonate belt deposits that existed along the edge of the "cliff" of the escarpment during the time the Burgess Shale was being deposited in deeper water at the escarpment's base.[2] The escarpment can be traced southeast to northwest through this region, and many fossil sites appear at the base of the ancient feature. There are also several that are within the relatively shallow water deposits of the Stephen Formation.

The fossil deposits at the foot of the Cathedral Escarpment were probably on a smooth, gentle slope of mud, within a few yards of where the older limestone outcropping rose steeply toward the shallow water on top of the escarpment. Fossils at sites along the base of the escarpment were likely buried quickly by repeated episodes of submarine debris flows, underwater "landslides" formed of mud and silt slurries that flowed down the slopes of the sea bottom. Sometimes these flows started up on the top of the escarpment and came down the cliff, fanning out at the base and burying animals living along the gentle slope at the foot of the cliff. Other flows might start from slumps of mud at the foot of the cliff itself. In either case, the quick burial of the species living in the area helped preserve their remains for fossilization. Animals appear to have been buried at various angles to bedding. Although it once was thought that the animals were trapped in the debris flows and came down from upslope with the sediment, it now seems that the Burgess Shale animals were buried in shorter slides close to where they lived at the foot of their underwater cliff. And recent work has also suggested that some of the areas where they lived may have been close to underwater brine seeps, localized areas where sulfide-, ammonia-, and methane-rich brine works its way up into the water column through the sediment. Experiments at modern brine seeps suggest that organismal soft tissue may not degrade in this setting for years – up to a decade.

These seeps are unlike hydrothermal vents of mid-ocean ridges in that they are cold – usually only slightly warmer than the surrounding water – rather than extremely hot. But they also may support chemosynthetically based communities at great depth. Such is the case at a parallel for the Burgess Shale brine seeps off the Gulf Coast of Florida at what is called the Florida Escarpment; here the contact between deepwater sediments at the base of the escarpment and the steep, eroded limestone of the escarpment itself provides the origin of the brine that works its way up through the mud to the water column above and supports the community living near the seep on the bottom of the Gulf of Mexico. Similarly, the Burgess Shale biota may have, in some cases, clustered close to the seeps, and these communities may have been based on chemosynthesis of the producer-level organisms.

Another phenomenon that may have assisted preservation of the fossils at these sites was occasional low levels of oxygen in the water and sediments, something that would have slowed decomposition of the remains of the species. It is uncertain, however, to what degree oxygen may have been depleted in these settings.[3]

The water along the base of the Cathedral Escarpment was probably somewhere around 100 m (328 ft.) deep, the bottom end of the **euphotic zone,** the uppermost, well-lit level of the ocean in which nearly all the sea's photosynthesis takes place. Below this depth is a level at which the ocean is dimly lit (**disphotic zone**) down to about 1000 m (3280 ft.), the depth below which no light penetrates at all. So the Burgess Shale deposits along the Cathedral Escarpment placed their resident organisms about as deep as they could get and still have a shot at having a food web based on photosynthesis. The depth is important, however, because being situated at the boundary between the euphotic and disphotic zones, and in some cases near brine seeps, suggests that the communities individually may have been either photosynthetically or chemosynthetically based.

No doubt Richard McConnell had been directed to the fossil beds above Field by locals who had noticed the remains up on the hillside sometime before. A rail workman had found "stone bugs" up that hill, and McConnell certainly knew that probably meant fossils. But McConnell ended up making the first collection from a deposit almost ludicrously rich in trilobite fossils and strange, detached shrimp-like fossils that were soon after named *Anomalocaris*. Surrounding Field are the large mountains and glaciated valleys of Yoho National Park, and most of these features are composed of Cambrian-age rocks from the western (or at the time, northern) offshore shelf of ancient Laurentia. On the talus slopes high above town and below the western side of the mass of Mount Stephen are the Trilobite Beds (fig. 6.2). The hike up to the exposure is a short but thigh-burning affair straight up a now-forested pile of cobbles from the ice age known as a moraine. What formed as a pile of debris deposited by a glacier during the time of mammoths now provides a direct route up to a slope of shale slabs containing fossils of marine animals that lived in the area during a period 5,000 times further back in Earth history. The

Mount Stephen Trilobite Beds

6.2. The slopes of Mount Stephen with abundant fossiliferous slabs of the Mount Stephen Trilobite Beds.

6.3. Trilobites of the Mount Stephen Trilobite Beds. (A) *Ogygopsis klotzi.* (B) *Olenoides serratus.* (C) *Bathyuriscus rotundatus.* (A) and (B), Scale in centimeters. (C) Scale bar = 1cm.

Field photos.

talus slope of the Trilobite Beds is perched like a lookout spot far above the town of Field and within view of Fossil Ridge to the north, where more finds were made later. The view off to the west deeper into British Columbia is impressive, too, and it is an easy spot to sit all day, surrounded by fossils and great scenery.

The Mount Stephen Trilobite Beds are within what was originally called the "*Ogygopsis* Shale," and they are located low in the Burgess Shale Formation in the Campsite Cliff Shale Member. Abundantly preserved on the shale slabs of that slope are fossils of large trilobites such as *Ogygopsis* (fig. 6.3a) and *Olenoides* (fig. 6.3b; plate 14a), along with smaller forms such as *Bathyuriscus* and *Elrathina* (fig. 6.3c). The biota also includes other forms such as algae, chancellorids, eocrinoids, brachiopods, crustaceans, unmineralized arthropods such as *Naraoia* and *Marrella*, sponges, and the spiny *Wiwaxia*.[4]

The trilobite *Olenoides serratus* is a very well known form, thanks to its soft-body preservation in the Burgess Shale. *Olenoides* lived on the seafloor and was a predator on small, soft prey. It may have hunted in part by digging through the top levels of the substrate, and perhaps pulled worms and other benthic animals out of the sediment. Food was probably crushed with the coxae of its limbs and passed forward to the mouth, a feeding mode probably typical of a lot of trilobites.[5]

Terror of the Seas? *Anomalocaris*

Anomalocaris was named as a shrimp several years after McConnell's report on his findings at Field, and it was based on specimens he found at the Mount Stephen Trilobite Beds; these fossils are still very abundant at the site (fig. 6.4).[6] But it was not a shrimp; it was the feeding appendage of a larger animal. What was thought to be an open-center jellyfish from the Burgess Shale was in fact the mouth. Other parts of this animal were initially identified as worms, sponges, or sea cucumbers. Eventually,

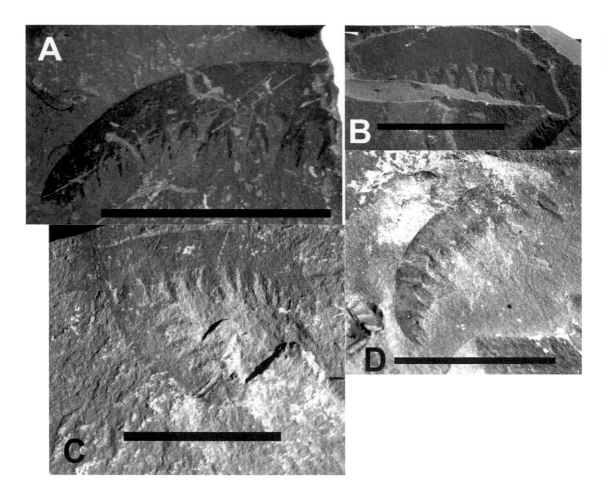

6.4. Feeding appendages of *Anomalocaris* from the Mount Stephen Trilobite Beds. (A)–(B), (D) Isolated appendages. (C) Two appendages on one slab. All scale bars = 5 cm.

Field photos.

it was discovered what the full animal looked like, thanks to a nearly complete specimen, and it is obviously one of the largest animals at the time and clearly a predator (fig. 6.5). *Anomalocaris* had long feeding appendages at the front of the head, a ventral mouth, dorsolaterally placed compound eyes, and a body with a fantail and individual flap-like lateral folds, which flapped in waves to make the animal move, much like modern cuttlefish. New finds from Australia have shown that the compound eyes of *Anomalocaris* each possessed about 16,000 individual lenses on the visual surface![7] This suggests a visual acuity rivaling the best of modern arthropods. So much for being primitive.[8]

Trilobites with massive but healed injuries to their exoskeletons suggested that *Anomalocaris* made meals of these iconic denizens of the Cambrian seas (sometimes unsuccessfully). But maybe not all anomalocaridids were feeding on the poor trilobites. Although *Anomalocaris* is popularly thought of as *the* tyrant predator of the Burgess Shale, there are actually several genera and up to seven species of anomalocaridids known from the formation, most of them contemporaneous. Certainly among such a diverse group there had to have been those that made a living in some way other than chomping on trilobites. J. W. Hagadorn, along with others previously (and since), has suggested that our view of

ANOMALOCARIS

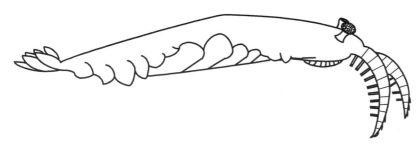

6.5. Three genera of anomalocaridids from the Burgess Shale Formation. Adult lengths ~30–50 cm.

HURDIA

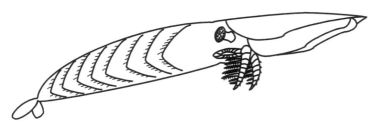

LAGGANIA

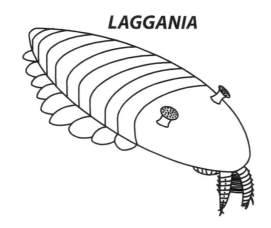

Anomalocaris itself, as a terror of the seas ripping apart trilobites, may be off the mark. We do find healed bite injuries in trilobites but we have no direct evidence that they were made by *Anomalocaris*. In fact, *Anomalocaris* appears to have had rather soft mouth plates. How might apparently soft, unmineralized mouth plates that never are preserved with breaks or wear have managed to injure the calcified exoskeletons of trilobites? Might the mouth plates have been better adapted to eating soft prey? Would at least some anomalocaridids have instead used the feeding appendages as filter scoops for combing soft prey from the sediment, prey they then "sucked" up with a soft mouth? There is such diversity of form preserved in the feeding appendages just in the Burgess Shale that it seems likely that these predators had a range of ecologies and feeding modes. Trilobites may have sometimes been on the menu of *Anomalocaris* but probably not as often as we originally thought, and perhaps more as scavenged carcasses.

The greatest diversity in anomalocaridids is seen at the Mount Stephen Trilobite Beds, where there are seven species known, each unique at least in feeding appendage morphology, and in some cases we know how different the bodies (for example, the heads) could be. The feeding appendage of *Anomalocaris canadensis* was long with ventral spikes on the podomeres, but the new form *Caryosyntrips serratus* had feeding appendages that were straighter, more distally tapered, and had very short, simple spikes on each podomere. The genus *Hurdia*, for which there are two morphologies in the feeding appendages, and probably *Peytoia* (*Laggania*) *nathorsti* had shorter feeding appendages in which each podomere had elongate extensions, each with many spikes (fig. 6.5). *Amplectobelua stephenensis* had short, stout feeding appendages with spikes on the tip and only small spikes on the ventral side of each podomere, but most interesting, it had a basal podomere with an especially sharp, elongate spike that gave the entire appendage the appearance of a can opener. Perhaps this was more a predator of hard-shelled animals than was *Anomalocaris*? All of these anomalocaridids are found at the Mount Stephen Trilobite Beds. These appendages fall into two groups: flexible, long, and with short spikes (*Anomalocaris*, *Caryosyntrips*, and *Amplectobelua*); and shorter, less flexible, and with elongate spikes with short secondary spikes (*Hurdia* and *Peytoia*). These five forms may have split their feeding strategies into two broad categories. The first group with flexible feeding appendages may have been more active hunters, tracking down and grabbing prey and manipulating them into the mouth. This doesn't necessarily mean hard-shelled prey, just active predation on mobile forms. The second group (*Hurdia* and *Peytoia*) may have used the more elaborate feeding appendages as effective nets or filters to trap or sift abundant, small prey items in a kind of predatory grazing approach with less active hunting. To use a modern marine mammal analogy, the feeding appendages of the hunting group might be thought of as the toothed jaws of an *Orca* (killer whale), grasping prey and working it toward the gullet, and those of the *Hurdia/Peytoia* group, in comparison, could be thought of as the baleen of baleen whales, filtering multiple individual animal prey items out of the environment. A recent study of two co-occurring species of *Anomalocaris* from the Emu Bay Shale of Australia suggests that even species within this genus demonstrated a range of feeding strategies. Some species within the genus seem to have preyed on soft-bodied organisms while others may have concentrated on animals with somewhat harder exoskeletons.[9]

The most important point in all of this is that we have moved far beyond the "terror of the seas" view of a world subject to the erratic hunger pangs of a voracious, lone *Anomalocaris* species—the Cambrian was a time with a full array of anomalocaridids—some predatory on soft-bodied organisms, some possibly on those with harder shells, and still others just unselectively, and almost passively, catching what they could. The ecosystem was far more complex, and more interesting, than we ever thought, even 20 years ago.[10]

6.6. Charles D. Walcott at his desk at the Smithsonian. Smithsonian Institution photo archives.

Courtesy of Smithsonian Institution.

In early August 2009, more than one hundred researchers from around the world gathered in Banff, Alberta, Canada, for three days of presentations on the latest findings in studies on Cambrian paleontology, and specifically on those cases of exceptional preservation known as Burgess Shale–type deposits. This was just weeks before the 100th anniversary of C. D. Walcott's discovery of the quarry that bears his name, the site north of Field that has produced so many thousands of specimens of soft-body preservation, the site that started it all in terms of showing us that there was a lot more in the Cambrian than previously met the trilobite-, brachiopod-, and hyolith-saturated eye. The Walcott Quarry on Fossil Ridge in Yoho National Park in British Columbia is now one of many very productive fossil localities for Burgess Shale–type preservation of less commonly preserved elements of the Cambrian biota. But as the first quarry of this type to be found, and one out of which so many type specimens have been collected, it has a special status still, more than 100 years after Walcott's serendipitous find.

Richard McConnell reported on his fossil find above Field, what became the Trilobite Beds, in 1887, and one of the readers of this report was the man who was everywhere in the Cambrian, Charles D. Walcott (fig. 6.6). It took Walcott 20 years, but he eventually did get up to Field, shortly after taking over as Secretary of the Smithsonian Institution in Washington, D.C. Walcott visited the area and did field work for several weeks in 1907, and although he did see material from the Burgess Shale around Mount Stephen, most of his work that summer was in older rocks of the Lower Cambrian and the older part of the Middle Cambrian. In 1908 Walcott worked mostly around what is now Glacier National Park in Montana, and he did not work in the Banff-Field area. But near the middle of July 1909 he arrived in Banff again and worked for some time

Walcott's Bonanza: Fossil Ridge

based out of Lake Louise. It wasn't until the end of the 1909 season, in late August, that he found the site that eventually was named after him.

The Walcott Quarry is on a north–south trending line between Mount Field and Wapta Mountain known informally as Fossil Ridge. The approach to the Walcott Quarry passes through several miles of forest, and the trail through these trees winds through ferns and moss ever upward, under the shadow of Wapta, and into the brush at tree line. Mountain lakes and glaciers are in view much of the time and one may encounter sun, rain, snow, marmots, or traces of grizzlies seemingly at any point and any time of year. The quarry itself is huge, having been worked by several serious expeditions since 1909, when it was first excavated for just a few days by Walcott.

Charles Doolittle Walcott was born in 1850 in New York and had only 10 years of formal education, but by age 13 he was developing an interest in geology due to chance encounters with trilobites and other fossils near his home, as well as contact with a local geologist. By age 17 he had resolved to study rocks and fossils of the Cambrian in North America, that period having only been named by Sedgwick some 32 years before. A vast, almost entirely unexplored system of rocks lay awaiting young Walcott across the continent, and during his career Walcott would visit much of it.[11] From age 21 to 26 Walcott worked as a farmhand in Trenton Falls, New York, being sure to arrange with his employer plenty of free time so Walcott could prospect for fossils and study geology. He amassed a significant collection of fossils in this time, and he eventually became an assistant to the state geologist of New York, James Hall. By 1879, Walcott had a job with the newly created United States Geological Survey (USGS), a post he was offered thanks to a recommendation from Hall. He stayed with the USGS for the next 28 years.[12]

Walcott was a longtime researcher on the Cambrian and trilobites, and he had worked throughout the northern Rockies and other regions studying the Cambrian rocks of the western side of the continent. In fact, Walcott originally named many of the species and rock formations that we have already discussed in this book. His role only becomes larger now.

Stories as to how the Walcott Quarry was actually discovered in 1909 vary. Exactly when he noticed it is also unclear. The first mention of what probably were specimens from what became the Walcott Quarry came on August 31, but he seems to have been aware of the site a day or two before this. The common story is that Walcott moved a rock out off the trail to Burgess Pass so that the path would be clear for the rest of his crew coming up with the pack train a couple days later; fossils were found in the rock the next day when he broke it open. This may not be correct, however, and according to Walcott's daughter, Helen, it may have been Walcott's wife who first found the fossils on the trail while her husband was prospecting.[13] In fact, Walcott never wrote of exactly how the site was found. Notes in his field notebook from late August 1909 are somewhat cursory. He writes that on August 31 he and part of his small crew (composed mostly of family members, a packer, and camp cook

6.7. Charles D. Walcott (left) digging at the Walcott Quarry in the Burgess Shale. Photograph undated. Smithsonian Institution photo archives.

Courtesy of Smithsonian Institution.

Arthur Brown) were up in the area collecting and found several types of animals, which he then sketches. These small sketches are recognizable as the arthropod *Marrella*, a crustacean, and possibly a *Naraoia*. He remarks on little else. However, his notebook was fairly small and was predated, so he had little room for expansive monologues on everything that happened on a given day.[14]

Before this work, however, the crew was camped to the north at beautiful Takakkaw Falls. Previously, on August 28, Walcott had ridden up to Burgess Pass on a day trip and noted the presence of the Stephen Formation. The next day he set up camp at Burgess Pass and seems to have collected along Fossil Ridge on August 30, the day the rest of his team came up from the falls.[15] The first mention of significant fossils, and drawings of such, as noted above, was on August 31. On September 1 Walcott drew several sponges that the crew found at the quarry, for the first time noting next to one of the sponge sketches "in situ," meaning it was found in place. Up to this point, it appears possible the specimens were found loose on the talus slope. Loose or in place, Walcott was finding what he recognized were some important fossils. Within a few days, though, he had moved back to working around Mount Stephen.[16]

The truth is that no one is quite sure anymore exactly how the Walcott Quarry was found. It appears from Walcott's notes that fossils were found at least by August 30, several hours before the main group arrived, and that well-preserved fossils appeared on the 31st. The quarry level and in-place material was found at least by September 1 but possibly earlier. The impression one gets is that the significance of the site came to the group only gradually as more and better material appeared and as they closed in on where it was eroding from. It was less the "Eureka!" moment

6.8. The Walcott Quarry on Fossil Ridge in the Burgess Shale Formation. (A) The quarry looking southeast in 2009. Note height of quarry wall; paleontologists for scale in lower right. (B) The layers of the Phyllopod Bed, with shovel for scale. (C) Close-up of the quarry argillite shale. Vertical marks are chisel scores.

of a single stupendous discovery and more the revelation of great things coming slowly from seemingly nicely preserved but otherwise unremarkable initial finds. But it really doesn't matter that much *how* the Walcott Quarry was found because the significance of the find is undisputed and, regardless of who actually found the first fossil there, the Walcott family and their crews packed tens of thousands of fossils out of the site over the following years. Walcott came back to collect from the quarry every summer from 1910 to 1913 and again in 1917. He also made shorter visits to collect in the talus in 1919, 1921, and 1924 (fig. 6.7), by which time he was 74 years old.

This quarry and immediate area along Fossil Ridge was visited in later years by other crews. Percy Raymond and a crew from Harvard University excavated Walcott's pit and opened one of their own (the Raymond Quarry) just upslope in 1930. The Geological Survey of Canada worked the Walcott Quarry in 1966 and 1967 with a crew of 17 that included Bill

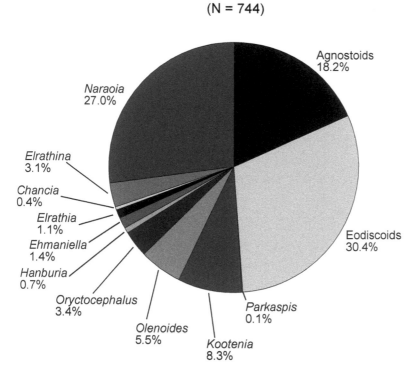

Burgess Shale
(N = 744)

Agnostoids
18.2%

Naraoia
27.0%

Elrathina
3.1%

Chancia
0.4%

Elrathia
1.1%

Ehmaniella
1.4%

Hanburia
0.7%

Oryctocephalus
3.4%

Olenoides
5.5%

Kootenia
8.3%

Parkaspis
0.1%

Eodiscoids
30.4%

Fritz, Jim Aitken, and Harry Whittington. In the years since Walcott had worked there, the quarry had filled with a fair amount of debris, which had to be cleared before crews could work. Still, in 1967 alone, the blasting crew removed about 180 cubic meters of overburden rock in the early days of excavation that summer. Results of this work showed that there was more to be found in Walcott's bonanza. The Royal Ontario Museum worked the same site and numerous others all around Mount Stephen and Fossil Ridge from 1975 to 2000 and significantly increased the total collection.[17]

However it was found, the Walcott Quarry along Fossil Ridge is one of the most incredible fossil deposits in the world. The Walcott Quarry (fig. 6.8a; plate 13) is in the Walcott Quarry Shale Member just above Wash Member, both of the Burgess Shale (fig. 6.1). Phyllopod Bed is about 2.3 m (7.5 ft.) thick and is near the middle part of the quarry (fig. 6.8b), consisting of dark argillite shale.[18] Preserved in these rocks is one of the most spectacular fossil biotas of any age anywhere in the world. Let us take a look at some of these fossils now. It is one of the most complete single-site pictures we can get of the Cambrian ocean world.

Elsewhere the Stars: The Trilobites of the Walcott Quarry

There is a respectable diversity of very well preserved trilobites in the Burgess Shale, and in any other formation they might be in the center spotlight, but with such a diversity of strange soft-bodied organisms preserved in this almost unique window on the Cambrian, the poor trilobites take a bit of a back seat to the rest of the fauna. To make up for that, we will take a look at the trilobites of the Burgess Shale Formation first. The formation at the Walcott Quarry contains at least a dozen species of trilobites, including the corynexochids *Olenoides* (plate 17; plate 14b), *Oryctocephalus*, and *Kootenia*; the ptychopariids *Elrathina*, *Elrathia*, and *Ehmaniella*; the eodiscoid *Pagetia*; and the agnostid *Peronopsis*.[19] *Ogygopsis*, a deepwater form so common at the Mount Stephen Trilobite Beds just a few miles south and a little downsection, is absent. Most abundant among the trilobites at the Walcott Quarry are the agnostids, eodiscoids, and the unmineralized trilobite *Naraoia*, followed by *Kootenia* and *Olenoides* (fig. 6.9). There are unmineralized trilobite body parts preserved in some specimens from the Burgess Shale Formation; we know what the antennae and biramous limbs of *Olenoides* looked like, for example, and in quite good detail (plate 17; fig. 4.14).

Seaweeds: Algae and Similar Forms

In previous chapters (such as chapter 3) we discussed cyanobacteria as the microorganisms that probably built the structures known as stromatolites. Cyanobacteria show up in the rock record of the Burgess Shale as the wispy *Marpolia*, a filamentous fossil that can cover bedding surfaces with what look like tufts of shed cat hair. Closer to plants, green algae are found in the Burgess Shale in the form of the genera *Yuknessia* and *Margaretia*, forms that are known from a number of other Cambrian formations also. Red algae are represented in the Burgess Shale by *Dalyia* and *Waputikia*. The algae are comparatively uncommon in the Burgess Shale, being known from up to just a few dozen specimens each; *Marpolia*, however, is known from hundreds of specimens cataloged just at the Smithsonian Institution's National Museum of Natural History in Washington, D.C.

Sponges

The Burgess Shale Formation has produced numerous species of sponges, the simplest metazoans known, a group of filter-feeding animals so diverse even in today's oceans. We may use their remains as art tools or to wash our cars, but we must respect a group with such a deep history and great diversity even in the Cambrian. There are about 40 species of sponges just in the Burgess Shale! Yes, that is a lot by any measure in Cambrian rocks. A few of the genera present include *Hamptonia*, *Vauxia* (fig. 6.10; plate 15a), *Diagoniella*, *Choia*, and *Hazelia*. We will discuss sponges more in chapter 7, but for now we can simply appreciate the fact that the ocean floor in the area of the Walcott Quarry (and other sites) was likely peppered with a number of individuals of many species of humble sponges at any given time.[20]

Ctenophora (Comb Jellies)

Although they superficially resemble jellyfish, these animals are in a separate phylum from the Cnidaria (see chapter 8). The comb jellies (phylum **Ctenophora**) are known from well-preserved specimens from China and the Burgess Shale in British Columbia. Modern forms are generally planktonic, although some species are epibenthic. They are biradially symmetrical, and it is unclear if they are **diploblastic** (most likely) or possibly **triploblastic.** Like jellyfish, they have a netlike nervous system with no central system, and no excretory organs or respiratory system, but they have a more developed digestive system. Unlike cnidarians, they are only medusoid (i.e., jellyfish-like) with no polyp stage of the life cycle, and they are never colonial. Many species are **hermaphroditic** and can self-fertilize. Ctenophores have eight ciliated comb rows at some stage of their life history and often possess two elongate tentacles. Locomotion is achieved through the movement of the cilia in each comb row, which all work together.

There are about 100 modern species of ctenophores known, and most are pelagic, with species living at depths ranging from the surface down to more than 2500 m (8200 ft.). All known ctenophores are pelagic

6.11. The indeterminate animal *Amiskwia* from the Burgess Shale (USNM 198670). Scale bar = 1 cm.

Courtesy of Smithsonian Institution.

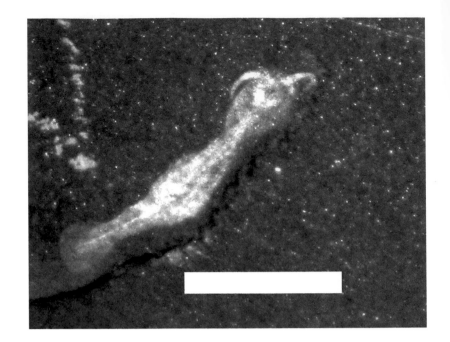

predators of small zooplankton. Prey items are caught by adhesive material attached to spiral filaments that burst out of what are called lasso cells, upon contact with the item.

The Burgess Shale has two genera of ctenophores, *Fasciculus* and *Ctenorhabdotus*, that appear to be ctenophores, based on the structure of what appear to be comb rows. Unlike modern species, however, there are many comb rows in these species, numbering about 80 and 24, respectively. The Lower Cambrian ctenophore *Maotianoascus* from the Chengjiang fauna of China has 16 comb rows, although an apparent embryonic specimen of this genus has 8 rows.[21]

Ribbon Worms? *Amiskwia*

Little *Amiskwia* is known from a few specimens in the Burgess Shale (fig. 6.11), and its structure still leaves us guessing as to what it actually is. Its body is elongate and short for its cross-sectional diameter; the head appears to have a pair of anterior tentacles, the mid-body appears to have two lateral flaps or fins, and the posterior has a flattened, fanlike tail structure. The animal does not appear to be segmented. *Amiskwia* has been compared to the Nemertea (ribbon worms), but is different in a number of ways from that group.[22] It does appear to have been an active swimmer, however.

Segmented Worms: The Annelida

Two well-known species in the Burgess Shale belong to a familiar group of modern worms, the bristly forms *Canadia* (fig. 6.12) and *Burgessochaeta* (fig. 6.13). These small worms have tufts of hairlike **chaetae** along their sides and belong to the Annelida, or segmented worms; within the annelids these two species are within the Polychaeta. Paleontologist Martin

6.12. The polychaete worm *Canadia* from the Burgess Shale (USNM 198724). Scale bar = 1 cm.

Courtesy of Smithsonian Institution.

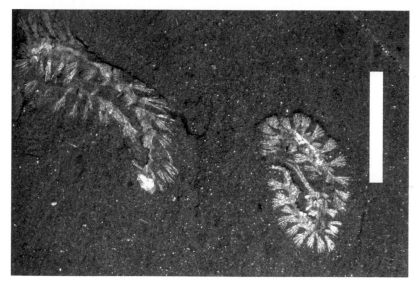

6.13. Two specimens of the polychaete worm *Burgesso-chaeta* from the Burgess Shale (USNM 198705 and 198698). Scale bar = 1 cm.

Courtesy of Smithsonian Institution.

Brasier once referred to *Canadia* as a "worm in drag,"[23] as these two guys do look surprisingly like miniature feather boas. But they are ancient members (some of the first, in fact) of a rather taxonomically and ecologically diverse group of modern worms.

The Annelida today includes more than 16,000 species, among them the earthworms and the parasitic (but sometimes useful) leeches. The phylum is divided into two classes, the Polychaeta (of which *Canadia* and *Burgessochaeta* were members), which includes modern clam-, sand-, and tubeworms, and the Clitellata, which includes earthworms and leeches. Annelids are characteristically segmented, bilaterally symmetrical, and possess the bristle-like chaetae mentioned above (these are often reduced or lost in clitellates). Polychaetes have well developed chaetae (as seen in the Cambrian forms and the numerous modern marine species), chitinous jaws, and indirect development with free-swimming larvae. The clitellates reduce or lose the chaetae and are hermaphroditic with direct development into near–adult form juveniles (no larval stage).

Morphologically annelids are characterized by the segments, or annuli; each segment often has unjointed appendages (not really legs) called parapodia, each with a pair of chaetal bundles (dorsal and ventral) on each lateral surface. The chaetae are derived from the epidermis. The parapodia and chaetae are well developed in polychaetes but much reduced in clitellates. The segmentation consists of repeated structures on the inside of the body as well; it is not just an external feature, and in this sense annelids are similar to the arthropods. The body is covered with a thin, soft cuticle, but annelids do not molt. Like most bilaterians, but unlike chordates, the main circulatory vessel is dorsal and the nerve cord (or cords) is ventral. The head is not enlarged but contains a well-developed nervous system and, in some species, small single-lens eyes. The nervous system includes a dorsal cerebral ganglion or "brain," which in some species is developed enough to be divided into a fore-, mid-, and hindbrain. The ventral nerve cord appears to have started out as paired, lateral cords that ran down the body in a ladder-like fashion, but this seems to have gradually closed into a single, medial nerve cord with lengthened lateral nerves in each segment in derived groups. The species with ladder-like nervous system arrangements have lateral connections and segmental ganglia in each segment. Some forms possess tentacles and other similar appendages on the head around the mouth.

The gut can be straight or sometimes coiled, and it is often regionalized, with a rectum, mid-gut (intestine), and sometimes a crop and gizzard. The digestive system passes undigested material after nutrients have been absorbed and also sediment (in deposit feeders). But paired **nephridia** ("kidneys") get rid of waste from each body segment cavity and regulate fluids in the body, a setup similar to arthropods and onychophorans.

The circulatory system is closed and often is without a heart or specific pumping organs. There are a variety of forms and locations for gas exchange in different species; in some, vessels in the parapodia serve a gill function, while in others the body surface conducts gas exchange. In still other species, anterior, tentacle-like branchial vessels act as gills, or trunk filaments on each segment serve the gill function.

Locomotion is achieved largely by muscular movement of the body segments (extension and contraction), assisted by hydrostatic expansion of these structures. Various forms can crawl well or swim somewhat; others can burrow but swim better, and still others can "walk" on the parapodia. The forms of locomotion among annelids are quite variable. The Clitellata lack parapodia so they tend to use hydrostatic pressure and chaetal movement to crawl, as in earthworms.

Modern annelids range from as small as 1 mm (0.04 in.) up to nearly 3 m (10 ft.) in length and live in fresh- and marine waters as well as terrestrial environments. Reflecting their range of habitat, annelids run a wide range of feeding ecologies also. They demonstrate great morphological

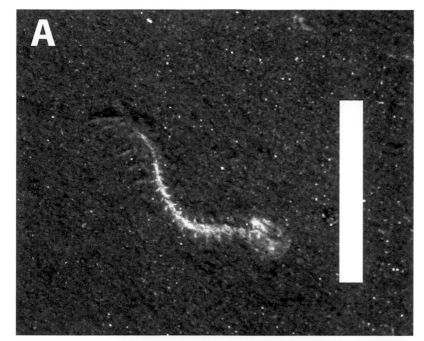

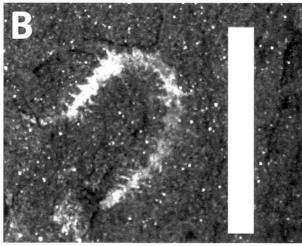

6.14. Specimens of the tiny polychaete worm *Peronochaeta* from the Burgess Shale. (A) Small specimen (USNM 198716). (B) Slightly larger specimen (USNM 83936a). Both scale bars = 1 cm.

Courtesy of Smithsonian Institution.

disparity along with their high species diversity. Some polychaetes and clitellates are predators on small invertebrates, with both active hunters and ambush predators known; some of these can capture prey with an evertable proboscis tipped with hook-like jaws (some possess poison glands, too). Many of the active hunters are epibenthic forms, and many of the ambush predators are infaunal forms, living in burrows. There are also among the annelids deposit-feeding species that ingest sediment for the contained organic matter and excrete the sand or mud. Others separate out the organics from the sediment without ingesting it. Some of these deposit-feeding annelids are mostly stationary and live in burrows, pulling water and organic material in through the sediment, and others burrow through and ingest the sediment. Other annelids are tube-dwelling suspension feeders and filter seawater for organic material. A number of species are symbiotic with or parasitic on other animals (sometimes including other annelids); many of these species are little modified from non-symbiotic or non-parasitic annelids except in being much smaller in body size. These symbiotic and parasitic annelids may live in, on, or around other animals. Among the Clitellata, earthworms are familiar terrestrial deposit feeders; other forms are detritivores, and of course there are the bloodsucking parasitic leeches (although some leech species are in fact predator-scavengers).

The reproductive range of annelids is, once again, vast and reflective of the amazing morphological and ecological diversity of the group. In polychaetes, reproduction can be asexual and complete individuals can be grown from fragmented adults,[24] or reproduction can be sexual with internal or external fertilization and with free-swimming larvae. Among Clitellata, many are hermaphroditic and engage in mating in which both individuals are inseminated and separate, acting essentially as impregnated females. (The name Clitellata comes from a Latin word for "saddle" and refers to an enlarged, ring-like sleeve on many species; this is a familiar structure on earthworms. The **clitellum** serves a number of functions, but one of them is to help pairs stick together during mating.) The fertilized, hermaphroditic individuals then deposit "cocoons" in the environment, each containing one to several embryos; the cocoon is a tough casing produced by the clitellum. Embryos are direct developing and hatch as juveniles in adult-like form; there is no larval stage.

Cambrian fossil annelids from North America include five genera of polychaetes from the Burgess Shale in British Columbia. These genera are *Burgessochaeta* (fig. 6.13), *Canadia* (fig. 6.12), *Insolicorypha*, tiny *Peronochaeta* (fig. 6.14), and *Stephenoscolex*. Each of these had between 18 and 34 segments, and on each segment were laterally projecting pairs of bristle tufts (the chaetae), which made these animals look a little like fuzzy caterpillars, although of course they are not. *Canadia* may have been a scavenger or carnivore and *Burgessochaeta* and *Peronochaeta* selective deposit feeders; *Insolicorypha* may have been a free-swimming (pelagic) animal.[25] The mode of life of *Stephenoscolex* is not well understood. Possible annelids have been found at the Chengjiang locality in China, but they are poorly known.[26]

The Goblet Stalk: *Dinomischus*

Dinomischus isolatus was described in 1977 from three specimens of a metazoan with a cup-shaped calyx; a long, narrow stem; and short, flat, petal-like blades (called bracts) surrounding the upper surface of the calyx. The calyx contains a U-shaped gut with a sac-like stomach, and there are twenty bracts on the calyx. The stem has a slightly expanded attachment point at the bottom end opposite the calyx. The mouth and anus are both on the upper-facing surface of the calyx, and the gut and stomach are held in place within the hollow calyx by fibers that suspend the structures. *Dinomischus* probably lived attached to the bottom sediment by its stem with tiny, hairlike cilia on the bracts capturing microscopic food particles in the water and transporting them to the mouth on the calyx's upper surface.

The closest modern relatives of *Dinomischus* may be members of the phylum Entoprocta, a group of about 150 species (most marine) of small, stalked animals with a cup-shaped body (calyx), a U-shaped gut with the mouth and anus both on the upper surface, and with a crown of tentacles around the calyx. Entoprocts can be either solitary or live colonially, and many are found attached to other benthic marine animals such as sponges; they are known to live from within the intertidal zone down to depths of about 500 m (1640 ft.). Despite an overall similarity to *Dinomischus*, including having a stalk and having the mouth and anus on the same surface surrounded by tentacles or bracts, entoprocts differ in being significantly smaller (most are 5 mm long or less) and in having actual tentacles rather than the flat, blade-shaped bracts on the calyx; also, few entoprocts have the gut and stomach suspended within the body. Although *Dinomischus* was probably not a Cambrian entoproct, it may well have been related to them;[27] perhaps it was a paleontological great uncle of the first entoprocts that shared a similar body plan and ecology and differed in a few details.

A Second Goblet

Recent work in the Campsite Cliff Shale Member of the Burgess Shale Formation at Mount Stephen has produced more than 1000 specimens of a new goblet-shaped, stalked animal up to 20 cm (8 in.) tall, a form named *Siphusauctum*. This was a benthic filter feeder that sometimes lived in groups like a field of flowers (but it was an animal, of course). The animal consisted of a holdfast, stalk, and hexaradial calyx with comb-like filtering rows inside an encasing sheath. Water was pumped in and across the filter rows through holes near the bottom of the external sheath. The mouth and anus were arranged somewhat as in *Dinomischus*, and it is unclear to what phylum *Siphusauctum* belonged. It is not necessarily a weird or unusual animal, it is just that it has some characters in common with a range of phyla including entoprocts and tunicates; most likely it is a stem group taxon, which would explain its less-than-obvious phylogenetic identification at this point. The most we can say at this point is that it is a bilaterian.[28]

Arrow Worms

Oesia disjuncta was described by C. D. Walcott in 1911 based on a handful of specimens from the Walcott Quarry. At the time, *Oesia* was identified as an annelid worm; it was later proposed to be a tunicate, or sea squirt. It then was recently identified as an arrow worm, a member of the phylum Chaetognatha. And then it was even more recently identified as a possible hemichordate but definitely not a chaetognath.[29] Notice a trend? Flux. It is more the rule than the exception in science. (In this case, the taxonomic shuffle of *Oesia* is mostly due to imperfect preservation and a dearth of specimens.)

Modern arrow worms (chaetognaths) are elongate animals up to about 10 cm (4 in.) in length, and they are mostly marine predators living in the water column as plankton, although some live on the bottom. They are long, bilaterally symmetrical, and have sets of elongate lateral fins and a fan-shaped tail fin. Some have eyes and tentacles, and most have relatively large and sharp sets of grasping spines in the mouth. Most species seem to feed on small planktonic crustaceans. Chaetognaths are generally small, about 2 mm or a little bigger and are second only to copepod arthropods in abundance among the zooplankton. They serve as both consumers and prey food for other animals.

Although *Oesia* itself may not be a Cambrian chaetognath and instead may be a hemichordate (a relative of echinoderms), paleontologist Simon Conway Morris identified a specimen from Walcott's Burgess Shale collection that may in fact belong to a chaetognath. The specimen is an isolated, tiny pair of grasping spines from the mouth of one individual. Although *Oesia* is yet another difficult animal to classify, its study has resulted in the discovery of another form new to the Burgess Shale Formation, a Cambrian arrow worm.

Walking Hallucinations: The Lobopodia

Perhaps no animal of the Burgess Shale is more iconic than *Hallucigenia*. This small, spiked creature gained notoriety in 1989 with popularization in Stephen J. Gould's *Wonderful Life*, at a time when it was still reconstructed in what turned out to be an upside-down position.[30] What have since been found to have been dorsal spikes were originally thought to be stilt-like legs, with the animal walking around on the pairs of appendages in such a bizarre mode of life that it evoked a sense of hallucinatory dreams. When more specimens indicated that we needed to flip the animal over it became only a little less Carrollesque, but it did then appear to belong to a more familiar group. In fact, *Hallucigenia* (fig. 6.15) and a number of related Cambrian forms now appear to be members of a group known as the Lobopodia, an extinct conglomerate of stem taxa related to today's Onychophora, or velvet worms, and tardigrades ("water bears"), and more distantly to arthropods.[31]

Modern onychophorans are all terrestrial animals living in humid environments, and there are more than 100 species known, although some estimates suggest we have only identified half of the possible diversity of this group today. Many species are colorful. Onychophorans are segmented and are basal to stem group arthropods. The modern species range from about 5 to 50 mm in length, and their bodies are covered with a soft, thin, flexible cuticle of chitin over the epidermis that is not separated into articulating plates, is not calcified, and possesses no sclerites. Like arthropods but unlike annelids onychophorans molt this cuticle regularly. The "velvet" part of their common name comes from their external texture, which is caused by the body being covered with small tubercules bearing tiny scales.

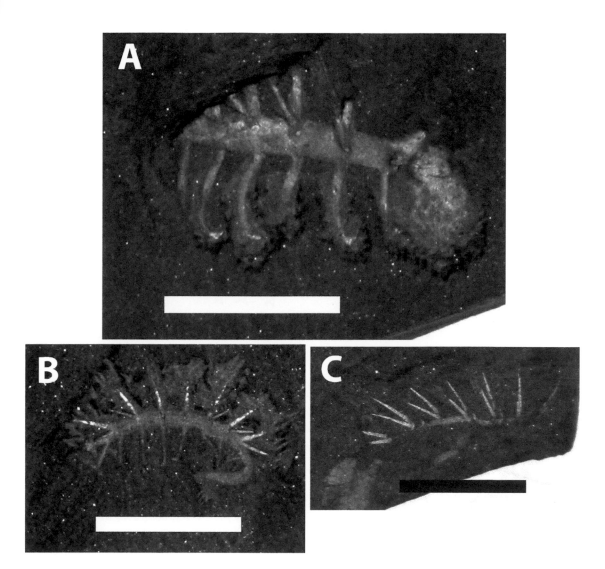

The legs of onychophorans (known as "lobopods") are short, un-jointed, and often wrinkly textured, with ends possessing three to six pads and a paired, terminal claw. The animals walk on the legs through a combination of hydrostatic extension and contraction of the body to move the leg bases, along with muscular movement of the legs themselves.

The head is rarely expanded and possesses paired antennae, sharp, curved spiky jaws surrounded by hook-like lips, and oral **palillae** lateral to the mouth. In some modern forms the oral palillae contain the exit tubes of slime glands that can shoot adhesive strings to ensnare prey or entangle would-be predators. The head also has small, single-lens eyes below the cuticle near the bases of the antennae. Eyes are even known from some Cambrian fossil lobopodians; they appear to have been tiny, but should have been effective.

The gut of onychophorans is straight and large, occupying much of the body cavity. The heart is dorsally placed and is tubular in shape, open

6.15. Specimens of the lobopodian *Hallucigenia* from the Burgess Shale. (A) Specimen showing anterior half of animal including head, first six legs, and first five pairs of spikes (USNM 83935). (B) Nearly complete specimen (USNM 198658). (C) Specimen consisting mostly of spike pairs and part of body trunk (USNM 53918). All scale bars = 1 cm.

Courtesy of Smithsonian Institution.

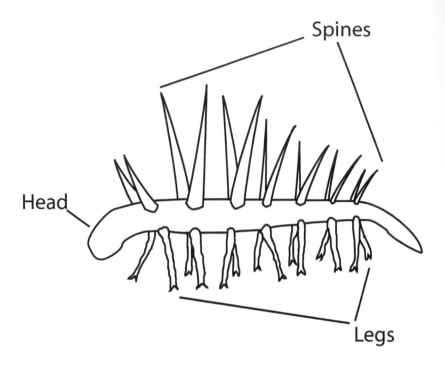

6.16. Reconstruction of *Hallucigenia sparsa*.

Spines

Head

Legs

at both ends and running an open, arthropod-like circulatory system. Nerve cords are ventral and lateral. The males are often smaller than females, sometimes with fewer pairs of legs. Onychophorans are sexually reproducing and may be either oviparous or viviparous depending on the species, and there is no larval stage; rather, the animals are direct developing and are hatched or born as juveniles with adult-form organs and the full number of segments.[32] In viviparous forms, gestation can last up to 15 months, and females may have several embryos in various stages of development in their uterus at one time.

Nearly all modern onychophorans are predators on small invertebrates such as insects, snails, and worms; fossil marine forms probably fed on small benthic invertebrates. The jaws are used to grasp and chop up prey, and the secretions from salivary glands can be used to partly digest prey before the soft material is sucked up into the mouth.

An onychophoran invasion of the land may have occurred during the Ordovician period about 488–443 million years ago,[33] but definitely by the Carboniferous, as terrestrial lobopodian fossils are found in those rocks. In addition to a diversity of Cambrian fossil forms, there are Tertiary fossils as well.

Cambrian lobopodians include *Hallucigenia* (figs. 6.15 and 6.16), and also *Aysheaia* from the Burgess Shale, plus *Antennacanthopodia*, *Microdictyon*, *Cardiodictyon*, and a slew of others from China. There are several strange lobopodians from the Burgess Shale that also are still undescribed. All of these forms were marine species as far as we know. Recent studies have identified a new species of *Hallucigenia* in China (*H. hongmeia*), found along with the original species also known from British Columbia (*H. sparsa*). *Hallucigenia hongmeia* has short, stumpy

6.17. Cap-shaped shells of the monoplacophoran-like mollusc *Scenella* from the Burgess Shale. Largest specimen is just under 1 cm across.

dorsal spines quite unlike the long, spiky needles known from *H. sparsa*. Study of the microstructure of the dorsal spines of complete *Hallucigenia sparsa* specimens from the Burgess Shale has recently permitted the identification of these elements isolated in several other Cambrian rock units, including the Mount Cap and Forteau formations in Canada and other units in Siberia and Europe. Hallucigeniids indeed appear to have been widespread in their biogeographic distribution[34]

Also known from China is *Diania*, the "walking cactus," a recently described species with spiny armor and (contentiously) jointed legs, suggesting a close relationship to arthropods. Indeed, *Diania* looks like a toy animal made of twisted pipe cleaner by an anonymous Cambrian balloon artist. Another study of *Diania* has found the legs to be unjointed and this lobopodian to be relatively basal within its group.[35] Lobopodians are still related to the origins of arthropods, but *Diania* may not be as close to illustrating that origin as it first appeared. In any case, in recent years it has become clear that *Hallucigenia* was just the beginning; the diversity of lobopodians in the Cambrian was impressive.

Among Burgess Shale lobopodians, *Hallucigenia* in North America is known only from the Burgess Shale (although indeterminate hallucigeniids are known from other formations, as mentioned above), whereas *Aysheaia* is known from the Burgess Shale and Wheeler Formation. Many lobopodians were probably mobile, benthic predators; because of a taphonomic association found in the fossil record, it has been suggested that *Aysheaia* may have specialized in feeding on sponges.[36]

Molluscs

Very common on some beds in the Burgess Shale are shells of *Scenella*, a small mollusc preserved as an often nearly flattened, cap-like shell (fig. 6.17).[37] Members of the phylum Mollusca today are divided into seven classes and include forms as morphologically diverse as clams, snails, chitons, and squids. The group's fossil record stretches back well into the Early Cambrian, but the first molluscs probably appeared late in the Proterozoic.

Molluscs are characterized by having a thick **mantle** composed of epidermis and cuticle; sometimes a calcium carbonate shell secreted by the mantle; a fleshy, muscular foot; a three-chambered heart; relatively complex kidneys; a tooth structure called a **radula;** and by being unsegmented (mostly) and bilaterally symmetrical. Molluscs live in a wide variety of marine, freshwater, and terrestrial environments. Most molluscs are herbivorous grazers, predators, or suspension (filter) feeders.

Among the classes of Mollusca are the Aplacophora, which consists of wormlike, bottom-dwelling marine molluscs that lack shells and rather secrete calcareous spicules embedded in the mantle. The Monoplacophora (or class Tergomya) are small molluscs that have a single, cap-shaped shell on the dorsal surface and move about on a flat, muscular foot. They generally lack eyes and sometimes have tiny tentacles around the mouth. Monoplacophorans are bilaterally symmetrical with pairs of muscles, gills, and nephridia along the body; this serial repetition of body parts has led to speculation that monoplacophorans are similar to ancestral molluscs and that mollusc origins may lie among the segmented animals. Most monoplacophorans are extinct and lived during the early part of the Paleozoic era, but a living form was discovered in 1952, and there are now six genera of modern monoplacophorans known. The Polyplacophora are flat, elongate molluscs that possess eight dorsal shell plates and comprise today's chitons. These forms are marine and live from the depths of the ocean into the intertidal zone. The Gastropoda includes marine, freshwater, and terrestrial snails and slugs; they have a secondary asymmetry (formed by a process known as torsion), coiled shells (lost in slugs), eyes (often reduced or lost), and a well-developed radula. Some forms have tentacles or a proboscis. Gastropods today are characterized by torsion, a twisting of the body forward along the right side that brings the back of the animal over the head; this occurs in the larval stage and may be secondarily lost in adults in some species. The reason that torsion evolved in the first place is not known, but it may have postdated the origin of shells in gastropods, so very early gastropods likely did not do this. The Bivalvia includes clams, oysters, and mussels and is characterized by having paired shells (called valves) with a plane of symmetry between them and with muscles and shell articulations to hold them together.[38] Most bivalves have lost their eyes and reduced or lost the radula. Bivalves live in freshwater or marine environments and often are suspension feeders. The Scaphopoda are tube-shelled, infaunal marine molluscs; they lack eyes and a heart but have a radula and proboscis. The Cephalopoda includes today's octopuses, squid, and cuttlefish and is covered in chapter 8.

Monoplacophorans and aplacophorans appear to be the most primitive among the modern groups of molluscs, and these groups have often been suggested to be closely allied, either as a grade of basal molluscs or as a clade that is the sister group to other molluscs.[39] Thus, monoplacophorans may be similar to what we might imagine as ancestral molluscs. However, one recent analysis that differs substantially in its results from most others seems to indicate that the Monoplacophora and Polyplacophora form a clade that together is closer to some bivalves and all gastropods than it is to other molluscs.[40] This would imply that monoplacophorans are relatively derived and are not necessarily close in form to ancestral molluscs. This remains to be supported further, however, and a majority of studies still indicate that monoplacophorans are probably basal among molluscs.

The molluscan body is covered by a non-chitinous cuticle over the epidermis, both of which overlie the muscles. The body also includes a mantle: a thick, sheet-like organ that forms the dorsal body wall and forms a cavity above the organs that houses the gills. The epidermal part of the mantle also secretes a calcareous shell in most species, and the shell can be either single, paired, or plated, or of course can be lost, as in slugs. The foot is a flat, muscular structure that secretes mucus and moves over it by way of muscle contractions and movement of cilia. Circulation in molluscs is open and driven by a dorsal, three-chambered heart; gills are generally well developed except in the Aplacophora. The nervous system is ventral and includes a pair of lateral nerve cords with anterior cerebral ganglia, and is developed to varying degrees in different classes. There are no segmental ganglia, however. Eyes occur in some species, including cup eyes in abalone, for example, and single-lens eyes in some gastropods and bivalves. The radula is a strip possessing many recurved chitinous teeth arranged in rows. The radula is shaped differently and is employed differently depending on the diet of the animal. The gut is often regionalized and includes a stomach and coiled intestine, and the kidneys are well developed. Most molluscs have one pair of kidneys (as do we, of course), but nautiloid cephalopods have two pairs; monoplacophorans have three, six, or seven pairs; and aplacophorans have none. Reproduction is sexual with internal or external fertilization and direct or indirect development, again depending on the species.

Among classes of molluscs that were around during the Cambrian, monoplacophorans today are mostly herbivorous grazers; gastropods may be predators, grazers, or suspension feeders; and bivalves are generally suspension feeders. Most species move around by muscle contractions that move in waves down the foot, but the foot is compressed laterally (not flattened) in bivalves and is used to anchor or burrow.

Cambrian molluscs include monoplacophorans, bivalves, probable gastropods, and (much later in the period) cephalopods. The shell of *Scenella* is similar to those of monoplacophorans, and it may be an early member of that class of molluscs (fig. 6.17). These fossils are common in the collections from the Burgess Shale, with more than 1000 specimens known just from Walcott's collection at the Smithsonian Institution. Although they occur in a number of formations in North America, usually in varying abundance, molluscs are particularly abundant in Early Cambrian rocks in Siberia. Worldwide the mollusc record is fairly diverse in the Cambrian (though nowhere near as diverse as today). Cambrian bivalves are represented by a number of genera of tiny fossils (often less than 1 mm across) that look, appropriately, somewhat like miniature clamshells. Other Cambrian molluscs include the gastropod-like forms *Strepsodiscus*, *Scaevogyra*, and *Aldanella* and the somewhat cone-shaped *Cambrioconus* and *Ilsanella*, the latter of which has a shell that looks like a ribbed ski cap. Overall, Cambrian molluscan taxa remained small (usually less than 15 mm for shelled forms) for most of the period, and then the upper size range shot up in the Late Cambrian (to about 100 mm), although there were still plenty of small taxa. The size increase may in part have resulted from the mollusc invasion of shallow, high-energy environments.[41]

Odontogriphus is another form found in the Burgess Shale and was named in 1976. It has been found to possess the characteristically molluscan structure of a radula, a toothed feeding structure deep in the mouth. This genus is a long, strap-like, soft-bodied form with a small, round, ventral mouth and a large mollusc-like foot. Preserved fossils of it are elongate and ovoid in shape, looking a little like a Cambrian shoe insert. The specimens are often preserved along with a cyanobacterium, so it is believed *Odontogriphus* may have fed on that algae-like material that grew on the ocean bottom. *Odontogriphus* was about 48 mm (2 in.) long, on average, although it apparently grew up to 125 mm (5 in.) long, and it appears to have been from a series of molluscs of the stem group that were far more primitive than the bivalves, gastropods, and monoplacophorans. This position and the fact that apparently more derived molluscs occur quite early in the Cambrian both suggest that similar ancestors of *Odontogriphus* must date back nearly to the beginning of the period and probably back into the late Proterozoic.[42]

Wiwaxia appears to be yet another stem group mollusc and looks just a little like a hellish bear claw breakfast pastry. The small, oval body is strongly convex and is covered with leaf-shaped, striated, hollow sclerites, all pointing posteriorly.[43] Each dorsolateral edge of *Wiwaxia*, however, is lined with about seven long, spine-like sclerites that must have served a defensive function (fig. 6.18). The ventral surface was devoid of sclerites and may have had a muscular foot used for locomotion. *Wiwaxia* appears to have had a radula associated with its mouth but is unlike other molluscs in having sclerites. This animal was described initially in 1899 based on one spine found on Mount Stephen, but it later showed up in much greater numbers at Walcott's quarry on Fossil Ridge. It probably was a slug-like animal living on the seafloor, moving on its foot by muscular contractions and feeding on algae and other organics on the bottom (plate 19). Specimens of *Wiwaxia* have been found recently in shallow-water deposits of Middle Cambrian age in the Czech Republic, and these occurrences, along with others from Siberia, China, and Australia,

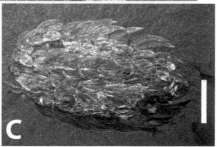

6.18. Specimens of the stem mollusc *Wiwaxia* from the Burgess Shale. (A) Specimen in slightly dorsolateral view showing body sclerites and longer defensive spines (USNM 198745). (B) Specimen in dorsal view showing length of defensive spines (USNM 198669). (C) Specimen in dorsal view showing body sclerites (USNM 83938). All scale bars = 1 cm.

Courtesy of Smithsonian Institution.

reinforce an association of *Wiwaxia* with tropical to subtropical climate zones during the Cambrian. *Wiwaxia* was once thought to be related to annelid worms, but it now appears to have been a stem group mollusc.[44]

Orthrozanclus makes *Wiwaxia* look soft and fuzzy. As a recently named, small, stem mollusc from the Burgess Shale, *Orthrozanclus* possesses similar sclerites along the body and seems to have had a bare ventral surface with a muscular foot, but the 15–20 spines along each side of the body are curved and much longer and more numerous than those in *Wiwaxia*. *Orthrozanclus* had a fringe of outward-pointing sclerites, the long spines, and a dorsal surface covered with tiny sclerites and an anterior shell; it was only about 6–10 mm (0.25–0.5 in.) long, and like *Wiwaxia* it probably moved along the ocean floor feeding on organics. This genus helped clarify the relationships of *Wiwaxia* and its relatives to *Halkieria* and its kin; these two groups and *Orthrozanclus* are all related stem molluscs, now called the Halwaxiida. Another recent study also supported the position of *Odontogriphus* as a basal-most mollusc, with the halwaxiids *Wiwaxia* and *Orthrozanclus* as slightly closer stem group molluscs. Other research has suggested that halkieriids at least are within the crown group of molluscs, perhaps related to chitons. As with many

6.19. Specimens of *Opabinia* from the Burgess Shale. (A) Specimen in side view showing proboscis, eyes, and body segments (USNM 57683). (B)–(C) Specimens showing body segments and proboscis (USNM 155600 and 155598). (D) Specimen showing the eyes and distal end of proboscis particularly well (USNM 57684). All scale bars = 1 cm.

Courtesy of Smithsonian Institution.

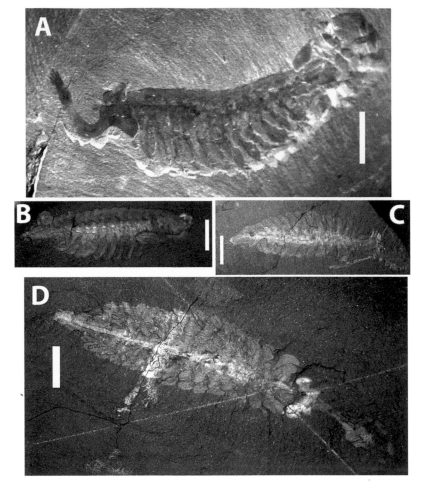

other disciplines in science, things can go back and forth between several camps on some topics, and as we have seen this is just one of many in Cambrian paleontology.[45]

Five Eyes and a Proboscis: *Opabinia*

Perhaps no animal from the Burgess Shale has been the subject of as much controversy or has been as maligned as little *Opabinia regalis*. Named by Walcott in 1912 in the series *Smithsonian Miscellaneous Collections*, poor *Opabinia* was literally laughed at when it made its more public debut in a life-restoration slide in a talk given to the Palaeontological Association by paleontologist Harry Whittington in 1972.[46] Segmented and possessing five eyes and a long proboscis, *Opabinia* is odd looking, and it has gone on to be as contentiously reinterpreted (still) in its morphology as any animal from the Burgess Shale.

Opabinia is a segmented, bilaterally symmetrical animal that most agree (surprisingly, given the disagreement about its structure) belongs among a stem group somewhere between onychophorans and basal arthropods. The front end of the animal had a long proboscis that in life was probably hollow and fluid filled; the anterior end of the proboscis had

Plate 1. View of the Grand Canyon from the South Rim looking northeast, showing Cambrian Tonto Group along with overlying formations. Cambrian formations are labeled on Sumner Butte east of Phantom Ranch.

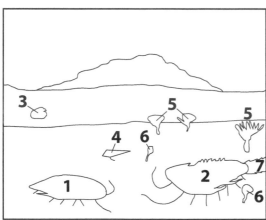

Plate 2. Life in the shallows of a Middle Cambrian expansive epicratonic estuary. At low tide in the shallow subtidal zone, trilobites move along the rippled, muddy bottom scattered with brachiopods, eocrinoids, and hyoliths, while a tiny bradoriid arthropod swims nearby. In the distance, the muddy bottom drops off into a deeper channel, and beyond, an exposed island of Precambrian schist is lightly patched with algae and mosses. Key: (1) Ptychopariid trilobite *Amecephalus althea*. (2) Corynexochid trilobite *Anoria tontoensis*. (3) Bradoriid arthropod. (4) Hyolith. (5) Eocrinoid *Gogia*. (6) Inarticulate brachiopods. (7) Trace fossil *Cruziana*.

Based on data from the Bright Angel Formation in Grand Canyon National Park, Arizona. Painting by Terry McKee.

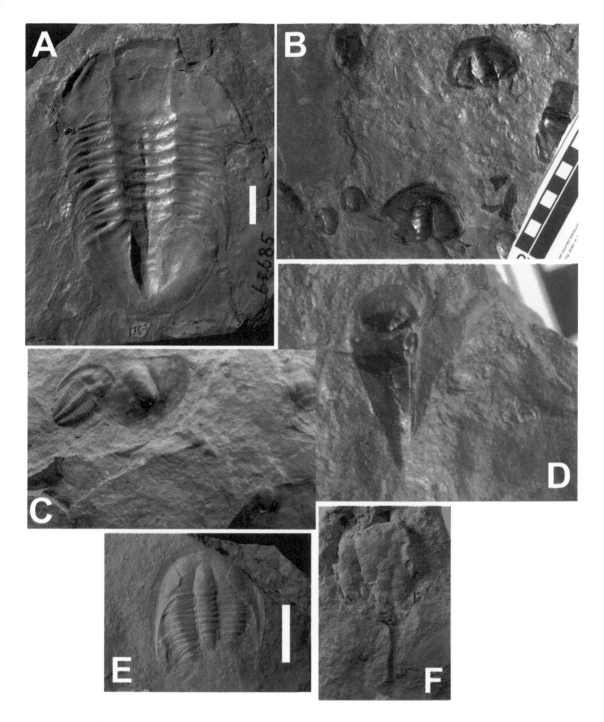

Plate 3. Fossils of the Middle Cambrian part of the Bright Angel Formation in Grand Canyon, Arizona. (A) The dolichometopid trilobite *Anoria tontoensis* (USNM 62685). Scale bar = 1 cm. (B) Pygidia of dolichometopid trilobites on a slab (GRCA 11802). Scale bar in centimeters. (C) Complete specimen of the ptychopariid trilobite *Amecephalus althea* (length ~3 cm.) next to a pygidium of a dolichometopid. (D) A hyolith fossil consisting of conch and operculum (GRCA 21400). Length ~3 cm. (E) *Amecephalus althea* (USNM 61573). Scale bar = 1 cm. (F) The eocrinoid *Gogia ?longidactylus* (GRCA 2641). Length ~4 cm.

All specimens from the museum of Grand Canyon National Park, except (A) and (E), which are courtesy of Smithsonian Institution.

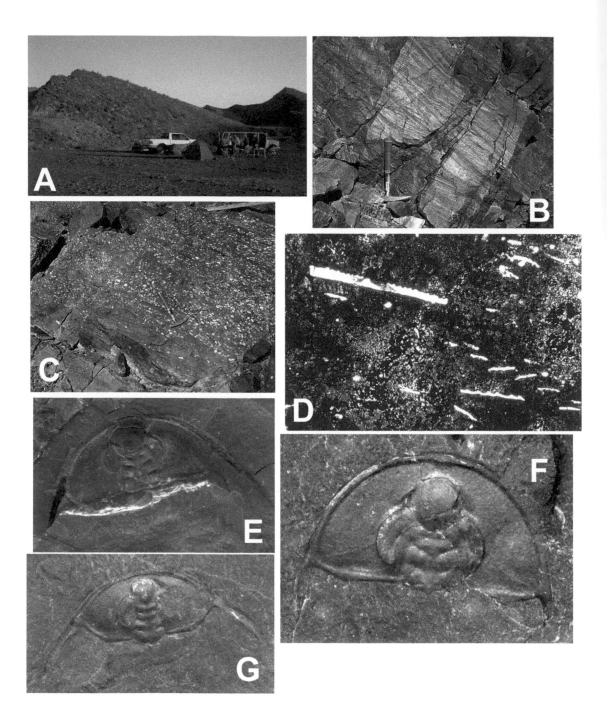

Plate 4. Rocks and fossils of the Cambrian units of the Marble Mountains, California. (A) Camp near the Zabriskie Quartzite. (B) Crossbedding in the Wood Canyon Formation. (C) Conglomeratic sandstone in the Wood Canyon Formation. (D) Thin-section micrograph of a carbonate layer in the Latham Shale with abundant, flat fossil fragments seen in cross section. Width of view = 2 mm. (E)–(G) Trilobites of the Latham Shale. (E) *Bristolia* aff. *fragilis* sp. A. Cephalon width ~4 cm. (F) *Olenellus clarki.* Cephalon width ~5 cm. (G) *Bristolia mohavensis.* Cephalon width ~3 cm.

Plate 5. Life on the shallow shelf of the Early Cambrian seas. A muddy bottom about 50 m (164 ft.) down is covered in algae, inarticulate brachiopods, and eocrinoids. Scattered on the surface are a number of hyoliths, conulariids, and an articulate brachiopod. Also present are olenellid trilobites and burrowing worms. Surveying the scene from above are two *Anomalocaris.* Key: (1) Anomalocaridid arthropods *Anomalocaris.* (2) Eocrinoid *Gogia.* (3) Olenellid trilobite *Mesonacis fremonti.* (4) Olenellid trilobite *Olenellus clarki.* (5) Conulariid cnidarians. (6) Hyoliths. (7) Articulate brachiopod *Nisusia.* (8) Olenellid trilobite *Bristolia bristolensis.* (9) Palaeoscolecidan worms. (10) Clusters of inarticulate brachiopods *Paterina.* (11) Alga *Margaretia.*

Based on data from the Latham Shale in the Marble Mountains of California. Painting by Terry McKee.

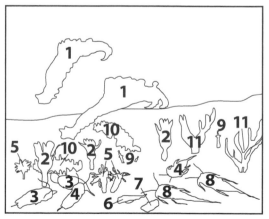

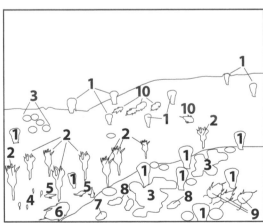

Plate 6. Life in the shallow carbonate shoal environment of the Early Cambrian seas. In less than 15 m (50 ft.) of water, carbonate shoals are covered in ovoid algal growths and burrowed lime-mud, with silty inter-shoal areas containing interference ripples indicative of tidal and storm currents. The shoals and inter-shoals are covered with algae and cyanobacteria, archaeocyathid sponges, eocrinoids, and a scattering of brachiopods and hyoliths. Trilobites include olenellids, dorypygids, and indeterminate ptychopariids. The shoal in the foreground contains all these trilobites moving around on a surface covered with algal growths and large archaeocyathids. Key: (1) Archaeocyathid sponges. (2) Eocrinoids *Gogia*. (3) Clusters of ovoid algal structures (oncoliths). (4) Inarticulate brachiopods. (5) Hyolith *Novitatus*. (6) Olenellid trilobite *Olenellus puertoblancoensis*. (7) Ptychopariid trilobite. (8) Dorypygid corynexochid trilobite *Bonnima*. (9) Olenellid trilobite *Bolbolenellus euryparia*. (10) Olenellid trilobite *Bristolia* sp.

Based on data from the Chambless Limestone in the Marble Mountains of California. Painting by Terry McKee.

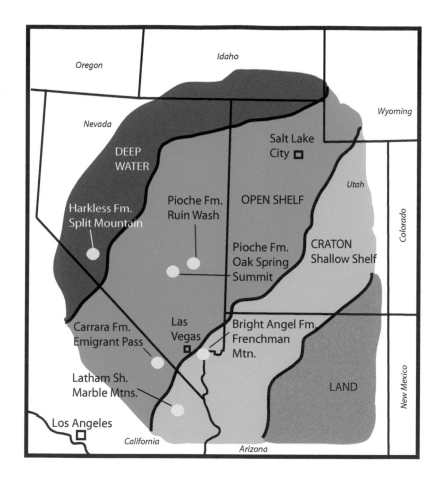

Plate 7. The coast line of western North America during the latest Early Cambrian (Dyeran). Localities mentioned in the text are marked by yellow dots. Position on shelf and water depth are marked by shades of blue. Map does not account for Cenozoic extension of the Great Basin, which exaggerates east–west distances compared to original paleogeography.

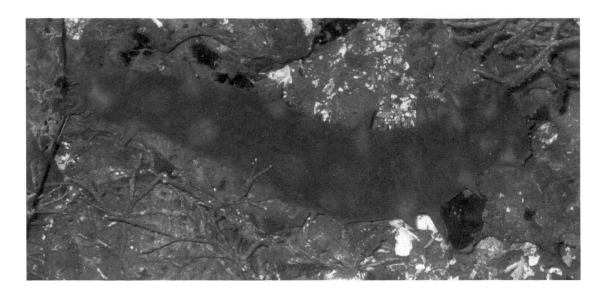

Plate 8. A modern sea cucumber (Holothuroidea, Echinodermata). Length about 15 cm (6 in.).

Plate 9. Modern marine bristle worms (Polychaeta, Annelida) among bottom sediment gravel. Length of each about 1.5 cm (0.6 in.).

Plate 10. A modern horseshoe crab (Chelicerata, Arthropoda). Length about 30 cm (1 ft.).

Plate 11. *Overleaf:* Life on the shallow shelf of the Middle Cambrian seas. A muddy bottom about 40 m (131 ft.) down is covered in algae and cyanobacteria. Also living on the bottom are brachiopods, hyoliths, holothuroid echinoderms, sponges, and eocrinoids leaning into the current. On the left, the arthropod *Meristosoma* winds its way along the bottom; center, *Utahcaris* feeds on an anomalocaridid carcass, and a worm emerges from its burrow. The water column is patrolled by anomalocaridids and bivalved arthropods. Key: (1) Anomalocaridid arthropod *Anomalocaris.* (2) Sponge *Vauxia.* (3) Alga *Acinocricus.* (4) Eocrinoid echinoderm *Gogia.* (5) Dolichometopid corynexochid trilobite *Glossopleura similaris.* (6) Arthropod *Meristosoma.* (7) Ptychopariid trilobite *Amecephalus althea.* (8) Alga *Yuknessia.* (9) Inarticulate brachiopods *Micromitra.* (10) Arthropod *Utahcaris.* (11) Palaeoscolecidan worm. (12) Bivalved arthropod *Isoxys.* (13) Dolichometopid corynexochid trilobite *Glossopleura yatesi.* (14) Hyolith *Haplophrentis.* (15) Holothuroid echinoderm *Eldonia.* (16) Cyanobacterium *Marpolia.* (17) Ptychopariid trilobite *Amecephalus idahoense.* (18) Corynexochid trilobite *Zacanthoides liddelli.* (19) Articulate brachiopods *Wimanella.* (20) Medusoid cnidarians (jellyfish). Some species shown are only found in either the Wellsville Mountains or the Bear River Range; all are shown together here, in a setting modeled on data from the Wellsville Mountains, because it is plausible that these species had overlapping geographic ranges, even though this has not yet been proven in most cases.

Based on data from the Spence Shale in the Wellsville Mountains of Utah and Bear River Range of Idaho. Painting by John Agnew.

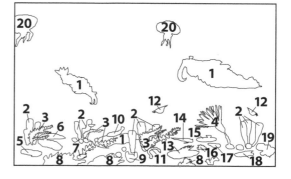

JOHN N. AGNEW ©09

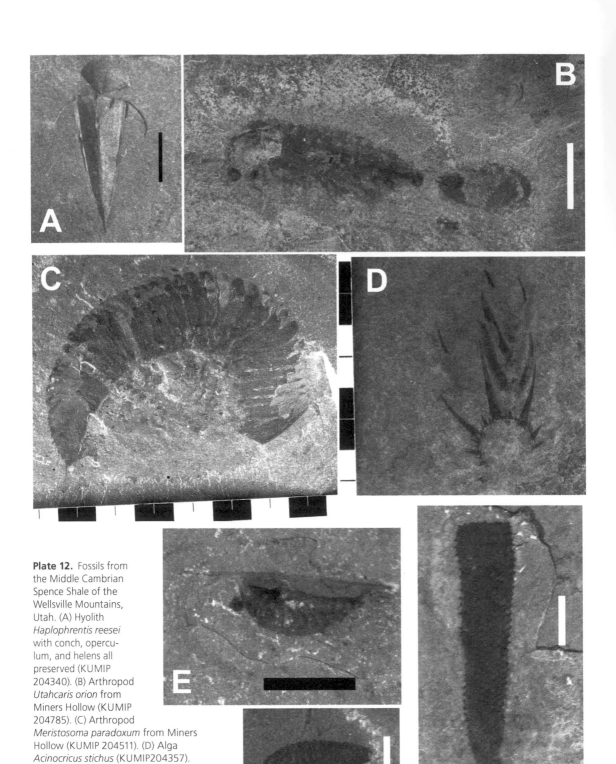

Plate 12. Fossils from the Middle Cambrian Spence Shale of the Wellsville Mountains, Utah. (A) Hyolith *Haplophrentis reesei* with conch, operculum, and helens all preserved (KUMIP 204340). (B) Arthropod *Utahcaris orion* from Miners Hollow (KUMIP 204785). (C) Arthropod *Meristosoma paradoxum* from Miners Hollow (KUMIP 204511). (D) Alga *Acinocricus stichus* (KUMIP204357). (E) Bivalved arthropod *Isoxys* sp. from Miners Hollow (KUMIP 312404). (F) Arthropod *Mollisonia symmetrica* from Miners Hollow, collected by the Gunther family in 1990 (KUMIP 314041). (G) The sponge *Vauxia magna* from Miners Hollow (KUMIP 111763). Scale bars = 1 cm.

All University of Kansas specimens.

Plate 13. View of the Walcott Quarry in the Burgess Shale, right foreground, looking north toward Wapta Mountain and the President Glacier.

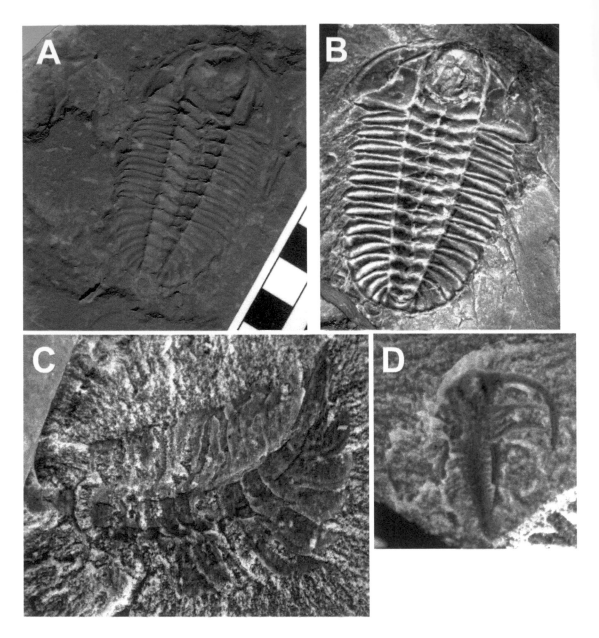

Plate 14. Fossils from the Burgess Shale. (A) *Olenoides serratus* from the Mount Stephen Trilobite Beds. (B) *Olenoides serratus* from the Walcott Quarry. (C) Arthropod *Leanchoilia* from the Walcott Quarry. (D) Arthropod *Marrella* from the Walcott Quarry.

Field photos.

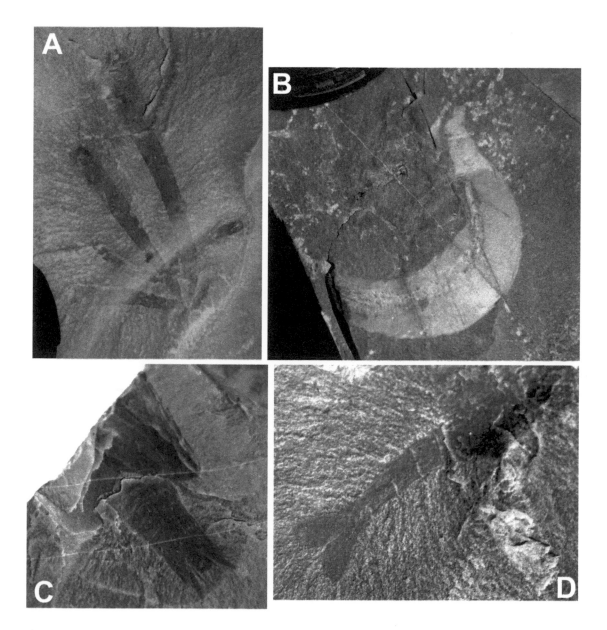

Plate 15. Fossils from the Burgess Shale at the Walcott Quarry. (A) Sponge *Vauxia*.
(B) Priapulid worm *Ottoia*. (C) Posterior half of bivalved arthropod *Canadaspis*. (D) Posterior
end of arthropod *Waptia*.

Field photos.

Plate 16. Diorama showing the setting of the Burgess Shale at the foot of the Cathedral escarpment with algae, filamentous cyanobacteria, and sponges growing in abundance. Living among these are numerous arthropods, lobopodians, brachiopods, and at least one *Opabinia*.

Courtesy of Smithsonian Institution.

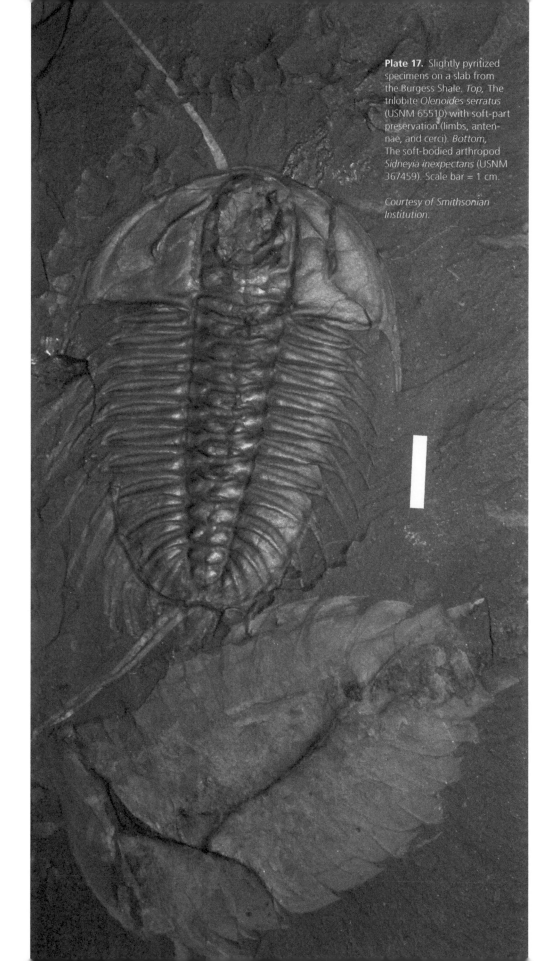

Plate 17. Slightly pyritized specimens on a slab from the Burgess Shale. *Top,* The trilobite *Olenoides serratus* (USNM 65510) with soft-part preservation (limbs, antennae, and cerci). *Bottom,* The soft-bodied arthropod *Sidneyia inexpectans* (USNM 367459). Scale bar = 1 cm.

Courtesy of Smithsonian Institution.

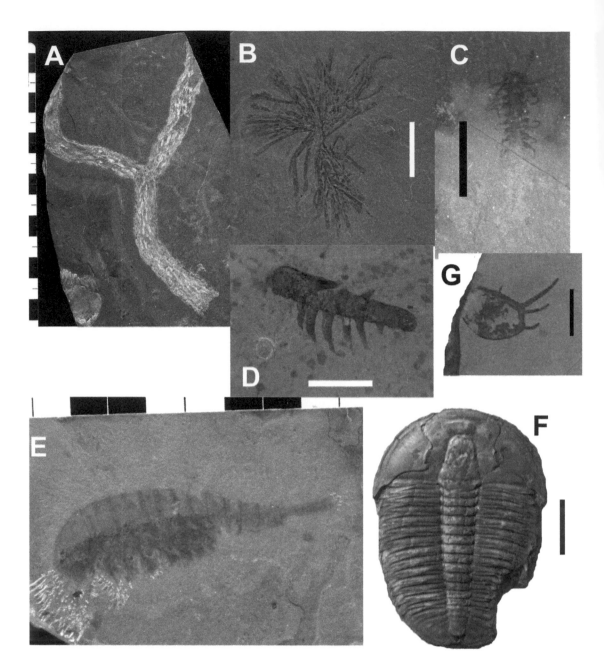

Plate 18. Fossils from the Middle Cambrian Wheeler Formation of Utah. (A) Alga *Margaretia dorus* (KUMIP 127811). (B) Alga *Yuknessia simplex* (KUMIP 204380). (C) Arthropod *Cambropodus gracilis* (KUMIP 204775). (D) Lobopodian *Aysheaia prolata* (KUMIP 153923). (E) Arthropod *Dicranocaris guntherorum* (KUMIP 135148). (F) Trilobite *Elrathia kingii* with healed bite mark (KUMIP 204773). (G) Bivalved arthropod *Tuzoia? petersoni* (KUMIP 134799). Scale bars = 1 cm.

All University of Kansas specimens.

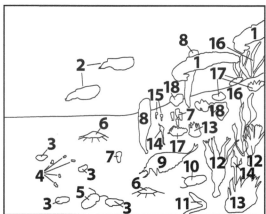

Based on data from the Wheeler Formation in the House Range of Utah. Painting by Terry McKee.

Plate 19. Life on the bottom of a deep intra-shelf basin of the Middle Cambrian seas. A muddy bottom about 91 m (300 ft.) down is on the edge of having enough light for photosynthesis, so the algae are mainly on the slopes in shallower water. Filter-feeding bottom dwellers include the eocrinoids, brachiopods, the enigmatic *Chancelloria*, and sponges. Also living on the bottom are carpoid echinoderms and trilobites. The arthropod *Dicranocaris* swims just above the sediment attempting to stir up a meal. In the right foreground an indeterminate worm is about to move out of view, and a grazing stem-mollusc moves across the sub-strate. In the water column, swimming bivalved arthropods and anomalocaridids patrol for food. Key: (1) Anomalocaridid arthropods *Peytoia* (*Laggania*). (2) Bivalved arthropods *Branchiocaris*. (3) Ptychopariid trilobites *Elrathia kingii*. (4) Agnostid trilobites *Peronopsis* and *Ptychagnostus*. (5) Corynexochid trilobite *Asaphiscus wheeleri*. (6) Sponge *Choia*. (7) Sponge *Diagoniella*. (8) Enigmatic sponge-like animal *Chancelloria*. (9) Arthropod *Dicranocaris*. (10) Soft-bodied trilobite *Naraoia*. (11) Indeterminate worm. (12) Eocrinoid echinoderm *Gogia*. (13) Stem-mollusc *Wiwaxia*. (14) Carpoid echinoderm *Castericystis*. (15) Inarticulate brachiopods. (16) Alga *Margaretia*. (17) Alga *Yuknessia*. (18) Cyanobacterium *Marpolia*.

Plate 20. Post-storm sunrise along the rocky shoreline of islands in the Late Cambrian sea of what is now Wisconsin. As large surf continues to pound the Precambrian rocks of the islands and seastacks, a tidepool provides a home to green algae and animals living on the margins of the ocean world. Key: (1) Green algae in and around tide pools. (2) Edrioasteroid echinoderms. (3) Indeterminate shelled mollusks.

Based on data from Sauk County, Wisconsin. Painting by Karen Foster-Wells.

Plate 21. A modern coral (Anthozoa, Cnidaria).

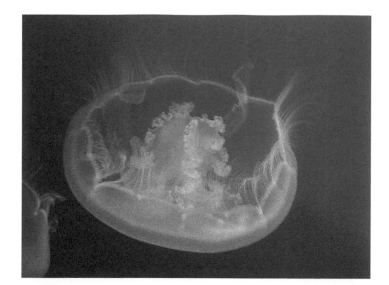

Plate 22. A modern jellyfish (Scyphozoa, Cnidaria).

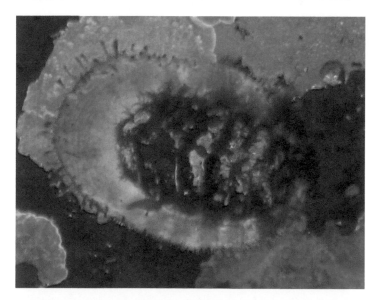

Plate 23. A modern chiton mollusc (Polyplacophora, Mollusca).

Plate 24. A modern crustacean, the peppermint shrimp (Malacostraca, Arthropoda).

Plate 25. A modern starfish (Asteroidea, Echinodermata).

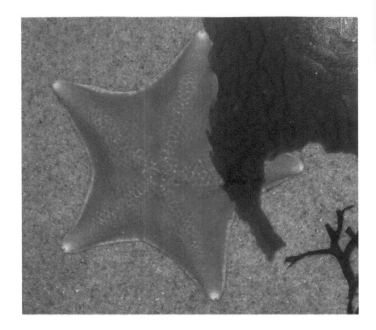

Plate 26. A modern sea anemone (Anthozoa, Cnidaria).

Plate 27. Fossils from the Latham Shale (Early Cambrian), San Bernardino County, California. (A) Nearly complete trilobite *Bristolia harringtoni* (UCR 10/7). Found by Douglas Morton. Scale bar = 1 cm. (B) Cephalon of the trilobite *Mesonacis cylindricus* (UCR 7897.2). Found by Larry Pearce. Scale bar = 1 cm. (C) Cephalon of the trilobite *Olenellus nevadensis*. Width ~4 cm. (D) Cephalon of the trilobite *Bristolia bristolensis*. Width ~3 cm. (E) A palaeoscolecidan worm (UCR 7003.1), one of two halves (part and counterpart) found years apart by Jack Mount and students, and Simon Conway Morris, respectively. Scale bar = 1 cm. (F) Partial feeding appendage of a relatively large individual of the anomalocaridid *Anomalocaris* (UCR 7002.1). Found by Jack Mount and students. Scale bar = 5 cm.

(A), (B), (E), and (F) from the University of California–Riverside, Invertebrate Fossil Collection.

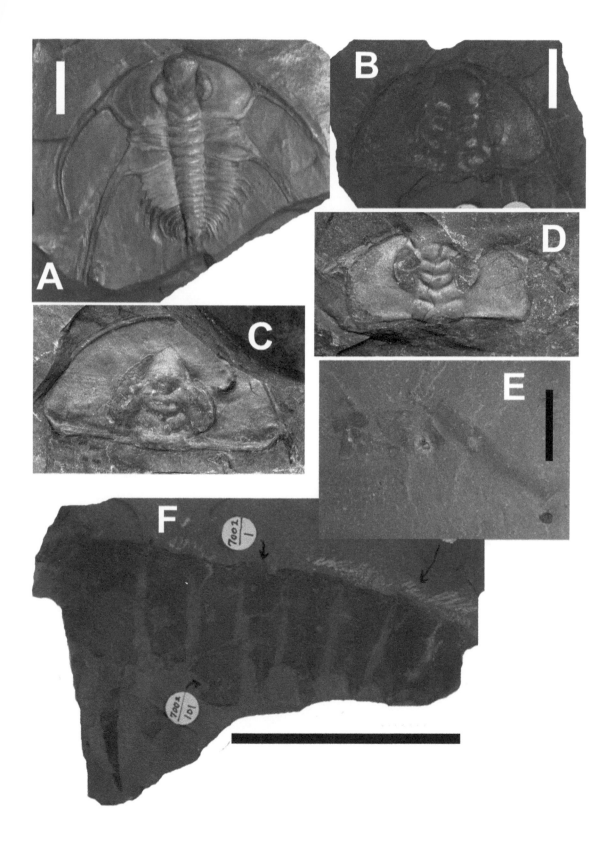

Plate 28. Fossils from the Early and Middle Cambrian of Nevada and Idaho. (A) Early Cambrian olenellid trilobite *Olenellus gilberti* from the Combined Metals Member of the Pioche Formation, Ruin Wash, Nevada (MWC 7466). (B) Feeding appendage of a small, unidentified anomalocaridid arthropod from the Middle Cambrian Comet Shale Member of the Pioche Formation, Chief Range, Nevada (MWC 7912). (C) Small, rare trilobite *Oryctocephalites palmeri*, Early Cambrian, Ruin Wash (MWC 7422). (D) Middle Cambrian ptychopariid trilobite *Amecephalus idahoense* from the Spence Shale, Spence Gulch, Idaho (MWC uncataloged). Scale bars (A), (B), and (D) = 1cm. Scale bar (C) = 5 mm.

All Museum of Western Colorado collection. (A)–(C) collected and donated by Andrew R. C. Milner.

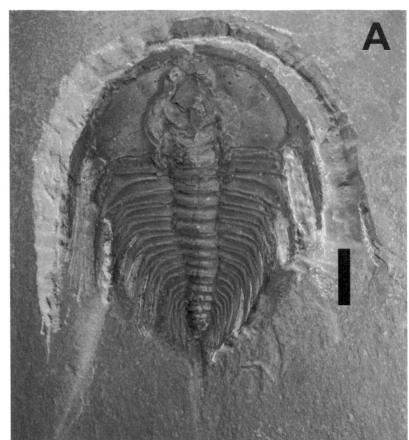

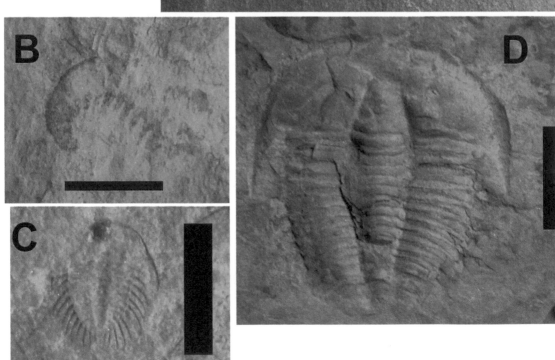

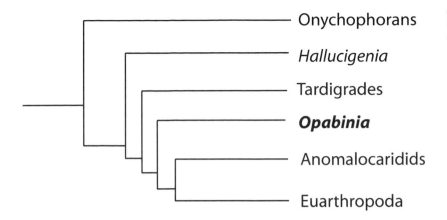

Onychophorans

Hallucigenia

Tardigrades

Opabinia

Anomalocaridids

Euarthropoda

two crescent-shaped pincers, each with six or so spikes of various lengths facing forward and inward. The head of *Opabinia* had a ventral mouth that faced posteriorly, and there were five eyes, each probably compound, and all but the central one (possibly) were stalked. Most specimens are about 40–70 mm (1.5–3 in.) in length, and the body between the head and posterior tail is divided into 15 segments (fig. 6.19). Each segment consisted of an axial ring and two lateral lobes, with a row of flattened blades associated with the lateral lobes that may have functioned as gills. The posterior end consisted of an elongate segment (possibly 3 fused) with a tail fan of six dorsolaterally directed lobes, three on each side.[47] And that is about where the agreement about *Opabinia*'s morphology ends.

There are two main aspects of the animal's structure that are debated: Are triangular, lateral extensions visible off the central part of the body gut diverticulae or lobopod-like limbs? And do the flattened gill blades attach along the anterodorsal edge of the lateral lobes or along the posterior edge? If the triangular extensions are not gut diverticulae and are lobopod-like limbs, they would be ventrally placed, below the lateral lobes, and the currently debated positions for the gill blades are just two of four that have been proposed in the past 35 years. Part of the reason that these structures have proven so difficult to interpret is that, despite beautiful soft-body preservation in most specimens, details of the structure in any one individual are often still obscured by decay, folding of parts of the body over other regions, and vagaries of the geochemical preservation process of different body tissues. In most of these soft-bodied species the tissues are preserved as thin films of different reflective capacities; often these films are composed of silica, aluminum, and potassium, and analysis of the chemical concentration of these elements in parts of the body can sometimes tease apart different structures better than one can just by looking at them in natural light. But even these results can be interpreted differently by different researchers, and so the questions above remain unresolved (at least to a reasonable level of satisfaction).[48]

Despite the disagreement over morphology, there seems to be some consensus that *Opabinia* falls somewhere between onychophorans (velvet worms) and arthropods. In several phylogenetic hypotheses, *Opabinia* is

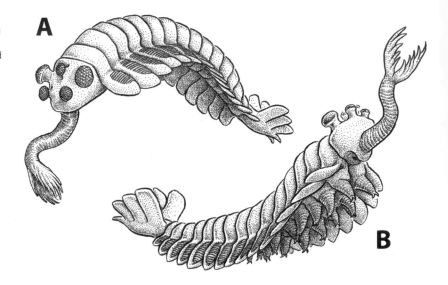

6.21. Reconstructions of *Opabinia.* (A) Based on Zhang and Briggs (2007). (B) Based on Budd (1996) and Budd and Daley (2012) with lobopod-style limbs on ventral part of body. Length 4–7 cm.

Drawings by Matt Celeskey.

most closely related to anomalocaridids + arthropods, with the tardigrades ("water bears") as the next most closely related sister taxon and then the lobopodians and onychophorans (fig. 6.20). Despite its segmentation and almost trilobite-like appearance in the trunk region, *Opabinia* has very un–arthropod like features in the unsegmented, tube-shaped proboscis and in its lack of jointed legs. So it is clearly not an actual arthropod, but the tail fan is reminiscent of anomalocaridids. If the debated triangular structures are lobopod-like limbs this would add to the mix of characters in the animal that make its classification complex.

The structural debate about *Opabinia* affects less its general classification between onychophorans and arthropods and more the nature of the early evolution of the arthropod limb. Remember that trilobites and many other arthropods have a biramous limb composed of a ventral walking leg and a more dorsal filamentous branch. The interpretation of *Opabinia* in which the medial, triangular areas are gut diverticulae and the gill blades attach along the posterior part of the lateral lobes implies a model in which the arthropod walking leg differentiated, segmented, and split from the rest of the original lateral lobe, which became the dorsal filamentous branch of the biramous limb.[49] Conversely, the other interpretation of *Opabinia* holds that the gill blades lay parallel to the dorsal surface of the lateral lobes and attached anteriorly and that the triangular reflective areas were lobopod-like limbs; in this model, the lateral lobes and gill blades evolved into the dorsal filamentous branch and the lobopod-like limbs evolved into the ventral walking leg of the biramous limb of arthropods.[50] This latter model also implies that the anomalocaridids, occupying a position between *Opabinia* and arthropods, secondarily lost the lobopod-like limbs.

But regardless of its role in study of early arthropod limb development, *Opabinia* is also interesting as a rather peculiar animal, one that was once alive and moving about the depths of an ocean in what is now

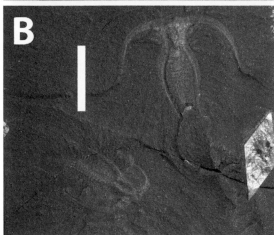

6.22. Specimens of the arthropod *Marrella* from the Burgess Shale. (A) Single well-preserved specimen (USNM 57674). (B) Two specimens (USNM 57670). Both scale bars = 1 cm.

Courtesy of Smithsonian Institution.

British Columbia about 505 million years ago (fig. 6.21). What ecological niche would this animal have occupied? *Opabinia* was probably a predator that grabbed prey with its proboscis and used that organ to bring the food down and back to the mouth. The proboscis may have been hollow and fluid filled and moved by some type of hydrostatic pressure. There has been some speculation that maybe it hunted burrowing animals and used the proboscis to access the tube-like dwellings. It is believed to have lived near the seafloor and to have moved along the bottom and perhaps swum above it in search of food. As a hunter and as a small animal that was likely on the menu of other, larger predators, *Opabinia* would have valued visual input from its environment, something that is apparent from its five compound eyes.

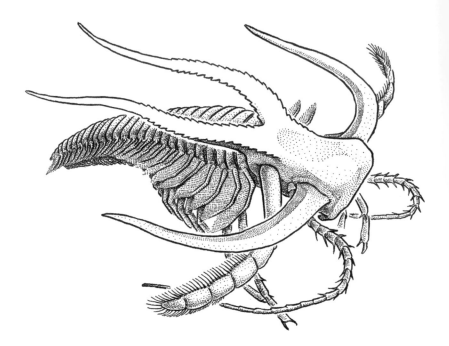

6.23. Life restoration of the arthropod *Marrella*. Total length ~2.5 cm.

Drawing by Matt Celeskey.

Given its unique morphology, its continued appearance as a subject of reinterpretations in new life restorations, and its role in the debate over arthropod limb evolution, it appears that *Opabinia* has had the last laugh—and at our expense. It was, after all, an audience of paleontologists that laughed at this animal when it was presented to the world by a colleague decades ago, and yet, ironically, we now recognize how important and interesting an animal it actually was and that perhaps we owed it a little more respect.[51] At least *Opabinia* seems to be getting that respect now.

A Parade of Chitinous Knights: The Myriad of Arthropods

Burgess Shale–type deposits often produce soft-bodied arthropods, animals related to today's insects, spiders, scorpions, lobsters, and shrimp that had an articulated exoskeleton but one that was not calcified like that of trilobites. Although these were hard, protective exoskeletons for the living animals, because the exoskeletons were not mineralized they did not fossilize with nearly the frequency that trilobite elements did. So it is rare and exciting when a fossilized non-trilobite arthropod appears in a formation and gives us a glimpse of the diversity of other arthropods from the time period. Soft-bodied, non-trilobite arthropods of the Cambrian seem to have larger geographic ranges and longer stratigraphic ranges than their trilobite cousins.[52] We have seen a few of these animals in older formations such as the Buen and Spence, and we will see more later in the Wheeler Formation, for example. But one aspect (among many) in which the Burgess Shale Formation really shines is the production of non-trilobite arthropods. There is a surprising morphological diversity of them.[53] Let us take a look at a few.

Marrella splendens

C. D. Walcott's first mention of fossils from what became the Walcott Quarry, in his notebook on August 31, 1909, included a sketch of what is clearly a specimen of the arthropod *Marrella*, so this was one of the first fossils found from this deposit. There are now more than 9000 specimens known from the Burgess Shale Formation at six localities ranging through the Kicking Horse Shale, Campsite Cliff Shale, Walcott Quarry Shale, and Raymond Quarry Shale members. These sites range from the top of the *Glossopleura* into the *Ehmaniella* trilobite zones.

Marrella is a small fossil arthropod about 2.5–25 mm (0.1–1 in.) in length with 26 body segments (fewer in young individuals) (fig. 6.22; plate 14d). The long tapering processes curving off the head include two pairs of cephalic spines, a pair of antennae, and a pair of what are believed to be swimming appendages. The latter two sets of processes are modified appendages and were flexible; the cephalic spines did not move. As in trilobites, the mouth is ventral and faces backward, with the stomach anterior of this and the gut passing over dorsally and heading back through the thorax. The heart is dorsal to and slightly anterior to the stomach and is long and tapering posteriorly. The legs on each segment were jointed and contained hairlike filaments, as in many arthropods, and the limbs were biramous, with a gill branch dorsal to each walking leg. The swimming appendages are believed to have functioned like paddles, thrusting the animal forward on the power stroke and rotating to reduce drag on

---

Diego García-Bellido is a researcher at the Institute of Geosciences of the Spanish Research Council and specializes in the paleobiology and anatomy of Cambrian animals, particularly arthropods. Born and raised in Madrid, Spain, Dr. García-Bellido was never much into rocks or fossils as a kid, but more often was an observer of animals wherever he encountered them. He also was unavoidably influenced by the scientific household in which he grew up. Both of his parents were developmental biologists studying the fruit fly. Dinner table conversations sometimes revolved around experiments on *Drosophila*. Francis Crick, co-discoverer of the physical structure of DNA in the 1950s, was a dinner guest. It is no surprise that most of the García-Bellido children grew up to be biologists. Diego went only slightly afield from the others by becoming a paleontologist, which he credits in part to the additional influence of his grandfather, who was an archaeologist of Roman, Greek, and Iberian history. "My attraction towards paleontology is kind of a compromise between my interest in evolution and my interest in archaeology," he says. Biology, but with lots of digging.

With the additional influence of three sabbatical years spent with his parents at Cal Tech and in Australia, Dr. García-Bellido did his

The Cambrian Corps 5 – Diego C. García-Bellido

college and graduate work at Complutense University in Madrid, studying biology and geology. His early interests in paleontology related to paleoanthropology, but he eventually became fascinated with the Cambrian explosion, which has been at the core of his research ever since.

It was during a visit to Cambridge University that Simon Conway Morris suggested to Diego that he contact Desmond Collins at the Royal Ontario Museum and offer to volunteer as a field crew member. He did so, was accepted, and spent three summers working at the Walcott Quarry in the Burgess Shale as part of the ROM crews. "There are few things," he says, "as awe-inspiring as splitting a rock and laying your eyes on a new type of animal that no human—or even dinosaur, for that matter—has ever seen." His fieldwork eventually led to research on many of the animals he collected and to redescriptions of some of Walcott's beasts based on the newer material; among the animals he helped redescribe was *Marrella*, the odd little arthropod that is also Dr. Garciá-Bellido's personal favorite among the Burgess Shale animals. Thousands of specimens of this arthropod are preserved in museum collections, but one individual was actually buried and fossilized in the process of molting its exoskeleton, and this individual is one that Diego described in an article in the journal *Nature*. "We know from modern arthropods that [molting] takes minutes to occur, so the chances of it being preserved were one in millions," he says.

Dr. Garciá-Bellido's recent work has involved excavation and description of new material from the Early Cambrian Emu Bay Shale in Australia. This material includes some of the most detailed preservation of eyes from the Cambrian that paleontologists have yet found, and has shown that the visual acuity of some arthropods was at nearly modern levels of effectiveness—and was better than many contemporary trilobites.

Paleontology is not the most lucrative career path, nor does it have the healthiest job market, but Dr. Garciá-Bellido is like many of us and recognizes the value of loving what you do and encourages aspiring paleontologists to follow their dreams—and to beware listening too intently to your own inner voice that may favor practicality over everything else. "There is a wonderful world out there waiting to be discovered," he says, "telling us how nature came to be what it is today." Diego is among the dreamers who get to pursue a passion as a career.

---

the recovery stroke. *Marrella* is believed to have been a swimmer low in the water column just above the seafloor (fig. 6.23); it may have swum in groups. This arthropod may also have fed on small prey items in the water and to have filtered its food with a meshwork formed by its smaller posterior limbs. One unique specimen from the Burgess Shale was found preserved in the actual process of molting. *Marrella* has also been found in the Early and Middle Cambrian of China. It has been suggested that

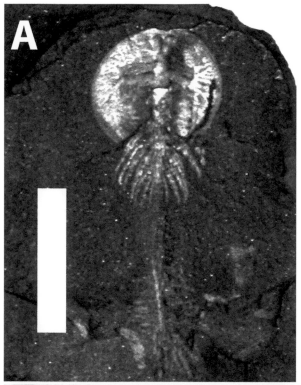

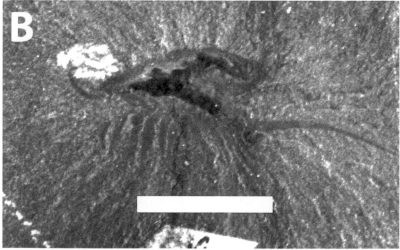

6.24. Specimens of the arthropod *Burgessia* from the Burgess Shale. (A) Specimen in dorsal view showing carapace, legs, and telson (USNM 57676). (B) Specimen in lateral view showing same (USNM 57680). Both scale bars = 1 cm.

Courtesy of Smithsonian Institution.

Marrella, along with *Canadia* and *Wiwaxia*, was an iridescent to silvery color, an adaptation to deter predators. *Marrella* is an arthropod within a group known as the Marrellomorpha, but how this group relates to other arthropods is still debated.[54]

Burgessia bella

Burgessia is a small arthropod with a single, round carapace over the head and thorax, a carapace that had a triangular notch in the back (fig. 6.24). Two delicate antennae protruded from the front end of the carapace, but

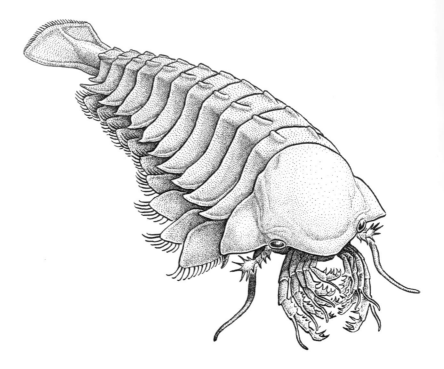

6.25. Life restoration of the stem crustacean *Sanctacaris* from the Burgess Shale. Total length up to ~10 cm.

Drawing by Matt Celeskey.

the animal apparently lacked eyes. The digestive system of *Burgessia* is complex, with a central gut and branching diverticulae spreading out to near the edge of the carapace on each side of the animal. In some specimens the walking limbs are hidden under the carapace, but others show three pairs of appendages on the head and seven sets of biramous appendages on the thorax. The telson is very long and consists of a single spine. In some specimens the telson is straight like that of a horseshoe crab, but in many others it is bent. This telson may have been flexible in life and controlled by the animal through internal hydrostatic pressure.[55] *Burgessia* is a very abundant arthropod in the Burgess Shale Formation (more than 1800 specimens), and it is classified near the base of the arachnomorph clade, making it a primitive relative of spiders, scorpions, horseshoe crabs, trilobites, and a whole host of other Cambrian forms. It probably crawled along the bottom sediments in the ocean, but exactly how it fed is unclear.

Sanctacaris uncata

One arthropod from the Burgess Shale appears to be a true chelicerate. This is *Sanctacaris*, the "sacred crab," named in honor of its informal field name assigned after its discovery, "Santa Claws." This is a rare, relatively small, soft-bodied arthropod with 10 feeding limbs on the head (5 on each side), 11 thick body segments, and a flat, fan-shaped telson. In addition to the feeding limbs there are antennae and a sixth pair of limbs on the wide, almost hammerhead-shaped cephalon. There are also two eyes, one near the base of each antenna. Each body segment had a pair of relatively long, biramous limbs, the dorsal branch of which was paddle-like and fringed with filaments that may have been involved in respiration also. *Sanctacaris* was probably an active, swimming predator (fig. 6.25).

Sanctacaris was first discovered from the basal Burgess Shale Formation in 1983. The few specimens found were well preserved but were in the *Glossopleura* trilobite zone (see chapter 5) and were therefore a bit older than the Walcott Quarry specimens. *Sanctacaris* has a number of characters that suggest a close relationship with chelicerates. Modern chelicerates like the horseshoe crab have six pairs of appendages on the head, a characteristic that *Sanctacaris* shares with these arthropods.[56] If *Sanctacaris* is considered an ancient chelicerate, then it indicates that this modern group's origins date back well into the Cambrian.

Emeraldella brocki

Emeraldella was named by Walcott in 1912 and has a large, semicircular cephalon with antennae, thirteen tapering thoracic segments, and a long telson, or tail spine. The limbs under each thoracic segment were biramous. *Emeraldella* may have been a mobile predator and scavenger that walked along the bottom sediment; it is classified as an arachnomorph, which is a large division of the arthropods that includes chelicerates and trilobites.[57] *Emeraldella* may also occur in the Marjum Formation in Utah (see chapter 7).

Sidneyia inexpectans

This arthropod was named after one of Walcott's sons. This wide-bodied arthropod had been interpreted as having an anteroposteriorly short cephalon with antennae and two stalked eyes (plate 17). It has recently been shown that the head was not as short as we originally thought, most previous specimens having preserved the head in a slightly deformed state. There were nine thoracic segments and two ring-like abdominal segments, and the telson was fan-shaped. The first four anterior appendages were uniramous walking legs, but the posterior five appendages were biramous and included paddle-shaped branches that may have served a gill function in part. *Sidneyia* is classified as an arachnomorph arthropod and was probably a bottom-dwelling predator on small trilobites, hyoliths, and small bivalved crustaceans (as indicated by gut contents preserved in some specimens).[58]

Leanchoilia superlata and L. persephone

These two species of *Leanchoilia* may be sexually dimorphic forms of the same species, although *L. superlata* is much more abundant. It is found in nine quarries along Fossil Ridge down to Mount Stephen, and over the years more than 1500 specimens have been collected (fig. 6.26; plate 14c). Stratigraphically, this arthropod ranges from the base of the

6.26. Specimens of the arthropod *Leanchoilia* from the Burgess Shale. (A) Specimen in right lateral view showing great appendages (USNM 250219). (B) Two specimens on one slab (USNM 250221). Scale bars = 1 cm.

Courtesy of Smithsonian Institution.

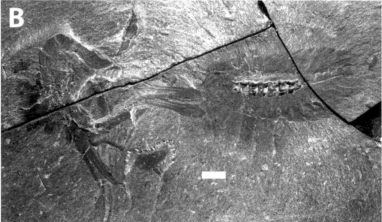

Burgess Shale Formation, in the Kicking Horse Shale Member, up into the upper middle of the formation in the Raymond Quarry Shale Member. The cephalon was translucent and had on the underside four small, compound, unstalked eyes; the head also had three sets of cephalic appendages, the first pair of which consisted of a "great appendage" on each side, each possessing three very long, whip-like flagella that may have functioned like antennae. The main body had 11 thoracic segments; the short telson was fringed with short spikes. Each thoracic appendage was biramous, with a leg branch and a paddle-like upper branch with gill filaments. *Leanchoilia superlata* also had paired axial ridges on the dorsal surface of each thoracic segment. Some specimens of *Leanchoilia* have paired mid-gut glands preserved three dimensionally; these structures can be associated with a predator's diet of occasional high-calorie meals (and intervening lean times) in modern species. *Leanchoilia's* leg appendages may have been adapted for walking or perhaps were better suited to swimming, but either way it appears these animals were predators and scavengers living on and near the bottom.[59]

Yohoia tenuis

Yohoia was named by Walcott in 1912 and was a small, somewhat elongate arthropod (fig. 6.27). The details of the head are not clear, but there

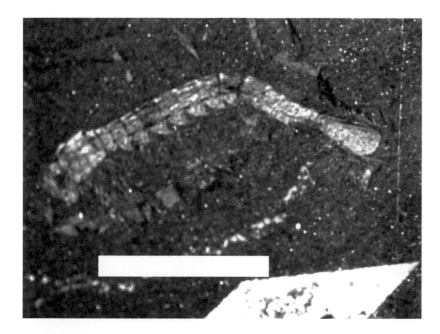

6.27. Specimen of the arthropod *Yohoia* from the Burgess Shale (USNM 57699). Scale bar = 1 cm.

Courtesy of Smithsonian Institution.

6.28. Life restoration of the arthropod *Yohoia*. Note eyes and grasping, anterior appendages. Total length ~3 cm.

Computer-generated image by and courtesy of Joachim Haug.

appear to have been several walking legs under a head shield and possibly relatively large eyes. The thorax consisted of 10 segments with pairs of filament-lined, paddle-like appendages underneath, and below these were spindly walking legs; there were 3 ring-like segments without appendages behind the thorax and then a flat, paddle-shaped telson. The front set of appendages was relatively long and jointed, with four spines on the ends (fig. 6.28). These raptorial appendages have recently been hypothesized to have been used to thrust outward to capture prey items (possibly bradoriid arthropods and agnostid trilobites) and bring them to the mouth. This use of the front appendage in feeding, as an ambush predator, is somewhat similar to that of modern mantis shrimps. *Yohoia* probably spent most of its time walking along the bottom, and occasionally swimming just above the sediment, capturing prey items with its elongate anterior appendages.[60]

Courtesy of Smithsonian Institution.

6.29. Specimens of the arthropod *Alalcomenaeus* from the Burgess Shale. (A) Holotype specimen (USNM 155658). (B) Smaller specimen (USNM 155659). Scale bars = 1 cm.

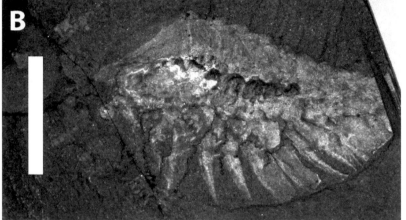

Alalcomenaeus cambricus

Alalcomenaeus had five eyes on its head, two large ones, each placed anterolaterally on the cephalon, and three median ones. The cephalon also had three pairs of cephalic appendages, two biramous posterior ones, and a pair of anteriorly placed great appendages that were generally similar to those of *Leanchoilia*. The mouth was ventral and posterior facing on the head, and the gut curved up dorsally and over the mouth and then ran posteriorly through the thorax. The 11 thoracic segments included biramous limbs, with a paddle-shaped outer branch with gill-like filaments on the posterior edge. The paddle-like telson was short with a posterior fringe of spikes.

This arthropod ranges through most of the Burgess Shale Formation, from the basal Kicking Horse Shale Member up to the Emerald Lake Oncolite; it is now known from more than 300 specimens (fig. 6.29). A recently described specimen of this genus from the Chengjiang biota in China shows details of the neural anatomy and indicates that among modern arthropods it was most like that of chelicerates. *Alalcomenaeus* was probably a benthic walker and swimmer and an active predator.

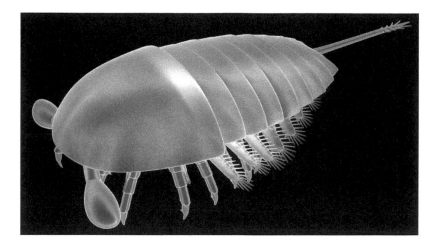

6.30. Life restoration of the arthropod *Sarotrocercus* from the Burgess Shale. Note large eyes and spiked telson. Total length ~2 cm.

Computer-generated image by and courtesy of Joachim Haug.

Individuals would have scanned the environment with the antenna-like flagellae of the great appendages, and the leg branches of the appendages were capable of propelling the animal along the bottom sediments. The paddle-shaped branch of the appendages may have aided in swimming, along with the telson. The predatory habits of the animal are indicated by the stout and spiked inner segment of the walking legs, which were used to crush prey and move pieces of food forward to the mouth.[61]

Helmetia expansa

Helmetia is a giant among Burgess Shale arthropods. *Alalcomenaeus* is fairly small, *Leanchoilia* can be midsized, but opening a drawer and coming across a *Helmetia* fossil gets your attention. It was of a size rivaled by many of the larger specimens of *Anomalocaris*. Some fossils of *Helmetia* come close to 30 cm (1 ft.) in length. *Helmetia* had carapaces over the head and tail sections of its body and six thoracic segments also; the corners of nearly every element of this multipart shield were sharp, almost spines. But the shield appears to have been flat and relatively thin. The eyes were relatively small. Not a terror of the seas, *Helmetia* appears to have been a large, fairly slow-swimming, nektonic filter feeder.[62]

Sarotrocercus oblitus

Sarotrocercus was a small arthropod from the Burgess Shale just about 1 cm (0.4 in.) long; it was named by Harry B. Whittington in 1981. It possessed a large head shield, a thorax with 10 or 11 thoracic segments, and possible biramous thoracic appendages with paddle-like exopods and walking-leg endopods at least on the anterior segments (fig. 6.30). The long telson had small spikes on the posterior end, and the antennae were so short they did not extend beyond the rim of the head shield. Whether *Sarotrocercus* actually swam upside down in the water column, as it was reconstructed by Whittington and subsequently was shown by Stephen Jay Gould in *Wonderful Life*, is unknown. It may have been a benthic

6.31. Life restoration of the Burgess Shale arthropod *Waptia*.

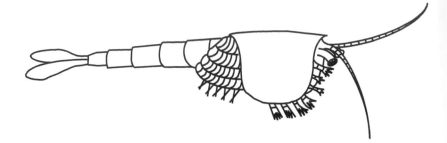

6.32. Lateral view of the fossil arthropod *Waptia* from the Burgess Shale (USNM 83948a). Scale bar = 1 cm.

Courtesy of Smithsonian Institution.

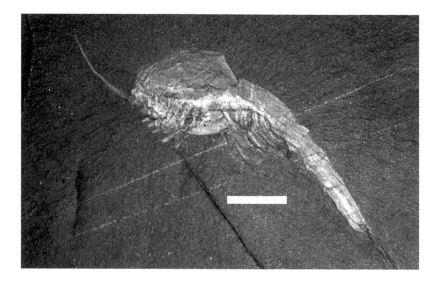

walker.[63] We simply do not know. In either case, one of the distinctive aspects of this arthropod's appearance is its dramatically large eyes (fig. 6.30), which may have been an adaptation for low light levels in the relatively deep Burgess Shale paleoenvironment.

Waptia fieldensis

Waptia is a small, bivalved arthropod with a distinctively long posterior section of six ring-like segments; the posterior-most segment is tipped by a pair of apparently segmented flaps that probably served as a type of tail fluke (fig. 6.31; plate 15d). There were two "stalked" compound eyes as well as antennae on the head,[64] a thorax surrounded by two relatively small carapace valves (fig. 6.32), a set of three paired head appendages ("jaws" or mandibles) behind the antennae, five walking thoracic appendages, and six gill appendages that were also flap-like and may have helped with swimming. The three pairs of jaws appear to be small and not adapted for large or hard food items. The head appears to have cerebral ganglia indicative of a significant brain, and the wide eye surfaces look mostly forward and laterally. *Waptia* probably lived on and near the bottom as a walker and swimmer that fed on small organics. It appears to have been a primitive crustaceomorph (stem crustacean), possibly close to the malacostracans (shrimps, etc.).[65]

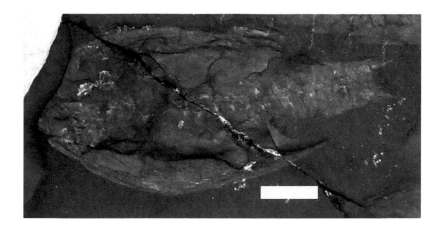

6.33. The bivalved arthropod *Canadaspis* from the Burgess Shale (USNM 57708). Scale bare = 1 cm.

Courtesy of Smithsonian Institution.

Canadaspis perfecta

Canadaspis is a relatively common bivalved crustacean—not just in the Burgess Shale (fig. 6.33; plate 15c), but in a number of other formations of the Cambrian in North America and around the world. The eyes were small and the antennae short; the dozen or so limbs were attached to a body mostly covered by the carapace; and the segmented abdomen was short and thick, with a pair of short spikes on the telson.

Perspicaris dictynna

Perspicaris is a rare bivalved crustacean from the Burgess Shale Formation, less abundant—and also less primitive—than *Waptia*. Its head had two large, stalked eyes that gave the animal its name as the "sharp sighted shrimp"; the antennae were relatively short. The abdomen had six segments without appendages, with a posterior forked telson segment fringed with small spikes.[66] The limbs of the main body were covered by the carapace valves and were short and paddle shaped. Considering the eyes and limbs preserved in *Perspicaris*, it appears that this animal was a swimmer in the water column in relatively deep parts of the ocean, ones in which light was still present but in short supply. Indeed, the estimated depth of the Burgess Shale deposits at the Walcott Quarry is right at the edge of good available light—below it light was still technically present for many more meters but photosynthesis no longer would have been effective. Where light is available but not abundant, the large eyes of the sharp sighted shrimp would be highly valued by their owner. The swimming habit of *Perspicaris* would partially explain its rareness in the fauna, as species living in the water column would be less likely to be caught up in the sea-bottom slides that helped bury other taxa.

Isoxys acutangulus and I. longissimus

Isoxys is a soft-bodied, bivalved arthropod that is widely known throughout Cambrian deposits in North America and elsewhere. The first species of this genus was named by Walcott from a specimen found in the Cambrian of Tennessee, and the first specimen of *I. acutangulus* was named from the Mount Stephen Trilobite Beds. *Isoxys* was later found at the Walcott Quarry and several other sites in the area. The distinctive carapace valves (or head shield) of *Isoxys* have anterior and posterior extensions along the dorsal articulation, none longer than those of *Isoxys longissimus*. For many years we knew only the carapace of this animal, but recent specimens have finally shown us what the rest of the body looked like. *Isoxys* had a head with large, bulbous eyes and – anteriorly – two long, grasping appendages; most of the rest of the head was covered by the head shield. The thorax consisted of 13 segments with long, paddle-like appendages, and there was a short, tail fluke–shaped telson. Interestingly, the head shield covered most of the body segments but lifted clear of them posteriorly so that there was a significant space between the last thoracic segment and the posterior "spike" of the head shield.

Isoxys acutangulus is known from about 300 specimens from the Burgess Shale, but there are only a handful of the long-spined *I. longissimus* known. *Isoxys* was probably a free-swimming genus, as indicated by its paddle-like limbs; the large eyes and long grasping appendages of the head suggest that it was a predator of small prey items living in the water column.[67]

Nereocaris exilis

This recently described bivalved arthropod from the Burgess Shale on Mount Stephen was about 14 cm (5.5 in.) long and had round eyes and a head and thorax mostly covered by the carapace. The abdomen comprises more than half the length of the animal and consists of about 60 segments, with a "tail" section (telson) composed of three sets of spiny processes. *Nereocaris* is a primitive bivalved arthropod distantly related among arthropods to *Branchiocaris*, *Perspicaris*, and *Canadaspis*. *Nereocaris* and its associated but paraphyletic fellow bivalved arthropods appear to demonstrate that the jointing of the exoskeleton in arthropods first evolved to facilitate swimming, and that benthic (bottom-walking) habits are a more derived condition within the group.[68]

Priscansermarinus barnetti

Hundreds of millions of years before they would be able to attach themselves to seemingly everything fixed (or not nailed down) in the ocean, from pier pilings to rocky shores, to ships and the snouts of whales, barnacles may have been present in the Burgess Shale. *Priscansermarinus* was a medium-sized genus that often lived in groups and looked somewhat like a bivalved mushroom. These fossils were described as possibly being goose barnacles, the oldest ever recorded. This identification has met, and was proposed, with some caution, and in general most paleontologists and

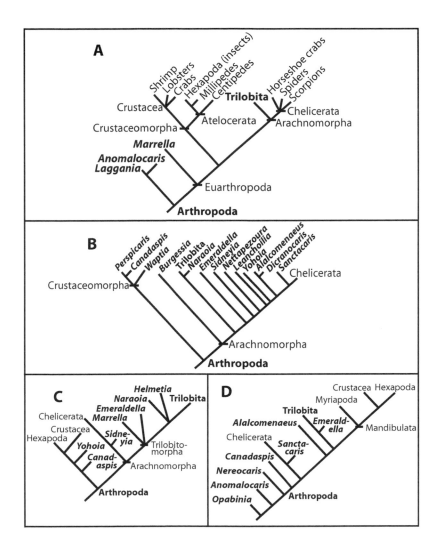

6.34. Hypothesized relationships of Cambrian and modern arthropods. (A) Classification of arthropods into crustaceomorphs and arachnomorphs, with trilobites allied with chelicerates in the Arachnomorpha. Note differences with traditional classification in figure 4.11. (B) Relationships of some Cambrian taxa within the Arachnomorpha and Crustaceomorpha. (C) Another classification with arachnomorphs but with crustaceomorphs broken down. (D) Classification of arthropods into chelicerates and Mandibulata, with trilobites allied with the mandibulates. In (A)–(D) bolded taxa are specifically Cambrian forms.

(A) and (B) based on data in Wills et al. (1998) and Hendricks and Lieberman (2008); (C) and (D) based loosely on data in Scholtz and Edgecombe (2005, 2006) and Edgecombe (2010).

barnacle workers do not seem sure what to make of *Priscansermarinus*. The possibility of barnacles in the Burgess Shale, however, is intriguing.[69]

Arthropod Systematics

We first met arthropods in chapter 4 when we encountered the trilobites and several other forms of the Early Cambrian. A few more non-trilobite arthropods appeared in chapter 5 (along with plenty of trilobites, too). But we have really gotten into a crawling, spindly mass of non-trilobite arthropod taxa in this chapter with all the representatives from the Burgess Shale. How these forms relate to each other and to modern arthropod groups with which most of us are familiar is worth a few comments. First, arthropod systematics have been a source of some controversy for some time, and in recent years, with an acceleration in the number of fossil forms being found, and with molecular studies of modern forms contributing insights as well, while some issues appear to some extent more resolved, others are becoming more contentious. Some arthropod groups have been allied at some point with nearly every other group in their clade. One point almost universally agreed on is that arthropods are monophyletic: the groups of arthropods share a common ancestor and did not emerge from separate lines.

As we saw in chapter 4, in relatively recent traditional classification the arthropods appear to split into two main clades or groups defined by characters shared by all (or most) members. These clades include the Crustaceomorpha and the Arachnomorpha. The crustaceomorphs include the Crustacea, which includes shrimp, lobsters, and crabs, among many others; and the Hexapoda, or insects (fig. 6.34a). The arachnomorphs include the Trilobita and the Chelicerata, the latter of which includes scorpions, spiders, and horseshoe crabs, possibly along with pycnogonids (sea spiders).[70] The Cambrian forms that we have been discussing fit into a variety of spots on this cladogram. Figure 6.34b shows hypothesized relationships of a number of Cambrian arthropods as they fit within this scheme. Most of the bivalved arthropods, genera such as *Waptia*, *Perspicaris*, and *Canadaspis*, appear to be crustaceans or more generalized crustaceomorphs, although some seem to be basal euarthropods. Anomalocaridids such as *Anomalocaris* and *Peytoia* (*Laggania*) may be basal arthropods; some workers place them just outside Arthropoda, but either way they are very close to the base of the arthropod tree. *Marrella* is often classified as a basal euarthropod or as near the Arachnomorpha (fig. 6.34a). Many other forms, such as the Burgess Shale's *Yohoia*, *Alalcomenaeus*, *Emeraldella*, and *Sidneyia*, appear to be arachnomorphs related to but more primitive than chelicerates; *Sanctacaris* from the Burgess Shale is a chelicerate (fig. 6.34b). There are also other classifications with several Burgess Shale taxa in different positions within and outside the Arachnomorpha (fig. 6.34c).

An alternative classification has two groups of arthropods, but they are the Chelicerata and the Mandibulata, another name for crustaceans, insects, and their allies. This classification (fig. 6.34d) has many of the bivalved arthropods (such as *Canadaspis*) as more primitive arthropods, outside the Crustacea. This hypothesis of relationships (fig. 6.34d) still has Hexapoda (insects), Myriapods (millipedes and centipedes), and Crustacea closely related, as they are in figure 6.34a, but notice that crustaceans and insects are more closely related to each other than either is to millipedes or centipedes; this is opposite of the other classification (fig. 6.34a), in which myriapods are closer to insects than crustaceans but within that millipedes are closer than centipedes. Most significant about the classification in figure 6.34d, however, is the position of the trilobites. This classification does away with the concept of the Arachnomorpha and the idea that trilobites are related to the chelicerates; instead, comparisons of the structures of the head and the origins of limbs and antenna in all these various groups appear to indicate that the trilobites are more closely associated with crustaceans and insects than they are with chelicerates.

This upends years of association of chelicerates and trilobites. At least one published but general classification of arthropods leaves the relationship of trilobites to chelicerates and mandibulates unresolved.[71]

The Creeping Stomach

A species just recently described from the Burgess Shale Formation, and apparently not found previously, is *Herpetogaster collinsi*, whose generic name, roughly translated, means "creeping stomach."[72] This was a truly otherworldly looking animal with an elongate, bag-like body with long, finely branching tentacles at the anterior end and an elongate process called a stolon protruding from the mid-posterior end of the trunk. The pharynx and stomach of the animal are large, the intestine short in comparison; the trunk appears to be segmented; and the stolon has a distal attachment disc that helped anchor the animal to the substrate. The trunk appears to have been free just above the seafloor sediment, secured to it through the stolon, with the tentacles and mouth facing upward, capturing food items from the water.

What are we to make of such an animal as *Herpetogaster*? A sock-like body attached to a stem with tentacles so branched they look like parts of a bush, all attached to what looks like a stem? This is an animal that is not easily classified at first and, if it were not extinct, it would appear to have potential as an incredibly expensive delicacy – an off-the-menu sushi dish for the extra adventurous. But it probably would have tasted terrible. It turns out we do have some idea where this animal sits in the scheme of animal classification. It appears that *Herpetogaster* is related to echinoderms, the group that includes modern starfish, although it is not an echinoderm itself. In fact, in its official description in 2010, paleontologists Jean-Bernard Caron, Simon Conway Morris, and Degan Shu proposed that *Herpetogaster* and several other taxa, including the Burgess Shale's *Eldonia*, form a clade that would be the sister taxon to echinoderms and the hemichordates. Or it could be the sister clade to one of those groups individually and not the other. In any case, *Herpetogaster* is closely related to both echinoderms and hemichordates and is one of the Burgess Shale's most morphologically interesting animals.[73]

Banffia: Old Taxon, New Phylum

The Burgess Shale form *Banffia constricta* was named by C. D. Walcott in 1911, so it has been known for more than 100 years. But its classification has remained tough to crack, and it has spent time believed to be a worm and eventually listed as "Problematica," perhaps a label that would be striven for by species if they were bent on stumping us, because that is essentially what that classification status is, where we place (temporarily, we hope) taxa we just cannot quite figure out. Of course, for decades all we knew of *Banffia* was the slightly more than a dozen specimens that Walcott collected in his time up on Fossil Ridge. But in the 1980s and 1990s the Royal Ontario Museum returned to those sites and turned up

6.35. Reconstruction of the Burgess Shale vetulicolian *Banffia*. Length ~10 cm.

more than 300 additional specimens. *Banffia* has also been turning up in decent numbers recently in the Spence Shale at Miners Hollow in Utah. What we found out from these fossils and from others turning up around the same time in China made us realize *Banffia* was even more problematic than we thought.

Banffia is an unusual animal about 5 cm (2 in.) long with a carapaced anterior section and a segmented posterior structure (fig. 6.35). The front end of the anterior section has a soft, round mouth, and the posterior section appears to be twisted slightly. The midsection between the anterior and posterior sections may be slightly constricted. *Banffia* appears to have lived on the bottom of the ocean, perhaps in the sediment, and likely filtered food out of the water or from the sediment; its occurrence in large numbers on some shale slabs suggests that it may have lived in large groups at times.[74]

Banffia's unusual morphology proved to be similar to several other forms that appeared in Cambrian rocks in recent decades, and in 2001 it was proposed that these species constituted a separate phylum, the Vetulicolia, whose relationships to other phyla are even today somewhat unclear. We will discuss the vetulicolians more in chapter 7.

Sea Lilies?

A rare form in the Burgess Shale is *Echmatocrinus*, a stout, stalked, filter-feeding animal that lived on the bottom of the ocean. It has been debated what type of animal *Echmatocrinus* was, and it may have been an octocoral, but it has also been described as a basal crinoid.[75] Paleontologists James Sprinkle and Desmond Collins recently listed the pros and cons for each case (*Echmatocrinus* as crinoid or octocoral), and determined that the evidence was inconclusive but that the weight of more pros and fewer cons was on the side of a stem-crinoid interpretation. Crinoids are known as sea lilies today, and they are plated echinoderms somewhat similar to eocrinoids, but the history of the Crinoidea extends through a spectacular period in the Paleozoic all the way to today. We will cover these and other echinoderms more in the next chapter.

Ancient Written Rocks: Cambrian Graptolites

A specimen collected from the Burgess Shale Formation was described in 1931 as a hydrozoan cnidarian,[76] although it was initially and later thought to be a member of the **graptolites,** a group of fossils common in marine rocks of later in the Paleozoic. This Burgess Shale fossil is *Chaunograptus scandens*, a series of long filaments with alternating buds that does superficially resemble a primitive graptolite (fig. 6.36). Graptolites are small fossils often preserved on dark shales, and they can look a lot like small hacksaw blades drawn on the rock with a graphite pencil—thus their name, which roughly translated means "written stones." Some graptolites grew on the bottom of the ocean, whereas others were free-floating pelagic forms. It appears that graptolites, which became common by the

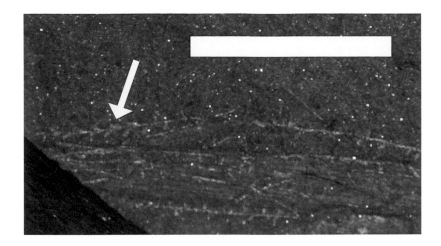

6.36. Fossil of the possible graptolite from the Burgess Shale, *Chaunograptus* (USNM 83484a). Note small cup-shaped structures alternately extending from the main strand (arrow). Scale bar = 1 cm.

Courtesy of Smithsonian Institution.

Ordovician but appeared at least by the Middle Cambrian, are fossils of budding, chitinous colonial structures of animals of rather tiny stature that may have been hemichordates, a group that includes modern acorn worms and **pterobranchs.** Pterobranchs are also known from Cambrian rocks, but graptolites are an extinct group.

The modern hemichordate pterobranchs comprise three genera and are colonial, or at least live in aggregates of individuals. Colonial forms such as *Rhabdopleura* have individuals living each in a separate, tube-shaped living chamber that grows off a main horizontal tube structure. Each animal is sac-shaped but has a mouth, tentacles, cephalic shield, single gill slit (in some forms), stomach, and U-shaped gut, and the individual is attached to the colony structure with a fleshy stalk with which it can retract itself into its individual living chamber for protection. These pterobranchs filter food from the seawater using their tentacles to strain food (capturing it with a sticky mucus covering the tentacles, which sometimes form a net) and move it to the mouth with small cilia on the tentacles. Although graptolites and pterobranchs are closely related, they are separate biological lines, and each appears in the Middle Cambrian. *Chaunograptus* may have been a graptolite, but it clearly was not a ptero-branch. Could the fact that it is very rare in the Burgess Shale Formation suggest that it was one of the free-floating pelagic graptolites? Cambrian graptolites are also known from a number of other deposits, including Upper Cambrian rocks in Tennessee and Colorado.

True pterobranch hemichordates are known from at least the Middle Cambrian. In fact, the genus *Rhabdopleura*, our modern example from above, is even known from one species found fossilized in Cambrian rocks in Siberia—making today's *Rhabdopleura* a living fossil that has survived relatively unchanged in 520 million years! Pterobranchs are also found in other Middle Cambrian sections, including ones in the Czech Republic and in the Wheeler Formation of Utah, the latter formation of which we will see more of in the next chapter. This Middle Cambrian pterobranch from the Wheeler Formation was described as *?Cephalodiscus* sp. and

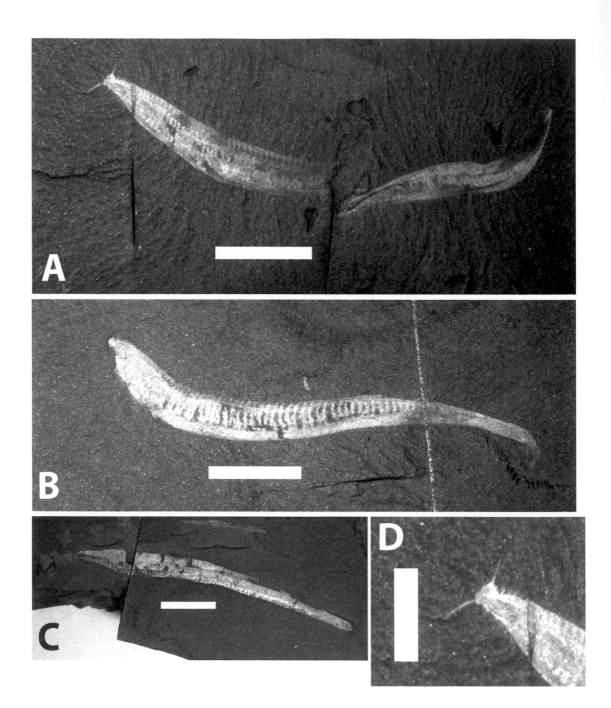

consists of many elongate and upright tubes branching off from the base, a form rather different from most graptolites.[77]

Roots of the Vertebrate Tree: The Cambrian Chordates

Given the diversity present in the fossil biota of the Burgess Shale, it may not be surprising that so many modern groups of animals can trace their lineages back at least that far. Even the origins of our own phylum, the Chordata, lie deep in the Cambrian. We are a bit biased in our fascination with the chordates of the Cambrian period. This is, after all, a rather minor group with a relatively spotty fossil record, but as it represents the origins of our own line, we cannot help but pay a little extra attention to these few species. Although the chordates were nowhere ecologically dominant at the time, their occurrence in any formation of the Cambrian is of interest to us, if for no other reason than that we are looking at the oldest members of our own phylum, ancestors that laid the genetic and morphological groundwork for things as diverse as tuna, gophers, and turtles. And in North America two representatives of the chordate phylum are found in the Burgess Shale. These are *Pikaia* and *Metaspriggina*. Although we know of plenty of Cambrian chordates and probable vertebrates worldwide,[78] they are not very abundant at any given locality; in the Burgess Shale there are only 114 specimens of *Pikaia* (fig. 6.37) and just two of *Metaspriggina*. The vertebrates known from rocks in China are effectively Cambrian-age jawless fish, not far from lampreys and hagfish, as far as we can tell.

Chordates originate in the Early Cambrian, with specimens known from China; the Burgess Shale chordates, then, are slightly later in the game. The chordates are united by a short list of characters that includes a **dorsal hollow nerve cord** (forming the brain and spinal cord), a type of nerve cord that contrasts with the solid and ventral one in most non-chordate metazoans; a **notochord** (a stiffening rod between the gut and the nerve cord); and a muscular **postanal tail.** Many chordates are also characterized by chevron-shaped muscles called **myomeres** along the body. Vertebrates are chordates but go one step further by developing cartilaginous or bony vertebrae, teeth, and a dorsal hollow nerve cord that expands anteriorly into a more complex brain.

Chordates include the cephalochordates like *Branchiostoma* (lancelets); the urochordates, which are also known as tunicates or sea squirts; and the vertebrates, beginning with jawless fish and later including bony fish and tetrapods. Cambrian urochordates are represented by *Shankouclava* from the Maotianshan Shale in China.[79] Urochordate adults are stationary and only the juveniles possess some of the characteristics of chordates. The hemichordates are now recognized to be a sister taxon to the echinoderms. Among modern vertebrates, the most primitive (or basal) are the lamprey and hagfish. The fossil conodonts (see chapter 8) are more derived than lampreys and hagfish but are basal among jawless fish, which are paraphyletic.[80]

The forms from China include the oldest chordates, both apparent basal chordates of a grade similar to the cephalochordate *Branchiostoma* and true vertebrates. *Haikouella* is an Early Cambrian chordate and appears to have had gills, a heart, brain, and arteries preserved, and it is known from more than 300 specimens. The myomeres are relatively straight and there are 25 of them.[81] *Haikouichthys* is also Early Cambrian in age and is a true vertebrate, at the level of a jawless fish; it has eyes, possible ear capsules, possible nasal capsules, gill slits, a dorsal fin, and

6.37. Fossils of the Burgess Shale chordate *Pikaia*. (A) Specimen showing body shape and myomeres along with head and tentacles (USNM 57628). (B) Another specimen showing myomeres (USNM 83940b). (C) Less well preserved specimen (USNM 202217). (D) Close-up of USNM 57628 showing head shape, tentacles, and lateral anterior appendages. Scale bar in (A)–(C) = 1 cm. Scale bar in (D) = 5 mm.

Courtesy of Smithsonian Institution.

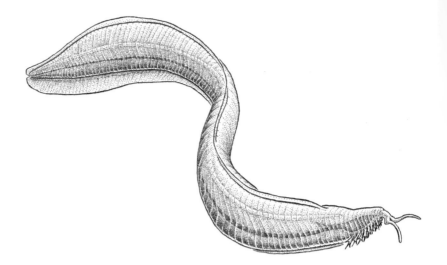

6.38. Life reconstruction of the chordate *Pikaia*. Length ~4 cm.

Drawing by Matt Celeskey.

simple vertebrae, along with chevron myomeres. More than 500 specimens have been found.[82] The Early Cambrian *Myllokunmingia*, also a vertebrate, has chevron-shaped myomeres, gills, vertebrae, a separate notochord, and a dorsal fin.[83]

Chordates are rare in Cambrian rocks in North America, and the oldest specimens are Middle Cambrian, not Early Cambrian in age as in China. *Pikaia* is known from 114 specimens in the Middle Cambrian Burgess Shale, and it was named as a polychaete worm in 1911 by C. D. Walcott. It was later informally suspected to be a chordate, with official published confirmation of this appearing not until 2012. *Pikaia* has about a hundred chevron-shaped myomeres, a nerve chord and notochord, an elongate dorsal fin, a long ventral caudal fin, and a pair of short tentacle-like structures on the head. The small head appears to be laterally bilobed, and behind the head are nine laterally paired appendages with openings near their bases that may be related to a gill-like function. Perhaps unexpectedly, there is no evidence in the specimens for eyes. A gut trace appears to show a small mouth at the posteroventral end of the head and an anus at the posterior end of the body just above the end of the ventral caudal fin. *Pikaia* is a stem chordate that probably swam freely but close to the bottom feeding on small, soft organic material (fig. 6.38).[84]

Metaspriggina is known from just two specimens from the Burgess Shale, and it is about 55 mm long. It is apparently a basal chordate and not a vertebrate, and it has well-developed myomeres and a linear trace that may be either the gut or the notochord. The cranial area is not well preserved and is tough to interpret.[85]

It is possible that some of the taxa we have interpreted as being basal chordates could be vertebrates. One recent study of modern chordates discovered that many of the characters shared by chordates (both basal and vertebrate) last the longest after the animal dies, and thus are most likely to preserve; conversely, some of the key structures of vertebrates are among the first to decay. This leads to what the authors called "stem-ward slippage," the tendency for a decayed chordate to retain as a preserved fossil more primitive characters and to lose its more derived ones and thus appear to be a more primitive or stem group animal than it actually is.[86] Thus, something like *Metaspriggina*, which is known from few specimens and appears to be a stem chordate, could actually be something more derived, like a vertebrate, and just have decayed enough not to preserve its gills, eyes, and heart. So the picture of the Cambrian chordates may be more complex than we realize. Most Cambrian chordates were probably swimmers in the water column, but we don't fully understand how they fed.

Beyond Yoho

Recent work by the Royal Ontario Museum has demonstrated that Burgess Shale–type preservation and soft-bodied taxa occur not only at the base of the Cathedral Escarpment in deeper water. At least one site in Kootenay National Park, British Columbia, is in the "thin" Stephen Formation sediments that accumulated on top of the Cathedral Formation, above and slightly landward from the escarpment itself. The Stanley Glacier site has yielded a biota of algae, hyoliths, brachiopods, ptychopariid trilobites, sponges, worms, and soft-bodied arthropods such as *Tuzoia*, *Sidneyia*, *Hurdia*, and the new anomalocaridid *Stanleycaris*. The recently named great-appendage arthropod *Kootenichela deppi* from the Stanley Glacier site is a relative of *Yohoia*, *Alalcomenaeus*, and *Leancoilia* from the Burgess Shale. *Kootenichela*'s specific epithet is also in honor of actor Johnny Depp for his titular role in the movie *Edward Scissorhands*. *Kootenichela* and some of its relatives do in fact have "great appendages" with elements that collectively look surprisingly like shears, although they would not have functioned as such.[87]

Common Worms

The priapulid worms, particularly *Ottoia* (fig. 6.39; plate 15b), are abundant in some layers of the Burgess Shale. These predators actually appear to form a significant proportion of the total biovolume of the biota and so are an important part of the fauna. They are not the most unusual or eye-catching elements of the fauna, but neither should they be easily dismissed. Priapulids play a role in more than just the Burgess Shale's ecosystem; we have already seen them in the Spence Shale. They are abundant here in the Burgess. And we will see more of them in formations in Utah, as we cover priapulids in more detail in the next chapter.

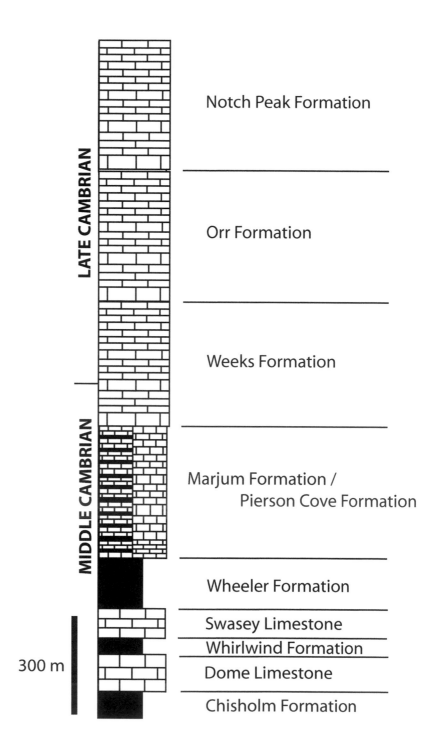

7.1. Stratigraphic section of geologic formations exposed in the House Range of western Utah. Solid black pattern = shale; brick pattern = limestone.

LATE CAMBRIAN

Notch Peak Formation

Orr Formation

Weeks Formation

MIDDLE CAMBRIAN

Marjum Formation / Pierson Cove Formation

Wheeler Formation

Swasey Limestone

Whirlwind Formation

Dome Limestone

Chisholm Formation

300 m

WE NOW TRAVEL BACK TO THE EASTERN PART OF THE BASIN AND Range Province in the western United States. In the Burgess Shale we spent time in the *Ehmaniella* zone of the Middle Cambrian (fig. 4.19), above the numerous *Glossopleura* zone sites we had seen in chapter 5. We will now move into the upper part of the *Ehmaniella* zone and the overlying *Bolaspidella* zone, into rocks just younger than the Burgess Shale. We are nearing the last part of the Middle Cambrian.

Driving down south, back into the arid terrain of the Great Basin, we find ourselves a ways west of the farming town of Delta in western Utah. We turn north off the "Loneliest Road in America" (U.S. Highway 50) and travel about half an hour into the heart of the House Range. This north–south strip of mountains, covered in piñon and juniper trees, is surrounded, as are many ranges in the area, by plains of very sparse brush and at least one nearby dry lake. The House Range has long been known to collectors drawn here for the beautifully preserved, articulated, and abundant trilobite fossils. In fact, most of the formations here were named by Walcott in the early twentieth century. As we've seen, there are few areas that Walcott didn't get to. There are a number of quarries along the eastern flank of the House Range, and many are in three of the most productive formations in the area: the Wheeler, Marjum, and Weeks formations (fig. 7.1), which generally dip east off the range. Relating back to the Marble Mountains in chapters 4 and 5, these three formations in the House Range are approximately equivalent in age to the upper Bonanza King Formation, the uppermost unit in that outcropping in California.

Wheeler Formation

The Wheeler Formation is 145 m (476 ft.) thick and consists mostly of dark to medium gray, calcareous shale with some minor interbedded limestone. These layers are deposited in repeated cycles, suggesting patterns of sea level rise and fall over millions of years. The Wheeler is largely within the *Ehmaniella* and *Bolaspidella* trilobite zones (fig. 4.19). The Wheeler Formation is part of the outer detrital belt and comprises dark gray, relatively deepwater shales; geologists have traced the lateral contact with the carbonate belt to the east of the Wheeler's exposure in the House Range. The shales here were deposited in deep water adjacent to the shallower water of the carbonate belt because they were within a tectonic structure called the House Range embayment. This was a

House Range Motherlode

7.2. Digging in the Middle Cambrian Wheeler Formation of the House Range, Utah. (A) Wheeler Amphitheater. (B) Shallow pit that produces many agnostids and *Elrathia*. (C) A site that produced several soft-bodied taxa near Mockingbird Gulch.

fault-bounded basin of deeper water, about 120 km (74 mi.) north to south, which projected from the open ocean toward the (modern day) east to as much as 121–162 km (75–100 mi.) into the carbonate belt. Although not necessarily an escarpment of the steepness seen in the Burgess Shale area, the slope of the House Range embayment's south edge indicates that the Wheeler Formation was also deposited in a relatively deep setting (plate 19) next to a shallower carbonate bank. To the southwest of the deep water of the House Range embayment were adjacent parts of the surrounding shallow carbonate shelf, represented by the Highland Peak and Bonanza King formations.[1]

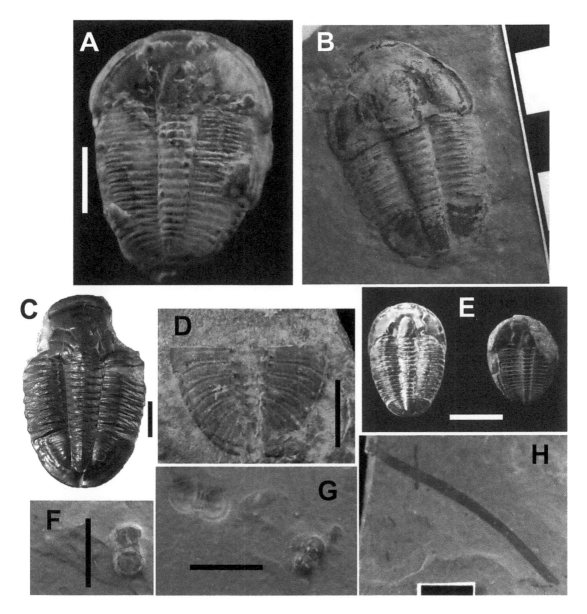

And preserved in this deep setting in the Wheeler Formation are some of the most abundant, well-preserved, and frequently complete trilobite specimens in North America. Trilobites are so abundant here it becomes almost boring to collect them. Particularly interesting is the proclivity for many of them to separate intact from their encasing shale (after a period of weathering) so that all one must do to find them is to walk or crawl a talus slope and pick them up like so much loose change. It is an embarrassment of riches but one that has provided us with some of the best data from this time in the Cambrian and made many of the trilobites from the formation among the best known in the world. This is the kind of boredom that one can get used to. But it also pays to put the effort into excavation in the Wheeler Formation (fig. 7.2). Not everything

7.3. Middle Cambrian trilobites and associated biota from the Wheeler Formation in the House Range. (A) Complete *Elrathia kingii*. (B) Complete *Asaphiscus wheeleri*. (C) Nearly complete *Asaphiscus wheeleri* missing only free-cheeks. (D) Pygidium of *Bathyuriscus*. (E) Two complete specimens of *Elrathia kingii*. (F)–(G) Agnostid specimens. (H) Fragmentary strand of algae. Scale bars = 1 cm.

All Museum of Western Colorado specimens.

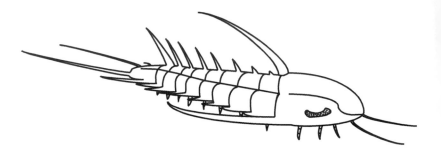

7.4. Reconstruction of *Olenoides vali,* based on data in Robison and Babcock (2011). Length ~10 cm.

can be found just lying there, and some of the best specimens one finds often come from a few hours of labor with a pry-bar and shovel.

There are at least 13 species of trilobites found throughout the Wheeler Formation. This was the time of trilobites such as the corynexochids *Asaphiscus wheeleri* and *Bathyuriscus fimbriatus,* the agnostid *Peronopsis interstricta,* and arguably the all-time most widely sold trilobite on the planet, the ptychopariid *Elrathia kingii,* which stocks the shelves of seemingly every rock shop in North America (fig. 7.3).[2] Characteristics of the Wheeler Formation and the preservation of the material suggest that burial of these trilobites was relatively rapid. Preservation at some sites may have involved burial by loose, fine-grained submarine debris flows. The new Wheeler trilobite *Kootenia randophi* has not only pleural spines and marginal pygidial spines (typical of the genus) but also axial spines on the occipital ring and thoracic segments. Even more dreamlike, *Olenoides vali* has an extra elongate spine on the occipital ring – one so long that it has the appearance of a CB radio antenna, as they used to adorn Jeeps back in the 1970s. In addition, *Olenoides vali* has axial spines on the thoracic segments that increase in length posteriorly, long genal spines, healthy pleural spines, and two to four extra-long pygidial spines (fig. 7.4). It was a spiny world during Wheeler times, and we are seeing that now, thanks to these very well preserved new specimens and very skilled preparation of them.

The trilobite *Elrathia kingii* (fig. 7.3a,e) is very abundant in the Wheeler Formation, having been preserved often intact by rapid burial at the base of the steep slope coming down from the carbonate belt to the deep water of the House Range embayment. So many specimens are available that it has been possible to study the paleobiology of this trilobite species more than most. Rare species are interesting to find, but we have a better chance of learning about trilobites as living animals from the yawn-inducing common ones. Some of what paleontologists have learned about *Elrathia kingii* is that the number of thoracic segments can vary between individuals from 10 to 13, that it does not occur at all sites in the Wheeler but often dominates those at which it is found (or is the only trilobite species), that it feared at least one predator of the Cambrian seas, as indicated by healed bite marks in one specimen (plate 18f),[3] and that it appears to have been an opportunistic scavenger living in deep water near the transition to low-oxygen bottom conditions.[4] There, it was a consumer in an ecosystem possibly based on chemosynthetic (rather than photosynthetic) production.

Robert Gaines is associate professor of geology at Pomona College in Claremont, California. His research concentrates on the Cambrian radiation and the reasons and timing behind it, particularly regarding why exceptional preservation of Burgess Shale–type deposits was so common compared to other intervals of geologic time. He also is interested in the microbial-mineral interactions that sustain the most primitive life forms on Earth. After undergraduate work in geology at the College of William and Mary and a master's in geology and paleontology from the University of Cincinnati, his PhD project at the University of California at Riverside focused on the paleoecology and Burgess Shale–type preservation of the Wheeler Formation of Utah.

The Cambrian Corps 6– Robert R. Gaines

As a youth in Montgomery, Alabama, Dr. Gaines spent time "plucking shark teeth, oysters, and ammonites from Cretaceous marls found widely around the area," he says. He went through the dinosaur phase that many kids do, and that some paleontologists never quite leave behind, but it was the direct influence of his parents that helped set his future path. The first event occurred in the late 1970s era of disco and *Star Wars* when he was 5 years old. His parents returned from a trip out west, and his mother gave him a specimen of the trilobite *Elrathia kingii* from the House Range of Utah. "The power of that small gift was transformative to me," he recalls, "and, when I am working in the field, I always collect fossils to give to children as a result." The next important influence was his parents' taking him as a 13-year-old to hear a lecture by *Wonderful Life: The Burgess Shale and the Nature of History* author and high-profile paleontologist Stephen J. Gould, at Auburn University.

In college, thanks to his accommodating geology mentor, Dr. Jerre Johnston, who detoured a geology field trip for his obsessed student, Bob finally got to visit the Wheeler Formation in the House Range, the source rock for his mother's gift trilobite. "I was blown away," he says, "and these are still my favorite rocks in the world, bar none." Even then he didn't realize that these rocks, from which his parents got his *Elrathia*, would end up being the subject of his dissertation and some subsequent research. That much of this work results in a fairly direct way from his parents' simple vacation souvenir is important to him. "I have my parents to thank entirely," he says, "because they always encouraged me to follow my interests without a second thought towards choosing a practical profession."

Research on the Wheeler Formation and others has led him to the conclusion that most Burgess Shale–type deposits of the Cambrian share a similar mode of preservation. This process depends in part on the slowing down of decay of soft-bodied taxa courtesy of the unique seawater chemistry of the time, which was low in sulfate, relatively low (at least at depth) in oxygen, and more alkaline than today. Alkalinity may have allowed a calcium carbonate barrier layer at the sediment-water interface that helped seal buried carcasses away from destruction, and low oxygen and sulfate would have impeded the work of microbes working to consume the carcasses as well. The microbes "were unable to 'eat' the organic matter of the fossils because they were unable to 'breathe' because of oxidant deprivation," Dr. Gaines says. Thus sequestered from decay, Burgess Shale–type fossils were free to preserve the exquisite, rare detail we know and whistle admiringly about today.

And it may have been changes in this unique seawater chemistry that has seemingly prevented Burgess Shale–type preservation from extending beyond the early to mid-Paleozoic. Some of the most important findings of Bob Gaines's research, and some of the most surprising to him, are the recognition that the Cambrian radiation and exceptional preservation in Burgess Shale–type deposits "may both be tied to a single, common cause – transient chemical conditions in the Earth's ocean and atmospheric system." This finding was, he says, "a real shocker."

The Wheeler Formation is another example of a geologic unit exhibiting Burgess Shale–type preservation, with numerous soft-bodied forms preserved in its calcareous gray shale in particular. Among the soft-bodied material are not only animals with unmineralized exoskeletons such as bivalved arthropods (*Branchiocaris*), the arthropod *Mollisonia*, the anomalocaridid arthropods *Anomalocaris*, *Peytoia* (*Laggania*), and *Ecnomocaris* (fig. 7.5), and the trilobite *Naraoia*, but also fully soft organisms like cyanobacteria and algae (*Marpolia*, *Margaretia*, *Yuknessia*; plate 18a,b), the lobopodian *Aysheaia* (plate 18d), and the undetermined worms that are not uncommon in these layers (fig. 7.5c). Also found in the Wheeler Formation are forms such as the eocrinoid *Gogia* (fig. 7.6), the sponges *Choia* and *Diagoniella*, the odd echinoderms *Ctenocystis*, *Castericystis*, and *Cothurnocystis*, and the unusual genera *Chancelloria*, *Wiwaxia*, and *Selkirkia*.[5]

NON-TRILOBITE ARTHROPODS

Only recently described from the Wheeler Formation is the "pitchfork shrimp" *Dicranocaris guntherorum* (plates 18e and 19), a moderately large arthropod with a large rounded cephalon with three pairs of cephalic appendages. The body has nine thoracic segments with biramous appendages, and there are three ring-like abdominal segments in front of an elongate, forked telson.[6]

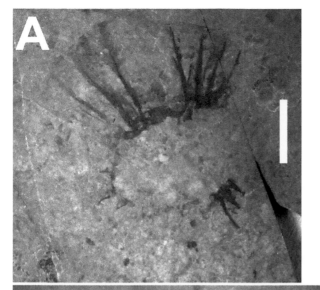

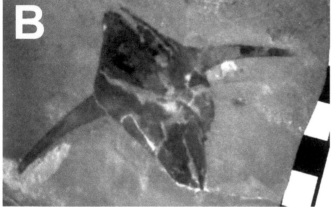

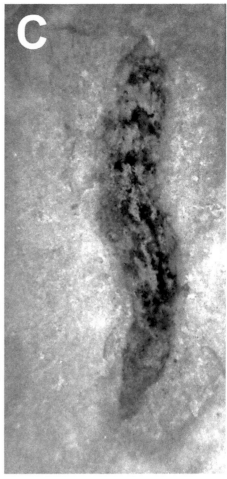

Cambropodus gracilis was found in the Wheeler Formation of the Drum Mountains in Utah, and is a long, multi-segmented arthropod with many uniramous legs (plate 18c). Each segment is short and the head is small, with short antennae and two other pairs of appendages. At least in general appearance, it is similar to modern myriapods, the arthropod group that includes centipedes and millipedes. But *Cambropodus* appears to lack mandibles, which differentiates it from the myriapods and their allies, the insects and crustaceans. The *Cambropodus* specimen is incomplete, so it is unclear how long the animal would have been, but there are at least nine pairs of uniramous thoracic appendages. This is a unique animal among Cambrian faunas of North America, but there is still no consensus on how *Cambropodus* relates to other arthropods. With luck, more specimens will be unearthed in the future which should tell us more about this rather interesting taxon.[7]

An unusual, unmineralized, bivalved arthropod was described from the Wheeler Formation in 1956. *Pseudoarctolepis* had a large, chitinous carapace with a long, curved projection on each side, originating from

7.5. Soft-bodied forms out of the Wheeler Formation. (A) Mouth cone of an anomalocaridid (KUMIP 153093). (B) Carapace of the bivalved arthropod *Pseudoarctolepis* (KUMIP specimen). (C) Undetermined worm (MWC 6919). Scale bars in centimeters.

(A) and (B), University of Kansas; (C), Museum of Western Colorado.

the lateral edge of the valve (fig. 7.5b). These spine-like processes may fold under or splay out when preserved, but they often give *Pseudoarctolepis* a very distinctive appearance in the rock when they are found. What this arthropod needed the spines for is unclear, but based on related forms from the same age it was presumably a free-swimming animal, so the processes would not have impeded movement along the bottom, for example. Perhaps they were developed to try to deter would-be predators.[8]

Speaking of predator deterrence, a bivalved arthropod named *Tuzoia? petersoni* is one of the more unusual out of any Cambrian formation (plate 18g). You may recall that the valves of *Tuzoia* have short spikes all around their perimeters; *T.? petersoni* has three extremely long spikes on at least one end near the dorsal articulation. Any predator would get a mouthful of as much spine as prey from this arthropod.

ECHINODERMS: EOCRINOIDS AND STRANGER BEASTS

Among the fossils relatively common at some sites in the Wheeler Formation are eocrinoids, mostly of the genus *Gogia* (fig. 7.6). This is a genus we have run across at many sites along the journey so far, starting in the Latham Shale (and Chambless Limestone) in the Marble Mountains and continuing through the Bright Angel, Pioche, and Chisholm formations to the Burgess Shale, and now the Wheeler and Marjum formations. *Gogia* and other members of the Eocrinoidea are echinoderms, the "spiny skins" that include species we know today as starfish, sea urchins, and sand dollars, along with a number of other forms including sea cucumbers and crinoids. Echinoderms seem to have increased in diversity from the Early to Middle Cambrian and then taken a slight hit in diversity in the Late Cambrian.

The Echinodermata today includes about 7000 species, but there are close to 13,000 known from fossils. The modern species are almost all marine; there are a few that live in brackish waters, but none inhabit freshwater. Almost all modern echinoderms are benthic animals, and they occupy a range of feeding niches from grazers of algae to predators to suspension feeders. As any pier-hugging starfish indicates, echinoderms have five-lobed radial symmetry and are the only animals known with such a design. Also unique to echinoderms is their **water vascular system,** a network of fluid-filled canals and chambers running out each appendage from a central ring canal. This system includes the ends of the canals, called podia of the **tube feet,** which function in sensation, feeding, movement, and gas exchange; podia are what you see on the bottom of the arms of a starfish helping the animal "walk," very slowly, along the sides of a fish tank, for example. Despite their apparent strangeness, however, echinoderms are a sister group to our own phylum, the chordates (fig. 1.10), the shared similarities of the two phyla being apparent in the early development of the cells and tissues of the respective embryos.

Echinoderms usually have about five body radii called **ambulacra** arranged around a central mouth. Although echinoderms are

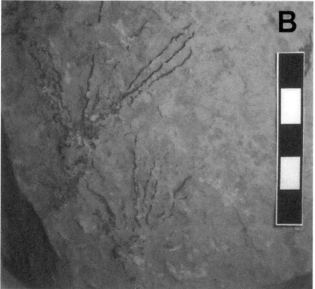

7.6. (A) Site in the Wheeler Formation of the House Range of Utah rich in eocrinoids. (B) Pair of *Gogia* eocrinoids from site in (A) (MWC 6837). Scale bar in centimeters.

(B), Museum of Western Colorado.

characteristically pentaradiate as adults, they are, interestingly, bilaterally symmetrical as larvae, suggesting that the group's origins lie with a bilateral ancestor. As already mentioned, the water vascular system of echinoderms rings the mouth in the central disc area and runs out along the ambulacra. Plates, mostly of calcium carbonate, form an endoskeleton embedded in the skin under the epidermis; in some echinoderms or parts of the body these plates are well spaced, but in many they are tightly packed, forming a type of armor protecting the animal. Some of the plates may develop spines for protection as well.

The **Eocrinoidea** were echinoderms that appeared in the Early Cambrian and were one of the blastozoan lines, relatives of the ancient and modern crinoids. As we have seen throughout this book, *Gogia* was a common genus. The eocrinoids consisted of a stalk that attached to the seafloor, and this was topped by an enlarged and rounded **calyx,** equivalent to the central disc of modern echinoderms. The calyx contained three to five ambulacra, and from these extended several feeding appendages (**brachioles**) that stretched up into the water. The mouth was situated on the top of the calyx, ringed by the brachioles, and faced up. The entire animal (stalk, calyx, and brachioles) "stood" on the seafloor, possibly attached to the substrate by the stalk (at least oriented vertically by it), filtering food particles from the water with podia in its brachioles and moving the food then to the mouth. Small, irregularly shaped plates covered the stalk and brachioles, and larger star-shaped plates covered the calyx in most Cambrian eocrinoids. It is believed that eocrinoids like *Gogia* attached to hard skeletal parts of other animals when possible; some Cambrian eocrinoids have been found attached to brachiopod shells and trilobite fragments.

There are a number of species of *Gogia* named from the Early and Middle Cambrian in North America. Among the differences between species are the form and arrangements of the plates covering the calyx and the number of brachioles. Most species of *Gogia* appear to have lived in a low-energy (sometimes deep) marine environment on a muddy bottom. There has been some speculation, based on comparisons with modern crinoid echinoderms, of a correlation between high numbers of brachioles and shallow paleoenvironments and between low numbers and deep settings; this remains unconfirmed, however. Some species of eocrinoids had coiled brachioles, which also may have been an adaptation to low-energy environments.[9]

The odd fossil forms *Ctenocystis*, *Castericystis*, and *Cothurnocystis* from the Wheeler Formation are all members of the para-echinoderm group **Homalozoa,** also known as the carpoids (plate 19). These relatives of the earliest echinoderms were not quite echinoderms themselves, but were interesting—and not extremely rare—animals in their own right. *Castericystis* is known from more than 200 specimens. Homalozoans (carpoids) were covered in an irregular armor of calcareous plates, like many echinoderms, and they had a single brachiole leading to the mouth, but they were not radially symmetrical and it is unclear if they had a water vascular system. What appears to be a tail in carpoids is called the **aulacophore** and may have been employed to orient the animal. Between the brachiole and aulacophore was the main part of the body, also covered by irregular plates. Carpoids probably lived on the bottom of the ocean as sometimes mobile, epifaunal suspension feeders that may have occasionally attached themselves to the bottom. In rare cases, juvenile carpoids have been found anchored to adult individuals.[10]

The first true echinoderms were likely the helicoplacoids, which we met briefly in chapter 4 in the Poleta Formation. Helicoplacoids appear in the Early Cambrian but do not last a lot longer than that. It is not that they were poorly adapted, but they appear to have been transitional between the bilateral design of the ancestral proto-echinoderms and

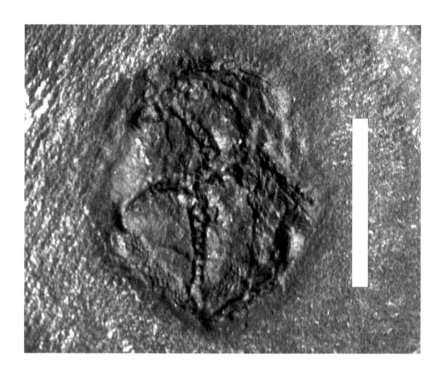

7.7. The edrioasteroid echino-derm *Walcottidiscus,* from the Burgess Shale (USNM 90754). Scale bar = 1 cm.

Courtesy of Smithsonian Institution.

the pentaradially symmetrical design of later echinoderms such as the eocrinoids and modern forms. Helicoplacoids were shaped somewhat like potatoes with many rows of spirally arranged skeletal plates covering them and three (not yet five) ambulacra. The mouth was placed not at the apex like in many echinoderms (eocrinoids, for example) but rather off to one side. The animals were benthic and probably fed on suspended food particles or detritus; it is unclear if they were mobile or sessile.[11]

Yet another class of echinoderms known from the Cambrian are the **edrioasteroids.** These were moderately convex, disc-like animals with calcareous plates that lived attached to the bottom of the ocean (fig. 7.7); there do not appear to have been any brachioles or an elongate stalk. There were five ambulacra radiating from the centrally placed mouth. In overall appearance, edrioasteroids look a bit like sand dollars, except that the ambulacra often curved away from the mouth (giving them collectively a spiral pattern), and the protective plates did not compose a single shell as in sand dollars. In some ways the edrioasteroids looked like sea urchins devoid of spines. Some forms could be only mildly convex, whereas others were very bulbous. Cambrian edrioasteroids in North America include, for instance, rare individuals from the Burgess Shale (fig. 7.7) and a number of specimens from the Chisholm Formation in Nevada. Most later edrioasteroids probably attached to hard substrates, and some have been found attached to the shells of brachiopods. Edrio-asteroids ranged into the Permian period but then became extinct.

If we ever find life on other planets, I would be willing to bet on two things: that the first and most abundant species found will look nothing like a vertebrate or arthropod, despite what our limited human imagina-tions seem to consistently produce in the movies; and if the alien species even proves to be multicellular,[12] it might just remind us of a sea cucumber in its unfamiliarity.[13] The **Holothuroidea** includes the modern sea cucumbers, and fossil forms of the group are known from the Burgess Shale, possibly in the form of *Eldonia* (see below). Like most other echinoderms, modern sea cucumbers (holothuroids) have a water vascular system arranged around a central ring canal; most have an elongate body, stretched out along the oral–anal axis, and the water vascular system has arranged itself to accommodate this. There is a single gonad, rather than several as is seen in most echinoderms. The podia are retained and used for locomotion and attachment. The holothuroids today demonstrate a wide range of body types, from wormlike species to pelagic ones that look superficially like an inverted, free-floating, web-limbed octopus (*Pelagothuria*).

Holothuroids often have tentacles near the mouth to assist with feeding, and these are supplied by the water vascular system. Many holothuroids retain five-lobe radial symmetry but it is stretched out along the long axis, and the ambulacra are enclosed grooves. The gut runs through the long axis of the animal and exits the anus. Near the end of the gut is a canal leading to the respiratory tree, a complex assortment of diverticula used in gas exchange; water for gas exchange is pumped in through the anus. Also attached near the respiratory tree internally are the Cuvierian tubules, sticky strings that can be shot out of a break in the posterior body wall to entangle prospective predators. In this defensive phenomenon of evisceration, holothuroids may also eject some of the gut and other organs, but after the animal escapes the lost elements are usually regenerated. Holothuroids are generally softer than other echinoderms and are covered with fewer (and less extensive) calcareous plates. Most sea cucumbers are deposit feeders or epibenthic scavengers, eating organic material in or on the sea bottom sediments; others are suspension feeders.

The biology and affinities of the Cambrian fossil *Eldonia* (fig. 7.8) are not well understood, but it is possibly a holothuroid. *Eldonia* is a disc-like animal overall, but its gut is coiled, and it possesses tentacles around the mouth (fig. 7.8b), as do many holothuroids. The disc-like shape is similar to that of jellyfish, but the coiled gut is unlike any cnidarian. C. D. Walcott suggested that *Eldonia* was a holothuroid, and perhaps it had evolved a coiled gut and jellyfish-like disc shape in order to become a pelagic suspension-feeding form. Did holothuroids start out this way? It is doubtful. *Eldonia* is probably just a holothuroid with an interesting adaptation, although that adaptation may in fact be something different. It appears now that *Eldonia* lived upside down in the sediment with its tentacles facing up; this is based on comparisons with another eldonioid from China. But even the free-floating *Eldonia* idea wasn't so strange, as there are modern pelagic echinoderms such as *Pelagothuria*, and there are forms with the mouth and anus close to each other as in *Eldonia*. But it is fascinating how diverse echinoderms are in body form and ecology, and *Eldonia* shows that this morphological diversity may have begun early in the group's history. Of course, just to make things interesting,

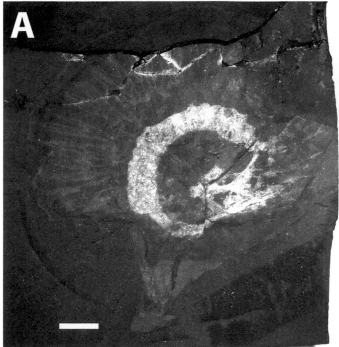

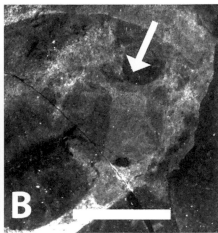

7.8. The Burgess Shale fossil *Eldonia*. (A) Full fossil showing disc shape and coiled gut (USNM 188555). (B) Close-up of mouth (arrow) surrounded by finely branched tentacle pattern in another specimen (USNM 201692). Scale bars = 1 cm.

Courtesy of Smithsonian Institution.

it has been suggested also that *Eldonia* is not an echinoderm at all, but a lophophorate, related to brachiopods, bryozoans, and phoronids (see chapter 8).[14]

In China, a number of ancestral echinoderms have appeared in recent years, suggesting a complex and diverse origin of the modern phylum. Diversity of echinoderms, particularly eocrinoids and edrioasteroids, appears to have increased steadily through the Cambrian period, and by the middle of the Ordovician there were more than three times as many classes of echinoderms as there were in the Early Cambrian. Early and Middle Cambrian species seem to have attached mainly to available firm skeletal debris in muddy substrates on the bottom of the ocean, with later species attaching to hard surfaces such as rocks or the shells of other living animals. This latter trend was probably related to the increase in sediment burrowing that occurred after the Middle Cambrian. When such burrowing activity took off in the Late Cambrian, it turned firm, muddy, algal mat-covered substrates of previous times into mixed up, soupy mud layers unsuitable for echinoderms that wanted to attach to something stable. It is believed that this change in conditions of the bottom of the sea, driven by diversification and changes in the behavior of animals that lived in the sediment, affected the evolution of the echinoderms that were just trying to find a solid place to attach and filter food from the water.[15]

CHANCELLORIA

One odd animal preserved in the Wheeler Formation (and a few other units in the Cambrian of North America, including the Burgess Shale)

7.9. The enigmatic animal *Chancelloria* (USNM 66528) from the Burgess Shale. Scale bar = 1 cm.

Courtesy of Smithsonian Institution.

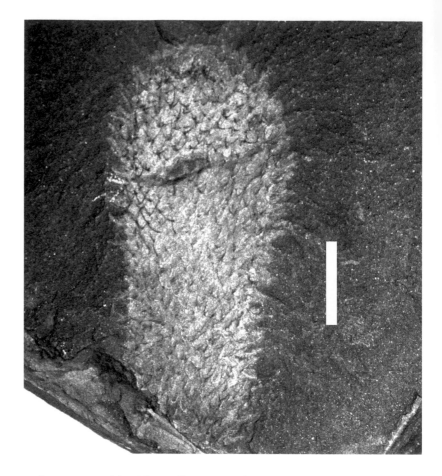

is the spiny, sac-like *Chancelloria*. First described as a sponge by Walcott, *Chancelloria* indeed has a sponge-like body shape and spiny structures superficially not unlike sponge spicules in shape (fig. 7.9). But unlike sponges, which contain solid spicules as internal structural supports, *Chancelloria* possessed hundreds of tiny, hollow, six-rayed (or seven-rayed) sclerites as an external, protective layer. The sclerites generally consist of six radially arranged spikes around a central disc, and they are often composed of calcium carbonate. There are several other chancellorid genera now, including *Allonia* and *Archiasterella*, known from Australia, China, and other regions.[16]

WIWAXIA

Occasionally, fossils of the unusual *Wiwaxia* have shown up in the Wheeler Formation, often single sclerites but sometimes partial individuals. This strange animal was first found in the Burgess Shale and named *Wiwaxia corrugata* (see chapter 6). It appears that *Wiwaxia* crawled around on the sea bottom, feeding on organics that existed on and in the top layer of sediment, a feeding mode known as epifaunal deposit feeding (plate 19). Although *Wiwaxia* appears to have been slug-like, except for the covering of sclerites, its relationships are not well understood. It may

be part of a unique group of stem molluscs, and it has sometimes been compared to a group known as the halkieriids, which include flattened, slug-like forms coated with *Wiwaxia*-like sclerites and two shells loosely resembling those of brachiopods.[17] Regardless of its true relationships, *Wiwaxia* is an intriguing animal to find in the shale.

Marjum Formation

The overlying Marjum Formation is 420 m (1378 ft.) of gray to tan laminated limestone with silty interbeds and is in the *Bolaspidella* zone; the Marjum Formation also contains some units of shale and mudstone (fig. 7.1). The dark gray limestone beds are generally about 2–10 cm (0.8–4 in.) thick and may be finely laminated. The cyclical layering of the Marjum Formation is even more pronounced than in the Wheeler Formation. The Marjum Formation was named by Walcott for outcrops near Marjum Pass in the House Range, and the amount of shale in the formation lessens as one moves away from this type of area. The Marjum Formation, like the Wheeler, represents deposition in the House Range embayment. The layering within the Marjum is a result of frequent, relatively minor rises and falls in sea level (which would have occurred on the order of every 10,000–100,000 years), with silty or shaly layers being deposited in relatively deeper water when higher sea levels moved the carbonate

7.10. Work in the Marjum Formation in the House Range. (A) A trilobite quarry in the Marjum Formation east of Marjum Pass. (B) Clearing a bedding plane for excavation. (C) Close-up of interbedded limestone and shale layers in the Marjum Formation, scale bar = 5 cm.

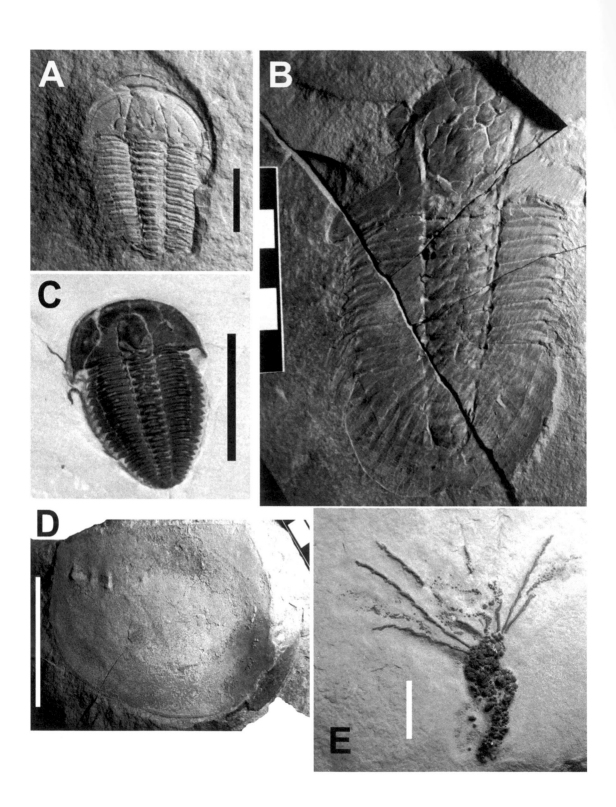

belt farther away toward shore, and the limestone layers being deposited when sea levels were lower and carbonate production was closer to the area. The sediments were probably deposited in quiet, possibly poorly oxygenated water below 50-m (164-ft.) depth, out of reach of the influence of even storm currents.[18]

Buried in these quiet-water deposits of the Marjum Formation are not only trilobites but soft-bodied animals, making this yet another formation with Burgess Shale–type preservation. There are quarries (fig. 7.10) in the type section near Marjum Pass that produce the large trilobite *Hemirhodon amplipyge* and abundant ptychopariids such as *Modocia laevinucha* (fig. 7.11); other sites in the Marjum contain trilobites such as agnostids, *Elrathia marjumi*, *Marjumia typa*, and the unmineralized trilobite *Naraoia compacta*. Also found among these trilobites are bivalved arthropods such as *Branchiocaris*, some of which get quite large (fig. 7.11d). *Branchiocaris* was identified only as a bivalved arthropod for years, poor preservation having prevented us from being able to refine its identification further. But recent finds of well-preserved specimens from

7.11. Fossils from the Marjum Formation in the House Range. (A) Complete specimen of the ptychopariid trilobite *Modocia*. (B) Nearly complete specimen of the large trilobite *Hemirhodon*. (C) Complete ptychopariid *Bolaspidella*. (D) Unmineralized shell of the bivalved arthropod *Branchiocaris*. (E) The eocrinoid *Gogia*. Scale bars for (A), (C), and (E) = 1 cm. (B) Scale bar marked in centimeters. (D) Scale bar = 5 cm.

(A), (B), and (D) from Museum of Western Colorado collections; (C) and (E) from Raymond M. Alf Museum of Paleontology collections.

7.12. Specimen of the arthropod *Emeraldella* from the Marjum Formation. University of Kansas specimen (KUMIP 204791). Specimen ~5 cm.

7.13. Specimen of the priapulid worm *Ottoia* from the Marjum Formation (KUMIP 204770). Scale bar = 1 cm.

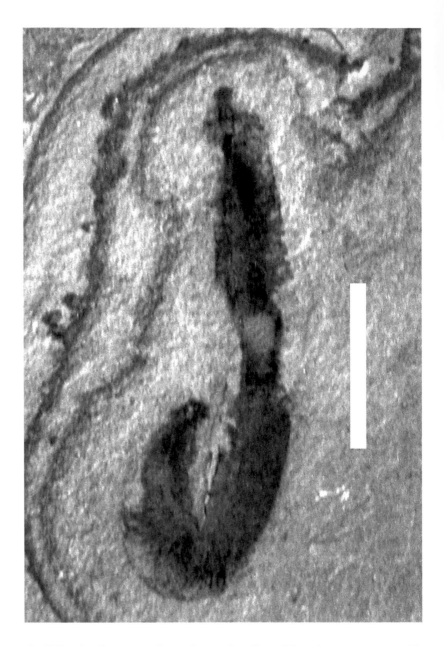

the Wheeler Formation have shown that *Branchiocaris* was more specifically a bivalved crustacean.[19] The alga *Yuknessia* has been reported from the Marjum Formation as well. Other animals found in the Marjum include an anomalocaridid (possibly *Peytoia*), eocrinoid echinoderms (*Gogia spiralis*; fig. 7.11e), hyoliths, molluscs, *Selkirkia*, *Eldonia*, *Emeraldella* (fig. 7.12), and six species of brachiopods.[20] Also known from the Marjum Formation is the priapulid worm *Ottoia*.

PRIAPULID WORMS

Ottoia is found in the Marjum Formation (fig. 7.13) and in the Burgess Shale Formation (plate 15; fig. 6.37). The form from the Marjum appears

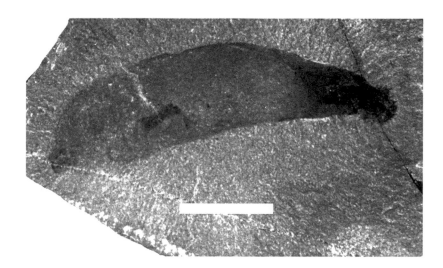

7.14. The priapulid worm *Ancalagon* from the Burgess Shale (USNM 198608). Scale bar = 1 cm.

Courtesy of Smithsonian Institution.

to be a bit more slender than that from British Columbia. Priapulids such as *Ottoia* are unsegmented, bilaterally symmetrical worms that possess a proboscis (or introvert) with hooked spines and a thin or sometimes bulbous "neck" that is also covered with short spines. The proboscis is used in acquiring food. The main body (trunk) is elongate in most forms, but generally thicker in *Ottoia* (at least the Burgess Shale specimens), and is sometimes annulated, though not structurally segmented. Some forms have a tentacle-like caudal appendage.

There are sixteen species of priapulid worms still around today, and they have a thin cuticle that forms the spines on the head (and tubercles, if present). The pharynx is also cuticle lined and has hook-like teeth, which are used in feeding. There is no specialized circulatory system and the nervous system is radial, without large cerebral ganglia. The gut is complete and straight or slightly coiled. Most modern species are benthic, infaunal, burrowing predators on soft invertebrates; all are marine animals. They move over or through the bottom sediments through extension and contraction of parts of the body. Fertilization is external and development indirect in all but one modern species.

Priapulids lie within a much larger group of vermiform metazoans called the Nemathelminthes. This group also includes the nematodes (roundworms) and nematomorphs (hair worms), and priapulids may be even more closely related to a group of tiny (~1 mm) sand- and mud-dwelling wormlike modern animals called kinorynchs. The nemathelminths live in a variety of habitats and have a diverse range of ecologies; many are parasitic. Nemathelminths other than priapulids are known from Cambrian deposits in China, and embryos probably belonging to this group have been found there recently.[21]

Ottoia has not been found in much abundance in the Marjum Formation, but in the Burgess Shale it is quite common in some layers. In some Cambrian paleoecosystems priapulids appear to dominate the numerical abundance of individuals, sometimes comprising more than 40% of the fauna.[22] The Burgess Shale also has yielded at least four other

7.15. Sketch of a USNM specimen of *Ottoia* from the Burgess Shale preserved with three hyolith conches in the gut. Length ~8 cm.

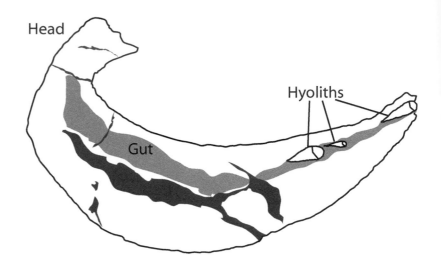

Head

Hyoliths

Gut

genera of priapulids in addition to *Ottoia:* these are *Ancalagon* (fig. 7.14), *Fieldia, Louisella,* and *Selkirkia. Ottoia* has also been found in the Spence Shale in Utah, and *Selkirkia* has been identified in the Wheeler Formation, just below the Marjum. *Ottoia* specimens in the Burgess Shale are often preserved in a U-shaped manner, and it is believed they probably lived in burrows of a similar shape, with the proboscis poised at one end, ready to shoot out and capture prey items. Among the prey items of *Ottoia* were the hyoliths and articulate brachiopods. We know this from specimens found in the Burgess Shale that were preserved with up to three intact hyolith shells in the gut (fig. 7.15); others had brachiopods in their digestive tracts. It appears the priapulids consistently swallowed the hyoliths whole and opercular end first.[23]

SPONGES

Exposed in the high desert of western Utah, in one of those wonderfully ironic juxtapositions of geology so common in the North American Cambrian, the Marjum Formation has yielded 10 genera of very marine fossil sponges. You might expect that of a rock unit that contains a quarry (now mostly reclaimed) known as Sponge Gully. The sponges found in the Marjum Formation are *Choia, Hazelia, Hamptonia, Leptomitus, Diagoniella, Protospongia, Hintzespongia, Kiwetinokia, Testiispongia,* and *Valospongia.*[24]

The animals known as sponges (phylum Porifera) are the simplest of the Metazoa and include a colorful array of about 5500 species today. Although what we use in our sinks and to wash our cars now are thankfully (for these animals) synthetic sponges, the real thing can still be found in some art supply stores; of course, the best way to see sponges is live in the wild, snorkeling or scuba diving (you, not the sponges). Recall that sponges arose during the late Proterozoic, so their history goes back a ways, and sponges had been around for close to 100 million years already by Marjum times. Proterozoic sponges may have helped oxygenate the oceans by way of their pumping action; circulation may have driven

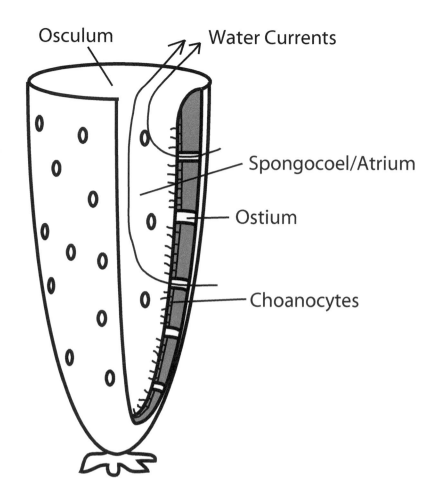

Osculum

Water Currents

Spongocoel/Atrium

Ostium

Choanocytes

water with higher oxygen content to levels deeper than those at which photosynthesis can occur.

Characteristics of sponges include: possession of flagellated cells called **choanocytes** that work together by the thousands to move water through the animal; a lack of true body tissues; body cells that are capable of changing form and function; adults are sessile but larvae may move or even swim; adults are often asymmetrical and shape may be plastic. Sponges are stationary, multicellular animals and are suspension feeders, filtering out of the seawater microscopic food particles such as organic molecules, algae, bacteria, and protists. Although different sponge species may be shaped like tubes, branching tubes, bushes, amorphous masses, or a number of forms, a typical generalized form is the shape of a squat, thick-walled jar. The thick walls of the sponge are known as the **cortex,** which contains a canal system for water flow; the open center of the jar is the **atrium,** and the top opening is the **osculum** (fig. 7.16). Water is pulled in through the walls of the sponge and filtered, then is pumped into the atrium and out the osculum. Water coming in from outside the sponge is often drawn in through openings on the external surface, each of which is called an **ostium** (in some sponges these openings are surrounded by cells called porocytes).

The choanocytes, so characteristic of sponges, line either the inner surface of the atrium of the sponge or the insides of chambers within the cortex called choanocyte canals; the choanocyte flagella work together to move water through the sponge, and filamentous collars around the base of each flagellum are what (in most cases) filter food particles out of the water. The food is then metabolized by each cell. Just to give you an idea of how many flagella are working together to move the water through a whole sponge, some species may have up to 15,000 choanocytes per cubic millimeter of cortex! The choanocyte cells of sponges are strikingly similar to the unicellular eukaryotic choanoflagellates, a group not surprisingly thought to be closely related to the Metazoa.

Structural support for the sponges is provided by **spicules,** tiny, rayed prongs of material embedded within the cortex. The shape and composition of the spicules characterizes the three classes of sponges that are known. Members of the class Calcarea possess spicules of calcium carbonate; those of the class Hexactinellida ("glass sponges") have six-rayed spicules of silica; and of class Demospongiae spicules of silica (not six-rayed) and/or collagen fibers. Of the 5500 living species of sponges, about 95% belong to the Demospongiae; all three classes were present in the Cambrian, and demosponges also were very diverse then. A high percentage of Cambrian sponges belonged to the Demospongiae, including many in the Burgess Shale, but in the Marjum Formation—for some reason—at least half (possibly more) of the sponge genera are hexactinellids, and only four genera are demosponges.

As a group, modern members of the Demospongiae range across most depths of the marine environment, down to thousands of meters. The main environmental requirement for sponges is that the water not be too choked with sediment, as this interferes with their pumping system. Calcareous sponges live at depths generally less than 200 m (656 ft.), whereas modern hexactinellids generally live deeper than 200 m. Ancient hexactinellids, however, have been found in deposits that represent environments that were shallower than 200 m, including units such as the Marjum and Burgess Shale formations, and in some settings that were near shore and quite shallow.[25]

Reproduction in sponges may be by either asexual or sexual processes, depending on the species. Two forms of asexual reproduction in modern sponges are by budding or by the pinching off of a previously existing branch; in either case, the piece falls off, regenerates, and forms a new individual. Sexually reproducing sponges are often hermaphroditic and alternate producing eggs and sperm at different times, sometimes just once each. Other species have male and female individuals. In some species, sperm are released into the water, and fertilization occurs there. In others, fertilization occurs in neighboring sponges; in these, embryos are eventually released as swimming larvae that either settle quickly or swim for a few days before attaching and growing into new adults.[26]

At least 16 genera of sponges are known from the Burgess Shale, and these animals generally are known from many formations in the

7.17. Reconstruction of the arthropod *Nettapezoura* from the Marjum Formation. Length ~14 cm.

Cambrian. Isolated sponge spicules are relatively abundant in some formations as well. The high diversity of sponges in the Cambrian indicates that filter-feeding organisms had achieved a separation by filtering height (above the mud of the seafloor), and were thus dividing their ecological niches, early in the Paleozoic.[27]

STRANGE ARTHROPODS

Among new arthropods described from the Marjum Formation is the predatory *Nettapezoura basilika*, the "regal duck-foot tail" (fig. 7.17). This non-trilobite, arachnomorph arthropod features a flattened and expanded telson that looks somewhat like a swim fin (or duck foot), although it may not have been used like one. There are two small eyes along the anterior rim of the head and six cephalic appendages. The appendages on each of the nine thoracic segments were biramous, and there are two ring-shaped abdominal segments anterior to the telson. It was probably a nektobenthic predator-scavenger.[28]

Also known from the Marjum Formation is a possible specimen of the Burgess Shale arthropod *Emeraldella* (fig. 7.12). This is a moderately large, bottom-dwelling predator and scavenger (see also chapter 6).

SOFT TRILOBITES

The species *Naraoia compacta* from the Marjum Formation is a trilobite, but it lacks the mineralized, calcitic dorsal exoskeleton that nearly all other trilobites have. It has an exoskeleton, but it is relatively soft and unmineralized. *Naraoia* also has a cephalon with antennae and three pairs of limbs (as in other trilobites), but it has a single shield over the posterior limbs equivalent to the thorax and pygidium. This species also occurs in the Burgess Shale, and the genus is also found in the Chengjiang fauna in China.

VETULICOLIANS: A NEW PHYLUM

A fossil found in the Pierson Cove Formation in the Drum Mountains, not far from the House Range, indicates the presence of members of the new phylum Vetulicolia in the Middle Cambrian of western Utah. The Pierson Cove Formation is a lateral equivalent of the Marjum

7.18. The vetulicolian *Skee-mella clavula* from the Pierson Cove Formation (KUMIP 310501). Scale bar = 1 cm.

Formation,[29] and in 2005 a form called *Skeemella clavula* was described from the lowermost part of this unit (fig. 7.18). *Skeemella* is a little different from other vetulicolians such as *Vetulicola* and *Banffia* (see chapter 6). *Skeemella* has a longer segmented posterior section and a smaller anterior section than does *Banffia*, for example.

The phylum Vetulicolia was named in 2001 by paleontologist Degan Shu and a number of others in a paper on *Vetulicola* and several other related forms from the Lower Cambrian of China. Although neither *Skeemella* nor *Banffia* are of similar proportions, some of the Chinese species of vetulicolians can very superficially resemble bivalved arthropods, so large are the anterior section carapaces and so short are the segmented posterior sections; in fact, *Vetulicola* was originally described as a bivalved arthropod. However, unlike those arthropods, vetulicolians may have four anterior plates (not two) and have a division into dorsal and ventral sections in the anterior region; additionally, vetulicolians lack limbs entirely. Complicating the classification of the vetulicolians is the fact that some have gill pouches, another non-arthropod character. So far, vetulicolians have been proposed to be possibly strange arthropods that lost their limbs and gained gill elements, or to be stem group deuterostomes (i.e., basal relatives of chordates and echinoderms) or possibly relatives of the urochordates. Another possibility is that the vetulicolians are further down from true arthropods, somewhere among the stem taxa where segmentation is well developed but limbs have not yet appeared. This hypothesis is complicated by the fact that it would require the loss of some characters in vetulicolians that would later have to be regained by arthropods. None of these ideas proves to be entirely satisfactory, and there is still little agreement on which hypothesis is preferred, although a cautious preference for the basal deuterostome classification seems to have a majority at this point. The gill structures argue in favor of a deuterostome affinity, but the lack of a clear notochord complicates this and their association with urochordates. One other possibility is that the vetulicolians are related to the modern phylum Kinorhyncha, which comprises about 150 species of tiny, marine animals that live in the sediment and feed on single-celled microorganisms. These animals are generally less than 1 mm long and have a round, anterior mouth, and a segmented trunk, often with a split terminal segment (as in *Skeemella*). The kinorhynchs do resemble vetulicolians except that they lack the distinct, carapaced anterior section and gill structures of the Cambrian forms.

Regardless of their relationships to members of other phyla, the Cambrian vetulicolians also seem to have ecologies about which we are not certain. Although their mouths suggest a filter-feeding or deposit-feeding habit, *Skeemella* and *Banffia* appear to have been capable of respectable mobility; their segmented posterior sections suggest a certain degree of movement was possible. Whether they were straining organic material from the water or from the sediment is not entirely clear. Similarly, it appears that they could move, but whether that movement was through the sediment, over the sediment, or through the water is not evident just yet.[30] With luck, future fossil finds will tell us; and it is entirely possible that individual species of vetulicolians fed and moved in different ways. In any case, this very old phylum, new to science though it is, should provide some exciting news in coming years as we slowly continue to decipher its fossil record.

Middle Cambrian rocks of the *Ehmaniella* and *Bolaspidella* zones do not occur only in the western part of North America. As we saw with the Early Cambrian, there are plenty of outcrops on the eastern side of the continent as well. In northwestern Georgia and over into Alabama and Tennessee is exposed the Conasauga Formation, which is Middle Cambrian in age and represents the south-shore Laurentia equivalent to the Wheeler-Marjum-Weeks layers in Utah. If those rocks in the House Range represent the shallow shelf off the north shore of Laurentia, facing the Panthalassic Ocean to the north, the Conasauga represents the southern shelf on the other side of the continent, facing down into the Iapetus Ocean (see figs. 1.3 and 1.4). The Conasauga Formation is dominated by shale and, in some areas, limestone and can be more than half a mile thick (up to 880 m [2886 ft.]). The formation occurs in areas often heavily faulted so the true thickness in many places is difficult to measure with certainty.

The Conasauga Formation contains some of the same trilobites we see in the House Range, including *Asaphiscus*, little *Peronopsis*, and

Middle Cambrian Dixie

7.19. The ptychopariid trilobite *Modocia pelops* from the Conasauga Formation near Centre, Alabama (USNM 94893). Scale bar = 1 cm.

Courtesy of Smithsonian Institution.

super-abundant *Elrathia* – plus other forms such as brachiopods, hyoliths, sponges (including *Brooksella*, originally described as a jellyfish), algae, soft-bodied trilobites (*Naraoia*), and priapulid worms (*Ottoia*). Some of the trilobite specimens found in the Conasauga Formation are quite complete too (fig. 7.19); the trick is finding outcrops, as things are not as wide open and sparsely vegetated as they are in Utah. The sponge *Brooksella* is often preserved as three-dimensional chert fossils that are nearly round and strongly convex on one side, with a radial pattern centered on the convex surface. Their appearance has earned them the informal name "star cobbles." These were described by Walcott as possible jellyfish but have been more recently interpreted as sponges.[31] *Brooksella* has also been identified in the Middle Cambrian Spence Shale in Utah.

Big Sky Trilobites

Cambrian fossils from approximately this age also occur in the mountains of western Montana in some of the wildest terrain in the region. In the depths of the wilderness just south of Glacier National Park lies Pentagon

Mountain, and on the slopes of this peak, in the Pentagon Shale, are trilobites of Middle Cambrian age. The mountain lies, appropriately enough, at the south end of the Trilobite Range and not far from Trilobite Lakes and Trilobite Peak. The trek in to this site consists of about 32 km (20 mi.) of trail travel by foot or horseback through the Bob Marshall Wilderness. The Middle Cambrian Pentagon Shale here contains trilobite genera found elsewhere in the West and East of North America in rocks of this age: *Bathyuriscus*, *Agnostus*, *Ehmaniella*, and *Kootenia*. The fossils may not be particularly difficult to find, but getting to them (and getting them out) requires some serious commitment.

We are almost through the Middle Cambrian. Time now to hit the road again and continue our journey ever forward through the period. Next stop is the Late Cambrian, 11 million years when our now-established animal faunas settle in, generate a few new forms, experience a few setbacks, but stage themselves for a takeoff.

Taking Off: The Late Cambrian

8

WE HAVE MOVED FORWARD THROUGH TIME NOW TO THE LATE Cambrian, the last 11 million years of the Cambrian period. Rocks of Late Cambrian age occur throughout much of North America, in part because by this time the ocean had risen high enough to flood even the low-lying parts of central Laurentia. No longer were the edges of the continent beachfront property; now there were shallow marine deposits and the life that lived in and around them as far inland as what would one day become Oklahoma, Wisconsin, and Minnesota.

Once again, after all this travel, we now find ourselves back in the House Range of Utah. We had visited here in our exploration of the late Middle Cambrian and the Wheeler and Marjum formations, but now we are back to see what rocks and fossils we can find from the very beginning of the Late Cambrian way out west, much farther offshore from the rocks of similar age in the Rockies and Black Hills. The rock unit we are interested in here lies just above the Marjum and is known as the Weeks Formation (fig. 7.1). It is exposed over the ridge from some Marjum Formation sites in a place called North Canyon (fig. 8.1a), where you can split open gigantic slabs of limestone with many brachiopods and a few trilobites on them.

Back to the House Range

Transition: The Weeks Formation

The Weeks Formation consists of 366 m (1200 ft.) of gray to tan, laminated limestone (fig. 8.1b) similar to the Marjum, but generally finer grained and sometimes splitting into large, wavy, tan-colored slabs. The Middle–Upper Cambrian contact appears to be in the lower Weeks Formation where we pass into the *Cedaria* zone, and upward from here the rocks are Late Cambrian in age.[1] Digging in the Weeks Formation and splitting these limestones you will often see complete trilobite specimens, but they are generally small and hard to see, usually showing under a thin layer of sediment rather than splitting right above the fossil. To really expose them you need to use an air-abrasion machine. This is like a miniaturized sandblaster that blows air and a light mixture of dolomite powder out of a tiny nozzle to knock the sediment off the fossil one grain at a time. It turns out that under the light veil of sediment, the Weeks Formation trilobites are often beautifully preserved, sometimes with details of the

8.1. Outcrops of the Upper Cambrian Weeks Formation in the House Range of Utah. (A) Quarry slope in North Canyon covered with silty limestone slabs. (B) Close-up of rock slabs and layering.

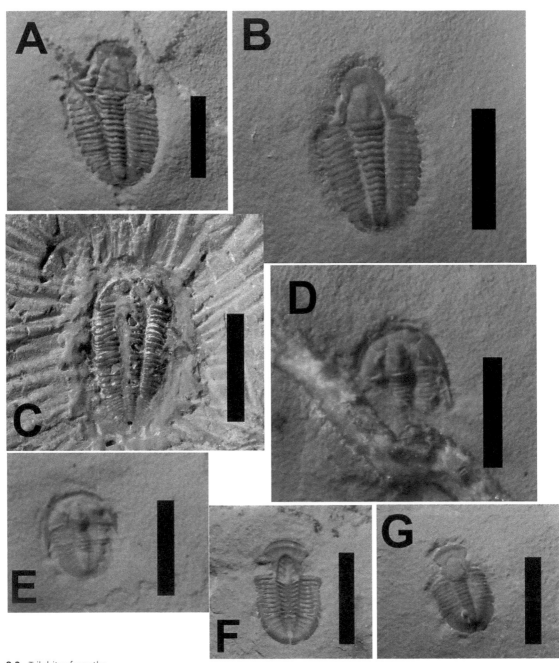

8.2. Trilobites from the Weeks Formation of North Canyon, House Range, Utah. (A) and (B) The ptychopariid *Modocia weekensis*. (C) The elongate *Menomonia semele*. (D) and (E) The raymondinid ptychopariid *Cedaria minor*. (F) and (G) *Cedaria* specimens missing the free cheeks. Scale bars = 1 cm.

All Museum of Western Colorado specimens.

gut indicated. Among the trilobites preserved in the Weeks Formation are little *Cedaria*, the elongate *Menomonia*, *Meteoapsis*, *Cedarina*, *Modocia*, *Genevievella*, *Meniscopsia*, and fluke-"tailed" *Tricrepicephalus* (fig. 8.2).[2] Recent work has found new species of trilobites in the Weeks Formation, including *Ithycephalus stricklandi*, *Coosella kieri*, and *Modocia comforti*, to name a few. Little *Norwoodia boninoi* has been found on large bedding planes preserving nearly 50 individuals; these trilobites have a relatively large, half moon–shaped cephalon with a wide flattened rim, long genal spines, a short posteriorly tapering thorax, and extremely long

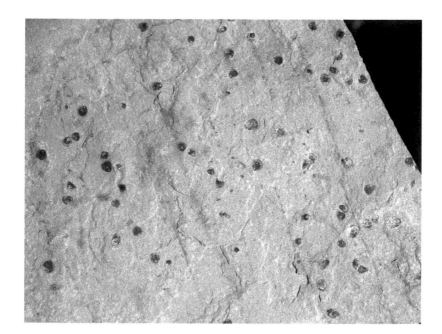

8.3. A slab of Weeks Formation containing abundant brachiopods, most about 5 mm in diameter.

spines on the occipital ring and fourth thoracic segment that collectively (if preserved in alignment) resemble the telson of a horseshoe crab.[3] They, of course, are no closer to horseshoe crabs – as far as we know – than any other trilobite, but the superficial similarity in shape is noticeable. Also preserved in the Weeks Formation is the arthropod *Beckwithia*. We will discuss that form shortly.

What you notice quickly in the Weeks Formation in North Canyon is that brachiopod shells are abundant (fig. 8.3), and many belong to the genus *Lingulella*. Some of these wavy limestone slabs at the site have been mined as flagstone, and if I needed to pave a back porch here in Utah, there is a local rock-supply yard I know of in Colorado where I could pick up some Weeks Formation for the project – brachiopods and all.

BRACHIOPODA

Slabs of the Weeks Formation can be pried out from outcrops in North Canyon, Utah, which is just south of some outcrops of the older Wheeler and Marjum formations that we visited in chapter 7. These limestone slabs can be quite large, and many contain small, articulated trilobites, but almost every piece you excavate from these rocks has small, black brachiopods preserved on it. Many of these, as noted above, belong to the genus *Lingulella*, which is a common brachiopod form of the Late Cambrian.[4] You may recall brachiopods being among the fossils found at nearly every site we have visited all the way back to those of the Early Cambrian in chapter 4; brachiopods are indeed seemingly everywhere in the Cambrian (fig. 8.4), and if there is one fossil type that you can almost count on to find at any given site of this age, the brachiopods would be it. You sometimes find them instead of trilobites (the other particularly common fossil type).

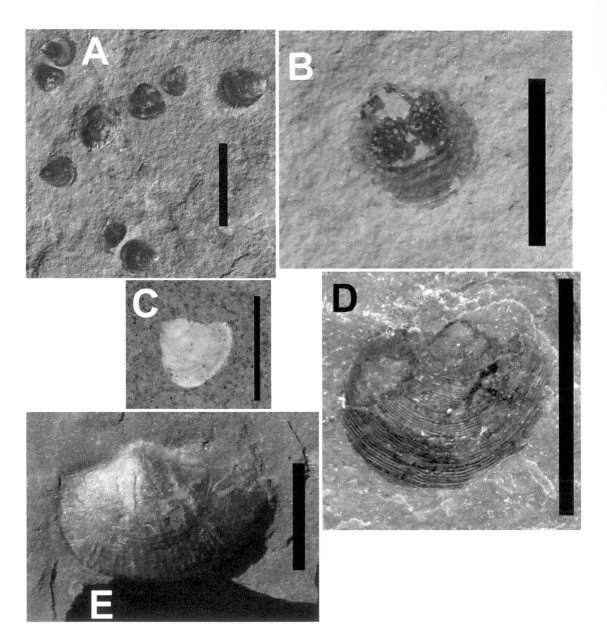

Although they look like clams, brachiopods are an entirely separate group from the molluscs, with very different internal anatomy. In fact, the brachiopods are most closely related to rare animals known as ectoprocts (or bryozoans) and phoronids (small tube-dwelling animals); these three groups together are known as **lophophorates,** animals possessing lophophores, which are structures of coiled rings of tentacles around the mouth that assist with filter feeding. Lophophorates are an interesting case in that it is still unclear (or at least no consensus has been reached) whether they are protostomes or deuterostomes; are they closer to annelids, molluscs, and others or to echinoderms and chordates? We don't know. We seem to have stem group brachiopods among a group of Early Cambrian small shelly fossils known as tommotiids in rocks in Australia, and these fossils are tiny, ribbed shells around the pedicle opening, belonging to a cone-shaped animal.

Brachiopods are shelled creatures that, again, are reminiscent (superficially) of clams. Modern types are known as lamp shells and are all marine bottom dwellers. They are quite different from the molluscs, however, not only in biology but also in their shells; for one, brachiopod shells have a plane of symmetry through the middle of the shell so that, viewed from above, left and right are mirror images of each other. The other valve of the animal may be totally different in shape, but left and right are mirror images.[5] In clams, in contrast, the shells are not symmetrical when viewed from above, but each of the two valves of one clam are mirror images of each other in side view. Thus, the planes of symmetry are totally different in the two types of organisms. (Clams also lack a lophophore, of course.)

There are about 300 species of brachiopods today, but there are as many as 12,000 fossil species, and they were particularly diverse in the Paleozoic era. The brachiopods appear in the Early Cambrian (early enough that the common ancestor of the lophophorates probably dates back into the Proterozoic), become very diverse throughout the rest of the Paleozoic, and then decline in diversity after that, probably because of competition with the bivalved molluscs. Most brachiopods are solitary, benthic, marine animals that live attached to the bottom and pump water through their shells to filter feed on organic particles. Some Cambrian brachiopod taxa are rather similar in overall appearance to the modern genus *Lingula*, which, despite showing little morphological variation (both intraspecific and compared to ancient species), demonstrates great intraspecific and intrageneric genetic variability.

The brachiopod body is contained within two symmetrical shells, or valves, composed of calcium carbonate or calcium phosphate (or both, in one case!). The body and its valves are attached to the bottom by a fleshy extension called the **pedicle,** which may extend into soft sediment as a type of "anchor" or may attach to a hard substrate. The two valves are often slightly different in shape, with the pedicle connecting with the rest of the body through a small opening in the hinge region of the ventral, or pedicle, valve; the other, dorsal shell is known as the brachial valve.

The way the valves articulate helps define two general groups of brachiopods (although evolutionary relationships are a bit more complex). The two valves are held together entirely by muscles in the non-hinged inarticulate brachiopods, such as *Lingulella* and *Paterina*; articulate brachiopods possess socketed hinges that keep the valves together – although these are, of course, still opened and closed by muscles.

The brachiopod body that lies protected between the valves includes the mantle, mouth, gonads, a stomach and intestine, and the lophophore (fig. 8.5). Structurally the lophophore consists of two arms with tentacles that extend from near the mouth out into the internal mantle cavity area between the two valves. Movements of cilia pump water into the mantle cavity from the opening in the slightly cracked valve contact, and the

8.4. Cambrian brachiopods from around the western United States. (A) Cluster of *Lingulella* from the Upper Cambrian Weeks Formation of Utah. (B) Isolated inarticulate brachiopod from the Upper Cambrian Dotsero Formation of Colorado. (C) Brachiopod from the Upper Cambrian Deadwood Formation near Lightning Creek, South Dakota. (D) Inarticulate *Paterina* from the Lower Cambrian Latham Shale in the Marble Mountains of California. (E) Articulate *Wimanella* from the Middle Cambrian Spence Shale at Spence Gulch, Idaho. Scale bars = 1 cm.

(A), (B), and (E), Museum of Western Colorado specimens.

8.5. Morphology of a brachiopod seen in cross-sectional side view. Thick black lines are the shell valves; gray inside is body tissues (with specified internal structures in white); the pedicle shows attachment to the substrate, which is shown as gray bed at bottom.

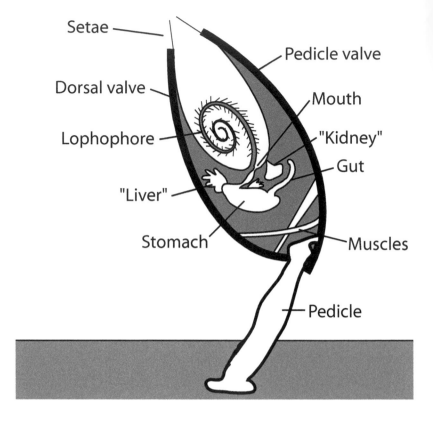

water flows through the lophophore, where food particles are caught in filaments and transported first to an axial food groove, then moved to the mouth.

The circulatory system of brachiopods is open with a heart and blood that mainly carries nutrients. Oxygen is probably transported around the body by the coelomic fluid, rather than the blood. Gas exchange to oxygenate the coelomic fluid seems to occur across unspecialized body surfaces. Reproduction in brachiopods is sexual, with external fertilization of eggs by the sperm in water. In some species, the eggs are retained inside the body area and are fertilized by sperm cells caught in the female's water current; in these species the eggs are brooded until released at a larval stage. The larval stage of the life cycle is free swimming. Articulate brachiopods have a three-part, linearly arranged larva that eventually sinks to the bottom, metamorphoses into its adult form, entrenches its pedicle, and then secretes its shells; inarticulate brachiopods, on the other hand, have a slightly longer free-swimming larval stage and then secrete their shells and sink to the bottom to attach their pedicle and become a benthic adult. Inarticulate larvae are already of essentially adult morphology and so they never metamorphose.

Many of the brachiopods of the Cambrian probably lived propped up slightly above the sediment on their deeply buried pedicle with their valves barely open, pumping seawater through the body area to feed. Fossils of Early Cambrian brachiopods from the Chengjiang biota in China show that other organisms liked to attach themselves to the shells

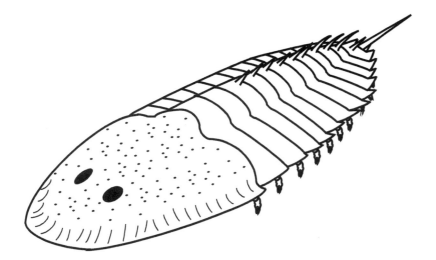

8.6. The arthropod *Beckwithia* from the Upper Cambrian Weeks Formation of Utah. This form is also known from rocks of similar age in Wisconsin. Total length ~10 cm.

of still-living brachiopods, not necessarily parasitically but often just as a solid base on which to grow. Brachiopods, in turn, are known to have attached their pedicle to the exoskeletal fragments of trilobites. In other cases, paleontologists have found brachiopod shells with circular borings in them and with healed breakage scars, indicating that predators fed on, or at least had bitten, the animals. In addition to attaching to solid bottoms or shell pieces by their pedicle, various species of brachiopods may have laid free on the bottom, have been nearly planktonic, or have lived in the sediment.[6] Although they are often small and rarely grab the attention of collectors in the same way trilobites do, especially in Cambrian rocks, brachiopods are really rather interesting animals, and their large numbers suggest they were an important part of their respective ecosystems at the time.

AGLASPIDIDA

Among the soft-bodied (unmineralized) arthropods in the Cambrian, some of the more diverse, if not necessarily abundantly preserved, are the aglaspidids. These non-trilobite arthropods have half moon–shaped heads and segmented bodies, and there are at least eight species of these animals known from Cambrian rocks of Wisconsin alone! Aglaspidids have been allied with the trilobites and naraoiids in the Artiopoda. The aglaspidid *Tremaglaspis* has now been described from the Cambrian Weeks Formation. The Weeks Formation of Utah has also yielded an aglaspidid-like arthropod called *Beckwithia* (fig. 8.6), and this genus also occurs in Wisconsin. *Beckwithia* from the Weeks Formation consists of a semicircular, textured head shield with two close-set compound eyes, about a dozen thoracic segments, and – presumably – relatively short legs set under the dorsal shield (fig. 8.6). It may have lived as a mobile animal on the sediment at the bottom of the ocean, but its feeding ecology is unknown. Could it have been a scavenger-predator? A deposit feeder? We don't know, but comparisons with true aglaspidids suggest that these modes of life are possible, and given the diversity of forms present in the Cambrian, we might also expect a diversity of ecologies among members of this group of arthropods. Aglaspidids are believed to have preferred relatively shallow water.[7]

Orr and Notch Peak Formations

Above the Weeks Formation in western Utah, thick limestone deposits of the Orr Formation and Notch Peak Formation contain trilobites of Late Cambrian age (fig. 7.1), including some of the same genera that we see farther east. We will see more of the Orr Formation later in this chapter as one of the formations in which evidence of an isotope excursion event has been observed in geochemical analyses of the rocks.

Also found in the Orr and Notch Peak formations are brachiopods, echinoderms, and sponges, and these were deposited in water around 100 m (330 ft.) deep in the deeper parts of the distal shelf. To the east was a shallow, higher-energy shoal of oolitic carbonate similar to what we see on carbonate banks today in the Bahamas, and much farther east were the sandy deposits of the Midwest. Among the trilobites preserved in these Upper Cambrian formations are *Dunderbergia* and *Pseudagnostus*.[8]

The Heartland of the Mississippi– St. Croix Region

Waves Lapping on the Midwest Shore

The type area for many of the fossils found in Late Cambrian rocks in North America is the region of the valleys of the upper Mississippi and St. Croix rivers along the border between Minnesota and Wisconsin. The cliffs along these rivers expose rocks of Late Cambrian age, first explored by early geologists and paleontologists and known to C. D. Walcott and his associates, and the exposures run well east into south-central Wisconsin. Anyone who has ridden the army "ducks" at the Wisconsin Dells, for example, has seen crossbedded sandstones of some of these Upper Cambrian formations along the waterways. The formations in this region include the Elk Mound Group, consisting of the Mount Simon Formation, Eau Claire Formation, and Wonewoc Formation; above these lie the Lone Rock Formation, St. Lawrence Formation, Jordan Sandstone, and Franconia Formation (fig. 8.7). These formations are largely sandstone and were formed mostly in shallow marine, beach, tidal flat, and dune environments during the Late Cambrian.

In these marine rocks are typical Late Cambrian trilobites such as *Dikelocephalus, Elvinia, Idahoia, Crepicephalus, Ptychaspis, Prosaukia, Wilbernia,* and *Ellipsocephaloides* (fig. 8.8). *Pemphigaspis*, an unusual trilobite with a double-bulb pygidium is found in this area as well. It was once thought that there were as many as 26 species of *Dikelocephalus*, but this is most likely an artifact of overly enthusiastic species splitting by paleontologists early on. In reality, there were probably only a handful of

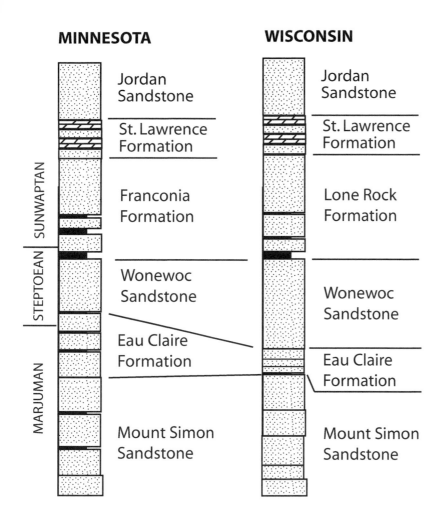

MINNESOTA

Jordan Sandstone

St. Lawrence Formation

Franconia Formation

Wonewoc Sandstone

Eau Claire Formation

Mount Simon Sandstone

WISCONSIN

Jordan Sandstone

St. Lawrence Formation

Lone Rock Formation

Wonewoc Sandstone

Eau Claire Formation

Mount Simon Sandstone

SUNWAPTAN | STEPTOEAN | MARJUMAN

8.7. Stratigraphic section of geologic formations from the Middle to Upper Cambrian of Minnesota and Wisconsin. Stipple pattern = sandstone; solid black pattern = shale; slanted brick pattern = dolomite.

Dikelocephalus species around during the Late Cambrian. Also found in the Upper Cambrian rocks of Wisconsin and Minnesota are fossils such as hyoliths, brachiopods, and conulariids.[9]

During the Late Cambrian in this region, Minnesota and Wisconsin were near the shoreline of the Laurentian landmass, which was at times just to the north in what is now Ontario, Canada, and at other times much closer in eastern Wisconsin. The area of modern-day Iowa and Illinois was a finer-grained, offshore marine setting, and the Mississippi–St. Croix region was as close as a few tens of miles and as far as 200 miles (324 km) from land. When land was near it was exposed due to a structural feature known as the Wisconsin Arch, which formed an uplift of a narrow strip of land, with marine waters on the eastern side. The land of the arch was rimmed with sand dunes and rivers along the shoreline beach, with sand being blown and washed in from land to the north and east. On that other eastern side of the arch there was a full 1134 km (700 mi.) of shallow marine shelf one would have to travel to reach the other shelf edge and deeper water to the southeast; most of this 700 miles of shelf, from land to deep water in the Appalachians region, was less than 100 m (328 ft.)

A

B

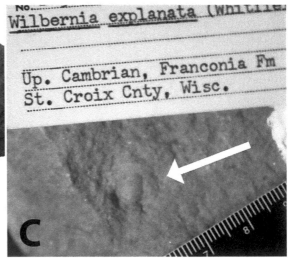

Wilbernia explanata (Whitile

Up. Cambrian, Franconia Fm
St. Croix Cnty, Wisc.

C

deep. Obviously, this was an incredibly large area of shallow water, but it was typical of Laurentia during the Cambrian.

In the Wisconsin area, a mere 20–200 miles offshore, there were islands of Precambrian rock that stuck up out of the ocean, some islands as much as 8 km (5 mi.) long and up to 183 m (600 ft.) high (plate 20). During storms, waves as high as 8 m (26 ft.) crashed on the shores of these islands—Waimea-sized breakers that moved large boulders that sat in the shallow water around the islands![10] Despite normally placid conditions, this was an environment that could become very high energy during ancient tropical storms.

Among the other fossils found in this area are non-trilobite arthropods and scyphozoan medusae, better known as jellyfish. Among the arthropods found were the first phyllocarid crustacean from the St.

Lawrence Formation of Wisconsin, *Arenosicaris*, and an unusual arthropod, *Mosineia*, from lowermost Upper Cambrian (or possibly upper Middle Cambrian) rocks, also in Wisconsin. *Arenosicaris* was a small swimming crustacean with a bivalved shell over its thorax (fig. 8.9), whereas *Mosineia* was a somewhat larger, primitive arthropod of a type known as euthycarcinoids, which walked on seven pairs of legs on the bottom of very shallow seas, as evidenced by mudcracks in the rocks in which the fossils are found. These soft-bodied arthropods are preserved in part in sandstones, which are rarely media for preserving such delicate fossils. Euthycarcinoid arthropods similar to *Mosineia* may have made some trackways recently described from Late Cambrian sand dune deposits in Ontario, Canada. Animals were moving around in air along the Late Cambrian shoreline, at least for short periods. These arthropod trackways represent the earliest evidence in the fossil record of animals on land.[11] Apparently, like crabs and some other arthropods today, Late Cambrian marine (and possibly freshwater) arthropods had already learned the amphibious art of walking on land when the need arose.

Even more evidence of arthropods walking on land has emerged in the form of trails in intertidal sand flat deposits in the Elk Mound Group of Wisconsin. What makes these trails interesting is that they seem to show that the arthropods were carrying mollusc shells around with them, held by the animals with a long tail inside the shell. Like hermit crabs today, the arthropods carried the shells with them, in this case probably in order to keep the gills wet so the animals could still breathe in the air. These animals probably went up on the exposed, sandy tidal flats at low tide so that they could feed on algal mats or other exposed organisms, and, in addition to carrying a shell, they may have reduced their own exposure to desiccation by leaving the water only for short periods of time and often at night.[12] Hermit arthropods of the Cambrian suggest that the hermit crabs we see at the beach today are just doing what their (indirect) ancestors were doing 500 million years ago.

Cambrian jellyfish have also been found in Wisconsin, and consist of impressions of multiple specimens that had probably become stranded on a beach, much like modern jellyfish do (fig. 8.10). These fossils are found on ripple-marked sandstone surfaces that were once intertidal, sandy shorelines of barrier islands.[13]

8.8. Trilobites from the Upper Cambrian of Minnesota and Wisconsin. (A) Pygidium of giant *Dikelocephalus oweni* from the Lodi Member of the St. Lawrence Formation near Prairie du Sac, Wisconsin (USNM uncataloged "85"). Full scale bar is 10 cm. (B) Spined pygidium of *Crepicephalus iowensis* from the Eau Claire Formation near Mountain Island, Minnesota (USNM 2728). Scale bar = 1 cm. (C) Cranidium (arrow) of idahoiid asaphid trilobite *Wilbernia explanata* from the Franconia (Lone Rock) Formation in St. Croix County, Wisconsin, shown with its museum collections tag. Scale in millimeters.

(A) and (B) courtesy of Smithsonian Institution; (C) from University of California Museum of Paleontology collection.

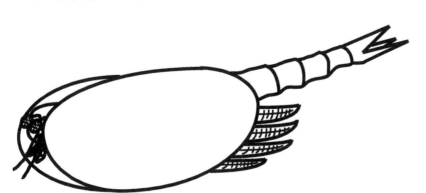

8.9. The Upper Cambrian bivalved arthropod *Arenosicaris* from the St. Lawrence Formation of Wisconsin. Length ~5 cm.

The Cambrian Corps 7 – James Hagadorn

A native of San Diego, California, James Hagadorn is curator of geology at the Denver Museum of Nature and Science in Colorado. He specializes in the Precambrian–Cambrian transition and early Cambrian ecosystems, particularly as indicated by the trace fossils left behind by animals and other members of these early biotas. "I am interested in how animals and microbes first impacted ancient Earth environments," he says. "By learning how they did this and what changes their activities fostered, this research informs me of how the outer membrane of our planet works – today, in deep time, and potentially in the future."

Although he never had a particularly strong interest in geology or paleontology early on (other than once digging an anti-burglar quicksand pit in his mother's garden), his path into Precambrian–Cambrian research was inspired in graduate school at the University of Southern California (USC); one of the reading assignments was Gould's *Wonderful Life*. Having majored in geology at the University of Pennsylvania, Dr. Hagadorn concentrated on land-based fieldwork in graduate school. "I realized I was not a seagoing or airgoing researcher." It was the influence of his professors at USC that led him into the questions of trace fossils and their evolution over the Precambrian–Cambrian boundary. "Trace fossils are like finding a deep-time crime scene," he says. "In deep time, trace fossils provide the same information, indicating how ancient animals moved, who they associated with, how they mated or nested or lived, who ate who, where they lived, and in some cases, they can even provide specific enough information to indicate who the perpetrator was."

Among Dr. Hagadorn's favorite surprises from the odd time period that is the Precambrian is the fact that some of the best-preserved fossil organisms from this time (the Ediacaran biotas) are found in sandstones, a lithology that is comparatively bad for preservation during subsequent Earth history. Also surprising, he says, is that most Ediacaran taxa appear now to be not basal animals or other organisms, but an entirely separate group of multicellular organisms. "Looks like Dolf Seilacher was right!" he says.

Among the most important lessons to be learned from the Cambrian, James believes, is to keep an open mind regarding such ancient ecosystems that were not yet entirely like our modern analogs. The Cambrian was a time from which we should expect some surprises. Existing beliefs and hypotheses should be readily thrown out the window if tests of data don't turn out as we expect. "For that matter," he says, "take all of your own interpretations with a giant grain of salt."

CNIDARIA

Anyone who has seen jellyfish washed up on the sand at a beach can picture the scene 500 million years ago when such stranding events happened in what is now Wisconsin and fossil impressions of these animals ended up in the rocks. But what seems (and is) so unlikely is the preservation of even impressions of animals that are composed entirely of very soft tissues. Jellyfish are members of the Cnidaria, a group of soft-bodied animals that includes today's sea anemones and corals as well. The cnidarians are more complex than the sponges but are less so than most other animal groups. One way in which other animal groups (above sponges and cnidarians) are more complex is that developmentally they form from three tissue layers rather than the two seen in cnidarians.

There are about 11,000 species of modern cnidarians. Many live colonially, and many others are solitary, free-floating, or swimming pelagic species. There are two general body forms, known as **polyps,** which are usually sessile and benthic (sea anemones, for example), and **medusae,** which are pelagic (jellyfish, for example). Cnidarians may reproduce sexually or asexually; polyps often produce medusae through budding. The polyp and medusa forms are both usually part of the cnidarian life cycle, one or the other being larval, but individual species may only

8.10. Ripple-marked surface in Upper Cambrian Wonewoc Sandstone near Mosinee, Wisconsin, showing multiple impressions of beach-stranded ancient jellyfish (circles).

Photo courtesy of J. W. Hagadorn.

take one form in many cases. Among the characters of cnidarians are: radial symmetry; ciliate, motile larvae; tentacles; stinging or adhering cells called **cnidae;** and a lack of a central nervous system (possession of a nerve net instead), a circulatory system, a respiration system, and excretory organs. Many cnidarians are predatory or are suspension feeders. Some cnidarians—some corals, for example—have symbiotic algae that provide a secondary source of energy.

There are four classes of Cnidaria: the Scyphozoa, Cubozoa, Hydrozoa, and Anthozoa. The Scyphozoa consists of medusoid species with a dome-shaped bell (or hood), a central, ventrally placed mouth with a stomach just above it, tentacles ringing the outside of the bell, and oral arms in the center, arranged around the mouth. The bell is lined on the outside with the umbrellar epidermis. The Cubozoa is another medusoid group that includes the box jellyfish, which have a bell with a squarish cross section. The cubozoan sting is very toxic and can be fatal to humans. The Hydrozoa have both medusoid and polyp forms; many modern species live in freshwater environments. There are medusoid hydrozoans from the fossil record, including a jellyfish-like species from the Marjum Formation of Utah and a similar type known as a chondrophorine hydrozoan from the Cadiz Formation of California. The small, disc-shaped genus *Scenella*, usually regarded as a monoplacophoran mollusc, has also been interpreted as a chondrophorine hydrozoan;[14] it is known from a number of Cambrian formations, including the Burgess Shale and Chisholm and Langston formations. The Anthozoa includes mostly polyps that live sometimes as solitary individuals, but often in colonies. Sea anemones and corals are anthozoans. The Anthozoa have no medusoid stage, and their reproduction may be either sexual or asexual. Sea anemones appear to have been present in the Cambrian seas, although, like many of the first sponges and molluscs, some of them were surprisingly small.[15]

Cambrian fossil cnidarians are preserved in a number of formations in North America, in addition to the stranded Wisconsin jellyfish mentioned above. Some of the best preserved specimens are jellyfish-like species of scyphozoans, cubozoans, and hydrozoans found in fine-grained rocks of the Marjum Formation in Utah (fig. 8.11). These specimens are very well preserved, showing muscles and tentacles in some, and are (along with Cambrian specimens from China) some of the best and oldest jellyfish known from the fossil record.[16]

Other known fossil scyphozoans include the **conulariids,** a group of (usually flattened) cone-shaped fossils found in rocks starting back at least in the Early–Middle Cambrian (fig. 5.5), but their first appearance may have been as far back as the late Proterozoic. The shells are usually made up of four triangular faces and are composed of many rods of calcium phosphate. Ridges running perpendicular to the long axis of the shell were formed by addition of calcium phosphate rods as the animal grew. The cone-shaped shells appear to have attached to the ocean bottom at their pointed ends, with these ends buried to varying degrees in the soft

8.11. Cambrian jellyfish fossil from the Marjum Formation of Utah showing tentacles and hood. Scale bar = 5 cm.

From Cartwright et al. (2007).

sediment; conulariids also appear to have frequently attached in the sediment next to brachiopods and other conulariids.

So what is a jellyfish relative doing in a cone-shell on the seafloor? It appears that the conulariid schyphozoans lived much like modern sea anemones, with the mouth and tentacles exposed to the water at the open end of the cone shell (plate 5).[17] The animals caught and ate prey and food particles from the water. Perhaps one way to imagine conulariids is to picture a small pseudo-jellyfish upside down, living in a cone-shaped shell on the bottom rather than free floating, but with the mouth and tentacles functioning much the same way.

In the Midwest, conulariids are found in the St. Lawrence Formation in the upper Mississippi River valley, a formation we met early in this chapter. The group is named after the genus *Conularia*, found in the same region and named by (who else?) C. D. Walcott back in 1890. Conulariids appear to be rare in most formations in which they occur. Possible conulariids have also been found in the Precambrian of Russia and, as we saw in chapter 4, in the Lower Cambrian Latham Shale in California.[18]

Also within the Anthozoa are the animals responsible for building so much of what snorkelers and scuba divers often travel so far to enjoy–coral reefs. It is modern coral reefs that provide the habitat for the many species that call shallow tropical seas home. Corals are tiny, colonial cnidarians that usually live in frameworks of calcium carbonate that the animals precipitate themselves. A branching coral, for example, may be a meter or so high and the entire surface will be covered with thousands of small pockets in the calcium carbonate structure, each of which contains an individual coral animal (polyp). Coral polyps are tube-like animals with the mouth and tentacles at one end (exposed to the water) and the bottom end, or pedal disc, attached to the bottom of the pocket in the coral colony's calcium carbonate framework. It is the framework that takes on such a variety of form in modern coral reefs, from branching stag horn corals to rounded "brain corals." In either form it will scrape your skin quite easily if you brush against it while swimming or surfing.

Corals are sessile, opportunistic feeders on tiny metazoans and other organic matter in the seawater; food particles are stung and brought in toward the mouth by the tentacles. The small size of the food items is necessitated by the generally small size of individual coral animals. The reefs that coral colonies build provide habitat for an entire ecosystem of other animals. There are corals and coralomorphs known from the Lower Cambrian in North America, Australia, and Russia (see chapter 4). The fact that a few fossil species of corals and coral-like animals have been identified in Cambrian rocks is interesting in that it shows that these animals, along with other organisms such as cyanobacteria, calcimicrobes, and archaeocyathids, were already building framework reefs in low- to high-energy environments at least 520 million years ago, when the ancestors of modern tropical reef fish (and indeed the ancestors of all modern vertebrates) were just appearing in the oceans.[19] Even at this early time it appears that reefs were high-energy environments crowded with species. In North America, the Burgess Shale genus *Echmatocrinus* is thought to have been either an octocoral or a crinoid (see chapter 6).

SPICE It Up

Evidence of an unusual event or series of events in the early Late Cambrian can be found in rocks of the upper Mississippi River valley in a sudden increase in one isotope of carbon over another. It appears that during the Steptoean stage of the Late Cambrian (Series 4) the isotope $\partial^{13}C$ increased about 4–5% relative to $\partial^{12}C$. That may not sound like much, but compared with typical variation it is a huge shift, and it suggests that there were big changes going on in the oceans. The shift even has its own name: the Steptoean Positive Isotopic Carbon Excursion (SPICE).[20] This trend in the rocks has been identified in the Upper Cambrian sandstones of the Mississippi–St. Croix region, and in an approximate stratigraphic equivalent of the Wonewoc Formation out in the Great Basin, the Orr Formation in Utah (fig. 7.1). Although the Wonewoc is sandy and the Orr is largely limestone and shale, they are of about the same age and appear to record some of the same events.

So what would cause a shift in what isotope of carbon was ending up in the rocks? It may have been that much more organic carbon was getting buried in the sediments of the time. This may have occurred due to an increase in total biomass of the tiny, planktonic, photosynthesizing organisms in the ocean water column, an influx of carbon from terrestrial algae washed into the oceans from the continents, or to a lowering of oxygen levels in the oceans due to vertical turnover of the water column near the coastlines, possibly caused by tectonic uplift of continental areas resulting in a significant drop in relative sea level.[21] The SPICE event appears to coincide with a great increase in the diversity of the phytoplanktonic acritarchs (see later in this chapter). A possible effect of the SPICE event was a significant trilobite extinction event, formerly called a **biomere,** one of several that occurred during the Cambrian. These "biomeres" were defined by A. R. Palmer and are different from the biostratigraphic trilobite zones (see below). The SPICE event, a trilobite extinction (and subsequent diversification), and the boundary between Sauk subsequences II and III all appear to occur at the same level, indicating a significant event.[22] What the SPICE event and biomeres show is that even trilobites, numerous as they were, were still susceptible to extinction at the hands of things causing rapid changes to their habitats, things such as sea level drops and oxygen level changes. The animals of the Cambrian were no less sensitive than species today.

Perhaps not surprisingly, the SPICE event is not the only isotopic variation known from the Late Cambrian. In fact, two other carbon isotope excursions identified in the latest Cambrian, during Stage 10, as it is known, are in fact temporary, negative carbon isotope changes, and these have their own acronyms. The earlier one, apparently the logical (but opposite) cohort of the SPICE event is known as HERB.[23] And a negative carbon isotope excursion in the Middle Cambrian (Drumian stage) has been identified as the DICE event.[24] Science isn't all lab coats and pure seriousness.

Extinction and Recovery

Repeated trilobite extinctions followed by diversifications (biomeres) were first identified by paleontologists working in the Late Cambrian of the Great Basin in the western U.S., but a number of the same events have now been identified in rocks in Oklahoma and the Midwest, for example.[25] Biomeres were defined by A. R. Palmer as "regional biostratigraphic unit[s] bounded by non-evolutionary changes in the dominant elements of a single phylum." Basically, this means that the boundaries above and below a defined biomere interval are marked by faunal turnovers that resulted from extinction and immigration.[26] This indicates ecosystem disruption rather than normal speciation change. Within a biomere, different species tend to be characteristic of different sub-environments on the shelf (near shore, lagoon, carbonate shoals, etc.); these associations have been termed **biofacies**.[27] We discussed these earlier in chapter 4.

Whereas stage boundaries are defined by the first appearance data of particular species, biomeres are defined by the broader faunal turnover of the extinctions above and below; in many cases, it has turned out that stage boundaries and biomeres end up being defined at the same horizons. And trilobite zones constitute a separate subdivision altogether. ("Biofacies" are more of a biological-sedimentological and taphonomic association and are not a zonation scheme; they can reoccur many times.) The bases of biomeres tend to have low morphological diversity, whereas the upper parts see an increase in the range of morphological diversities represented by trilobite species.[28]

There appear to have been three biomere-defining extinctions within about 10–15 million years during the Late Cambrian in North America. The extinctions are characterized just after the event by reduced overall diversity, less patchy (more geographically even) distribution of species diversity, and immigrations of species from more distal (shelf-edge) paleoenvironments into the shallower shelf settings.[29] As significant as these extinctions were, after the events in the Late Cambrian species diversity recovery appears to have been almost complete and relatively fast (within a few million years). More major extinctions, the kind that define period boundaries, for example, have generally greater loss and slower recovery.

An interesting finding about extinctions has been that the degree of species loss does not necessarily equate to the severity of the ecological disruption that results. An ecologically disastrous extinction does not necessarily have to result from equivalently massive species loss, and massive species loss may result in a relatively minor ecological disruption; both types of extinction events have occurred in Earth history.[30] In part, the determining factor here may be what elements of a fauna are lost, with **keystone species** and common or dominant species being ones whose loss has a large impact on the resulting disruption. The Late Cambrian extinctions among North American trilobites line up with Laurentian stage boundaries but do not involve wholesale ecological turnover; rather, they are followed by rapid immigration replacement. This series of relatively minor extinctions, however, does mark an introduction to the beginnings of the diversification of the soon-to-be-new Paleozoic Evolutionary Fauna during the Ordovician.

Trilobites in the Rockies

In the western part of the continent, the Late Cambrian saw the encroachment of the sea far in from the earlier areas we have seen in Nevada and Utah; certainly there are relatively deepwater units in these areas still in the Late Cambrian (the Weeks Formation in Utah, for example), but in order to see deposits in the shallow, near-shore environments in the Late Cambrian we need to travel to the mountains of Montana, Wyoming, and Colorado.

The Cowboy State

Rocks of Late Cambrian age are exposed along the flanks of the Big Horn and Wind River mountain ranges in central and western Wyoming, and there are also outcrops in the vicinity of Grand Teton and Yellowstone national parks. These units are known as the Gallatin Group (Pilgrim Limestone and Snowy Range Formation), which consists of shale and a predominance of limestone, and the stratigraphically lower Gros Ventre Group (Wolsey Shale, Death Canyon Limestone, and Park Shale), which has plenty of limestone but is mostly shaly. Below these two groups lies the Flathead Quartzite. Between these three units, we have, as at other sites such as Grand Canyon and the Marble Mountains, basal sandstones overlain by shale-limestone interbedded units that become dominated by limestone higher up. It is the same pattern we have seen before in a few places on the craton, but younger here because we are farther east in an area the rising sea took longer to reach. As with some other Cambrian formations, the Snowy Range Formation has flat-pebble conglomerates that appear have been deposited near shore in significant storms of the time.[31]

Trilobites occur in these rocks, but–in a pattern not uncommon in Upper Cambrian rocks of the middle part of the continent–they are very rarely articulated, and it can take a fair amount of work to find them. You definitely earn your catch when working these sites. In some cases, you may not even see any fossils at the outcrop but rather must bring rocks back to the lab and process them before finally finding some trilobite pieces under a microscope. The process? "Break rock by heating in oven, cooling in cold water, and crushing," reads step number 1 in the procedure summary of one worker who has studied Gallatin Group trilobites in Wyoming.[32] Among the species found in these rocks are the ptychopariids *Cedaria*, *Weeksina*, *Meteoraspis*, and *Tricrepicephalus*; along with other trilobite forms such as *Kingstonia*, *Elvinia*, and *Coosina*. Nearly all are preserved as isolated cranidia or pygidia, however. The trilobites are generally preserved in an interbedded mix of limestone and oolitic limestone, with some sandstone, all of which collectively are indicative of shallow, relatively high energy carbonate shoals on the continental shelf. Some fossil deposits have been identified as being wave-carried debris accumulations. Also preserved with the trilobites, as usual, are hyoliths and brachiopods, but there was also a sponge found in the Gallatin Group. Trilobite zones well represented by the fossils in these Upper Cambrian rocks in Wyoming include the *Cedaria*, *Crepicephalus*, and *Elvinia* zones.[33]

CONODONTS

In the Upper Cambrian rocks of the Gros Ventre Group and Gallatin Group in the Big Horn Mountains of Wyoming one can find tiny (nearly microscopic) cone- and spike-shaped fossils of animals known

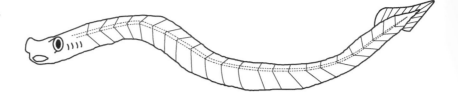

as **conodonts.** You do not generally find them on the outcrop but rather must go through an even nastier process than that with the trilobites mentioned above; the process involves heavy liquids and other things requiring a fume-hood. The conodonts were small, elongate, almost eel-shaped jawless fish that appeared in the Middle Cambrian and were particularly diverse later in the Paleozoic (fig. 8.12); they became extinct in the Triassic period. Not just chordates like the Burgess Shale's *Pikaia* and *Metaspriggina*, conodonts were full-blown vertebrates. There are about 550 genera known, and they are the main biostratigraphic group for defining age boundaries in the Paleozoic; in fact, the Cambrian–Ordovician boundary is defined on the first appearance of the conodont species *Iapetognathus fluctivagus*, as we saw in chapter 1.

The conodonts were ribbon like, long and laterally compressed, and the fossils found are the tooth-like elements of their feeding apparatus, which consisted of an array of sometimes elaborate conical, comb-, or molar-shaped "teeth" arranged in paired rows down the mouth/gullet region of the head. Like modern lampreys and hagfish, there were no jaws or bones to support the teeth but the conodonts were vertebrates.[34]

Conodonts were named as a group by anatomist Christian Heinrich Pander in 1856, and they range in size from 1.5–16 inches (40 mm–40 cm). As mentioned above, they are entirely soft bodied except for their teeth, which are composed of the characteristic vertebrate tissues dentine and enamel. Unlike some vertebrates, in which teeth are continually shed and replaced, it appears that conodonts retained their teeth throughout life. The teeth and feeding apparatus in general show unexpectedly complex occlusion (the interlocking that provides shearing and grinding functions in teeth). Cambrian conodont teeth were less complex than later types, generally being conical or spike shaped, but many were grooved, and it is possible that these conodonts were mildly venomous.[35]

For years, before more complete specimens were known, it was debated what type of animals produced the conodont teeth that were being found in Paleozoic rocks worldwide, and in general it was believed that they were either annelid worms or vertebrates.[36] Even when I was in college, right around the time the more complete specimens were coming to light, the traditional view was that conodont teeth represented "worm jaws." All that began changing in the early 1980s, however, when nearly complete, soft-bodied fossils of conodonts were found in rocks just younger than the Cambrian in Scotland and South Africa.[37] These fossils confirmed the identification of conodonts as vertebrates. Among the characters of these fossils that indicated they were in fact chordates and vertebrates were: enamel teeth, a cartilaginous head with large eyes and a differentiated brain, the chevron shaped muscles known as myomeres, a notochord, a ray-supported tail fin, and the possible presence of gills. The complete specimens were of two types in separate evolutionary lines of conodonts, which suggests that their anatomy is representative of conodonts as a whole, not just one branch of the conodont tree. Although their status as vertebrates seems established, the relationships of conodonts within Vertebrata is as confused as ever. They may be stem vertebrates, as seems logical, but it also appears possible that they are more derived than the modern, jawless hagfish.

The eyes in conodonts appear particularly large. This fact and the presence of extrinsic eye muscles suggest that the eyes were actively used by the animals, and that vision was not of secondary importance to them in their life habits. Larger species, however, have relatively small eyes. The apparent large size of the eyes in very small conodonts is probably a result of the eyes functionally not being able to be any smaller. Smaller eyes simply would not have worked, so tiny conodonts have disproportionately large eyes.[38]

The teeth, eyes, and structure of the body suggest that conodonts were active predators. What they ate is less clear, however. Given their generally small size and free-swimming locomotion they may have fed on zooplankton in the water column or perhaps small, soft-bodied species near the sediment-water interface.

Conodonts appear in the Middle Cambrian, and by the Late Cambrian consist of two main evolutionary lines; they also appear globally to fall into two faunas, one low-latitude and warm-water adapted and the other high-latitude and cooler-water adapted. The two evolutionary lines appear to diversify beginning in the Late Cambrian, around the time of the *Elvinia* trilobite zone.[39]

The presence of conodonts in the Cambrian indicates that despite the marked rarity of chordates such as *Pikaia* and *Metaspriggina* in the Burgess Shale, and despite the presence of apparent vertebrates in the Chengjiang biota in China, early members of our vertebrate line were also present in respectable numbers in a several parts in North America by the Late Cambrian. That such diverse groups as sharks, bony fish, reptiles, birds, amphibians, and mammals evolved from the humble Cambrian beginnings of Vertebrata, as represented by conodonts and *Haikouichthys*, is a testament to the ability of species to build on a **bauplan** when presented with ecological and evolutionary opportunities. This is far from the only example of such diversification, but it is one of the most easily relatable for us—our origins also are tied to the vertebrate cousins of the Cambrian conodonts.

ACRITARCHS

In the Gros Ventre Group and the Snowy Range Formation of Wyoming researchers have found microfossils representing unicellular, planktonic

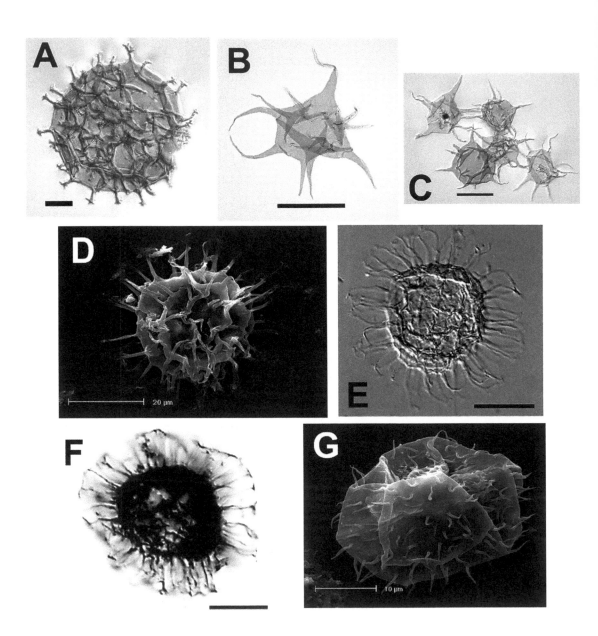

green algae of the Cambrian oceans.[40] These are the fossils known as acritarchs. These are not easy organisms to find, as they are of course tiny and you must break down bulk samples of rock and separate the fossils using, as with conodonts, an assortment of entertaining fluids that tend to have lots of warnings on the labels—things like hydrochloric and hydrofluoric acids. But we learn a lot about the ecosystems of the Cambrian from these small fossils. After all, as photosynthesizing unicellular green algae, many of the acritarchs were among the species that probably formed the bulk of the biomass of the primary producers of the ocean. Consumers are not going to get their energy out of the ether; it has to come from the foundation of the food chain. All the trilobites and arthropods, all the monoplacophorans and hyoliths, and all the brachiopods and sponges have to get energy from their food and that food ultimately got its energy from the primary producers, among which the phytoplankton (in the form of acritarchs in the Cambrian) were some of the most important. So those of us who like those larger fossils of the Cambrian owe a tip of the hat or two to the acritarchs for making the Cambrian world possible. For that matter, we owe them for making our own world possible, because they are among the ones who got it started so many millennia ago.

Technically the acritarchs are an informal, wastebasket group, and we first encountered them in chapter 3. Acritarchs comprise any small, non-carbonate and non-siliceous, organic structures that cannot be otherwise classified. By small, we generally mean less than 1 mm in diameter; large acritarch diameters are about 0.3 mm! Among the acritarchs are structures such as possible metazoan egg cases and resting cysts of green algae. However, most of these microfossils were probably free-floating, pelagic phytoplankton such as unicellular marine green algae (chlorophytes) (fig. 8.13), the origin of which is probably well back in the Precambrian;[41] some larger forms may have been benthic algae. Other acritarch species are similar to the mostly unicellular and largely photosynthetic dinoflagellates. As mentioned above, the acritarchs are not a biological group, but more of a holding tank for small marine planktonic and benthic taxa that cannot be more specifically identified. Recently, however, many acritarchs have been confirmed as green algae, so we are beginning to be able to classify many of these species.[42]

The acritarchs appear in the Precambrian 3.2 billion years ago, and their diversity increases starting 1 billion years ago. Their diversity spikes in the Ediacaran but then drops at the Ediacaran–Cambrian boundary. Diversity then increases through the Early and Middle Cambrian, reaching Ediacaran levels once again by about the Early–Middle Cambrian boundary. This Early Cambrian diversification was a recovery after the Ediacaran–Cambrian extinction event of acritarchs, and it preceded not only the diversification of trilobites and other complex metazoans later in the period but even the diversification of shelled animals in the Early Cambrian. In the Late Cambrian acritarch diversity really takes off, and by the end of the period it has more than doubled over its Early Cambrian peak to nearly 250 known species.

Beginning in the Early Cambrian, acritarchs appear to have become smaller and more spiked, and perhaps more species were becoming planktonic as opposed to benthic. Throughout the early part of the Cambrian, there were regular episodes of diversification of acritarchs every few million years, with a few species becoming extinct but an overall trend of increasing diversification. This Early Cambrian acritarch diversification has been hypothesized to have been related to the origin of marine phytoplankton, which was an important development in the oceanic ecosystem. Acritarchs also seem to have been going from exclusively benthic in the Precambrian to mostly pelagic in the Cambrian.[43]

8.13. Some Cambrian acritarchs. (A)–(C) Upper Cambrian forms from the Snowy Range Formation in Shoshone Canyon, Wyoming. (A) *Timofeevia phosphoritica*. Scale bar = 5 microns. (B) *Polygonium* sp. Scale bar = 25 microns. (C) *Polygonium* sp. Scale bar = 25 microns. (D)–(G) Lower Cambrian forms from the Lükati Formation of Estonia. (D)–(E) *Skiagia ornata*, scale bars = 20 microns. (F) *Skiagia scottica*, scale bar = 20 microns. (G) *Globosphaeridium cerinum*, scale bar = 10 microns. *Skiagia ornata* and *S. scottica* also occur in the Lower Cambrian of Greenland.

(A)–(C) courtesy of Brian Pedder. (D)–(G) courtesy of Malgorzata Moczydlowska.

8.14. The Cambrian of the Colorado Rockies. (A) Outcrop of the Upper Cambrian Sawatch (€s) Formation high above the Eagle River south of Vail. (B) Two asaphid trilobite pygidia from the Upper Cambrian Dotsero Formation just up section from the locality in (A). (C) Trilobite pygidium from the Dotsero Formation, same locality. (B) and (C), Scale bars = 1 cm.

(B) and (C), Museum of Western Colorado collection.

The Centennial State

To the south of the Cowboy State's mountain ranges, in Colorado, Cambrian rocks are exposed in canyons and high plateaus of some of that state's mountainous terrain. Near and in parks and ski towns, the Colorado Rockies produce almost unexpected Cambrian treasures, although they take some work to find. In the central part of the Centennial State, gray quartzites and quartz-rich sandstones of the Sawatch Formation sit on Precambrian metamorphic rocks and mark the appearance of the Late Cambrian beach and shallow marine environments in the area (fig. 8.14a). Sedimentary structures in the Sawatch Formation indicate the influence of occasional storms on these shallow marine sands. Trilobites and other body fossils are tough to find in the often-vertical outcrops of the Sawatch Formation, but there are trace fossils of trilobites (or at least trilobite-like arthropods) in the *Rusophycus* resting traces found in these rocks. *Skolithos* tubes indicate wormlike animals living in high-energy, sandy seafloor environments.

Above the Sawatch Formation, the Upper Cambrian Dotsero Formation represents slightly more distal shallow marine deposits with green, gray, and red shales and fine sandstones, light gray siltstones, tan quartzites, and orange carbonates interbedded for a thickness of about 30 m (100 ft.) total. A thick bed of stromatolitic limestone is also found near the top of the Dotsero.[44] The Dotsero Formation is present but not always exposed in a number of the mountainous areas within the Colorado Rockies; fossils are more common in this unit than in the Sawatch Formation, but they are not particularly abundant. Brachiopods, rare eocrinoids, and trilobites similar to forms found in the St. Croix and Mississippi valleys in Minnesota and Wisconsin are found in the Dotsero Formation with some determined searching of limestone and fine sandstone. Outcrops of Sawatch and Dotsero formations are exposed at approximately 2744 m (9000 ft.) elevation, high above the Eagle River in the Rockies, not far from the ski town of Vail, Colorado, and here one can find not only *Rusophycus* in the Sawatch, but also relatively abundant brachiopods and rare trilobites of the order Asaphida in the Dotsero Formation (fig. 8.14b,c). Also found in the Dotsero Formation are microbe fossils and the asaphid trilobites *Ellipsocephaloides*, *Idahoia*, and *Ptychaspis*. In the 1930s, several species of graptolites were identified in the Dotsero Formation near Glenwood Canyon in Colorado.[45] (See chapter 6 on Cambrian graptolites and related forms.)

Deep in the heart of Dinosaur National Monument, straddling the Utah-Colorado border and in the paddle-strokes of John Wesley Powell along the Green River, lie rocks of Late Cambrian age that lack abundant body fossils but contain many traces of animals, indicating–despite a near lack of body fossil remains–that 500 million years ago life was abundant here. In an area now frequented by mountain bighorn sheep rest small, flat rocks containing trails and other traces of trilobites and other animals.

Passing the towering Weber Sandstone cliffs of Steamboat Rock on the Green River you float north for a while, deep in a canyon 823 m (2700 ft.) below Harpers Corner Overlook. As you pass over the Mitten Park fault at river level you pass from Pennsylvanian-age rocks to the Precambrian sedimentary rocks of the Uinta Mountain Group (which we first encountered in chapter 3). A little farther downstream in Whirlpool Canyon you reach, at river level, rocks of the Upper Cambrian Lodore Formation. Most interesting here are the sea stacks, columns of 1.1-billion-year-old Uinta Mountain Group rocks around which the Lodore Formation sediments were deposited. At the time the Late Cambrian Lodore sea existed here, this was a shoreline with Uinta Mountain Group rock formations (themselves already close to 500 million years old) sticking up out of the coastal waters–not unlike the sea stacks seen today along parts of the Pacific Coast highway in California and Oregon. The traces indicate the activity of worms and arthropods on the seafloor in the shallows.[46]

From Colorado we head northeast to an island in the plains. The Black Hills of South Dakota and Wyoming were formed by uplift in the middle of the Great Plains, but are related to uplift of the main Rocky Mountains. The Black Hills are famous for gold and its associated mines, granite mountains turned into monuments (Mount Rushmore and the Crazy Horse memorial), and caves (such as Jewel Cave, Wind Cave, and many other smaller ones).[47] But the Black Hills may also be well known as a center of paleontological discoveries; at least it is surrounded by sites containing the large and flashy. "Sue" the *Tyrannosaurus* was found in the Hell Creek Formation northeast of the Black Hills and spent a short stint at the Black Hills Institute of Geological Research in Hill City; giant fossil brontotheres and other Oligocene-age mammals were found by the thousands at what is now Badlands National Park just east of the Hills; mosasaurs and other marine reptiles have been found all around the plains around the eastern and southern Hills; *Triceratops* and possibly the first partial *Tyrannosaurus* ever found were excavated from the plains on the Wyoming side of the uplift; and Late Jurassic dinosaurs such as *Camarasaurus, Brachiosaurus, Barosaurus,* and *Allosaurus* have been found at sites ringing the edge of the Black Hills since as far back as 1889. With all this going on, it has largely been overlooked by the general public that the heart of the Black Hills is home to many sites that have yielded humble but important Late Cambrian trilobites and brachiopods.

The Deadwood Formation was named after the once wild (now less so) mining and gambling town of the same name in the northern Black Hills of South Dakota. Outcrops of these Upper Cambrian rocks are exposed just a block or two from downtown Deadwood, within sight of the hotels, casinos, and saloons that comprise the town that once hosted Wild Bill and Calamity Jane. The Deadwood Formation consists of mostly sandstone, with basal conglomerate and interbedded limestone, and ranges from as little as 1.2 m (4 ft.) thick in the southern Black Hills up to approximately 152 m (500 ft.) thick in the northern Hills (fig. 8.15). It gets as thick as 274 m (900 ft.) in the subsurface of the Williston Basin below North Dakota. Some areas of the Deadwood Formation also contain minor amounts of shale. Although it occurs in outcrop as far north as the Bear Lodge Mountains in the northwestern part of the Black Hills uplift, the Deadwood Formation also occurs in the subsurface all the way up into (or rather below) Saskatchewan and Alberta in Canada. The formation is exposed more or less in a ring around the central Black Hills, the core of which is composed of Precambrian rocks. Thus, the Deadwood Formation sits on late Archean and early Proterozoic metamorphic and igneous formations, and the contact between the Cambrian and the Precambrian units indicates that the Deadwood Sea advanced from the west over an irregular surface of Precambrian rock that had valleys and ridges with (in some places) several hundred feet of topographic relief. The Deadwood Formation is mostly Late Cambrian in age, and it ranges through much of the epoch, but it also ranges up into the basal Ordovician in its northern reaches.[48] In at least one spot the contact consists of

8.15. The Upper Cambrian Deadwood Formation in Little Elk Creek Canyon, South Dakota. (A) Bedded outcrops of the Deadwood Formation (arrow) above Little Elk Creek. (B) Close up of interbedded shale and trilobite-producing bioclastic limestone of the upper Deadwood Formation. Scale bar = 25 cm.

Deadwood Formation sandstone resting on a flat, horizontal unconformity below which the metamorphic Precambrian rock shows signs of having been deposited as flat beds of shale, metamorphosed into slate, deformed into countless, tight folds and uplifted to a completely vertical orientation, and then eroded off flat–all before Deadwood Formation deposition began. This is yet another great example of an angular unconformity that lets us gaze into the "abyss of time."

The Deadwood Formation represents shallow marine deposition in the Late Cambrian at a time when the beach shoreline was just to the south and east. Basal conglomeratic sandstones probably were along the beach and shoreline, sandstones were in shallow environments just offshore, and the limestones were deposited on shallower shoals scattered along the continental shelf.

The Deadwood Formation contains many trilobites of genera seen at other Late Cambrian sites across the west and Midwest: the ptychopariids *Cedarina, Kingstonia, Modocia, Tricrepicephalus, Coosina, Coosia,* and *Crepicephalus* (fig. 8.16) occur in the lower part of the formation. The upper Deadwood contains trilobites such as *Ellipsocephaloides, Ptychaspis, Dikelocephalus, Idahoia,* and *Dartonaspis.* As at many other sites in the Late Cambrian, you pound a lot with your rock hammer to find trilobites in the Deadwood Formation–they don't just jump out of the outcrop, as they seemingly do at some older sites. Although several genera are often shared by different formations across the continent, in most cases the

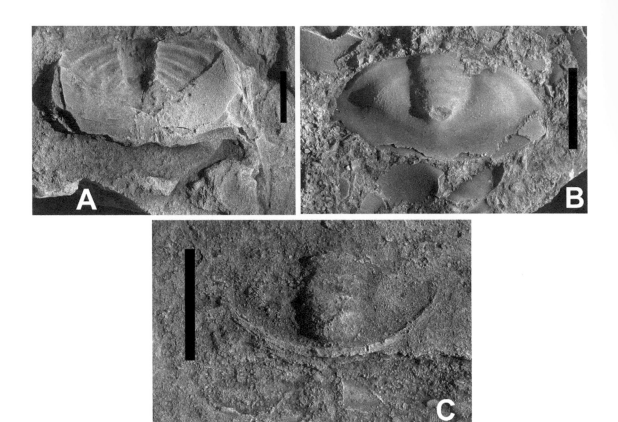

8.16. Trilobite specimens from the Upper Cambrian Deadwood Formation of the Black Hills, South Dakota. (A) Pygidium of *Crepicephalus buttsi montanaensis* from near Deerfield, South Dakota. (B) Indeterminate pygidium from the upper Deadwood in Little Elk Creek Canyon. Possibly a worn specimen of *Prosaukia*. (C) Indeterminate pygidium from the upper Deadwood in Little Elk Creek Canyon, possibly *Parabolinoides*. All scale bars = 1 cm.

species are different. *Tricrepicephalus* is found in the Weeks Formation in Utah (as we have seen), in rocks slightly older than the Deadwood Formation, but the forms belong to different species of *Tricrepicephalus*, and the Utah species is clearly different from that of the Black Hills. This is not entirely surprising, however, given the time and environmental differences between the two sites.

Preserved within the Deadwood Formation is a turnover of trilobite species characterized by older marjumiids such as *Crepicephalus* being replaced by pterocephaliids, including *Aphelaspis*. This change may have resulted from a decrease in water temperature that allowed deeper, cooler-water species to invade the shelf during Deadwood Formation deposition. The water temperature change does not need to have had anything to do with climate changes, but may rather have resulted from combined factors relating to sea level changes, shelf topography, and variation in the oceanic thermocline.

The Deadwood Formation also yields, at some localities, abundant brachiopods and some trace fossils. There are at least a dozen species of brachiopods known from the Deadwood Formation, and some of these were found in well cores taken from as deep as 2650 m (8700 ft.) below the ground surface east of Calgary and west of Saskatoon. Other specimens can be found in outcrop around the Black Hills (fig. 8.17). Among the trace fossils from the Deadwood Formation are rare trails of trilobites that appear to have sometimes "hopped" along the bottom, swimming

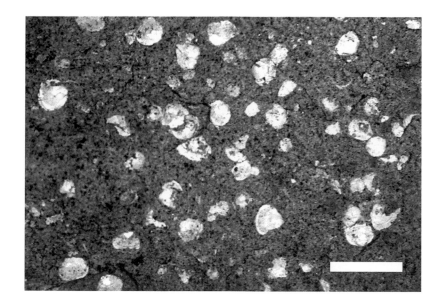

8.17. Brachiopods from a single sandstone bed in the Upper Cambrian Deadwood Formation near Lightning Creek, South Dakota. Scale bar = 1 cm.

in the water above the sand and contacting the sand itself only briefly before bouncing up again. This is a behavior we expected but see only in these traces.[49]

Recent drilling finds from the subsurface of Canada indicate the presence of limbs and mandibles of tiny branchiopod, copepod, and ostracod crustaceans in the Deadwood Formation. These finds suggest that at least three elements of the modern zooplankton had already appeared by the Late Cambrian.[50]

By the Late Cambrian, the sea had advanced to a point that it flooded nearly the entire continent of Laurentia, and fossils of trilobites and other Cambrian animals are now found all over North America as a result. In addition to the areas mentioned above, such fossils have been found in the east in New York, Vermont, Pennsylvania, Virginia, Maryland, and Quebec; in the northwest in British Columbia, Montana, Alberta, and Alaska; in the southeast in Tennessee and Alabama; and in the southern Midwest, Great Plains, and Southwest in Arkansas, Missouri, Oklahoma, Texas, New Mexico, and Nevada. Those are just a few examples. The Late Cambrian trilobite fauna is widespread and very similar in many areas.

In the Little Belt Mountains of Montana, the Upper Cambrian Pilgrim Formation has yielded some of the same trilobite genera as other sites in Utah, Wyoming, Colorado, and the Wisconsin-Minnesota region: *Crepicephalus* (fig. 8.18), *Idahoia, Ptychaspsis, Arapahoia,* and *Tricrepicephalus.* The Nolichucky Formation of Tennessee and Alabama has also produced *Tricrepicephalus* (fig. 8.19), and *Idahoia* and *Ptychaspis* are also found in the Wichita Mountains of Oklahoma. Articulated Late Cambrian trilobites are found in deepwater facies in the McKay Group of British Columbia. At Late Cambrian sites in Arkansas and in Nevada

Trilobites Everywhere: More Late Cambrian Deposits

8.18. Pygidium of *Crepi-cephalus* sp. from the Upper Cambrian Pilgrim Formation of the Little Belt Mountains, Montana (USNM 127063).

Courtesy of Smithsonian Institution. Scale bar = 1 cm.

fossils of shallow-water trilobite species have been found in deepwater basin deposits; these had arrived in deep water by way of submarine sand-mud flows that cascaded down the slope from the shallow continental shelf. Skeletal elements of trilobites that had already been buried seem to have been carried down to deep water and redeposited. Just as marine faunas of the Cambrian can be pervasive in North America, geologic mechanisms of today, such as submarine slides, seem to have operated to preserve these fossils even 500 million years ago. Some things are everywhere, and some things never change.[51]

Last-Minute Appearance: Squid Ancestors

Near the end of the Cambrian a group appeared that became quite significant in the oceans of later times. Their fossils are rare and they occur at few sites, but their presence is important for recognizing the emergence of modern groups of marine animals. What is this surprising group that appeared quietly in scattered parts of the globe near the end of the Cambrian? The cephalopod molluscs – ancestors of today's squid, octopuses, cuttlefish, and nautiluses. Cephalopods are molluscs with a body that occupies the most recent of sequential linear (but sometimes coiled) chambers in a calcitic shell, which in many species is lost. The body is connected to the original chamber by a filament of tissue called the **siphuncle.** In nautilus species at least, buoyancy may be in part controlled by regulation of fluid and gas levels within the chambers. The mouth is surrounded by external, prehensile (grabbing) tentacles that have small suckers for grip, and there is often a pair of strong can opener–shaped jaws (a beak) and a rasping radula for breaking up and processing food before it is swallowed. Cephalopods have eyes of a type similar to, but separately derived from, those of vertebrates. The muscular funnel can be used to force water out and provide jet-like propulsion of the animal. Among marine invertebrates, the brains of octopus are some of the largest. There are about 900 modern species of cephalopods, and all are mobile marine animals; most squid, cuttlefish, and nautilus are pelagic

8.19. Complete specimen of the fluke-"tailed" trilobite *Tricrepicephalus cedarensis* from the Middle–Upper Cambrian Nolichucky Formation of Cedar Bluff, Alabama (USNM 94955). Scale bar = 1 cm.

Courtesy of Smithsonian Institution.

and live in the water column at various depths, and many octopuses are mobile and live near the bottom. All cephalopods are predatory carnivores and some can inject **neurotoxins** into prey.

Cambrian cephalopods occur in rocks as widely scattered as China, Antarctica, Texas, New York, and Nevada. In general, Cambrian cephalopods are found as shell fossils, usually less than 4 inches (10 cm) long, and shaped like a cornucopia. In eastern New York, the Little Falls Formation has yielded latest Cambrian cephalopods, but only a handful. The specimens consist of slightly curved, conical shells about 3.6-cm (1.4-in.) long, with closely spaced individual chamber walls seen in cross section.

Many more Cambrian cephalopod specimens have been found in China, and the diversity there appears to be at the level of almost 150

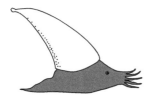

8.20. Reconstruction of the Late Cambrian shelled cephalopod *Plectronoceras*. Soft-part anatomy is conjectural but assumes eyes and some incipient tentacles. Length ~3 cm.

species! The first cephalopods appear in China as a single genus in the Late Cambrian, and within five million years have diversified into nearly 40 genera on at least two equatorial paleocontinents. These Cambrian cephalopods appear to have had partially gas filled chambers in the shells, and may have compensated for the resulting positive buoyancy by adding weight to the shell in the form of closely spaced chamber septa (often about just 1 mm between septa). The Cambrian cephalopods also appear to have occupied a wide range of habitats, indicating significant ecological diversity.[52] Interestingly, cephalopod diversity increases gradually through the Late Cambrian and then appears to drop at the end of the period. It then recovers during the Ordovician.

It is unclear what the cephalopod animals looked like occupying these shells, but they have been reconstructed as slug-like, epibenthic molluscs. Others suggest that they were pelagic and were capable of vertical migration within the water column (buoyancy controlled through gas changes in the shell chambers), although not necessarily jet swimming. I have reconstructed a Cambrian cephalopod (fig. 8.20) based on the shell, and have shown the animal as less like a slug. And to think that from these simple ancestors came creatures as impressive as the 10-meter-long giant squid!

And then there is the monkey wrench of *Nectocaris*. This unusual fossil from the Middle Cambrian Burgess Shale was originally collected by Walcott but not described until 1976, and then as an indeterminate arthropod. It has recently been redescribed based on 90 new specimens, this time as a primitive, unshelled, swimming cephalopod, generally similar to cuttlefish. The reason this is unexpected is that until now we had only seen cephalopods from the later part of the Late Cambrian, and they had calcitic shells. These were also assumed to be bottom-dwelling animals without (or possibly with) tentacles. *Nectocaris*, if it is a cephalopod, would be a much earlier, free-swimming (but near the bottom), unshelled species and one with two tentacles, camera-type eyes, a funnel near the mouth, and long lateral fins. *Nectocaris* looks more like a modern nektonic cephalopod than the cephalopods of the Late Cambrian do, but this is in part due to differences in preservation—we're simply not finding the soft parts of the Late Cambrian shelled species. At least some researchers are not convinced of the cephalopod affinities of *Nectocaris*, and it appears there is no consensus just yet,[53] but if it does prove to be a cephalopod it makes the origins of this group a bit more interesting.

The End of the Beginning: The Cambrian–Ordovician Boundary

It has been a nice ride for 54 million years, but the Cambrian had to end sometime. Not that the Earth or its animals knew this or would have observed anything but another sunrise, but we humans, in our need to understand through classification, had to define a boundary somewhere to mark the change in faunas that we saw in the fossil assemblages. So the end of the Cambrian is officially marked by the first appearance of a single species of conodont vertebrate: *Iapetognathus fluctivagus*. This species's first appearance actually marks the beginning of the Ordovician period, which by implication, of course, ends the Cambrian. There is nothing special about *Iapetognathus fluctivagus* itself in the history of either period. It was chosen as a boundary marker like any other key fossil simply due to widespread geographic occurrence, abundance, short time range, and identifiability. The Ordovician was recognized as being distinct many years ago, but the definition of the lower boundary has to be precise and its marker was defined relatively recently. Someone had to serve as the boundary-marking species and *I. fluctivagus* got the votes. Perhaps it was absent from the deciding committee meeting.

The Ordovician was a time of diversification and expansion of the marine biotas, a time when life cranked up the complexity of its ecosystems more than one notch; in retrospect, it begins the *rest* of the Paleozoic, a time when the marine realm made the Cambrian look, well, primitive in comparison. But none of that would have been obvious to observers of the last years of the Cambrian since there was no indication of what was to come. It is only through hindsight that we can see that not every niche in the benthic marine setting was occupied; the organisms of the Cambrian did the best possible at the time. They did diversify and expand their ecologic ranges during the 54 million years of the Cambrian. And, as we have seen, life was flourishing at the time.

Although there are a few species that become extinct near the end of the Cambrian, there are no wholesale losses, no major wipeout, and most groups continue into the Ordovician unscathed, even if a few individual species turn over. Rather than a mass extinction, the end of the Cambrian marks the end of the prelude. The stage is now set. The Ordovician marks the beginning of the taking off of many groups and the construction of a more diversified and complex marine ecosystem. The appearance of little *I. fluctivagus*, therefore, should not spell a black-shrouded end to the Cambrian with dirges and laments so much as it marks the triumphant graduation of Earth's "baby boom" generation of Cambrian explosion species into the wild, crowded world of intricate ecological interactions of innumerable taxa that the world saw after 488 million years ago. There would not have been a Paleozoic world and there would not be the modern seas without the Cambrian having set the stage for all the metazoan and other groups that developed and diversified during that time. The groups that spent their formative years learning in the school of the Cambrian exploded out of the gates or hung on in an increasingly competitive world after the Cambrian ended. The end of the Cambrian is when these groups really took off and started forming the world we know today.

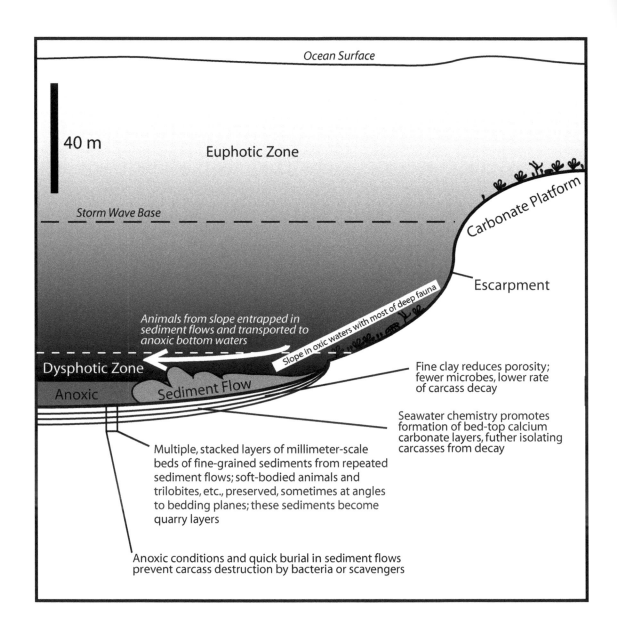

Ocean Surface

40 m

Euphotic Zone

Carbonate Platform

Storm Wave Base

Escarpment

Animals from slope entrapped in sediment flows and transported to anoxic bottom waters

Slope in oxic waters with most of deep fauna

Dysphotic Zone

Fine clay reduces porosity; fewer microbes, lower rate of carcass decay

Anoxic

Sediment Flow

Seawater chemistry promotes formation of bed-top calcium carbonate layers, futher isolating carcasses from decay

Multiple, stacked layers of millimeter-scale beds of fine-grained sediments from repeated sediment flows; soft-bodied animals and trilobites, etc., preserved, sometimes at angles to bedding planes; these sediments become quarry layers

Anoxic conditions and quick burial in sediment flows prevent carcass destruction by bacteria or scavengers

9.1. Generalized model of factors contributing to Burgess Shale–type preservation, based in part on the Wheeler and Burgess Shale formations.

Based on data in Gaines and Droser (2005); Gaines, Briggs, and Zhao (2008); Gaines, Hammerlund, Hou, et al. (2012); Gaines, Droser, Orr, et al. (2012).

Home by the Sea: A Closer Look

THE CAMBRIAN PERIOD MAY BE MANY THINGS, BUT MORE THAN anything it is a goldmine of information regarding how our modern biological world got underway. Times have changed since the days of *Anomalocaris*. The cast of characters in the oceans has changed in the details, and the ecological functioning of the entire marine realm has increased in its complexity, but the roots of the modern biota and the biota's modes of interaction lie in the eruption of life that occurred between the first appearance of a complex trace fossil around 542 million years ago and that of a particular species of a toothy and eel-shaped member of our own phylum about 488 million years ago. This interval, the Cambrian period we know and love, may be considered a 54-million-year key moment in the biological history of Earth. Although the Cambrian accounts for only about 1.2% of Earth history, and although by far most of our planet's history has consisted of the very different world of the Precambrian, this Cambrian 1/100th of our story set the stage for the next couple hundred million years and, importantly for us, the world we know around us today.

So, what can we learn from the fossil record of the Cambrian? Much that is important, it would appear. This fossil record has served as the key data in debates about the nature of evolution itself. But there are details in this record that can range from the curious to the critical, from merely interesting and inspiring to unexpectedly revealing about the world of the time.

Taxonomic Diversity

One of the aspects of this record that we should perhaps tackle first is the revealed patterns in the diversity preserved in some sites and formations of Cambrian age. The typical formation or locality in the Cambrian in North America produces brachiopods, trilobites, and sometimes hyoliths. This is from fossiliferous levels; there are plenty of thick sections of shale or limestone in any given rock unit that you can dig and not find a thing. But within the fossiliferous parts, the above three fossil types are almost guaranteed to be there at some level of abundance. Unfortunately, these of course tell only part of the story. Thanks to the fact that few taxa have the mineralized hard parts that trilobites, brachiopods, and hyoliths have, most of the diversity that lived in these environments, like worms and unmineralized arthropods, rarely fossilizes. In fact, paleontologist Simon Conway Morris once estimated that as much as 86% of the diversity and 98% of the individuals were not preserved at most Cambrian sites.[1] This is a huge hole in our knowledge of the local ecosystems that we look at,

Table 9.1. Cambrian lagerstätten of Laurentia (including true Burgess Shale–type deposits [Hagadorn (2002)], as well as those with rare soft-bodied preservation of taxa shared with the Burgess Shale and those preserving any unmineralized Cambrian taxa).

Formation	Location	Age
Burgess Shale	British Columbia	Middle Cambrian
Wheeler Formation	Utah	Middle Cambrian
Spence Shale	Utah	Middle Cambrian
Pioche Formation	Nevada	Early/Middle Cambrian
Latham Shale	California	Early Cambrian
Kinzers Formation	Pennsylvania	Early Cambrian
Conasauga Formation	Georgia/Tennessee	Middle Cambrian
Poleta Formation	Nevada	Early Cambrian
Buen Formation	North Greenland	Early Cambrian
Parker Slate	Vermont	Early Cambrian
Mount Cap Formation	Arctic Canada	Early Cambrian
Marjum Formation	Utah	Middle Cambrian
Weeks Formation	Utah	Late Cambrian
St. Lawrence Formation	Wisconsin	Late Cambrian

and this is where Burgess Shale–type deposits come to the rescue. These BST deposits, as they are sometimes called, range from the Lower Cambrian into the Ordovician (and by broader definition into the Silurian and Devonian) and preserve entirely soft-bodied organisms (e.g. worms and cnidarians), unmineralized arthropods, and other soft forms, along with the standard trilobites, brachiopods, and the other taxa bearing mineralized shells and exoskeletons. Unlike other sites, the soft-body parts of trilobites (limbs and antennae) and other unmineralized taxa are sometimes preserved at BST sites. The Burgess Shale itself is the prototypical deposit, of course, and its preserved diversity is truly impressive, as we have seen (and will see again). There are other Cambrian deposits in China and other countries that rival the Burgess Shale in diversity and preservation quality, but most BST deposits are not quite as productive. In fact, many of the now-known BST deposits, at least in North America, were until a couple decades ago known mainly as trilobite formations. And although there are plenty of formations that lack soft-bodied taxa entirely still, a surprising number of the traditionally productive trilobite formations of the region have been found to contain at least a few soft-bodied taxa (table 9.1).

Many Burgess Shale–type deposits are in shale from relatively deep water and are dominated by non-trilobite arthropods, with many types of sponges, a few lobopods, and an abundance of priapulid worms; plenty of other taxa may occur at these sites as well. Strictly speaking, true BST deposits share a confirmed and generally similar mode of preservation and similar taxa. Some other sites may preserve a few taxa in common with the Burgess Shale, but the mode of preservation is not necessarily the same as at major BST sites; still other sites preserve other soft-bodied taxa not occurring at these main sites. (All three types of localities are included in table 9.1 and might more generally be considered lagerstätten, regardless of preservation mode.) Lagerstätten (including BST deposits) are also known to occur in Cambrian rocks around the world, as widely distributed as Spain, England, Russia, China, Sardinia, Australia, Sweden, and

Poland; in fact, they are found on every continent except South America, Antarctica, and Africa – so far.[2]

Leaving the Burgess Shale and other sites with extensive soft-bodied records aside for the moment and concentrating on sites dominated by mineralized taxa, let us take a look at the diversity preserved at so-called average Cambrian localities. We are not excluding soft-bodied taxa from our analysis of these sites, we are simply removing, for now, the incomparable in the Burgess Shale fauna so that we are not comparing the proverbial apples and oranges. Citrus only, for this one. We will take a look at five localities in the late Dyeran of the Lower Cambrian and five localities in the Delamaran of the Middle Cambrian, and from each stratigraphic level the sites range from near shore on the craton to well offshore near the shelf edge. Our Lower Cambrian sites, arranged from nearest to shore to most distal, will be the Bright Angel Formation at Frenchman Mountain, the Latham Shale in the Marble Mountains, the Pioche Formation at Ruin Wash, the Carrara Formation at Emigrant Pass, and the Emigrant Formation at Split Mountain. The Middle Cambrian sites, similarly arranged, will be the Bright Angel Formation in the Grand Canyon, the Cadiz Formation in the Marble Mountains, the Chisholm Formation at the Half Moon Mine, the Spence Shale at Spence Gulch, and the Rachel Limestone in the Groom Range.

Our mode of comparison can be the total species diversity represented by the samples from these sites. This is a very pure measure, but unfortunately it is subject to high variability due to sample size; the more specimens that have been collected from a quarry, the more likely it will have many species – similarly, even a very truly diverse quarry can only have a few known species when only a few specimens have yet been collected. There is also the issue that counts of straight species diversities assume that all species are equally abundant. This is, of course, not the case, and in biological studies a fauna in which one species dominates in abundance is considered less diverse than a fauna in which species are all equally abundant, *even* when both faunas contain the same number of species.

So, we need some metric by which to compare faunas based on different sample sizes and at the same time account for different relative abundances of taxa regardless of total species diversity. Several of these exist, but the one we will use is the Shannon index, based on the work of mathematician Claude Shannon. This index will give us a single number by which to compare the different samples without having to correct for different sample sizes, different total species diversities, and different relative abundances in our heads.[3] Instead, relative abundances are represented by proportions of species within the sample. The Shannon index is expressed as

$$H = -\sum p_i \log(p_i)$$

H is the symbol for the Shannon index, and p_i is simply the relative abundance proportion of each species in the fauna. The proportion of each

Site	Cambrian Age	Shelf Setting	Formation	Number of Species	Shannon Index	Effective Richness
Frenchman Mtn.	Early	C	Bright Angel	4	0.7556	2.13
Marble Mtns.	Early	DC	Latham Shale	7	1.6755	5.34
Ruin Wash	Early	IMS	Pioche	11	1.5009	4.49
Emigrant Pass	Early	IMS	Carrara	5	1.1035	3.02
Split Mountain	Early	OS	Emigrant	2	0.1761	1.19
Grand Canyon	Middle	C	Bright Angel	6	0.8300	2.29
Marble Mtns.	Middle	DC	Cadiz	10	1.9165	6.80
Half Moon Mine	Middle	IMS	Chisholm	10	1.9750	7.21
Spence Gulch	Middle	IMS	Spence Shale	11	1.6824	5.38
Groom Range	Middle	OS	Rachel Limestone	5	1.1161	3.05

Table 9.2. Species diversity, Shannon index, and effective richness of five Early Cambrian and five Middle Cambrian sites distributed on the coast of western Laurentia, arranged from most proximal to most distal sites within each age. Shelf setting: C = proximal craton; DC = distal craton; IMS = inner and middle shelf; OS = outer shelf. Note that effective richness (and other diversity measures) is highest on the distal craton and inner- and middle- shelf settings and lower in proximal cratonic and outer-shelf settings.

species in the sample is multiplied by the natural log of that proportion. We then add all these individual species products (represented by the Σ in the equation) and reverse the sign to get the Shannon index.[4] However, the index value is just a number to compare; it has no units or specific meaning. So we are going to make one adjustment to our numbers for each fauna and convert them to **effective richness**, the number of species that the sample would produce if all of their relative abundances were even. This is a way of going back to comparing straight species diversity (theoretically at least). How many species would a particular quarry produce if there were only one specimen of each species rather than multiple specimens of abundant taxa and one of rare taxa? Effective richness (S_{eff}) is expressed as

$$S_{eff} = e^H$$

where e is the natural log base and H is the Shannon index. This is a simple calculator move, taking the constant e and raising it to the power of your calculated Shannon index. The result is a number ready to compare with other samples, but unlike the Shannon index, the effective richness can be thought of in terms of an actual number of species (even though it is rarely a whole number). This is how we can compare wildly different sample sizes and relative abundances.

Now that we have done this for all our sites, what do we find? Well, there are hints that during the Cambrian the moderately shallow depths of the open continental shelves were areas with the greatest diversity (as represented by effective richness; table 9.2). The very near-shore deposits of the proximal craton appear to have been somewhat depauperate in terms of species, and the most distal sites may be even lower in diversity. This is a pattern that has been noted, if not specifically quantified, in the past. But what does it mean? Perhaps the fluctuations of the nearly brackish environments close to the expansive epicratonic estuaries near shore were too hostile for a number of species, as were the depths of the distal, outer shelf (and its possible anoxia) for others. Maybe the Cambrian "Goldilocks Zone" of the open shelf was more to the liking of the majority of critters living in the shallow seas around Laurentia.[5]

It has also been suggested that ecological innovations and perhaps some evolutionary developments occur first in near-shore environments and expand later to the offshore settings. Patterns of species turnover also appear to indicate new appearances near shore and subsequent expansion offshore with older, more primitive faunas also pushed offshore.[6] Such patterns might suggest a reason that relatively shallow open-shelf settings might be more diverse relative to deeper-water environments, but the lower diversity in nearest-shore craton settings is a more surprising finding. Earlier studies may not have included sites as close to shore. Future studies of the potentially more hostile close-shore and estuarine settings may fill in the pattern here.

Taphonomy is the study of fossil preservation, of the remains of animals from the time of death to fossilization, and involves disarticulation of the carcass and skeletal elements, transportation (by currents etc.), burial, and diagenetic and other effects of fossilization. More than just the fossils are important to a taphonomist. The *context* in which they are found is critical. Not surprisingly, there is a myriad of ways in which the fossil record can deceive us, and taphonomy helps us to figure out what effects of the paleoenvironments and paleobiology may have skewed the sample that is ultimately preserved. The fossil record can tell us not just about the animals but also about the environments in which they lived and were ultimately entombed.[7] And all of this puts us closer to a more realistic understanding of the world of the past.

The Grand Illusion: Decoding the Record through Taphonomy

Preservation

That trilobites, hyoliths, and brachiopods dominate many Cambrian fossil samples is understandable given their mineralized hard parts. Some estimates are that up to 90% of the original, full complement of mineralized taxa that existed in local environments are represented in the fossil record on small geographic scales, thanks to their hard parts. Interestingly, however, it appears (counterintuitively) that within a group of mineralized species, taxa with small, thin or unreinforced shells were as commonly preserved as those that had larger, thick shells, sometimes with ribs, folds, or spines reinforcing them.[8] This suggests, for example, that the relative abundances of brachiopods we see preserved at many sites may be a reasonable approximation of the ratios of species that lived in each area. Thus, it would appear that in many Cambrian settings, articulate brachiopods were less abundant than their inarticulate cousins. Would the same taphonomic finding suggest that hyoliths were as rare, and trilobites as dominant, compared to brachiopods as the raw fossil record indicates? It is difficult to say, but that is possible. There is plenty of morphological diversity in trilobite exoskeletons; the size and thickness of the individual elements, and the number and shape of spines, for example, can vary greatly between species. Brachiopods seem to be a little less morphologically diverse in the Cambrian; most are oval in shape, and their size range does not appear to be as wide as with the trilobites. Molluscs, although mineralized, are generally tiny and rare. But does this mean that what we see is representative of the original populations? Possibly not, but our view of the Cambrian of North America would consist mostly of these mineralized taxa without the BST deposits now known from so many other areas.

The soft-bodied preservation at most Burgess Shale–type deposits consists of carbonaceous films representing the soft flesh and unmineralized exoskeletal sclerites. In some cases preservation can vary from one stratigraphic level to the next, in a matter of centimeters or meters vertically. Within the Pioche Formation in Nevada soft taxa are often preserved with a coating of red hematite in Lower Cambrian rocks, whereas just above in the Middle Cambrian the taxa occur as gray or black carbon films.[9] By whatever path, the preservation of soft-bodied taxa is rare, but we are learning more about the processes every year.

Several factors are believed to have contributed to this soft-part preservation. First, a carcass would need to be buried relatively quickly. In some cases, this may have been almost instantaneous, by means of entrapment in an ocean-bottom sediment flow. Quick burial need not involve high sedimentation rates for the whole deposit. In some cases, one can accumulate just 10 cm (4 in.) of mud every 1000 years but still have material buried very quickly in thin, fine-grained sediment flows.[10] In many Cambrian deposits, including the Burgess Shale, the sediment flows could be started by simple slumps in mud on low-angle slopes, or they could be flows fanning out from narrow canyon notches in the upslope carbonate ramps.[11] Second, low or nonexistent oxygen levels in the water in a deep-bottom setting could have slowed or prevented microbial decay of the carcass and prevented animal scavengers from being able to access either the bottom waters or the sediments just below, thus protecting the body from scavenger damage. These two factors result in physical and chemical shielding of a carcass from the action of enzymes that decompose the bodies. In general, carcasses must be protected from surface scavengers, infaunal burrowers, and decomposing bacteria and other microbes; anoxia should protect from all of these, except some anaerobic microbes. Field experiments with modern shrimp carcasses show that scavengers were the most destructive of the above three factors, even for buried specimens. In the lab, interestingly, oxygen levels caused only minor differences in decomposition rates, and in general all soft tissue was gone within about two weeks. That, again, was with shrimp. In other experiments, decay of a modern polychaete worm, however, was significantly slower under anoxic conditions. So decomposition rates—and thus likelihood of soft-tissue preservation in the fossil record—probably vary by taxonomic group. Whether annelid or shrimp, another recent lab study found that soft-tissue decay was impeded by low-sulfate (SO_4) conditions in the surrounding water, and this is a seawater condition that appears to have been present in Cambrian oceans.[12]

Other key factors in preserving soft-bodied fossils include fine grain size, which makes for sediments with low permeability that can prevent oxygen from working in to an entombed carcass; lack of burrowing of the sediment (often thanks to anoxia), which prevents both scavenging of the carcass and aeration of the sediments that would let in microbes or later scavengers; and proximity to well-oxygenated water, which provides a source biota for preservation (fig. 9.1).[13] Burgess Shale–type deposits also tend to be tropical to subtropical in paleolatitude, to have apparently variable oxygen levels and salinity, and, as we saw earlier, some may be associated with brine seeps. Although BST deposits are known from the Ordovician,[14] most are Cambrian in age, and their restriction to the early Paleozoic may be a result of an increase in sediment burrowing depth or other factors, as we will see below. By the middle of the Paleozoic, so many taxa may have developed an ability to burrow so deep that it was simply difficult for carcasses to last intact long enough to be preserved. BST preservation seems to be characteristic of Lower and Middle Cambrian deposits but is not restricted to these epochs.

More so than restriction of burrowing, what may have set Cambrian BST deposits apart was the unique seawater chemistry of the time. As mentioned above, low sulfate concentrations in Cambrian seawater may also have helped decrease the rate of carcass decay, thus increasing chances for soft-part preservation. Weathering of continental basement rocks during the Cambrian may have increased the alkalinity of the ocean, which may in turn have facilitated the formation of impermeable calcium carbonate layers near the sediment-water interface at the time, further sequestering carcasses from scavengers and decay and enhancing preservation of soft parts. The change in seawater chemistry would then have shut this mechanism down and helped explain the rarity of BST preservation after the Middle Cambrian.[15]

Chemically, there are probably several taphonomic pathways to BST preservation, some possibly involving iron- and sulfur-reducing bacteria and iron oxide replacement later in the process. The Burgess Shale itself preserves soft body parts mostly as aluminosilicate films, probably formed by replacement of organic material during post-depositional metamorphism of the shale. Aluminosilicates have even replaced the shells of the trilobites in the Burgess Shale. Some deposits preserve soft-bodied taxa as slightly chemically altered organic remains.[16]

The Burgess Shale Formation, particularly the Walcott Quarry, has some individual layers dominated by particular taxa. Although overall priapulid worms are particularly abundant, some layers are dominated by crustaceans or by the arthropod *Marrella*. In the Wheeler Formation in Utah, beds with abundant specimens of the trilobite *Elrathia* tend not to contain many soft-bodied specimens, although rare specimens do occur; the more common soft-bodied occurrences in the Wheeler are in layers in which *Elrathia* is rare. The biofacies in which *Elrathia* is common probably was oxygenated, although at relatively low levels, and burrowing was more abundant, but the biofacies containing most of the Wheeler's soft-bodied specimens was probably anoxic, unburrowed, and the preserved material was washed into the area. In the Wheeler, one mechanism that helped lock soft-body specimens away from bacterial activity after burial was the reduction of host-sediment porosity. Sealed in fine-grained sediments in anoxic conditions, carcasses may have been further protected from decay by the chemical precipitation of carbonate in what tiny pore spaces there were in the bottom sediments; porosity and oxygen thus reduced, bacterial decay would have slowed greatly, allowing preservation of the soft-bodied taxa that we see in the Wheeler such as worms and *Dicranocaris*. The carbonate material for this process may have been derived from seawater and, like the Burgess Shale, been influenced by the alkaline seawater chemistry of the time.[17] Soft-bodied taxa seem to have preserved more often when bottom oxygen levels fluctuated within the paleoenvironment; fossil samples in these conditions also tend to be of higher diversity and to contain more complete specimens than those from settings in which the oxygen levels were consistently low or anoxic. Samples from paleoenvironments with variations in oxygen levels also tend to be less transported.[18] Overall, then, preservation is favored under anoxic conditions that are not hospitable to animals, so the most diverse fossil assemblages are those that are transported into an anoxic environment but only a short distance. These assemblages were likely the biotas that had been living close to the oxic-anoxic boundary and were at some point swept into the anoxic zone or had it encroach upon their habitat. Burgess Shale–type deposits tend to be laminated, unburrowed fine-grained sediments interbedded with burrowed (oxic) layers devoid of soft-body preservation; the interbedding was caused as the oxic-anoxic interface shifted back and forth on the seafloor.

Biostratinomy

Another aspect of taphonomy is **biostratinomy,** the study of the physical orientation of the remains of animals in a fossil deposit, which can reveal aspects about not only the paleobiology of the organisms but also about abiotic conditions of the paleoenvironment. Each can influence the other. In particular, as it was once stated by Carlton Brett and Gordon Baird, the differential effects of paleoenvironmental processes "on particular types of skeletal remains in different facies are related to variations in rates and modes of burial, and the nature and intensity of environmental energy."[19] And different types of animal remains may be influenced differently by the same environmental conditions, due to potential differences in the hydrodynamics or the density of the elements, for example.

Characteristics of trilobite deposits can be particularly revealing, and the abundance of trilobites in Cambrian deposits makes them particularly well suited to biostratinomic studies. Among the characteristics of these deposits that are important in these studies are the species preserved and their relative abundances; the size of elements; up–down orientation of molts, carcasses, and isolated elements; compass orientation of elements; articulation and completeness of elements; fragmentation and breakage; and **corrasion,** a combined assessment of the degree of chemical corrosion and physical abrasion of a skeletal hard part.[20]

Let us now take a look at the characteristics of the trilobite samples of a few of the sites we have visited in this book. We started off in the Marble Mountains, so the rich sample from the Early Cambrian Latham Shale is as good a place to begin as any. Overall, the Latham has more than 10 species of trilobites from all levels, although it is only about 20 m (65 ft.) thick. Some species only range through the lower part, and some only the upper; others are restricted to the middle, and still others range through the entire formation. We can only base our conclusions here on a particular sample, and in this case it will be one collected recently from several pits ranging from low in the formation to the upper 50 cm (20 in.), a total sample of more than 800 specimens that turned up nine species, including *Mesonacis fremonti, Bristolia bristolensis, B. insolens,* and *Olenellus nevadaensis.* Individual sample pits seemed to each have a different dominant species, and what species this was did not seem to be tied to any noticeable difference in paleoenvironment, nor did it obviously relate to particular biostratigraphic patterns. Perhaps it was simply a sampling effect; we don't know. The maximum diversity preserved in any one sample pit was six species. It is clear from the sample of molted cephala that their convex-up or -down orientation and compass alignment on the bedding planes is entirely random. Articulated exoskeletons are also extremely rare (1.5%), and 98.5% of the sample consists of isolated cephala or thoracic segments.[21] Most thoracic segments are broken, although genal spines on the cephala are mostly intact. There are a lot of ways in which a trilobite sample like this can become so disarticulated, randomly oriented, and broken, but one possibility is molts or carcasses of populations of trilobites of several species, accumulated over time, sitting exposed on the bottom of the ocean for a relatively long period of time. If the depth of the ocean bottom is about 50 m (164 ft.) or more, the skeletal elements will be at a depth unaffected by most waves, and thus will rest on the bottom in unmoving water, a water column that will move only a little at depth (or not at all) when strong storms rock the surface waters above. These quiet waters and long exposure times lead to complete disarticulation of most molts and carcasses (fig. 9.2), their random orientation (figs. 9.3 and 9.4), and the eventual breakage of many elements at the jaws of scavengers that were perhaps after the cuticular lining of the sclerites. (It is unlikely that currents mechanically broke the skeletal elements, as water movement at this depth even during storms was probably pretty light, and experiments with modern eggshells as trilobite sclerite proxies suggest that even tumbling in coarse sediment does not break down eggshell very well, as we will see later in this chapter.) This may be how the Latham Shale trilobite sample came to be preserved as it is; mixed in with the trilobites,

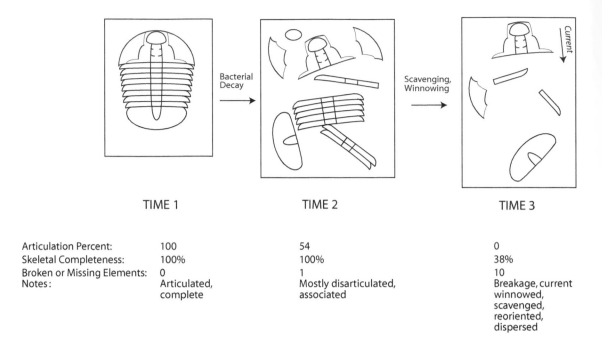

	TIME 1	TIME 2	TIME 3
Articulation Percent:	100	54	0
Skeletal Completeness:	100%	100%	38%
Broken or Missing Elements:	0	1	10
Notes:	Articulated, complete	Mostly disarticulated, associated	Breakage, current winnowed, scavenged, reoriented, dispersed

9.2. The disarticulation and winnowing of a trilobite exoskeleton, showing characteristics of deposits at each stage of process.

along with the usual brachiopods and hyoliths, are those rare soft-bodied specimens like the alga *Margaretia* and the arthropod *Anomalocaris*. The Latham appears to have been in shallower water than the Burgess Shale, and it was inboard of—not at depth below—a carbonate platform, and this may in part account for the rarer preservation of soft-bodied taxa. But the Latham is one of several formations in the region with abundant trilobites from the later Early Cambrian.

In contrast to the Latham, the slightly younger upper Combined Metals Member of the Pioche Formation demonstrates a higher degree of articulation and orientation of the cephala. One bed at Ruin Wash (see chapter 5) also has randomly azimuth-oriented specimens and contains eight species of trilobites, not far from the six known from one bed at several sites in the Latham Shale. But 9% of the specimens from this same Ruin Wash layer are articulated, versus 1.5% in the Latham, and a rather emphatic 93% of isolated cephala on this bed from Ruin Wash are oriented convex up, as opposed to the statistically 50:50 split in the Latham. Convex-up cephala dominate the samples from six other surfaces at Ruin Wash, from between 77% and 91%. These numbers suggest a strong preference for orientation of trilobite cephalon pieces in convex-up position,[22] a preference simply not present in the Latham Shale. What was causing these orientation and articulation differences? Was the setting of Ruin Wash during the late Early Cambrian higher energy, where trilobite pieces were flipped until hydrodynamically stable and whole carcasses were more quickly buried? There is little else in the sedimentology of the two formations to suggest a drastic difference in energy settings, but the above scenario is possible because a cup-shaped piece of exoskeleton is much harder to move when its convex side is up than when its concave side is up. Although the reasons for these differences between the Latham

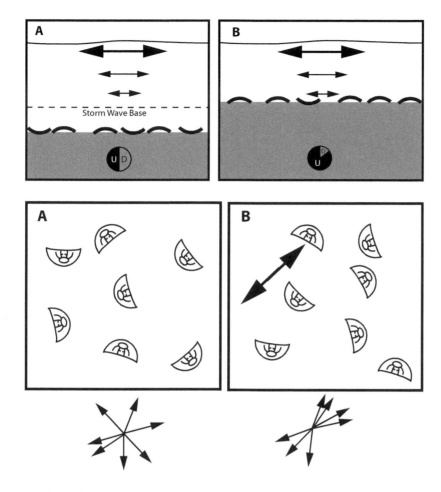

9.3. *Top,* preferential orientation of convex trilobite sclerites on ocean bottom. (A) convex trilobite sclerites (thick curved lines) rest on the bottom randomly oriented half up, half down in still water below storm wave base. (B) in shallower water above storm wave base, bottom currents cause convex sclerites to preferentially orient to a hydrodynamically stable convex-up position; note hypothetical 5:1 ratio of convex-up to convex-down sclerites in this setting. Double arrows indicate relative strength of wave movement. (Wave movement of individual water molecules would, in reality, be circular but still decreasing in strength with depth; only the lateral movement is shown in this figure for simplicity.)

9.4. *Bottom,* current orientation of trilobite sclerites on the ocean bottom. (A) randomly oriented olenellid trilobite cephala resulting from no current activity; azimuth orientations of anterior axes (arrows below) show random orientation. (B) oscillation current (large double arrow) reorients cephala to hydrodynamically stable alignments; azimuth orientations of anterior axes (arrows below) show bimodal, nonrandom distribution.

Shale in the Marble Mountains and the Pioche Formation at Ruin Wash are not immediately clear, the fact that such differences are apparent is tantalizing and leads us to want to understand more about these deposits.

Like the Latham Shale, the Bright Angel Formation at Frenchman Mountain, Nevada, consists of mostly disarticulated cephala and thoracic segments (combined, 99.3%). Articulated exoskeletons are even less abundant here than in the Latham and far less common than in the Pioche (~0.6%), and the broken thoracic segments tend to be even smaller. More revealing, 8.9% of the sample consists of cephalon fragments, a number of which are broken, possibly by predators. There also is a trend at this site for more cephala to be preserved convex down (56.7%), as opposed to the 50:50 split in the Latham sample; why more cephala are preserved in a hydrodynamically unstable position is open to speculation, but at some younger sites in the Paleozoic this is sometimes attributed to bioturbation. Studies at Frenchman Mountain are still in preliminary stages,[23] so it is difficult to say what may account for the similarities and differences between the Bright Angel and Latham at these two sites.

The breakage of thoracic segments and cephala in the Bright Angel Formation at Frenchman Mountain is interesting in that it suggests that scavengers may have been in the area feeding on the remains (carcasses and molts) of trilobites and that remains were experiencing long bottom-exposure times. A study by Brian Platt on the Upper Cambrian trilobites of the Rabbitkettle Formation in the Northwest Territories of Canada found that a significant percentage of the cranidia and pygidia and thoracic segments in the sample were broken and that much of the damage might be attributable to predation and scavenging by non-trilobite arthropods. As part of the study, Pratt also experimented with eggshell pieces as a proxy for trilobite elements and subjected them to hours of violent agitation in sand and flowing water and found that none of the eggshell pieces broke. Despite acknowledged structural differences between eggshell and trilobite scleries, the toughness of thin plates of biotic construction was demonstrated. This may indicate that scavenger damage may be a more likely cause of broken trilobite elements in most Cambrian formations than physical tumbling, especially in fine-grained sediments that were likely deposited in low-energy settings.[24] And scavenger damage may be indicated by the breakage evident not only in the Rabbitkettle and Bright Angel formations, but also at other sites with breakage to a significant proportion of the elements, such as the Latham and Pioche Formations, for example.

Paleoecology of the Burgess Shale

Paleoecology involves the reconstruction of the biology of fossil organisms and their ecosystems in order to understand how the biosphere has functioned in the past. At times, the way our planet has worked has been rather different than it is now, but for much of it the rules have been the same and just the cast has differed. But understanding these similarities and differences and the interaction of the planet and its life forms, how each influences the other and how life responds to changes, is ultimately the fundamental contribution of paleontology to modern science. No view of modern ecology, biodiversity, functional morphology, extinction, or other aspect of biology is complete without the perspective of what those life forms (or, more precisely, their ancestors) have accomplished in the past. No modern ecosystem or its responses to change can perhaps be understood as well if not viewed through the lens of comparison with one or more of the precedent paleoecosystems that, throughout Earth history, have dealt with all manner of upheavals. Paleoecology aims to take into account all aspects of paleontology to reconstruct the world as it was and its ways of functioning, its responses to change through time, and the interactions of the organisms involved. Paleoecology considers the evolution of individual clades as part of the evolution of entire ecosystems as the biotic characters (species and groups) and abiotic settings come, go, and fluctuate. On a personal level, paleoecology is rewarding in that it is as close as we get to a time machine to transport us back and see times past in all their glory. But more significantly paleoecological studies strive to bring it all together for each time and location studied, whatever the scale, in order to add to our understanding of the functioning of our planet.

(When I say that paleoecology reconstructs the workings of past eco-systems, I should clarify that we as paleontologists are very aware of the fact that we are *striving* to reconstruct the workings of past ecosystems, as best we can, millions of years later. There are many things that we do not know, and that we cannot know, but the way to deal with that is to be rather explicit about what we don't know, what we are assuming, and to quantify such things as much as possible in our work. This is yet another example of how science works—the more open we are about what we don't know, the more likely we are to ask the right questions that eventually lead us closer to an answer.)

Paleoecology takes all the aspects of the paleontology of a fossil de-posit, always with a defined scale, such as the species known, total diver-sity, relative abundance, taphonomic characteristics, morphology of the organisms, and so on, and wraps it all into the story of how the ecosystem functioned. It is a reconstruction of the dynamics of an ancient ecosystem based on all the best evidence we have, with the recognition that parts of it (sometimes much of it) may be missing. But that in some ways is the fun of the detective work. As vertebrate paleontologist Michael Brett-Surman once said about working in the Mesozoic, "Being a paleontologist is like being a coroner except all the witnesses are dead and all the evidence has been left out in the rain for 65 million years."[25] Add about 450 mil-lion years to that time interval for the evidence to get buried, uplifted, faulted, metamorphosed, rained on, and otherwise pulverized and you have adapted the analogy for working in the Cambrian.

Modes of Life

Some of the aspects of fossil organisms that paleoecologists look at and try to define for each species are their preferred habitat, their mode of locomotion (if they moved), their type of reproductive cycle, and their feeding mode, as a few examples. In the Cambrian, when just about all animals were marine,[26] preferred habitats may have been near the surface of the ocean, near the bottom, on the bottom, or in the bottom sediments. Reproductive modes may vary; as we have seen, some species are direct developing, involving growth of an adult-form juvenile, whereas others that have indirect development have a larval stage during which the ecol-ogy of individuals of the species may be rather different from that of the adults. One example of the latter is the free-swimming larva of tunicates that eventually attaches to the substrate and becomes a sessile sea squirt adult. Locomotion in studied species may be by walking or swimming, for example, and feeding modes include carnivorous predators, grazers of algae, and filter feeders of microscopic organics, to name a few. In many cases, the analysis of the paleoecosystem dynamics is based not as much on specific species but on the abundances of the different feeding and habitat modes represented.

One aspect of the fossil record that can hinder paleoecological stud-ies is patchy or incomplete representation of species in an ecosystem. This is especially problematic in the Cambrian, where so many of the sites are so dominated by trilobites, brachiopods, and other species with mineralized hard parts. As we have seen, up to 86% of species may be missing from many Cambrian deposits, so there are only a few localities of this age that would really work for full paleoecological analyses. The Burgess Shale is one of these; the Chengjiang deposits in China and perhaps a couple others are about the only other ones at this point.

The Burgess Shale has been the subject of paleoecological analyses, primarily in separate studies by Simon Conway Morris in 1986 and by Jean-Bernard Caron most recently in 2009. These were based on samples of many tens of thousands of specimens from the Burgess Shale collected over the years by C. D. Walcott and by the Royal Ontario Museum. A few conclusions were drawn in these studies. First, unlike at most other Cambrian sites, trilobites are inconspicuous, accounting for less than 1% up to at most less than 5% of the arthropods in the population. (Now *that's* something to keep in mind when collecting trilobites anywhere!) And the diversity and relative abundance of the Burgess Shale sample seem to be dominated by arthropods. This relative paucity of trilobite individuals is a rather different picture, and probably ecologically a more realistic one, than that painted by the vast majority of Cambrian sites at which one may dig for years and find nothing but trilobites and their hard-shelled associates before eventually finding a single example of soft-bodied preservation. Among the BST deposits that show this latter pattern and are dominated by trilobites despite well-known soft-body preservation are other significant lagerstätten such as the Wheeler Formation, which also preserves up to 75% of the species known from the Burgess Shale, including soft-bodied forms, but which on any given outcrop day is dominated by trilobites.

Analysis of different layers in the Walcott Quarry, and in the overlying Raymond Quarry of the Burgess Shale 20 m (66 ft.) higher, indicates that relative abundances of different taxa varied over the scale of feet (tens of centimeters) and that the paleocommunities represented by the two quarries, separated by a more significant amount of time, were somewhat different. One aspect of this can been seen in the relative abundances of the anomalocaridids from these sites (fig. 9.5). Whereas the Walcott Quarry is dominated by specimens of two types of the anomalocaridid *Hurdia* (fig. 9.6), the Raymond Quarry (and to an even greater degree the Mount Stephen Trilobite Beds) is dominated by *Anomalocaris*. Each of the three sites above also has between four and six species of anomalocaridids preserved in it, not just one. As we saw in chapter 6, the feeding appendages of *Peytoia* (*Laggania*), *Hurdia*, *Anomalocaris*, *Caryosyntrips*, and *Amplectobelua* are all morphologically different from each other and likely were adapted for feeding on different types of prey. This is strong evidence for **niche partitioning** among these Cambrian predators, some of the oldest such evidence in the fossil record. Niche partitioning, which we can see today in many ecosystems, is nature's way of splitting up the workload and avoiding direct competition between closely related or

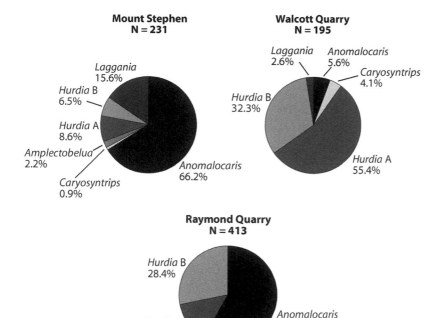

9.5. Compositions of the anomalocaridid faunas of the Walcott and Raymond quarries on Fossil Ridge and of the Mount Stephen Trilobite Beds, by numbers of specimens. Note total diversities of each site and difference in relative abundances of taxa at different sites. Based on data in Daley and Budd (2010).

morphologically similar species. It may have gone something like this. The role of large predator in the Cambrian may have initially been filled by a generalized anomalocaridid, but as a wide array of potential prey species appeared, this one anomalocaridid could not be an efficient predator of all of them. It would have been better catching some (likely groups it co-evolved with and by chance perhaps a few others) than it was at catching others (perhaps newer groups), but it still was the primary predator on all of them. With time, some sub-populations diverged from this initial anomalocaridid stock, with feeding appendages or other morphological adaptations that helped them feed more efficiently on the newer groups that the first anomalocaridids were less effective at catching; perhaps other, immigrant anomalocaridids appeared in the area that happened to be better at catching other prey. The point is that soon the ecosystem was relatively crowded with anomalocaridids and no one could afford to be a generalist and compete with other anomalocaridid species for the same food. Soon all had developed feeding appendages most efficient at catching *their* type of prey and had settled into niches that avoided competition with their cousins. This is evolution at work—finding a way to accommodate a crowd but still managing to keep everyone from stepping on each other. If selection and modification with descent were less effective there would be either fewer species and less efficient utilization of the resources of the ecosystem or there would be a lot of species wasting a lot of effort trying to feed on exactly the same things. As it is, the reality shows us what the most effective balance probably is, and was in the Cambrian (or close to it), given the peculiarities of each individual

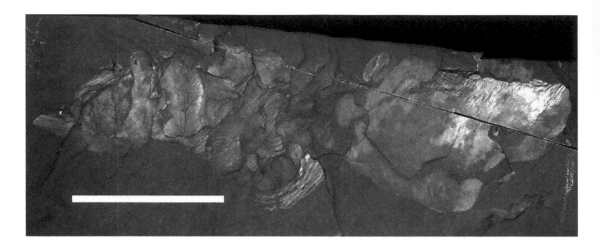

9.6. The Burgess Shale anomalocaridid *Hurdia* showing segmentation, mouth, feeding appendages, eyes, and anterior hood (USNM 274159). Scale bar = 5 cm.

Courtesy of Smithsonian Institution.

ecosystem. To be able to see such specialization, and probably niche partitioning, in animals from 505 million years ago shows how quickly after the Cambrian radiation such interactions began affecting paleofaunas in modern ways.

In the Walcott Quarry, the diversity of the biota was dominated by arthropods and sponges. Arthropods also dominated the relative abundance of individuals in the Burgess Shale, with arthropods, sponges, echinoderms, and priapulid worms making up the bulk of the **biovolume** in the populations (fig. 9.7a). Biovolume is another way of estimating the relative importance of species in an ecosystem. Due to the individual size differences between species, counts of individuals can be deceiving in terms of how much effect a particular species has on those other species it interacts with, in terms of ecological processes like energy flow within the ecosystem. Biovolume is simply the relative abundance of each species multiplied by its volume size (and often expressed as a percentage of the total calculated for the biota). One large member of one species can easily counterbalance many individuals of a smaller species, so biovolume gives us a more realistic picture of how, in energetic terms, the relative abundances of species are distributed in an ecosystem. Energy is important because more of it is needed by large organisms to grow to their large size, compared to smaller species, and because large individuals can serve as sources of more food energy to predators and scavengers. However, many, many individuals of a small species can dominate a few individuals of larger species, if there are *enough* to dominate the biovolume, and biomass (if we factor density into the calculation once we have the volume).[27] More biovolume (and biomass) means more total energy used and made available to others[28]—this makes the dominant species important, not sheer numbers or size. This is why counting individuals is not always entirely satisfactory. Factoring in the volumes allows us to account for both numerical abundance and size differences in one metric of relative abundance.

The Burgess Shale biovolume sample (fig. 9.7a) shows the abundance of arthropods, sponges, echinoderms, and priapulids, probably as a result

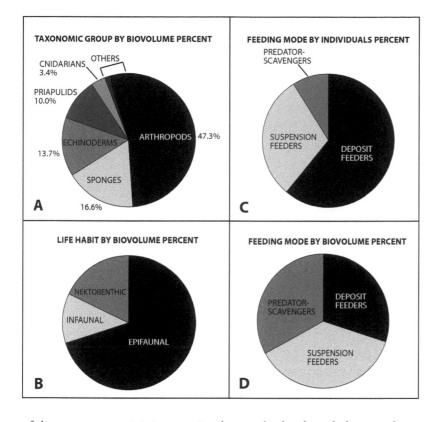

9.7. Composition of the Burgess Shale fauna. (A) Taxonomic group abundance as measured by biovolume percent. (B) Life-habit categories as measured by biovolume percent. (C) Feeding-mode categories as measured by percent of individuals. (D) Feeding-mode categories as measured by biovolume percent.

All based on data in Conway Morris (1986).

of these groups containing species that are both relatively large and numerically abundant in the fossil sample. If they are big *and* common, they will form a large percentage of the biovolume. Also recall that arthropods (even non-trilobite arthropods) and sponges dominate the diversity of the Burgess Shale. These two factors, biovolume dominance and high diversity among arthropods and sponges, suggest that these groups, along with echinoderms and priapulids, were ecologically important elements of the Burgess Shale biota. A significant amount of the ecosystem's energy probably flowed to and from individuals of member species of these groups. It is important to point out, however, that we cannot know this for certain because we only know the percent biovolumes of the groups, and energy flow is greatly affected by relative metabolic rates of the species in each group. For the same body mass, which species use more energy and which less? We don't know this, and so our biovolume relative abundance is only a very rough proxy for relative energetic importance. We need to be careful not to read too much into this, but the biovolume numbers at least give us a hint at what groups were playing key roles in the energy flow dynamics of the ecosystem.

It is also important that we not forget the source of energy flow within the ecosystem. We are so used to studying animals, many of us, that we have to remind ourselves that the energy used by all those so-called dominant groups of animals did not appear in the cycle by pure chance or magic. All the energy available to the arthropods, sponges, echinoderms, and priapulids that dominated our Burgess Shale system was only available thanks to the work of millions of photosynthesizing microorganisms that converted sunlight to energy, and it was that energy (part of it at least) that was ingested by herbivorous species for their growth and reproduction. When carnivores or scavengers ate one of the herbivores, it was a small percentage of *that* energy (and thus a tiny fraction of the original energy fixed by the photosynthesizers) that was utilized for that individual's growth and reproduction. This goes back to basic biology, but we heterotrophs—consumers of energy supply, from the meanest carnivores to the gentlest of grazers of algae or plants—need to remember that we have the photosynthesizing and chemosynthesizing autotrophs, and their ability to store energy in chemical bonds, to thank for our world. So remember that the real foundations of the Burgess Shale ecosystem were the cyanobacteria, the algae, and the phytoplankton that provided the base of the food chain and made the full menagerie of animals possible.

Paleoecologists can categorize individual species by a range of factors in order to analyze the structure of the paleoecosystem. As mentioned above, a few of the aspects that are often looked at include habitat, locomotion, and feeding mode. Three main feeding modes used by Cambrian animals—and still in use today, of course—are deposit feeding, suspension feeding, and predation/scavenging. **Deposit feeding,** in this case, involves removing organic matter from the sediment at the bottom of the ocean, either by selective picking of the material from the sediment or by wholesale ingestion of the sediment and its edible bits, and egestion of the sediment minus the utilized organics. A number of burrowing worms are deposit feeders. **Suspension feeding** involves straining or otherwise removing food particles from the water; this is the mode of feeding for many brachiopods, eocrinoids, and sponges, for example. Predators and scavengers are carnivores that feed on other animals; in the case of predators these animals are usually ones they have caught by hunting or ambush (although nearly all predators will scavenge when they happen across an already-dead prey item). Strict scavengers only clean up and eat carcasses. These two carnivore categories are often combined as **predator-scavengers** in paleoecological analyses. Many arthropods in Cambrian deposits, of both trilobite and non-trilobite affinity, are predator-scavengers. In the Burgess Shale Formation, the suspension feeders dominate the diversity, and among these suspension feeders sponges account for about half the genera. In relative abundance, selective deposit feeding arthropods dominate the sample of individuals (fig. 9.7c), although by biovolume, interestingly, the predator-scavengers, deposit feeders, and suspension feeders all are about equally abundant (fig. 9.7d). The nearly equal abundance of the three categories by biovolume may be accounted for by the relatively large size of the predator-scavengers and of the sponges among suspension feeders. Still, the evenness of the relative abundances here by biovolume suggests no one feeding group dominated the ecosystem.

Among the categories by which Cambrian animals can be classified, a combination of habitat and locomotion mode—a life habit—perhaps, includes several that we can outline here. **Vagrant infaunal** species are those that move around in the mud or sand of the bottom sediments below the ocean floor. This is basically a burrowing habit and might be best modeled by one or another species of worm. **Sessile infaunal** species are those that live in the bottom sediments but do not move, perhaps *Eldonia*. **Vagrant epifaunal** species move around on the ocean bottom at the sediment-water interface; many trilobites were probably of this life habit. **Sessile epifaunal** species are those that rested on or were attached to the substrate at the bottom of the ocean and did not move; sponges are a good example of a member of this group. Pelagic species are those that live in the water column at any depth of the ocean from near the surface to close to the bottom; pelagic species are generally either passive floaters (planktonic) or active swimmers (nektonic). *Canadaspis* and many other bivalved arthropods were probably pelagic. Finally, nektobenthic species are active swimmers (although, of course, they may float temporarily) that live in the water column but mostly just above the bottom; this group includes species that swam freely in the ocean but may have fed on bottom-dwelling species. The chordate *Pikaia* may have been **nektobenthic.**

Diversity and biovolume of the Burgess Shale paleocommunity was dominated by epifaunal elements, both vagrant and sessile (fig. 9.7b). Most of these epifaunal species and most of the biovolume were accounted for by arthropods and sponges. Similarly, the Chengjiang faunas in China are dominated by epifaunal arthropods and sponges (along with priapulids and brachiopods).[29] (Notice a bit of a theme here?) Relatively few species and little of the biovolume were swimming or living in the bottom sediments. Overall the environment seemed to be heavy with epifaunal, suspension-feeding species, with epifaunal deposit feeders and epifaunal predator-scavengers also accounting for a significant percentage of the population. Not everyone was running or crawling around on the bottom of the ocean, but many were. At a time when animals had not yet utilized infaunal lifestyles to the extent or depth that they do today, and at a time before active, swimming fish and other groups had really taken off and expanded into a multilevel food chain of an open ocean ecosystem, we might expect the epifaunal ocean world of the Cambrian to be crowded for its time. I say for its time, relative to the infaunal group and the nektonic/pelagic groups, because in an absolute sense the epifaunal crowd in the Cambrian had a bottom community that was still not all that densely populated. The descendants of these Cambrian animals managed to fill in the infaunal and open water column communities and still pack in more species and niches into the epifaunal setting. But during the Cambrian, the epifauna was where the action was.

If we look at a food web of the Cambrian ocean world, as represented by the Burgess Shale (fig. 9.8), we see that the primary producers, those taxa that turn sunlight and environmental elements into the energy

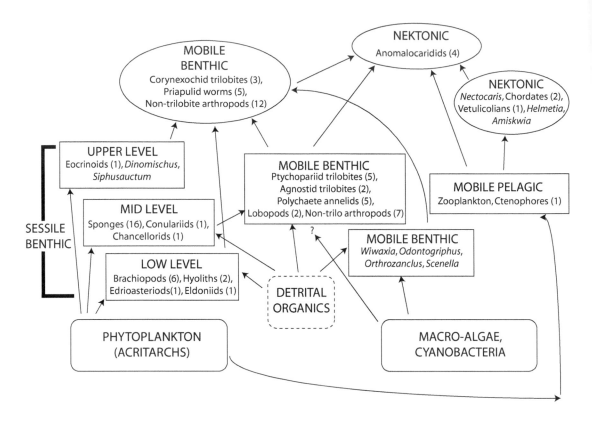

9.8. A hypothetical foodweb of the Cambrian marine realm, based mostly on the Burgess Shale. Primary producers in rounded rectangles. First-level consumers ("herbivorous" species) in rectangles. Second-level consumers (predators) in ovals. Numbers in parentheses indicate generic diversity within a group. There appear to be few grazers and few nektonic species. As previous studies have found, the bulk of the diversity is in benthic detritivores, predators, and sessile benthic filter feeders.

ultimately utilized by the whole ecosystem, include microscopic pelagic (and possibly some benthic) phytoplankton, macroscopic algae living on the bottom of the ocean as mats and filamentous forms (with some perhaps floating in the water column), and cyanobacteria. We also see that many groups fed on the primary producers; a few predator-scavengers fed on the primary consumers. Note that the diversity seems to be dominated by benthic forms, specifically epifaunal, sessile suspension filter-feeding taxa such as sponges and chancellorids and by mobile epifaunal detritivore-scavengers including ptychopariid trilobites (fig. 9.8). The main predatory groups appear to have been corynexochid trilobites, priapulid worms, anomalocaridids, and a host of non-trilobite arthropods. It is unclear if there was a true second tier of predator-scavengers, carnivores that specialized in feeding on carnivores; anomalocaridids may have fed occasionally on smaller predators, but most probably did not specialize in this.

It is difficult to estimate the diversity of phytoplankton that would have been present, although there are at least six genera of larger algae in the Burgess Shale. Similarly, the number of species of zooplankton cannot yet be precisely estimated. Disregarding producers for the moment (because we don't really know their diversity among phytoplankton) and just comparing the heterotrophic sample, first-level consumers (filter feeders, herbivores, and detritivores) account for about two-thirds of the genera in the food web; second-level consumers (predators) are the other one-third. An abundance of filter feeders and predator-scavengers seems to be a characteristic of many Cambrian ecosystems.[30] Overall, Cambrian food webs seem to be fairly similar to modern marine food webs in overall structure, if not necessarily component species and absolute diversity.[31]

Taxonomic Diversity

A few sections ago, we looked at the relative diversities of the paleofaunas from several sites in the Cambrian of North America. We used the Shannon index and its modification to effective richness to compare what these sites would represent in terms of straight taxa if all species were equally abundant and at a comparable sample size. Recall that the effective richness of these ten localities from the Lower and Middle Cambrian ranged from a little over 2 in the Bright Angel Formation at Frenchman Mountain and in the Grand Canyon (Lower and Middle Cambrian, respectively) up to just over 7 in the Middle Cambrian Chisholm Formation near Pioche, Nevada (table 9.2). In comparison, a sample from the Burgess Shale Formation of more than 16,300 specimens of 95 species has a Shannon index of 2.853 and an effective richness of 17.3,[32] more than twice that of the Chisholm Formation, which, although it is the most diverse among the typical sites we investigated above, preserves only trilobites, brachiopods, eocrinoids, and hyoliths. In straight species counts the Burgess Shale has almost nine times as many known species as the most diverse of the others (the Pioche Formation at Ruin Wash and the Spence Shale at Spence Gulch, each with 11 species in the studied samples). This is the difference that soft-body preservation makes in measuring diversity. It is difficult to compare the Burgess Shale with your average locality. And certainly in their time each of the typical sites that we looked at earlier had diversities and faunas more like that of the Burgess Shale, but the soft taxa simply were preserved much less frequently. We might well project an image of the Cambrian ocean world of eastern British Columbia, with a few adjustments here and there for local conditions, onto our views of sites like Ruin Wash and Wheeler Amphitheater, among others. Without our Burgess Shale–type deposit "windows to the Cambrian," our view of that time would be almost sadly deficient.

The primary event of the Cambrian, the unique aspect of this time, and the major source of debate about what this particular period is trying to tell us through its rocks and fossils from the incomprehensively distant past, is the Cambrian radiation–the relatively rapid diversification of the Metazoa that occurred mostly during the first half of the period. Also known as the Cambrian explosion, this event has, upon closer inspection, come to be seen as having actually started back in the latest Ediacaran and thus to have dragged out slightly longer than we thought at first. So perhaps it was more crescendo than explosion. Either way, it was geologically fast–especially compared to the time before and since. No animals for 85% or so of Earth history, and then suddenly they appeared in the

The Cambrian
Radiation

Ediacaran, and within 10–15 million years at the beginning of the Cambrian (0.2% of Earth history) animals had diversified into almost every phylum or body plan known today. And almost no new major body plans since then. It was all in that window. This of course brings up a whole host of questions such as Why then? Why not now? Why not before then? These are questions that paleontologists and biologists are still debating today. In this section we will take a look at this, the significance of the Cambrian radiation for our understanding of evolution and biology.

The Early Days

First, it is important to note that since the first days of the Cambrian as a period back in 1835 it was recognized that there were few or no fossils below the base of this period. In Charles Darwin's time the entire Precambrian was a black hole of zero fossil preservation. Even though far fewer formations and fossils were known from the Cambrian, and certainly the discovery of the Burgess Shale and its biota were 50 years off (C. D. Walcott was a child of about nine when *On the Origin of Species* came out), so little had ever even been seen, much less identified, from the Precambrian that even in 1859 the Cambrian appeared to be an explosion of animal forms on the scene. Darwin's view of evolution by natural selection was that it would have been gradual and taken quite a bit of time to develop new forms, so he recognized himself that the sudden appearance of animal fossils beginning in the Cambrian, and the contrasting emptiness of the Precambrian preceding it, was an argument against his view. As it turns out, Darwin probably would have been thrilled both with the appearance of animals and the diversity of the Ediacaran biota (whatever types of organisms they are that dominate those faunas along with the possible metazoans *Dickinsonia* and *Kimberella*) later in the Precambrian, and with the discovery of such previously unrecognized animal diversity early in the Cambrian. His excitement may have been tempered by a realization that, even though the Precambrian has proven to be far more fossiliferous than it appeared in his day, the diversification of animals in the Early Cambrian still appears to be rapid—and, yes, explosive. We simply are able to see enough detail now to recognize that the explosion had a fuse running back into the Precambrian and that not all animal body plans (phyla) appeared at the same time during the Cambrian—we know what order they appeared in and when it was. Some were in the Middle Cambrian. But the fact that all of today's phyla seem to have appeared around the time of the Ediacaran–Cambrian transition and not earlier or later has to be significant.

The debates about why this event occurred and its timing are still quite active. Was Darwin's worry about the Cambrian radiation potentially arguing against his view of life justified? For the time, perhaps. If evolution is always gradual and slow, as he assumed it must be, then the sudden appearance of animals in the Cambrian would be anomalous. But since his time we have seen from the fossil record and from biology that uniformity of tempo may be the exception rather than the rule in evolution, especially when it involves the origins of kingdoms and phyla and not just species. The more we know, the more we appreciate the complexity of the process as it has happened through Earth history. We may, in some (most?) of our studies, be left with as many additional questions as answers, but research since the mid-nineteenth century has shown us that much more was involved with the Cambrian radiation than we might have imagined; the details are so subtle and interesting. What follows is just an introduction to some of the questions that the Cambrian radiation raises and to some recent research findings about various aspects of these issues.

The first issue we must tackle, of course, is what precisely we mean by the Cambrian radiation (or explosion). In essence, the Cambrian radiation was the appearance of most metazoan phyla (major animal body plans) within a time span of about 10–15 million years in the Early Cambrian. I say most animal phyla premiered in the fossil record during the Early Cambrian because sponges may have appeared during the Proterozoic millions of years before the Cambrian, and the bryozoans appeared at the earliest (as far as we know) in the latest Cambrian, possibly in the Ordovician. In addition, there are a number of entirely soft bodied and mostly small, less diverse animal phyla that are only known from modern settings; perhaps many of these originated during the Cambrian, too, but we just do not have a fossil record to prove it.[33] The point is that other than sponges and bryozoans, just about anything with a preservable skeleton makes its first appearance in the first 10–15 million years of the Early Cambrian.[34]

Now, admittedly 10 to 15 million years is a very long time, but keep in mind that Earth had gone some 4 billion years by this point with no animal life at all, and that after billions of years of bacteria and eukaryotes and little change, it took these newcomer metazoans a mere few tens of millions of years to suddenly cause wholesale overhaul of Earth's ecosystems.[35] Simple ecosystems of the Proterozoic became increasingly complex as what we recognize now as more modern-style niches continued to be filled throughout the Cambrian, although ecospace even by the end of the period was not as packed as it became later.

In addition to the appearance of most modern phyla by the end of the Cambrian, the radiation also appears to have accounted for the appearance of many modern classes within these phyla during the Cambrian (and some just after, in the Ordovician).[36]

As mentioned above, sponges were probably the first animals to appear in the fossil record, possibly as early as the Cryogenian (Snowball Earth) period as much as 650 million years ago, and bilaterian metazoans may also have appeared during the Ediacaran.[37] Very rare and tiny (less than 1 mm in diameter) possible animal fossils are known from the Doushantuo Formation in China, fossils that appear to be of bilaterians,

Psycho Cambrian:
Qu'est-ce que c'est?

and these are from 582–597 million years ago (about 40–55 million years before the Cambrian).[38] Much of the initial, phylogenetic diversification of animals could well have been in the late Ediacaran, at least 10 million years before the Proterozoic–Cambrian boundary time,[39] with the ecological radiation and fossil record appearances not gaining steam until the Early Cambrian, and in some cases several million years into the Cambrian (~530–520 million years ago).[40] Although the major pulse of animal diversification may have been during the late Ediacaran, we have little fossil evidence so far of this series of events. The main evidence is from molecular studies of modern animals, which look at genetic differences between different groups and estimate how long ago they diverged. A lot of fossil evidence of the Cambrian radiation is from several million years into the period. What do seem to diversify and increase in abundance right at the Precambrian–Cambrian boundary at 542 million years ago are trace fossils, which were small and simple in the Ediacaran but become large and complex in the Cambrian.

The Cambrian radiation may consist of the appearance of most animal phyla, but it is interesting that the generic diversity within these phyla increases close to 300% during the course of the Ordovician, the period after the Cambrian.[41] The rise in generic diversity begins right at the end of the Cambrian and takes off in an increase that continues steeply and steadily into the early Devonian. Something else is going on there, an issue we will return to later in this chapter. After the Devonian there are a few peaks and valleys in diversity, but in general diversity stays roughly steady for the rest of the Paleozoic. After the Permian–Triassic boundary, however, another rise begins.[42]

Cambrian Radiation: The Big Questions

The big questions with the Cambrian radiation, now that we have a better idea of what that event was, revolve around the speed and timing of it. Once animals had appeared, why and how did so many body plans (phyla) appear in such a short amount of time, especially, again, considering the relative biologic stasis that prevailed for much of the Precambrian? Why then, at that point in metazoan history? And, an important question that researchers still struggle with, why did almost all animal phyla appear during the Early Cambrian and almost none since that time?

Explanations for the Cambrian radiation number close to two dozen proposed initial triggers or mechanisms of sustaining the event. Proposed triggers are generally either internal genetic factors or external abiotic factors, whereas the mechanisms that kept things going could be either of those or also external biological factors. Most proposed explanations for the Cambrian radiation revolve around two possible and non-exclusive mechanisms: an internal genetic tool kit influencing embryonic development, the full appearance of which defines what is possible in metazoans, and the early evolution of which determines fairly quickly what will become constrained in later species; and the crossing of an external

A native of Helena, Montana, Kevin Peterson is associate professor of biological sciences at Dartmouth College. His research centers on "the origin and early evolution of animals, specifically their dramatic and sudden appearance in the fossil record." He was taken by his parents to see a rerelease of *Fantasia* as a four-year-old. Enraptured by the dinosaur scenes, he determined to become a paleontologist at four and found his first fossil at age six. By age seven, however, he had changed plans to becoming a medical doctor. Paleontology was sent to the back burner. After an undergraduate regimen of premed courses, however, Kevin found that medical school itself now had less appeal. He worked in parcel delivery briefly, and for the highway department, and "after floundering for a few years" he relocated his first fossil finds, an act that rekindled the fascination he had felt as a child. So he started applying to graduate schools "the next day." This led him to a PhD in geology at the University of California, Los Angeles. Although the attraction to dinosaurs common to many paleontologists had its influence with Dr. Peterson as well, "I kept finding myself reading about the Cambrian and becoming more and more interested in the problem, especially as it seemed to be, at least in part, a problem of developmental biology," he says. And so this problem became his specialty. Given the apparent origin of animals at least several million years into back into the Ediacaran, the question of what then later triggered the explosion that began right at the beginning of the Cambrian is one that researchers debate today and may well for years to come yet. Dr. Peterson sees the triggers as potentially being several, possibly lying in an increase in oxygenation of the environment and in the opportunity created by the apparent extinction of most of the Ediacaran biota. Also, the discovery that a shared microRNA developmental tool kit existed in all animals from an apparently early stage was a surprise. "Here potentially," he says "was the source of morphological complexity, transcribed non-coding regulatory RNA molecules." Once this genetic potential was in place, animals may have been poised to take off and needed only to wait for the opportunity to do so (eventually provided perhaps by an increase in oxygen and the demise of many Ediacarans). Thus, possibly, the apparent delay between the origin of animals and their takeover of the planet. But Dr. Peterson also appreciates that we may still be a long way off from truly understanding what happened during the Cambrian radiation. "I think we should treasure," he says, "the sheer magnitude of the difficulty of the problem—the more we know about it, the harder the problem becomes!"

The Cambrian Corps 8– Kevin J. Peterson

ecological threshold (or thresholds) that set in motion a series of additional developments that fill in both ecomorphological space within Metazoa and niche space in Earth's marine ecosystems.

Triggers

Among possible triggers of the Cambrian radiation (those factors that are timing-dependent enough to suggest a critical role in initiating a series of events) are paleobiogeographic factors such as the breakup of the supercontinent Pannotia,[43] genetic factors within Metazoa, increasing oxygen levels in the world's oceans, the rise of predation, high sea levels, unusual ocean chemistry, and the diversification of pelagic acritarchs. And none of these factors is necessarily exclusive of the others; in fact, some are interrelated.

A genetic trigger to the Cambrian radiation is an attractive one because it would explain the timing of the diversification in that such an event cannot occur, of course, until the group (Metazoa) had the internal capacity to achieve it. In many cases the search for a trigger is as much a search for an explanation of the uniqueness of the radiation. Many researchers are more trying to explain why nothing like the Cambrian radiation has occurred since. The general conclusion regarding triggers has been that the metazoan ability to evolve more diverse and complex body plans than any group before it was by way of the genetic regulation of development and that it was in place early in metazoan history, not only because so many body plans were achieved so quickly but also because the gene tool kit that controls these functions is shared by almost all animals and is almost identical in most bilaterians. The question of why phyla have not continued to evolve since the Cambrian is another large issue that will be addressed in a later section.

In general, genetic factors come into play with two main questions: To what degree did the initial attainment of modern genetic capability in animals serve as a trigger for the Cambrian radiation? And did Cambrian animals have any degree of enhanced genetic capability relative to today's forms such that the Cambrian was a time of increased morphological evolution? Relating to the first question, there appear to be two genetic pathways by which animals may have evolved the morphological diversity expressed by the range of phyla that appear in the Cambrian. The acquisition of these genetic capabilities may have been necessary before the diversification of metazoan phyla could occur. Increases in the total number of genes would have been in place already when Metazoa appeared, but as new, more complex body plans appeared early in their history, animals increased the numbers of genes (with some variability). The overall trend is for higher gene numbers in more derived phyla, although this is a rather loose trend with its exceptions; humans have about 10 times as many genes as most fungi. Another pathway is in the increase in the number of *cis*-regulatory binding sites.[44] These are involved in the processing of developmental information in an embryo and are associated with sections of DNA, up to about 1000 base pairs long, that regulate the expression of nearby genes. The more binding sites, the greater the developmental power of the organism. These factors relate to the initial threshold animals passed giving them greater evolutionary capacity relative to their ancestors. We will address the question of higher capacity relative to later animals below.

Could high sea levels have triggered the radiation? Remember that when animals first appeared in the late Proterozoic, sea levels were relatively low, but they soon began to rise, reaching some of the highest levels in Earth history by the end of the Cambrian. So the radiation took place during an overall sea level rise that, other than some brief regressions here and there, was almost relentless. This is the Sauk Sequence transgression. Taken in concert with a relatively low-lying and flat continent, this sea level rise flooded vast areas of Laurentia, and this resulted in a massive increase in shallow marine habitat for animals, habitat that had not existed to the same extent a few tens of millions of years earlier. This may have opened up an opportunity for a more varied and taxonomically diverse animal fauna to expand into even more relatively empty ecological space than was available in the Ediacaran.[45]

Seawater chemistry may have played a role in triggering the Cambrian radiation as well. The widespread exposure of mostly unvegetated continental crust, along with very high sea levels during the Cambrian, resulted in the influx of erosional runoff to the shallow marine environment and an increase in calcium ions in the seawater. This may have helped promote the emergence of biomineralized shells in animals right at the time they were beginning to diversify. In addition to calcium, phosphorus may have played a role in the radiation as well. The emergence of animals, or at least their primary diversification, may have been delayed by low phosphorus levels in the ocean during the Precambrian. Particularly high or low levels of phosphorus in seawater, taken up into food sources, cause some marine animals to function below par. Studies of modern stromatolite-grazing marine snails found that they are most effective at feeding when phosphorus levels are intermediate, suggesting that the Cambrian radiation may only have happened once phosphorus levels in food sources for grazers and other animals had risen. The history of phosphorite deposits and their yields suggests that phosphorus levels were very low until about 600 million years ago, when there was a spike and then a decline, followed by another spike in the Late Cambrian. This pattern, and the fact that animals with phosphatic shells capable of sequestering excess phosphorus appeared around the very beginning of the Cambrian at 542 million years ago, indicate that the Cambrian radiation and the rise of animals and grazing and predation coincide with moderation of phosphorus levels in food sources.[46] This is one example of the interaction of life and the Earth's environment. The rise of animals might have been delayed by low P levels, but when those levels rose significantly animals adapted and then thrived when P levels declined a bit after the initial spike, changing the planet in the process.

A rise in phosphorite is correlated to some degree with a rise in diversity in planktonic acritarchs. This reflects the general increase in phytoplankton and nutrients, and might have been an additional trigger for the metazoan radiation, especially the rise in filter-feeding forms that would have benefited from a rise in nutrients and food supply in the water column. In fact, phosphorite levels, acritarch diversity and abundance, and metazoan diversity and abundance all increase almost simultaneously in the Early Cambrian. The phosphorite levels, along with high continental erosion rates and an initially low abundance of filter feeders, meant that nutrient content in the seas would have been high compared to the modern ocean.[47] This would have helped drive the Cambrian radiation with an increase in everything animals needed to diversify and expand.

Paleobiogeographic studies, combined with cladistic analysis, have found evidence that the breakup of the supercontinent Pannotia about 550 million years ago may have driven some early diversification at least among trilobites and possibly other bilaterians.[48] This, of course, suggests that arthropods and trilobites appeared by the late Ediacaran, and indeed some studies have placed the appearance of arthropods at around 546 million years ago, with trilobites presumably appearing almost immediately afterward.

Another possible trigger of the radiation could be the opportunity presented by the extinction of much of the Ediacaran biota at the end of the Proterozoic. Could animals have been around for millions of years but had no chance to flood into niches already occupied by Ediacarans? Was the explosion of ecology and body plans in the Cambrian simply a case of animals taking advantage of a chance misfortune to the Ediacarans? This has happened in the eras since (e.g., the rise of mammals) so there seems little reason this could not have played a part in the Cambrian radiation.

Mechanisms: Genetic

One proposed factor in the Cambrian radiation was that speciation rates, and by extension, the rate of appearance of new body plans, were higher during the Cambrian than during times since then. This relates to the question brought up earlier regarding whether Cambrian animals had greater genetic capacity than not just their ancestors, but also later metazoans. Some studies of marine invertebrates showed that speciation rates during the Early Cambrian were much higher than later in the Paleozoic and that these rates declined through time.[49] Later research has found, however, that Cambrian speciation rates were normal and not unusually high, at least among trilobites.[50] Trilobites were, however, at least apparently more prone to extinction in the Cambrian than they were in the Ordovician.[51] The appearance of trilobite fossils relatively late in the Early Cambrian is unexpected, if in fact they originated at or near the Ediacaran–Cambrian boundary as we suspect. Trilobites may have diversified in marginal environments or as tiny species and only appeared in the Early Cambrian Stage 2 as they achieved relatively large size and greater abundance. It appears that with trilobites at least, the main clades split from each other before they diversified in morphological range.[52]

Although speciation rates do not appear to be any higher during the Cambrian, trilobite specialist Mark Webster found that there was at least a higher level of within-species morphological variation in Early Cambrian trilobites versus those from later epochs in the Paleozoic.[53] The presence or absence of a genal spine on two closely related species of trilobites might be used as a distinguishing character defining pairs of the arthropods from the Early Cambrian and the Ordovician, for example. This study suggests, however, that the relative length of that genal spine in the one species from the Early Cambrian might have varied across a broader range (long to short) than would that of the Ordovician species. This suggests that natural selection, a driving force of evolution, had more to work with in the Early Cambrian than it did later. In this light, the apparently unremarkable speciation rates of many Cambrian clades are a bit perplexing.

If speciation rates were no higher during the Cambrian than today, was genetic capacity to create morphological diversity (in the form of phyla) any higher? Did this capacity become restricted after the Cambrian? It appears that the evolution of new phyla in the Cambrian did not occur as a result of enhanced genetic capacity, and that metazoan genetic capability may not have become any more constrained in subsequent time because it never was higher in the first place.[54] Rather, animals' genetic capacity reached modern levels very quickly and early in their evolution and never changed significantly. This all suggests that the Cambrian radiation was the result of typical processes.

Mechanisms: Ecological Thresholds

The transition from simple, microbial ecosystems of the Precambrian to typical Cambrian ecosystems with complex food webs – with second-level consumers and large size and predators, as driven by the appearance of metazoans, may have begun as early as 550 million years ago, at least 8 million years before the Cambrian.[55] Whatever triggered the radiation of animals, once multicellular organisms with the genetic capacity of Metazoa were on the scene it may have only been a matter of time before an ecological explosion came, in the form of a series of inevitable events triggered one after another. And perhaps the ecological first domino in the series was the development of **macrophagy.**

ALL THE BETTER . . .

The appearance of macrophagy, the eating of large food – whether by grazing or predation – appears to have been a key ecological threshold in the Cambrian radiation. It may well have initiated larger body sizes and the development of a large biomass of plankton in the oceans as organisms tried to escape the benthic feeding pressure. As one group of researchers put it, "The Cambrian explosion was the inevitable outcome of the evolution of macrophagy near the end of the Marinoan glacial interval [during the Ediacaran]."[56] Predation would come soon in the plankton as well.

Grazing, the consuming of marine algae in this case, appears – rather surprisingly perhaps – to have been fairly rare among Paleozoic and earlier ecosystems. This is based in part on the fact that herbivory in modern marine settings is typically a feeding mode of more derived animals that evolved from predators and detritivores, and appeared after the Paleozoic. Herbivory seems not to be the primitive starting point of feeding that it might initially appear to be. Certainly, more primitive Cambrian and Precambrian animals could have had this feeding mode, and some surely did. It is just that fewer than we might have expected were likely feeding this way. There may have been more detritivores in Cambrian ecosystems.[57]

Predation, although perhaps not in itself a trigger of the Cambrian radiation, may well have shaped the course of it. Although predation had probably existed for some time among microscopic prokaryotes, and perhaps had helped drive the origin and larger size of eukaryotes as much as 2.7 billion years ago, the appearance of macro-predators among the metazoans was a key event in biotic history.[58] The proportion of predator species in the marine fauna increased steadily throughout the Cambrian and into the middle of the Ordovician, from about 3% of genera at the beginning of the Cambrian to close to 15% by the end of that geologic period.[59] This was the initial diversification of metazoan predators.

It is not entirely clear whether the first macro-predators were benthic or pelagic, but they most likely were feeding on the diversifying zoo- and phytoplankton of the time. Indeed, paleontologist Nick Butterfield has found tiny filtering mouth parts of pelagic arthropods (possibly planktonic branchiopods) in Early Cambrian sediments in the Mount Cap Formation of Canada, suggesting that the predatory (filter-feeding) zooplankton of the world's oceans were operating early in the radiation.[60]

Responses to predation would have included the development of mineralized skeletons, the improvement of burrowing techniques, and the appearance and refinement of sense organs such as eyes and antennae.

I HEAR YOU KNOCKING

Possibly the most obvious response to predatory danger at your door is to block the adversary out. This is certainly the response of many species, ranging from clams and sow bugs to ankylosaurian dinosaurs and armadillos. The appearance of mineralized shells and skeletons in the fossil record (those beyond ones used just for support) dates back at least to the Ediacaran, and such protective shells explode in diversity and abundance in the earliest Cambrian in the form of the small shelly fossils, such as *Cloudina*, that we met in chapter 4. Caps, tubes, spirals, and two-part valves all appear in great numbers in many Early Cambrian deposits, and most are made of calcite, silica, or phosphate. Most are very small, but some get up to fairly large size. Hard shells of calcite—some with spikes and other elaboration—provided protection from predators and may also have been found by some species to provide additional body support and thus to facilitate even greater body size. Among the groups that developed these first mineralized skeletons are hyoliths, molluscs, brachiopods, trilobites, echinoderms, anabaritids, and at least seven other small shelly fossil groups.[61] The development of external skeletons and shells may have both responded to and driven other respective elements of the Cambrian radiation.

<h2 style="text-align:center">UP, UP AND AWAY</h2>

Another important threshold that seems to have been crossed during the Cambrian radiation was the invasion of the pelagic realm by metazoan zooplankton. Phytoplankton (acritarchs) had appeared early in the Proterozoic 1.8 billion years ago, and seemingly had the water column to themselves until about right at the base of the Cambrian. It appears then that small metazoans, perhaps driven upward by predation and competition, became pelagic and began feeding on the phytoplankton. This resulted in a simultaneous rise in spiked and ornamented acritarchs.[62] Acritarchs were abundant and diverse in the water column, and once tiny metazoan consumers began feeding on them, the pelagic and benthic realms of the oceans became interconnected and the world's marine ecosystems suddenly became more complex. This both resulted from earlier events in the radiation and fed additional developments. Again, this probably happened just before the Cambrian, but it was part of the same series of events.

Among the zooplankton known from the Cambrian are the chaetognaths, tiny predatory worms with spiky teeth that look like they were stolen from an alien movie nightmare. These predatory planktonic animals are important in today's oceans and may well have been important during the Cambrian as animals that helped start the complex food webs of the Phanerozoic and helped connect the pelagic and benthic realms.[63] Chaetognaths are known from probable examples from the Burgess Shale and from *Protosagitta* from the Chengjiang deposits in China (see chapter 6).

<h2 style="text-align:center">DIG DEEP</h2>

The Cambrian substrate revolution refers to the transition from relatively firm, microbe-matted sea bottoms with a sharp sediment-water interface during the Precambrian to a soft, burrowed sea bottom with a blurry sediment-water interface, and a significant mixed layer of bioturbated, well-oxygenated mud or sand at that interface, that developed throughout the Cambrian.[64] The ecological opportunities provided by the invasion of this new environment (the initiation of significant burrowing in the marine realm) helped drive the Cambrian's radiation. The depth and complexity of burrows in the seafloor increased dramatically through Cambrian time, probably driven by predation pressure. Driven into the sediment to escape predators, animals may have found habitats and modes of feeding that had not previously been exploited, such as deposit feeding and undermat mining.[65] Interestingly, the depth of burrowing in the marine realm only got deeper after the Cambrian.

SUPERSIZING

Trace and body fossils also record an increase in maximum body size of species among metazoans, starting at the base of the Cambrian.[66] Behavioral complexity also increases. There are of course large Precambrian para-animals in some of the Ediacarans, but by the Early to Middle Cambrian we have the likes of *Anomalocaris*, *Helmetia*, and *Tegopelte* among metazoans. These are far larger than the first animals of the Precambrian–Cambrian boundary.

EYES OF THE WORLD

The appearance of predation in Metazoa, the increase in body size, and the increase in the amount and depth of burrowing during the Cambrian all suggest an increase in the spatial complexity of their environmental habitats for most Cambrian organisms. In bilaterians this complexity may have driven the evolution of sense organs, particularly the eyes, and in some cases the antennae. This is what has been termed the "Cambrian information revolution."[67] The appearance of eyes is characteristic of Cambrian faunas and is indicative of this increasing complexity in the biosphere at the time. Development of better vision would have allowed animals to find food and avoid predators in deeper and deeper water as eye systems improved.

Indeed, eyes have evolved independently at least 40 to 60 times throughout animal history! There are about nine different types, but these generally can be classified as camera-type eyes and various forms of compound eyes. Thanks to the independent derivation of eyes in so many animal groups, a listing of taxa sharing one of the two basic eye types is almost humorous, as you might not otherwise associate such diverse lineages. For example, camera-type eyes have arisen independently (of course) in vertebrates, cephalopods, some spiders, some snails and slugs, some annelids, and some jellyfish. Compound eyes are known from arthropods, some bivalved molluscs, some annelids, and some echinoderms (brittle stars).[68] Note that among different species of molluscs and annelids, both types of eyes have appeared. The abundance of eyes and designs is perhaps evidence of their importance. In a newly dynamic and three-dimensional world, vision became crucial to survival.

And for a full-range appreciation of what is possible, at least in compound eyes, we can look to the trilobites and their huge eyes, tiny eyes, eyes with large lenses, eyes with many tiny lenses, eyes that look out in almost all directions but up, eyes that look up too, eyes that look all around for the pelagic species, eyes that fuse into one wraparound eye that looks like a pair of New Wave sunglasses—the list is seemingly endless. And all this within a few tens of millions of years of the first trilobites. Within trilobites as a group, eye size can vary by habitat depth as well; there were smaller eyes in those species that lived in well-lit shallows, huge eyes in those that swam in the deep ocean, and possibly blind species developing in depths even beyond light.

Whether predator or prey, Cambrian animals found themselves relying on vision in many instances for success. The appearance of the first eyes in Metazoa was a monumental event. Perhaps we should say the appearances were monumental events because it is entirely possible that eyes appeared in a couple of groups at nearly the same time. Either way, it probably started out in each lineage, unremarkably enough, with elaboration and refinement of a simple patch of photosensitive cells on the body of an unknown metazoan—and it ran from there. It makes me glad that chordates, too, wasted no time developing eyes during the Cambrian.[69]

The idea of the ecological threshold mechanism suggests that part of the reason no new metazoan phyla have appeared since the Cambrian is that as ecospace (niches) became filled during the Cambrian, any subsequent openings were now filled by members of existing clades. Speciation was enough to fill open spots as opposed to the land rush of opportunism that followed the initial appearance of numerous phyla in the open seascape of previously simple Precambrian ecosystems. The threshold passed may have been the appearance of large, complex metazoans, the radiation may simply represent the infilling of ecospace. To some degree this is suggested by the molecular data indicating Ediacaran divergences for most animal groups and the fossil data showing Early Cambrian explosion into a wide variety of ecological spaces.

Arguing against the threshold idea, however, at least as it relates to the realities of the post-Cambrian record, is the fact that no subsequent massive ecological opportunity has ever produced a new phylum. Smaller-scale extinctions may result in ecological infilling by speciation from a neighboring environment and an existing clade, but what of massive extinctions such as the Permian–Triassic event that ended the Paleozoic, when at least 95% of marine species seem to have become extinct? Surely this would have allowed at least one new phylum.[70] Apparently not. Perhaps the ecological threshold idea is too simple with regard to the post-Cambrian pattern. Or perhaps we underappreciate just how open the marine realm's ecosystems were at the Ediacaran–Cambrian boundary, compared with what had existed before. Perhaps it was equivalent to the recovery after a more-than-95% extinction.

Most likely, of course, the ecological threshold was operating in concert with the passing of a genetic developmental-mechanism threshold during the Cambrian, and this multipart influence is what made this time unique. It is the interconnectedness of Cambrian and later food webs, for example, that gives them their stability, whereas before the Ediacaran it was the simplicity of the food webs that likely made them stable.[71] In any case, ecospace in the Cambrian oceans got crowded quickly but did not really fill—that was a process for coming geologic periods.

What to Make of It

Flash in the Pan?

Getting back to an issue we covered briefly above, and to borrow the question wording from James W. Valentine, "Why no new phyla since the Cambrian?" Why does the appearance of new phyla seem to have been restricted to the Ediacaran–Cambrian interval and why after that golden era have things been rather quiet, no, silent, on the phylum front?

This question is often approached as the issue of what has been constraining morphological diversification since the Cambrian. As we mentioned earlier, one possibility is the infilling of ecospace such that any open niches are now filled by speciation of existing clades and that there are no longer entire open ecosystems for animals to flood into. Post-Cambrian diversification was a process of increasing specialization.[72] Another idea that has come up in recent years is genetic constraint. One possible operating mechanism here is the control of gene regulatory networks (GRNs), which are functional linkages among regulatory genes that control development in animals. The important aspect of these GRNs regarding morphological diversification is that the networks are hierarchical such that rewiring of the network is much easier at lower levels. Similarly, in your house it is a much easier task to change the lighting in the living room by changing a light bulb than it is to redo the entire electrical setup by changing everything around in the fuse box. The fuse box is generally left alone, but light bulbs are switched out all the time. In the GRNs of animals, the most frequent and least constrained modifications occur in the periphery of the network, and these result in speciation events. The axial parts of GRNs (what are called "kernels") are less changeable because they are higher in the hierarchy and are more interconnected. As researchers Eric Davidson and Doug Erwin put it, "[C]hange in [GRN kernels] is prohibited on pain of developmental catastrophe."[73] These kernels appear to have been in place during the initial diversification of Bilateria, and they do not seem to have changed much since then. (In fact, even individual regulatory genes may change so little over time that they may function similarly when substituted into a member of a separate phylum.) The idea is that the kernels, once assembled, can be built onto but cannot be rewired or taken apart. There also seems to have been ecological feedback from the direct anatomical developments of GRNs; as limbs appeared in Bilateria, new types of movement through the environment became possible, with the appearance of jaws came grazing and predation.

The early appearance of GRNs would have allowed hierarchical control of the developmental origins of parts of the embryo, and because the upper levels of the GRNs (the kernels) would be more difficult to change than the lower levels, most subsequent changes (later in the Phanerozoic) were on the lower levels, causing less significant morphological changes.[74]

Another factor that may have limited the emergence of new body plans after the Cambrian is the accumulation of microRNAs. MicroRNAs increase genetic precision; they reduce gene expression variability, effectively increasing heritability, and more are added to metazoan genomes through geologic time. By reducing the amount of variability in the expression of certain genes and in becoming more numerous through time, microRNAs might be central to the long-term reduction in disparity within clades.[75] This is one mechanism that may, just as an example, explain the decreasing amount of intraspecies morphological variation in trilobites from the Early Cambrian to the later Paleozoic.

The Appearance of New Phyla

Keep in mind that huge genetic changes are not impossible, just less likely with time. Morphological changes, the expressions of the rewiring of the genes, are similarly less dramatic through geologic time. Such large morphologic jumps are probably not the way successful new body plans (and thus phyla) appear. But this doesn't mean that the appearances of new phyla did not happen overnight. They only proceeded quickly on a geologic time scale. A lone individual of a suddenly new potential phylum, one born of a huge mutation on multiple dimensions, would be so unlike anything in its parent population that it would be unlikely to survive long enough to establish its new body plan as a phylum. "Almost inevitably," zoologist Richard Dawkins writes, "a megamutation of that magnitude will land in the middle of an ocean of inviability: probably unrecognisable as an animal at all."[76] Rather, the evolution of new body plans occurred by a relatively quick succession of small jumps, faster than later in geological time, but each step still small enough that the new, different individual, or growing population of individuals, was able to get by during its lifetime. You would not want the common ancestor of Lophotrochozoa to suddenly pop out a fully formed offspring mollusc or annelid – they would be so alien among the population of ancestral lophotrochozoans, and their adaptations perhaps so ill- or over-suited to contemporary conditions, that there would be no predicting their success or failure. So the appearance of new body plans and phyla in the Ediacaran–Cambrian, dramatic as it was, probably progressed in normal step-wise fashion over generations, just at a faster rate, relative to the amount of change involved, than in typical, later speciation events.

Diversity versus Disparity

Another issue raised by the Cambrian radiation is the respective patterns through geologic time of diversity on one hand and disparity on another. Diversity is simply the number of species of any specified higher-level taxon that inhabit a designated area. Disparity encompasses the morphological distance between species. Species are defined based on any of a few different criteria, but one of the most common is reproductive isolation of the respective populations. Morphologically, two species can be quite similar or very different, but their taxonomic separation (as long as they do not or cannot interbreed) is uniform: two different species. The species designation does not indicate how different they are, even within the same genus. This is also true at higher taxonomic levels; the fact that two families of animals are in the same class tells us nothing about how similar or different they are morphologically. Disparity measures this for us and is an important way of looking at the early history of animals.

Because diversity and disparity are not necessarily correlated through time at most scales, and because both appear to demonstrate different patterns through the Phanerozoic, it is worth looking at both. Diversity appears to increase significantly throughout the Phanerozoic, particularly during the Paleozoic. The numbers of species, genera, and families increase steadily and there are more now, and were later in the Paleozoic, than there were during the Cambrian. What of animal disparity? One might logically expect it to demonstrate the same pattern. To get right to the point, it doesn't. In *Wonderful Life* (and other venues), paleontologist Stephen Jay Gould suggested that while species diversity increases after the Cambrian, disparity (in the form of phylum numbers) actually *decreases* by a great pruning of the tree of life at the hands of comparatively random extinctions.[77] Gould argued that morphology or the developmental mechanisms controlling it were flexible during the Cambrian and less so afterward. Research over the past 20+ years has shown that this proposed pattern is not really correct either, in part due to the then-assignment of many Burgess Shale forms (now recognized as stem taxa in extant phyla) to unique, extinct phyla. Instead, it appears that while species diversity has increased gradually, disparity – the range of morphological distances between animal groups – expanded tremendously during the Ediacaran–Cambrian boundary interval, reaching modern levels very quickly, and has remained at that level ever since. Disparity, once we determined how to measure it, has proven to be approximately equal between Cambrian and modern groups, at least among arthropods and priapulid worms – not more, not less.[78] It appears that disparity, as measured by phyla, reached modern levels very quickly in the Cambrian and then plateaued (fig. 9.9). So the Cambrian radiation was a radiation of body form as well. The tree of life split off its main trunks (phyla) early and the smaller branches and twigs (species) have simply been filling things in ever since.

"You've Been Given a Great Gift"

Gould also argued that the survivorship of early Phanerozoic phyla was not predictable, that because some phyla were represented by so few species, extinctions weeded out some of the phyla by chance, thus limiting the range of body plans that later species could work with. He went on to

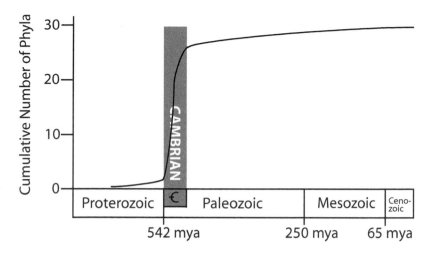

9.9. This is the Cambrian radiation. Curve of the cumulative number of metazoan phyla through time from the late Proterozoic to today showing origins of approximately 80% of known phyla during the Cambrian period (gray bar). Although generic and species diversities begin significant increases in the Ordovician, the great majority of animal body plans (phyla) appear during the Cambrian itself. Based on data in Erwin et al. (2011).

suggest that because the extinctions of these depauperate phylum experiments were by chance, and thus unpredictable, there is no reason why a second run of the Cambrian radiation would weed out the same phyla twice. Or three times. The idea was that successive experiments with "re-running" the Cambrian radiation would result in drastically different biological worlds today, each time you tried it.[79] This idea suggests that chance has as much influence on evolution as anything else.

An interesting recent experiment with *Escherichia coli* bacteria suggests that in fact chance may have less influence than we thought. This bacterium can produce about three generations an hour, and of course, being bacteria, they can be produced in high numbers and in multiple populations. So unlike the mind experiment of rerunning all of evolution, the researchers tried to miniaturize the process and run multiple identical populations through 1000 generations, each under different influences simulating chance, history, and other factors. The result was that regardless of chance influence and the conditions imposed, after 1000 generations of evolution each population ended up producing similar solutions. As the authors concluded, "Bacterial populations showed parallel and even convergent evolution in fitness."[80]

The important part of the experiment, even considering the many differences we might expect between evolutionary function in metazoans versus bacteria, is that the points arrived at were the same even though the routes taken were different. This suggests that, in fact, chance is less a factor in long-term evolutionary pattern than convergence or parallelism, the evolution of similar physical solutions (whether anatomical or physiological) to the same problem in completely unrelated groups of organisms. Convergence is why a dolphin (a mammal) looks so similar to an extinct ichthyosaur (a reptile); both are or were secondarily marine vertebrates of high swimming efficiency. It is also why many placental mammals have physical and ecological counterparts among marsupials, often in South America or Australia. Convergence in the form of mimicry has caused more than 2000 species of arthropods (mostly insects and spiders) to become antlike in appearance or behavior, and in fact ant-mimicry appears to have evolved separately at least 70 times.[81] This list goes on. Moles have short bones in their forelimbs with high-leverage muscle attachments and spade-like claws, adaptations for digging; and mammals such as the aardvark and the armadillo, that eat ground-dwelling insects, have peg-like teeth, often without enamel. Ingenious modern mammals? Perhaps, but also copycats. Little *Fruitafossor*, a Late Jurassic mammal from the same rocks that yield *Stegosaurus* and *Brachiosaurus*, had all the same features 150 million years ago, adaptations for a life of digging after termites and ants. Convergence is so prevalent in the animal world that paleontologist Simon Conway Morris has argued that Gould's experiment of giving evolution a do-over would in fact result in many similar types of animals emerging, ecologically if not necessarily taxonomically speaking.[82] There appear to be only so many ways of making a living, and life might well find them again on a second try.

Of course, part of the experiment of *Wonderful Life* seems less necessary now. After all, Gould was operating under the assumption that the number of phyla in the Cambrian was significantly higher than it is today. Because we seem to have found that Metazoa reached a level of phylum diversity about equal to today early on and then that level plateaued, there really does not seem to have been the decimation of phyla that would be needed to weed out body plans by chance. Might the "re-running of the tape" result simply in the same diversity of phyla that plateaus yet again? There would very possibly be some twists to what some of the phyla look like, but might convergence have worked its influence to generate many familiar forms again? Could the result of the Cambrian radiation, given a few minor surprises in casting, have been largely inevitable? We may never know, but it appears possible at least that the eventual destination of that rerun experiment could often have been something resembling a biologically familiar Bedford Falls,[83] simply with a few changes in the names and minor characteristics of the characters.

Evolutionary Faunas

Although disparity plateaus at the end of the Cambrian, overall diversity increases after the Cambrian. There is more to the story during the Paleozoic. The history of the Phanerozoic has been divided, at least for the marine macro-invertebrate realm, into three great associations of animals. We have been discussing throughout this book what is termed the Cambrian Evolutionary Fauna, consisting mostly of trilobites, inarticulate brachiopods, hyoliths, molluscs, polychaete annelids, and eocrinoids. Following this in the Ordovician and continuing through the rest of the Paleozoic to the end of the Permian is the Paleozoic Evolutionary Fauna (which, in fact, just begins diversifying in the Late Cambrian). This association includes trilobites, articulate brachiopods, true crinoids, corals, cephalopods, bryozoans, echinoderms, stromatoporoid sponges, ostracods, and graptolites. The rise in the early Mesozoic of the third association marks the beginning of the Modern Evolutionary Fauna.[84] This last fauna is devoid of trilobites, which disappear at the Permian–Triassic boundary, and is dominated by bivalved and gastropod molluscs, a variety of echinoderms, cephalopod molluscs, and crustaceans. The succession of these faunas is associated with an apparent overall rise in taxonomic diversity. From the Early Cambrian through the Paleozoic and then Modern evolutionary faunas, the species richness of reef communities increased from up to 80 species in the Cambrian to 400 and then 1200 in the subsequent evolutionary faunas.[85] But this is just one marine environment; other studies have found the question of continued taxonomic diversity increase through the Phanerozoic open to interpretation. There seems to be little argument, however, that the diversification during the Ordovician was real. Total time period diversity rose from about 250 genera at the end of the Cambrian to nearly 1000 in the Late Ordovician.[86] This rise in diversity is associated with an increase in ecospace packing—the marine environment is a much more complex place now than it was in the Cambrian, and that complexity began to increase right at the end of our favorite period, into the beginning of the Ordovician.[87] Did the Cambrian miss all the fun? No, it started the party. In fact, almost every major group involved in the Paleozoic and Modern evolutionary faunas first appeared during the Cambrian (cephalopods, echinoderms, molluscs, for example). And only a few Cambrian groups became extinct and did not live to be minor members of later faunal associations; although they are not characteristic of the Modern Evolutionary Fauna, several groups around today (mostly in deep environments) have their roots in the Cambrian (e.g., brachiopods and crinoids). So the evolutionary faunas represent not so much wholesale turnover of biotas but rather the rises and falls of groups dominating the diversity and ecology. As their time dominating the scene came to an end for one reason or another, the members of each evolutionary fauna appear to have been pushed into deeper, offshore settings by the diversification of each subsequent fauna.[88] This is perhaps why today's brachiopods and crinoids are often found in deeper water.

Another division of the faunas of the Phanerozoic finds 12 distinct associations called Ecologic-Evolutionary Units for level-bottom communities.[89] The first two of these, EEUs I and II, are in the Cambrian. EEU I includes the Early Cambrian time of olenellid trilobites and EEU II consists of the Middle and Late Cambrian.

The utility of the evolutionary faunas, despite the fact that they contain many of the same major groups, but in different levels of abundance, is that they reflect changes in the marine ecosystems—from the relatively simple systems of the Cambrian (which as we have seen were still many times more complex than those of the Proterozoic), to the complex ones of the rest of the Paleozoic, to the modern fauna that is so species rich and is dominated by such different taxa than was the fauna before it.

It Is Unavoidable?

It may be that the Cambrian radiation had its prerequisites, its triggers (more than one?), and its resulting effects that served to amplify the outcome. The following necessarily includes a lot of maybes, but it is worth outlining just to see how different prerequisites, triggers, and results interacted. We do not yet know what happened exactly, and we may argue about the details for decades to come, but this is one possibility.

About 600 million years ago the first animals appeared in the form of tiny sponges, with cnidarians appearing sometime later. These animals were around for several tens of millions of years until the rise of bilaterians. Bilaterians may have been around nearly as long as their less derived fellow metazoans, but sometime late in the Ediacaran they seem to have developed the genetic capability to generate diverse body plans and somewhat larger size. Around the same time, phosphorite and oxygen levels in the oceans were on the rise, as were levels of other minerals with increased erosion of the continents. Laurentia began to break off from Pannotia and approach the equator as the sea level began to rise, flooding large areas of the continent. The resulting increase in seawater nutrients and the expansion of shallow-marine habitat provided opportunities unprecedented in metazoan history. When some of the first experimentation in bilaterian body plans produced macrophagous species (benthic and planktonic filter feeders and possibly grazers), food webs immediately became multi-tiered and linked between the benthos and pelagic realm. Soon after, increasing primary consumers intensified feeding on phytoplankton, consumers of zooplankton appeared, and carnivorous predators after that. The diversity of body plans appeared rapidly as the genetic potential of bilaterians was exercised for the first time. The multi-tier food web explosion drove some species to burrowing (where some discovered deposit feeding), while others went up into the water column – both predators chasing prey and prey escaping their chasers. Many animals became larger. Mineralized shells and exoskeletons appeared quickly, as did eyes and antennae, and possibly auditory and olfactory systems. Behavioral complexity, the ability for which was genetically encoded in bilaterians, increased. Very quickly within the Early Cambrian, evidence of all of this became buried in the sediments of the oceans and was on its way to preservation in the rock record.

Soon environmental settings were a little more crowded with species, and the establishment of gene regulatory networks and accumulation of microRNAs made the generation of new, unique body plans less frequent if not impossible. As the Cambrian wound down, the body plans settled in, the ecosystems became established, and the stage was set for the next period of action, the dramatic increase in species diversity and ecosystem infilling that occurred during the Ordovician.

It is possible that once a few key developments happened early in the history of animals, the rest of the Cambrian radiation was only a matter of time. Once some thresholds were crossed this one-time-only primary radiation of Metazoa was inevitable. What thresholds were probably most important? Two big ones that stand out are that the genetic capability of bilaterians had to be in place – that perhaps is a prerequisite – but also the threshold that kicked things off may have been the development of macrophagy in animals. Once those two things happened, in concert with some serendipitous events occurring in the abiotic setting of the time, the cascade of biological developments that followed would seem almost unavoidable as the ecological tiering of the food web established itself.

Regarding this idea that much of what we think of as the Cambrian radiation was almost guaranteed after a few thresholds were crossed, Simon Conway Morris once wrote,

> Once the first animal appeared, and recall that on the metaphorical "day one" this was just another protistan experiment, then the ecological ball automatically began to roll. The complex ecologies which rapidly developed were subject to both continuing expansion and feedback, and as others have argued that in essence was the Cambrian "explosion," the reverberations of which continue until the present day.[90]

Given the factors that converged late in the Ediacaran, from genetic to biological to geological, the Cambrian radiation almost *had* to happen – and, as it was the Earth mother of all animal radiations, it *could* happen only once.

On and On: Legacy of the Explosion   10

AS THE THUNDEROUS, METAPHORICAL ECHOES OF THE CAMBRIAN
explosion rumble away into the distance, it is dawn in the Ordovician.
On a beach with light-blue tropical water reaching out to the horizon,
there is only the soft sound of small waves rolling in to the sand. It is a
scene that today might be punctuated by the cries of seagulls—only that
on the first day of the Ordovician there are none, of course.[1] There are no
land plants behind us; as in the Cambrian, the only organisms on land
are some algae, mosses, and an occasional arthropod venturing out of the
tidal zone. Life is still mostly in the ocean.

The arrival of the Ordovician has been announced somewhere out in
that vast expanse of ocean in front of us, by the appearance of a new type
of conodont. That's all—the great transition of one period to another. Indeed, little has changed. The Ordovician looks very much like the Cambrian. There is no post-apocalyptic wasteland of extinction, although a
few species have disappeared. The cast that appeared in the Cambrian
has begun to diversify. What is about to happen in the Ordovician is that
the newly graduated Cambrian animal groups will take off into young
adulthood and increase dramatically in diversity. Some will decrease in
diversity and abundance, and others will expand, but the overall diversity
and ecological complexity that got a toe hold in the Cambrian will fully
bloom in the Ordovician as animals mature into the Paleozoic Evolutionary Fauna.

As we close the Cambrian we see not an end but a beginning. A beginning of the rest of the story. Even old Paul Harvey would not assume
that we know this one, so now might be a good time to check the "Where
are they now?" file for what became of the cast of Cambrian characters in
the years since the Cambrian. For some animal taxa we know how they
fared through the Paleozoic up to today thanks to a strong fossil record;
for others we have little to go from other than what representatives they
may have had during the Cambrian and what the group looks like in
modern times—it is difficult to say much about what happened to them
in the intervening years of the post-Cambrian Paleozoic, the Mesozoic,
and the Cenozoic before now. What is important and encouraging is
that most Cambrian groups survive in some form or another today, except—sadly and ironically—for one of the most successful lines of the early
Paleozoic, the trilobites.

Sponges

Sponges, the most primitive of animals, were impressively diverse already in the Cambrian, particularly in the Burgess Shale in genera like *Vauxia* and *Diagoniella*. They sail right through the Phanerozoic with few major expansions or declines. They have reached a diversity of 5500 species in three classes today and live at a variety of depths in oceans worldwide. Sponges are still important filter-feeding animals in today's oceans.

Cnidarians and Ctenophores

The cnidarians, of which we saw a few jellyfish, anemones, corals, and possible sea pens during the Cambrian, become freshwater but stay mostly marine and expand to about 11,000 species today. Corals become established as important reef builders during the Paleozoic, and of course are important today as well. The coral structures disappear from the fossil record after the Permian, however, and do not reappear in a new form (the scleractinians) until 8–10 million years into the Triassic period. What happened during this gap? Where did our modern corals come from if we can trace fossil stocks only back to the Triassic? The corals of the early Mesozoic up to now probably did evolve from Paleozoic corals; the Triassic appearance was not likely a re-evolution of coral form from a basal cnidarian. More likely, some corals survived the Permian extinction and early Triassic recovery in a soft, non-skeleton secreting, almost anemone-like form and began building reefs once again when water conditions improved. Experiments have shown that this mode of survival appears to be a technique used by modern coral animals in acidifying seawater conditions that interfere with their ability to secret skeletons of calcium carbonate.[2]

Soft corals, sea pens, sea fans, sea anemones (plate 26), jellyfish (plate 22), hydras, and corals (plate 21) all of course make our oceans – from tropical reefs to temperate tide pools – more beautiful today than they would have been otherwise. Whole aquarium exhibits have been built around the aesthetic quality of jellyfish. Sea anemones occupy much of the space in many tide pools on rocky outcrops around some oceans. The comb jellies (Ctenophora) survive today also, in about 100 species. These animals were not diverse during the Cambrian, either, but they seem to have weathered the Phanerozoic just fine.

Bilaterians

PROTOSTOMES

Hair worms (Nematomorpha) are known from the Cambrian in several genera from Chengjiang in China. These worms today include about 320 species. Priapulid worms, represented in the Cambrian by such relatively common fossils as *Ottoia*, survive today and include just 16 species. But they don't seem ever to have been very diverse. If the Burgess Shale goblet-animal *Dinomischus* was in fact an entoproct, its modern relatives would include about 150 species, the large majority of them marine.

Nemerteans may be represented in the Cambrian by *Amiskwia* from the Burgess Shale. Today there are about 900 species of these mostly benthic marine worms, and they grow up to anywhere from 1 cm (0.4 in.) to several feet long.

The annelid worms, represented with such style in the Cambrian by polychaete genera such as *Canadia*, *Peronochaeta*, and *Burgessochaeta*, expand to include some 16,500 species today, many of them marine (plate 9), but annelids now live in just about any wet environment, including the soil and ponds. As we saw earlier, annelids such as earthworms and leeches are relatives of the Cambrian polychaete annelids, and there are plenty of feathery marine polychaetes in modern oceans. The modern genus *Notopygos*, for example, resembles the Burgess Shale's *Canadia* in overall appearance. Also impressive is the range of size and life habits of modern annelids: some annelids are as long as a car, others are almost microscopic; and while some are burrowing, marine deposit feeders, others may live in soils, whether parasitic, planktonic, or epifaunal benthic. One group specializes in being symbionts with crinoid echinoderms. It seems there are few ways of life for animals that annelids have *not* discovered.

Lobopods of the Cambrian, which we met in *Hallucigenia* and *Aysheaia* from the Burgess Shale, are represented in today's fauna by the Onychophora (velvet worms). There are about 110 species of modern onychophorans, and interestingly, all are terrestrial. The tardigrades ("water bears"), tiny sister taxa to onychophorans and arthropods, are eight-legged forms represented in the Cambrian by fossils from Siberia. Today, there are about 800 species of tardigrades, and most are so minuscule that they live in the moisture on plants and mosses and in soil. Others, still very small, are marine or freshwater species.

And then there are the arthropods. These ecdysozoans, so impressively morphologically diverse already in the Cambrian, in the trilobites and the unmineralized forms such as the bivalved and "great appendage" arthropods, diversify taxonomically throughout the Paleozoic and the rest of the Phanerozoic and take over even more environments. Members of some trilobite lineages become even more elaborate and large in the Ordovician, although trilobite diversity and abundance overall declines after that period. They slowly become less conspicuous in marine faunas up to the Permian–Triassic boundary, where they become extinct. The non-trilobite arthropods expand dramatically during the Paleozoic, both in the diversity of the crustaceans, myriapods, and chelicerates, and in the origin of insects. There are today well more than a million described species of arthropods!

The crustaceans, represented in the Cambrian by the phyllocarids and possibly some other bivalved genera, include about 67,000 modern species, and they have diversified to include familiar (and in many cases, delicious) groups such as crabs, shrimp (plate 24), lobsters, crayfish, barnacles, isopods (pill bugs), and a number of small aquatic forms such as branchiopods, ostracods, copepods, and amphipods (some of these latter groups probably were present in the Cambrian in the zooplankton—see the Deadwood Formation in chapter 8). Modern crustaceans may be either benthic or planktonic, some are parasitic; many are marine—such as some of the krill eaten by baleen whales—whereas others are freshwater. Mesozoic fossils indicate that many of the crustacean groups we know today were present in familiar forms and habitats at least 150 million years ago.

Myriapoda (centipedes and millipedes) may or may not be represented in the Cambrian fauna by little *Cambropodus* from Utah. This group is known from fossils starting at least in the Silurian in what may be marine forms, and myriapods increase in diversity through time. Another Paleozoic, terrestrial form left a fossil trackway in New Mexico indicative of an individual nearly 2 m (6 ft.) long. Modern myriapods include more than 11,000 species, and most are terrestrial, living among soil and leaves.

The group Cheliceriformes (chelicerates and pycnogonids) has diversified to include today's 70,000 species. Among the chelicerates are the horseshoe crabs (plate 10), the scorpions, the spiders, and (unfortunately) ticks. Pycnogonids are the rather alien-looking sea spiders. During the Cambrian the chelicerates were represented by some stem forms, including the Burgess Shale's *Sanctacaris*. Among fossil forms of the chelicerates are the eurypterids of later in the Paleozoic, a group known as the "sea scorpions," that in some cases grew to 2 m (6 ft.) long. Recent finds have shown that horseshoe crabs appear at least by the late Ordovician period.[3] Most modern cheliceriforms are terrestrial (ticks, spiders, and scorpions), although the horseshoe crabs and sea spiders are of course marine.

Hexopoda, the insects, did not exist in the Cambrian, although their arthropod ancestors must have—in whatever form they would have taken at that time. Insects appear as fossils in the Devonian, although the origins of the group may date back to the Silurian. They arise after the arthropod invasion of the land, so most insects are terrestrial forms today; known aquatic species are secondarily adapted to water. Winged insects appear shortly after the appearance of the group. There are more than 900,000 species of insects described today; estimates of how many species there are (described yet or not) in the world today run from several million to possibly tens of millions. In fact, insects are not only diverse, but so abundant today that many pastures probably have up to several million beetles per acre; at any one time there are, on the whole planet, perhaps as many as hundreds of thousands of billions of ants! Insects diversified during the Paleozoic, evolving the proverbial 2-foot-wingspan dragonflies during the Carboniferous, and had taken over the planet in numbers, diversity, and biomass probably by the Mesozoic, by which time most modern lineages had appeared. Dinosaurs lived with insects such as ants, bees and termites, for example, along with plenty of other groups. Whatever the Cambrian ancestors of insects were, the descendants of those ancient representatives in many ways rule the terrestrial environments of today.

The Mollusca was represented during the Cambrian by a number of tiny, millimeter-scale probable bivalves, gastropods, and monoplacophorans (or at least stem taxa of those groups) and, later in the Cambrian, the first cephalopods. Bivalves and gastropods diversify and become larger in average species size after the Cambrian. Cephalopods, after their Late (possibly Middle) Cambrian appearance, take off during the Paleozoic and become particularly important in the form of nautiloids and, by the Mesozoic, ammonoids; ammonoids are of great value in determining biostratigraphic correlations in the Triassic, Jurassic, and Cretaceous, but they do not survive the end-Cretaceous extinction. Today, the 93,000 known mollusc species include a great diversity of groups, some of which (like some arthropods) are of significant seafood importance. Clams, scallops, calamari – they all come from the ancestors known from Cambrian rocks. Among modern molluscs are bivalves, which include marine and freshwater clams and marine scallops; the gastropods, which include marine, freshwater, and terrestrial snails and slugs; cephalopods, the squid, octopuses, cuttlefish, and ammonoid-like nautiluses; aplacophorans; monoplacophorans; polyplacophorans (chitons; plate 23); and the tube-shelled scaphopods.

The Brachiopoda, you may remember, exists in the Cambrian in the form of two groups: the inarticulates (e.g., *Paterina*) and the articulates (e.g., *Nisusia*). Although brachiopods are a relatively minor group in today's oceans, they occupy nearly all depths in mostly benthic species that are as small as 1 mm and up to 9 cm across. There are about 330 species today, most of them articulate, and all are marine. *Lingula* is a textbook example of a modern inarticulate brachiopod. Brachiopods are abundant as fossils in the Cambrian, and their diversity and abundance increases after the Cambrian throughout the Paleozoic. In many Paleozoic formations in North America brachiopods are particularly abundant, and some forms, such as spiriferids and rhynchonellids, develop some interesting shell morphologies. Beginning in the Mesozoic, brachiopods decline in abundance and diversity as they seem to have been competitively replaced in most environments by bivalve molluscs. So it is happily due to brachiopod tenacity that the 300+ species survive today in oceans so dominated by molluscs. These holdouts from the Cambrian Evolutionary Fauna have done a great job of hanging in there for some 250 million years *beyond* what might be considered their heyday.

DEUTEROSTOMES

Echinoderms, as we saw, were relatively common during the Cambrian, as represented by the eocrinoids, such as *Gogia*, the rare possible crinoids like *Echmatocrinus*, and the carpoids, edrioasteroids, helicoplacoids, and holothuroideans. Echinoderms diversify during the rest of the Paleozoic, and the crinoids in particular become very abundant and diverse during that time. Many post-Cambrian Paleozoic formations in North America are packed with crinoids and crinoid parts; whether the formations are Mississippian or Permian in age, whether they are exposed in the roadcuts and rock quarries of the Midwest or in the desert canyons around Moab, Utah, it does not much matter, they often can be found to contain abundant elements of the crinoid echinoderms, all descended from Cambrian forms. And crinoids do not disappear at the end of the Paleozoic. They are still around today, along with 7000 other species of echinoderms. Crinoids are not particularly abundant in today's oceans, and many have been pushed into deeper water, but they are still there. The carpoids, edrioasteroids, and helicoplacoids of the Cambrian fauna are gone, but their indirect descendants – their nieces and nephews, you might say – are alive and well. And in the time since the Cambrian, echinoderms have diversified and today consist of starfish (plate 25), sea urchins, brittle stars, and sand dollars, in addition to the crinoid and holothuroidean (sea cucumber) descendants of the Cambrian fauna (plate 8). Echinoderms inhabit all depths of the seas today, from fully marine to brackish waters, and although most are benthic animals, some are pelagic. Some starfish can get up to 1 m (3 ft.) across. Without all of these offspring of the Cambrian echinoderms, our visits to modern tide pools would be much less interesting.

The hemichordates are related to the echinoderms and may be represented in the Cambrian fossil record by the possible graptolite *Chaunograptus* from the Burgess Shale, by other graptolites from Colorado and Tennessee, and by true pterobranchs from Siberia. Modern forms are infaunal benthic worms and include about 85 species.

The Chaetognatha, or arrow worms, are known from the Cambrian from unnamed elements in the Burgess Shale and from more complete material from China. Then, as now, they were small pelagic predators (with a few benthic species) that were nearly microscopic. Today's species number about 100 and some can get up to 12 cm (5 in.) long, although most are still tiny.[4]

Finally, the Chordata includes urochordates (tunicates, or sea squirts), cephalochordates, and vertebrates. There is at least one Cambrian tunicate known, from China; the modern cephalochordate *Branchiostoma* bears some overall resemblance to the Burgess Shale's *Pikaia*; and Cambrian vertebrates are represented in *Haikouichthyes* and others from the Chengjiang deposits of China. Tunicates are now – and were during the Cambrian – sessile filter feeders (as adults), and today the group includes some 3000 species. As entirely soft-bodied chordates, their fossil record is spotty. Cephalochordates include about 20 species and not a lot is known of their fossil history, either. The vertebrates, on the other hand, have a good fossil record that documents a radiation in the sea during the post-Cambrian Paleozoic and a rather spectacular terrestrial radiation after the invasion of land.

Vertebrates kept a low profile for a few million years after the Cambrian, but during the Ordovician and Silurian they diversified from the *Haikouichthyes*-like form into body plate–covered agnathans ("jawless fish"), which had a respectable diversity. It was not long before jaws, internal bones, and teeth evolved, and from there fish took off in several directions at once in the middle part of the Paleozoic: the bony ray-finned fish, the bony lobe-finned fish, and the sharks with their possibly secondarily cartilaginous skeletons. By the Devonian and Carboniferous, around the time fish diversity was reaching impressive levels, descendants of lobe-finned fish invaded the land and evolved into the first tetrapods. From here land vertebrates diversified into amphibians and amniotes, the latter including reptiles, mammals, and birds. Then, some reinvaded the water. Among the diversity of vertebrates that appeared among the amniotes during subsequent Earth history, the descendants of *Haikouichthyes* and its kin, were forms as different as the sail-backed pelycosaurs of the Permian, the large and dangerously predatory rauisuchians of the Triassic, the tiny dryolestid mammals and the massive sauropod dinosaurs of the Jurassic, the serpentine, secondarily-marine mosasaurs of the Cretaceous, and the horned and tusked uintatheres of the Cenozoic. The vertebrates truly pull out all the stops in their adaptive radiations as bony fish and tetrapods, much as insects and plants did in their land invasions just before the tetrapods. From the few vertebrates of the Cambrian have come probably about 55,000 species of vertebrates today, including more than 20,000 just among the bony fish. Wrasses, mackerels, butterfly fishes, surgeonfishes, puffers, remoras, eels, lungfish, and halibuts–all evolved from those first agnathans that developed jaws and bony skeletons.

Haikouichthyes might look with pride on its modern descendants, given that they include things as diverse as tuna and clownfish, leopard sharks and sting rays, tree frogs and tiger salamanders, komodo monitors and sea turtles, finches and pelicans, and aardvarks and bats. In a sense it might seem that, between the radiations of vertebrates and insects since the Cambrian, the Cambrian explosion has only continued. All this diversification has occurred within the body plan confines of single phyla, however many classes may have originated in the process. So, surprisingly, disparity has not increased as much in the diversifications of these groups as it might seem. Still, the morphological and ecological range achieved by these groups is striking. What observer of the arthropods of the Cambrian might have predicted the eventual appearance of katydids and walking sticks? What diver who caught a glimpse of *Pikaia* would have foreseen that its phylum would eventually produce aerial-acrobat mammals specialized to snap flying arthropods right out of the air (bats) and mammalian behemoths adapted to filter pelagic arthropods from the sea by the hundreds of pounds (baleen whales)? Convergence and parallelism suggest that the paths to these eventualities are not necessarily as unlikely as we might think, and yet the specializations achieved are impressive, given that they were evolved within the limitations of single body plans. If the Cambrian has shown us how much morphological diversity animals were capable of generating in a very short period of time, the eras since then have shown how many ecologically diverse niches animals can squeeze their way into occupying, given a bit of time.

Whatever we take from the Cambrian and its biota regarding the nature of evolution, it was this 54-million-year period that started our modern biological world on its way. Without the Cambrian radiation, today's world would not look nor operate as it does. And yet, as I and others have argued, once a small number of prerequisites were achieved and a few thresholds crossed, the radiation and its ecological consequences were almost inevitable. As I said at the beginning, the biological world that surrounds us today really began in those early years of the Cambrian, 542 million years ago. In this sense, the Cambrian is the most important period in geological history, at least from the perspective of us metazoans. The Cambrian is our origination story; the species fossilized in the rocks are our founding fathers. We can follow their story (and ours) through more than half a billion years of time.

What we can't predict is what may become of animal species from here. And as researchers profiled in this book have emphasized, the biological history of life on Earth since well back into the Precambrian has shown repeated interaction of life and planet. Abiotic conditions affecting living organisms and living organisms, thanks to collective biomass, influencing abiotic conditions on Earth. We and other organisms are capable of changing conditions on the planet, intentionally or not, and if historical geology is telling us anything in red bold print, it is that Earth, too, will change conditions itself, guaranteed, given enough time. Adaptability, as always, is the name of the game. We can only hope that we, and our animal cousins, have inherited that, too, from our Cambrian ancestors.

Glossary

Abathochroal eye Trilobite compound eye characterized by relatively few, larger lenses arranged on the eye with spaces between them.

Acritarch Any tiny, often microscopic, organic body or body shell of uncertain biological affinity; most are probably unicellular algae or other phytoplankton; particularly abundant in the Precambrian and Cambrian.

Ages Subdivisions of the geologic time scale below epoch. The Cambrian is divided into 4 epochs and 10 ages.

Agnathans Jawless vertebrate fish; modern forms include the lamprey and hagfish.

Agnostida The order of trilobites including agnostoids and eodiscoids, characterized generally by having: cephala and pygidia of essentially equal size, only two or three thoracic segments, and loss of eyes in some species; usually small, probably pelagic or benthic, and often cosmopolitan and characteristic of deepwater facies; agnostoids and eodiscoids are probably of separate origins (i.e., Agnostida is probably not a natural group); examples of agnostids from the Cambrian of North America include *Peronopsis, Ptychagnostus,* and *Pagetia.*

Ambulacra The body radii of echinoderms; for example, the five ambulacra of star fish are exemplified by the five arms on the animals.

Amoebae Microscopic, unicellular eukaryotes of irregular and variable shape; can move through the extension of parts of the body (pseudopodia).

Angular unconformity A geologic unconformity in which the underlying (older) rock has been uplifted and eroded before deposition of the overlying (younger) rock, resulting in an angle between the two sets of rock.

Annulation A ring-like formation on the outside of a worm body, for example; this is not the same as segmentation, in which the body internally and externally is compartmentalized in rings; annulation reflects an external texture and not an internal structure; annelid worms are segmented, other worms may be annulated.

Anoxic Characterized by a lack of oxygen.

Archaeocyathids Large (~5–15 cm), vase-shaped organisms from the Cambrian period once thought to be their own group but now recognized as a group within Porifera (sponges); were particularly abundant in the Early Cambrian but faded quickly in diversity and numbers in the Middle Cambrian.

Argillite Fine-grained sedimentary rock, consisting largely of clay particles, that has been very lightly metamorphosed. Less (or un-) metamorphosed rock of the same grain size is shale; more metamorphosed rock would be slate.

Arthropoda The largest phylum of animals as measured by modern species diversity and biomass; most species are characterized by segmentation, jointed limbs, and an exoskeleton (chitinous cuticle).

Articulating half ring A flange of the axial ring of the trilobite thoracic segment that projects forward and rests underneath the axial ring of the next segment forward; retains articulation of the elements (and maintains protection of the axial body) when the trilobite enrolls itself for protection.

Asaphida The order of trilobites characterized generally by having a smooth protaspid cephalon; examples of asaphids from the Cambrian of North America include *Dikelocephalus, Saukia,* and *Idahoia.*

Asthenosphere A layer of the Earth consisting of softer material of the mantle below the lithosphere; usually about 100 km (62 mi.) down to 350 km (217 mi.) in depth; magmas are often generated from this layer.

ATP Adenosine triphosphate, the chemical processing of which powers activities of cells.

Atrium The open, central part of a sponge.

Aulacophore The posterior extension of the body ("tail") in carpoid echinoderms.

Autotrophic Characterized by an ability to supply one's organic material through internal processes; autotrophic organisms usually manufacture organic compounds through photo- or chemosynthesis.

Axial lobe The middle lobe of the trilobite exoskeleton, arranged along a central, anteroposterior axis, and consisting of the glabella, thoracic axial rings, and the axial ridge of the pygidium.

Axial ring The raised, arch-shaped, central part of the trilobite thoracic segment.

Bauplan The combination of new and homologous original characters that defines a group of animals; similar to a body plan; from German for "blueprint."

Benthic Living on the bottom of the ocean (or other body of water); as opposed to those living in the water column (pelagic).

Bilateria Division of the Metazoa that includes most all animals except sponges and cnidarians; characterized by anterior–posterior orientation and bilateral symmetry.

Biofacies Associations of commonly co-occurring rock types and biotas.

Biofacies realms Different paleoenvironmental settings (e.g., shallow cratonic or deep shelf) in which the trilobite species are characteristically associated with particular rock types (e.g., one species in sandstones of the high-energy shoreface or another in black shales).

Bioherms Mound-shaped organic reefs composed of archaeocyathids, algae, cyanobacteria, sponges, and corals.

Biomere An interval defined by extinction and recovery events that bring entirely new faunas into paleoenvironments, probably mostly by dispersal from other regions; a term commonly identified with Middle and Late Cambrian trilobite turnover events, but less in use today.

Biostratigraphy The branch of geology involved with the correlation of sedimentary rocks by comparison of their contained fossil assemblages; because groups of species change through time, rocks of different ages contain characteristic fossils; by this means, distantly separated sedimentary rocks can be compared in age.

Biostratinomy A branch of taphonomy studying the orientation of the remains of organisms and what this reflects of the history of the material and the environment of deposition.

Biota An association of organisms (any combination of animal, plant, or other) living or preserved as fossils within a defined area, large or small.

Biovolume A measure of the relative abundance of species in an ecosystem based on their total numbers combined with body volume; more accurate than straight individual counts because it takes into account relative influence based on differences in size.

Biozones Intervals of Cambrian time characterized by particular genera or species of trilobites, the lower boundaries of which are marked by the first appearances of one of these species.

Brachioles The extending "arms" from the calyx of an eocrinoid or crinoid echinoderm; used in filter feeding.

Brood pouches Clusters of eggs kept by trilobites often in the anterior part of their cephalon before eggs are hatched.

Calyx The egg-shaped, central part of certain echinoderm bodies, such as crinoids and eocrinoids; rests on an attachment stem and contains extending brachioles (also the outer leaves of a flower).

Carbonates Sedimentary rocks consisting of often biogenically precipitated $CaCO_3$, usually in the form of limestone but also commonly (with an added element) as dolomite.

Cephalon The "head" of the trilobite dorsal exoskeleton, containing the eyes and glabella; the mouth is under the cephalon; consists of the cranidium and free cheeks in non-olenelloids.

Cerci Sensory appendages that extend from the posterior end of the pygidium in some trilobites (e.g., *Olenoides serratus*) and, more commonly, from the abdomen of some insects.

Cerebral ganglia Small masses of nervous tissue, connected by nerve cords, that form the central nervous systems of many invertebrates.

Chaetae Chitinous bristles characteristic of polychaete annelid worms; extend from the parapodia.

Chemosynthetic Characterized by generating energy through the oxidation of inorganic molecules or ions; some bacteria are chemosynthetic.

Chlorite A group of greenish clay minerals associated with some lightly metamorphosed rocks or occurring as products of altered minerals containing iron and magnesium.

Chloroplasts Organelles found in plant and algae cells that conduct photosynthesis.

Choanocytes Cells with a single flagellum that work collectively to move water through the body of a sponge; organic particles stick to the base collar of each flagellum, which is how sponges feed.

Cilia Hairlike tubular extension of a cell membrane.

Clitellum A saddle-like section of the segmented body of annelid worms.

Cnidae The stinging or sticking cells of cnidarians, as in the cells that cause jellyfish stings.

Cnidarians Members of the animal phylum Cnidaria, including corals, jellyfish, hydras, and sea anemones.

Conch The tapering, triangular tube-shell of a hyolithid.

Conglomerates Sedimentary rocks consisting of a mix of large (pebble-, cobble-sized, or larger) clasts of various sizes.

Conodonts Mostly small, eel-like vertebrates characterized by having multipart arrangements of multi-cusped teeth; appear first in Cambrian rocks but are particularly abundant later in the Paleozoic.

Continental shelf The shallow (generally less than 200 m deep) part of the margin of a continent between the shoreline and the slope to deep water; the slope of the continental shelf is only about 0.1 degrees; shelf settings are where many deposits of Cambrian fossils are found.

Conulariids Members of an extinct group of cnidarians that are preserved often as flattened, cone-shaped fossils; believed to have been similar to sea anemones in overall form; found at least as far back as the Early Cambrian.

Corrasion A taphonomic category assessing the degree of corrosion and abrasion present on skeletal remains of animals.

Cortex The thick wall of a sponge containing the water-flow canal system.

Corynexochida The order of trilobites characterized generally by large glabellas that extend nearly to the front of the cranidium and by pygidia often nearly as large as the cephalon; examples of corynexochids from the Cambrian of North America include *Olenoides* and *Glossopleura*.

Coxa The innermost segment on the walking leg of a trilobite; often very stout and spined in carnivorous taxa; may be used to move food toward the mouth.

Cranidium The central part of the trilobite "head" with the glabella and fixed cheeks.

Craton The relatively undeformed, stable part of the core of a continent; high sea levels during the Cambrian flooded much of the North American craton, and these deposits tend to be thinner and in paleoenvironmental settings shallower and closer to shore than the continental margin settings.

Crossbeds Stratification at an angle to the main bedding, formed by current flow and ripple or dune migration.

Crown group A lineage of organisms defined as the last common ancestor of a group of living organisms and all its descendants; this last part is important, as extinct species may or may not be within a crown group depending on how closely they are related to living forms.

Ctenophora Phylum of animals that comprises the comb jellies; outside Bilateria and very similar to jellyfish but outside Cnidaria.

Cyanobacteria Photosynthesizing bacteria sometimes called "blue-green algae," although they are not algae at all. Commonly make some stromatolite structures and other so-called algal structures in the rock record.

Deposit feeding Feeding by removing organic material from sediment on the bottom of the ocean; may be either by stirring up or filtering through sediment and removing organics, or by ingesting sediment and organics together and excreting just the sediment.

Detritivore An animal that feeds on detritus.

Detritus Organic debris from decomposing organisms (including animals and plants).

Deuterostomes Division of the Bilateria within Metazoa, characterized by an early embryo in which an initial infold of the dividing cells becomes the anus; includes echinoderms, hemichordates, and chordates.

Diploblastic Characterized by having the body composed of two cell germ layers (in animals).

Disconformity A geologic unconformity between beds that are parallel, marked by an erosional surface.

Disphotic zone The level of the ocean in which lighting is poor and photosynthesis does not take place; generally from 100 m to 1000 m; below 1000 m no light reaches.

Dorsal hollow nerve cord The characteristic spinal cord of chordates; most invertebrates have a ventral nerve cord, whereas that of chordates is dorsal and hallow.

Dysoxic Characterized by low oxygen levels; poorly oxygenated.

Ecdysis Molting; periodic shedding of the cuticle to accommodate growth, as in arthropods.

Ecdysozoa Subdivision of the protostome animals including priapulid worms, nematode worms, onychophorans, tardigrades, and arthropods.

Echinoderms Members of the animal phylum Echinodermata; includes starfish (sea stars), sea urchins, sand dollars, sea cucumbers, and crinoids.

Edrioasteroids An extinct group of echinoderms in overall appearance similar to sea urchins and sand dollars; benthic marine animals with five ambulacra and a generally oval shape in dorsal view.

Effective richness The species diversity of a sample as modified from Shannon's entropy, accounting for differences in total diversity, sample size, and relative abundance; given as a single number indicating the number of species that would be present if all taxa were equally abundant.

Endemic Characterized by a restriction (as in animal or plant species) to a particular geographic region.

Endopod The inner branch of a biramous appendage in arthropods; in trilobites it consists of the walking leg.

Endosymbionts Organisms that live within the body (or cells) of another organism.

Eocrinoidea A group of primitive echinoderms with a stem that attached to the sea bottom, a main-body calyx, and feathery arms, called brachioles, that were used in filter feeding; were covered in plates; a common Cambrian genus was *Gogia*.

Epochs The subdivisions of geologic time below periods.

Era Subdivision of geologic time above periods. The Cambrian is part of the Paleozoic era.

Euphotic zone The level of the ocean in which lighting is good and photosynthesis can take place; generally down to about 100 m.

Exopod The outer branch of a biramous appendage in arthropods; in trilobites it consists of the filamentous, fanlike structure above the walking leg.

Facies The characteristics of a rock unit, or an association of rock types, that is typical of a certain origin or paleoenvironmental setting for the rocks.

Faunal succession The principle that associations of animals change through time and are often characteristic of particular intervals; a key concept of biostratigraphy.

Faunas Associations of animals living or preserved as fossils at a single locality or area, however large or small.

Flagellae Whip-like projections on some cells, often used for locomotion; singular, flagellum.

Formation A geologic unit of distinctive rock type or associations of rock types that is mappable at a reasonable scale (often 1:24,000).

Free cheeks See librigenae.

Genal spines Spines on the posterolateral tips of the free cheeks (or librigenae) in trilobite cephala; in olenelloids, which lack free-cheeks, the genal spines are generally on the posterolateral corners of the cephalon (although the spines have moved forward in some species; e.g., *Bristolia anteros*).

Glabella The raised, axial ridge on the trilobite "head"; usually bulbous on the anterior end; organs that lie under the glabella include the stomach, "liver," heart, and brain of the trilobite.

Glauconite A mica-like hydrous potassium iron silicate mineral that is green in color and is common in some sedimentary rocks.

Graptolites Colonial marine animals of the phylum Hemichordata found fossilized in Paleozoic rocks; colonies consist of branches of lines of cup-shaped exoskeletons, each cup for an individual.

GSSPs Global Stratotype Section and Points. Markers indicating the bases of systems, series, and stages; usually based on biostratigraphic first appearances.

Helens Curved structures sometimes found between the conch and operculum in hyolithid fossils; function unknown; named after C. D. Walcott's daughter.

Helicoplacoidea Group of primitive, extinct echinoderms characterized by an elongate, oval shape, triradiate symmetry, and spiral ambulacra.

Hermaphroditic Characterized by producing both sperm and ova; can result in potentially self-fertilizing species.

Heterotrophic Characterized by the need to obtain organic material and energy through ingestion of other organisms or material; herbivores and carnivores are heterotrophic.

Holaspid The developmental stage in which trilobites have reached adult form in cephalon, pygidium, and number of thoracic segments; holaspid trilobites may still be small but the full adult form has been reached and from this point on growth involves only increase in size.

Holochroal compound eyes Trilobite compound eyes characterized by many small lenses packed tightly in contact with each other.

Holothuroidea A class within the phylum Echinodermata consisting of the sea cucumbers; although slug-like in appearance, they are echinoderms related to starfish (sea stars) and are not molluscs at all.

Homalozoa An extinct group of primitive echinoderms, also known as carpoids, characterized by a calyx with a short stem and a single brachiole; also covered with plates.

Hyolithida An extinct fossil group characterized by a tapered, tubular but flat-sided shell (triangular in shape when crushed); an operculum cover on the open end; and long, curved extensions (helens) connected to that end as well. They are probably molluscs or close relatives.

Hypostome An oval-shaped piece of the mineralized exoskeleton of trilobites that lies under the anterior end of the cephalon and just anterior to the mouth; serves as muscle support for mouth and protection for esophagus.

Instar The stage between two molts in arthropods, including trilobites.

Interference ripples Ripple marks caused by two separate currents, often at right angles to each other; forms a nearly checkerboard appearance; may be indicative of shallow, tidally influenced deposition.

Keystone species Any species whose influence on its ecosystem is so significant that removal (or subsequent reintroduction) of it causes a cascade of effects on the remaining species; most species have interconnected influences, but experiments have shown that some species have particularly large influence on others in their ecosystem.

Lagerstätte A fossil deposit characterized by exceptional soft-bodied preservation; from German; plural lagerstätten.

Laurentia The ancient continent, roughly equivalent to modern North America, that was equatorial during the Cambrian period.

Librigenae The free-cheeks of the trilobite "head" (left and right) that lie lateral to the cranidium; the eye is along the suture between the librigena and the cranidium; these are shed first during trilobite molting in order to allow the animal to more easily crawl out of the cephalon and the rest of the exoskeleton.

Lithosphere A layer of the Earth consisting of the crust and upper mantle, often about 100 km (62 mi.) thick, which is relatively strong and solid.

Lophophorates Animals such as brachiopods possessing a lophophore, a coil-shaped, feathery-textured filter-feeding apparatus through which the animals pump water to secure organic material for food.

Lophotrochozoa Subdivision of the protostome animals including flatworms, molluscs, brachiopods, and annelids.

Macrophagy The habit of feeding on large organic matter, as in grazing algal masses or plant material or predation on large (i.e., non-microscopic) animals.

Malacostraca The class within Arthropoda including shrimp, crabs, and lobsters.

Mantle The surface layer of the main body of a mollusc, which secretes the shell in shelled forms.

Medusae Pelagic forms of the phylum Cnidaria, as in a jellyfish, and as opposed to benthic forms such as sea anemones (polyps); singular is medusa.

Meraspid The developmental stage of a young trilobite when the cephalon, pygidium, and some thoracic segments are differentiated, but the full (adult) number of thoracic segments has not yet been attained.

Metazoa The kingdom of life made up of animals; metazoans are characterized by being multicellular eukaryotes that must get their nutrition through ingestion, by having cells that lack walls, and by having true tissues (except sponges in the latter character).

Mitochondria Organelles found in most eukaryotic cells that generate cells' chemical energy in the form of ATP; singular is mitochondrion.

Mollusca Phylum of animals including clams, mussels, scallops, snails and slugs, squids and octopus, as well as lesser known groups monoplacophorans, chitons, and aplacophorans.

Mudstones Sedimentary rocks consisting of mud- and clay-sized detrital grains; similar to shales but less platy in bedding characteristic.

Myomeres Blocks of muscles in a chevron or zigzag shape (in lateral view), arranged in a series from anterior to posterior, characteristic of chordates.

Nektobenthic Characterized by a free-swimming habit but living just above the sediment, close to the bottom.

Nektonic Living in the water column and swimming free.

Nematomorpha A phylum of long, thin worms resembling nematodes but with a single ventral nerve cord running from the brain and without excretory canals; larvae of modern forms bore into insects.

Nephridia Plural form of nephridium; a tube-shaped excretory organ of the invertebrate body that often occurs paired within each body segment and which performs a function similar to the kidneys in vertebrates, removing metabolic waste from the body.

Neurotoxins Biogenically produced chemicals that are toxic to the nervous systems of animals.

Niche partitioning The division of the ecological functioning of species that are either closely related or make their living in a similar manner, so as to reduce the amount of overlap in their roles.

Nonconformity A geologic unconformity between stratified rocks above and unstratified igneous or metamorphic rocks below.

Notochord A stiff but flexible rod located in a dorsal position in chordates; it is the developmental predecessor of the vertebral column in vertebrates.

Ommatidium The basic unit of the compound eye in arthropods, consisting of seven to eight sensory cells underneath a crystalline cone and the lens.

Oncoliths Concentrically laminated, ovoid sedimentary structures of calcium carbonate, generally about 5 mm to 10 cm in diameter, formed by the accretion of biofilms of cyanobacteria such as *Girvanella,* either by rolling in high-energy environments or by quiet-water stationary growth; common in limestone units of the Cambrian of the Great Basin.

Onychophora Phylum of animals including the velvet worms; today terrestrial predators but were marine in the past; lobopodians were very similar and probably closely related.

Ooids Small (~1 mm diameter) spherical accretionary bodies, usually of calcium carbonate and in limestones or dolomites; often formed by buildup of calcium carbonate around a nucleus of sand or fossil fragment in a high-energy environment where the ooids are rolled back and forth.

Operculum Fan-shaped shell piece that fit on the wide (open) end of a hyolithid conch; probably could be closed onto conch for protection of the animal (also a bone in fish skulls).

Optic nerve The nerve cord that connects the eye to the brain (in whatever form it takes) in animals.

Osculum The opening at the top of a sponge.

Ostium A small opening on the outside of the wall of a sponge, through which water is drawn for filtering.

Paleoecology The study of the relationships of fossil organisms (and groups of organisms) to their environments and to each other.

Papillae In anatomy, a fleshy protuberance; often nipple shaped.

Paraconformity A geologic unconformity between parallel beds that is not marked by a clear erosional surface.

Pedicle The long body extension used by brachiopods to attach themselves to the substrate.

Pelagic Living in the water column of the ocean (or lake), as opposed to on the bottom; pelagic species may be planktonic (free floating) or nektonic (swimming).

Periods The subdivisions of geologic time below eras and above epochs. The Cambrian is one of seven (or six) periods within the Paleozoic era.

Phaselus The first developmental phase of trilobites when the embryo consists of a tiny (< 1 mm) ovoid exoskeleton.

Phyla plural; singular is phylum; the Linnaean taxonomic rank below kingdom and above class; usually aligns with an overall body plan; from highest to lowest taxonomic rank the order is: kingdom, phylum, class, order, family, genus, species.

Phyllocarids Members of the subclass Phyllocarida, class Malacostraca, phylum Arthropoda; characterized by large, folded carapaces that cover the thorax, by five cephalon segments, eight thorax segments, six abdomen segments, and by five pairs of pleopods.

Phytoplankton Tiny, often microscopic, unicellular, photosynthesizing organisms that occur free floating in the water column.

Planktonic Relating to or characteristic of members of the plankton; plankton are small metazoans, microbes, and photosynthetic algae—for example, suspended in the water column and moved around more by currents and turbulence than by their own activity; ecologically very important in the oceans.

Pleomeres The segments in the abdomen of malacostracan arthropods.

Pleopods The abdominal appendages of malacostracan arthropods.

Pleurae The lateral sections of the trilobite thoracic segment, on either side of the axial ring.

Pleural lobe Either of the two (left and right) lateral lobes of the trilobite exoskeleton, as seen from dorsal view.

Pleural spines Spines on the lateral edges of the trilobite thoracic segment; sometimes short or absent; sometimes elongate; often projecting posterolaterally.

Podomeres The individual segments of the jointed walking leg in trilobites.

Polychaete A member of the Polychaeta, a class of the phylum Annelida; segmented and usually with a pair of parapodia on each segment; marine and with a head, jaws, and eyes.

Polyps Sessile, benthic forms of a cnidarian (phylum Cnidaria); e.g., a sea anemone or coral animal.

Porifera The phylum of animals that includes sponges, characterized by having flagellate cells that move water through the structure and filter out organics but also in lacking true tissues; probably includes archaeocythids.

Postanal tail An extension of the notochord and muscles posterior to the anus, characteristic of chordates.

Predator-scavengers Species that are carnivorous and hunt down prey or feed on carcasses.

Priapula Phylum of animals comprising the priapulid worms, which are burrowing marine predators; members characterized by a tubular body, spined anterior end containing the mouth, and an extensible feeding proboscis.

Protaspid An early stage in trilobite development when the animal has a cephalon but is still in the process of fully forming the pygidium and first-formed thoracic segment.

Protocerebrum Anteriorly placed, enlarged cluster of cerebral ganglia that forms the main part of the "brain" of trilobites (and some other arthropods).

Protostomes Division of the Bilateria within Metazoa, characterized by an early embryo in which an initial infold of the dividing cells becomes the mouth; includes most all bilaterian phyla except echinoderms (+ hemichordates) and chordates.

Pterobranchs Colonial, benthic marine animals of the phylum Hemichordata, class Pterobranchia; individuals live in tube-stalks and can extend to filter feed or retract for protection; see also graptolites for a fossil form that may be related to pterobranchs; both pterobranchs and graptolites are known from Cambrian rocks.

Ptychopariida The order of trilobites characterized generally by having glabellas that are short and tapering anteriorly and by having reduced pygidia; examples of ptychopariids from the Cambrian of North America include *Amecephalus, Modocia,* and *Elrathia.*

Pygidium The posterior element of the trilobite dorsal exoskeleton, consisting of a shield of small to rather large size, relative to the cephalon.

Radula The toothed strip in the mouth of molluscs that is used for rasping food.

Redlichiida The order of trilobites including olenelloids, a common Cambrian group from Laurentia; characterized generally by half moon–shaped cephala, crescentic eyes, often spiny thoraces, and tiny pygidia. Examples of redlichiids from the Cambrian of North America include *Olenellus, Bristolia, Mesonacis,* and *Nephrolenellus.*

Sabkha A supratidal environment in arid and semiarid settings that lies just outside (slightly above) the intertidal zone toward land; often very flat and characterized by evaporite-salt, carbonate, and eolian deposits; may contain mudcracks, dolomitization, and algal and stromatolitic laminae fabrics.

Sandstones Sedimentary rocks consisting of sand grains of any of a range of sizes larger than muds; often cemented with calcite or silica.

Sauk Sequence A sequence of early Paleozoic (largely Cambrian) rock in North America caused by a long-term overall transgression; characterized by several sea level rises and falls but with an overall trend of rising sea level.

Segmentation The division of a body into individual (and anteroposteriorly sequential) segments, as in annelid worms and arthropods.

Sequences Informal large-scale units that are bounded by unconformities and can be traced over large areas of a continent; may also be used on a smaller scale for any particular succession of rocks.

Series A subdivision of a system equivalent to an epoch. The second epoch of the Cambrian (time) includes rocks of Series 2.

Sessile Living, stationary, on the bottom sediment, and often physically attached.

Sessile epifaunal Characterized by a stationary and on-top-of-the-sediment lifestyle.

Sessile infaunal Characterized by a stationary and within-sediment lifestyle.

Shales Sedimentary rocks consisting of mud and clay sized detrital grains; often thin and platy bedding; may contain minor amounts of silt grains.

Siphuncle A cord of body tissue in shelled cephalopods that attaches the main part of the animal to the central part of the shell; extends through older body walls; also changes the gas and fluid levels in the chambers to regulate buoyancy.

Spicules The multi-axial support structures within a sponge, most often composed of silica or calcium carbonate.

Spreading center A plate tectonic margin at which two plates are moving away from each other; also known as a divergent margin; characterized by volcanic activity. The Mid-Atlantic Ridge is a spreading center.

Stages Subdivisions of series equivalent to ages.

Stem group A group of species outside (i.e., more distantly related to) the crown group of a lineage of organisms, but still within the lineage; stem taxa are less derived (i.e. have fewer advanced characters) than the common ancestor of the living species that defines a crown group; e.g., many Burgess Shale taxa clearly belong within Arthropoda but do not share characters that define the group that includes living forms, they are further down the "stem" and not part of the "crown."

Stem taxa See stem group.

Stipes The stems connecting the holdfast of algae to the wider thalli.

Strike-slip A type of fault (or plate boundary, on a large scale) in which the opposing sides of the fault line move laterally past each other with little up–down movement.

Stromatolites Layered, often bulbous, structures formed by the accretion of sediment by microbial biofilms; most often formed by photosynthesizing cyanobacteria.

Subduction zones Plate tectonic boundaries at which one plate (usually oceanic) is forced under another (often continental); such boundaries often result in frequent large earthquakes and active volcanoes.

Supratidal Characteristic of the area just outside (and above) the intertidal zone where tides flood and expose the substrate.

Suspension feeding Filter feeding by pumping water through or over a filter mechanism and removing any floating organics; so called because food particles floating in the water column are considered to be in suspension.

System The stratigraphic subdivision equivalent to a period. The time interval is the Cambrian period; the rocks are part of the Cambrian system.

Taphonomy The study of the remains of organisms from the time of death until fossilization; includes decay, disarticulation, modification, transport, and burial of skeletal or soft-part material.

Tardigrada Phylum of tiny marine and freshwater forms called "water bears"; close to arthropods; each has eight stubby legs.

Telson The last abdominal segment in some arthropods; sometimes a fan-shaped segment but may, as in horseshoe crabs, bear a long spine.

Thalli Flattened and ribbon-shaped filaments in algae (or plants).

Thermocline The point at which temperature in a body of water drops significantly, separating a warmer upper layer from cool, deeper waters; when measured from the surface, ocean temperatures, for example, will decrease gradually with depth, but at the thermocline the temperature will suddenly drop much more precipitously.

Thoracic segments The individual, articulating segments of the trilobite thorax.

Thorax The segmented section of the trilobite dorsal exoskeleton behind the cephalon and in front of the pygidium.

Tillites Sedimentary rock deposits composed of lithified glacial till, the rock debris left by retreating glaciers; including cobbles and gravel.

Trace fossil Any fossilized track or trace of animal activity.

Transgression The expansion of the marine environment inland towards or onto the continental craton caused by a rise in sea level.

Trilobite Any member of the Trilobita, a group of arthropods characterized by a central axial and two lateral, pleural lobes; also divided into cephalon, thorax, and pygidium; dorsal exoskeleton is mineralized with calcite.

Triploblastic Characterized by having the body composed of three cell germ layers (ectoderm, mesoderm, and endoderm) in animals.

Tube feet Echinoderm structures, connected to the water vascular system; hollow tubes that extend and retract and can move the echinoderm around, as in starfish.

Unconformity A break or gap in the rock record, representing a significant amount of time, caused by nondeposition or erosion.

Uniformitarianism The concept that Earth processes operating today are the same ones that operated in the past, and that most evidence in the geologic record reflects these consistent processes.

Vagrant epifaunal Characterized by an on-the-sediment and mobile lifestyle.

Vagrant infaunal Characterized by a mobile and in-the-sediment lifestyle.

Vendobionts Members of the Ediacaran biota that fall outside the metazoans and appear to be an extinct group (or groups) of multicellular non-animals that existed near the beginnings of metazoan diversification; based on the formal group name Vendobionta.

Ventral cuticle The unmineralized part of the trilobite exoskeleton that is under the body; opposite the commonly fossilized (and mineralized) dorsal exoskeleton.

Vermiform Having the shape of a worm.

Walther's Law The concept that paleoenvironments now seen in vertical stacks of sedimentary rocks were once laterally coexistent; the vertical stacking resulted from shifting of environments through time.

Water vascular system The water-filled tubular canal system inside the echinoderm body that connects with external tube feet; used in feeding and locomotion.

1. Natural Mystic

1. Darton's note on the Iron Mountain section is Darton (1907); also included in Darton et al. (1915); for a short biography of N. H. Darton, see Snoke (2003).

2. See Dott and Batten (1981) and Prothero (1990) for more on the history of stratigraphy.

3. Dinosaur-bearing deposits are almost always determined to be of their respective ages based on other lines of evidence besides the dinosaurs themselves, often invertebrate fossils included in interbedded marine rocks. Dinosaur biostratigraphy is generally what you would consider coarse.

4. Prothero (1990).

5. The conodont that marks the beginning of the Ordovician (and the end of the Cambrian) is *Iapetognathus fluctivagus*. There are complications with this definition, as the form sometimes does not actually occur at the boundary. See Terfelt et al. (2011).

6. For the Cambrian, Lower, Middle, and Upper and their time equivalents Early, Middle, and Late are no longer official, but they may be used, informally, just in lowercase; e.g., lower Cambrian.

7. See Geyer and Shergold (2000), Babcock et al. (2005) Walker and Geissman (2009), and Peng and Babcock (2011) for more on the Cambrian timescale. See also the International Commission on Stratigraphy's International Stratigraphic Chart.

8. North America is more correctly Laurentia when referring to the Cambrian continent. Laurentia included parts of what are now other continents, and North America now includes bits of other Cambrian continents such as Avalonia. But most of modern North America is essentially equivalent to Laurentia, the Cambrian continent.

9. See Palmer (1998a), as well as chapter 4, for more on the Cambrian stages of Laurentia.

10. Potassium-Argon dating (K-40/Ar-40) is a different process. In this there is conversion of a proton to a neutron through electron capture; the atomic weight stays the same. Also note that carbon has a half-life of only 5730 years so this process works only on (geologically very) recent material. U-235 converts to Pb-207 with a half-life of 713 million years, so the uranium-lead technique works better on older units. Also note that an entirely different technique dates original (rather than detrital) ash-derived zircon crystals in sedimentary rocks.

11. The most important part is that the decay rates of the isotopes are known, so as long as the ratios are measured accurately, and as long as you make sure you are measuring a mineral grain formed at the time (and not before or one that has been partially melted since), your results should be fairly accurate.

Radiometric dating has a certain error that is unavoidable.

It is usually a percentage (often about +/−0.2% to +/−0.3%), but obviously as you go back in time that becomes an increasingly large number of actual years. For example, the beginning of the Pleistocene epoch, which initiated our current series of ice ages, is dated as 2.588 million years ago, +/−< 1000 years. Not bad. If we go back to the end of the Cambrian period, however, the error is still less than 1% but the +/−on the date at that percentage accounts for more than half the length of the Pleistocene−1,700,000 years. But when one is talking about 488 million years ago, 1.7 million years arguably makes little difference.

12. See Scotese and McKerrow (1990) and Williams (1997) on Cambrian paleogeography.

13. The magnetic pole is relatively constant in position but varies a bit with time; depending on your position on the globe the magnetic pole is almost always a few to a dozen or so degrees off from the geographic pole. This is a separate issue from magnetic pole reversal, in which magnetic north flips toward the southern geographic pole, something that happens repeatedly through geologic time.

14. In fact, paleomagnetic work contributed to the formulation of plate tectonic theory; further reading on plate tectonics and paleomagnetism can be found in Cox and Hart (1986).

15. See Haq and Schutter (2008) on sea levels, and Holland (1984) on continental flooding area.

16. Sonett et al. (1996).

17. See Lamar and Merifield (1967) on the length of a day and days in a year, and Sonett et al. (1996) on lunar orbit.

18. Holland (1984); Wilhelms (1984).

19. Hargraves (1976).

20. Wilhelms (1984).

21. One scientist even referred to such a scenario for origin of the Moon as being "horrendously improbable" (W. M. Kaula, quoted in Holland, 1984).

22. The extremity of tides with a new and close Moon led Lamar and Merifield (1967) to suggest that this was the impetus for the Cambrian radiation; they were arguing that the Moon did not begin orbiting the Earth until late Proterozoic time. The sudden appearance of tides where none had existed previously, along with the resultant increase in current energy in shallow-marine environments, was suggested as a trigger for the hard parts (exoskeletons and shells) seen in some animals beginning around the Precambrian–Cambrian boundary; this idea does not seem to have generated a lot of support, however.

23. Riding (2009).

24. See Frakes (1979) on precipitation; Witzke (1990) on climate belts.

25. Riding (2009).

26. For example, Palmer and Gatehouse (1972) on trilobites from Antarctica.

27. I am referring to (mostly fossil) species that are lost not through human interference, such as habitat loss or other anthropogenic effects, but by other, more natural causes.

28. It is important to point out that most species, as defined by humans, can only trace their members back 1 million years or so, on average. Of course, regardless of species definitions, the lineages of all species are an unbroken continuum of individuals going

back hundreds of millions of years. The boundaries between species, at fine scale, eventually become arbitrary anyway (at what point is an offspring a different species than its parent?).

29. Immobile as adults, that is; sponges and some corals have mobile larvae, and corals have mobile relatives in the jellyfish.

30. Both chemosynthetic and photosynthetic organisms are considered autotrophic, as opposed to the heterotrophic condition. Chemosynthetic and photosynthetic autotrophy, as well as heterotrophy, are common among different species of single-celled organisms.

31. For more information on the diversity of Archaea, Bacteria, and Eukarya and their ecologies, see Runnegar (1992), Campbell (1993), and Knoll (2003a).

2. Into the Heart

1. Cooled lavas have been found in very rare cases to have enclosed logs and at least one rhinoceros, preserved as molds, the organic material long since rotted away.

2. Siliciclastic rocks are so named because they are often composed in large part of silica and the grains are detrital (clasts).

3. The key in erosion by water and wind is other grains of rock, however small, that they carry; both will beat less harshly alone but water or wind running with even sand grains in them will blast quite effectively. Other factors that help erode rock include repeated expansion and contraction through daily temperature flux or ice wedging and oxidation and other chemical weathering.

4. Although these non-siliciclastic rocks are often buried as deep as siliciclastic rocks, that is not always necessary to start the process of lithification, as some carbonates will actually begin to solidify to near rock hardness while still exposed at the surface.

5. Fossils are big in most cases, that is; some of course are quite small and fossils are not always tremendously larger than the surrounding sediments.

6. As we will see, conglomerates can also be formed by glaciers.

7. Nodular cherts are usually a result of diagenetic replacement.

8. For a summary of siliciclastic and carbonate rocks, see Boggs (1987).

9. Huntoon et al. (1996)

10. For more on the geology of the Toroweap, see Turner (2003).

11. See McKee and Bigarella (1979) and Middleton and Elliott (2003) on the geology of the Coconino.

12. Tracks and traces get names separate from those of body fossils; fossils have genus and species names, whereas traces are named by ichnogenus and ichnospecies names. *Laoporus* is an ichnogenus; we don't know what fossil species made this kind of track.

13. For details on Hermit geology, see McKee (1982) and Blakey (2003).

14. Geology of the Supai is also in Blakey (2003) and McKee (1982).

15. See Beus (2003) on the geology of the Redwall Limestone.

16. Noble (1914) named the Tapeats and further described it in Noble (1922); as mentioned earlier, a nice map of Grand Canyon geology with stratigraphy and faults is Huntoon et al. (1996).

17. For more on the geology of the Tapeats Sandstone, see McKee and Resser (1945), Hereford (1977), Middleton and Elliott (2003), Rose (2006, 2011), and Hagadorn et al. (2011). For general beach and tidal flat geology see McCubbin (1982) and Weimer et al. (1982).

18. E. Rose, pers. comm. (2011).

19. The Bright Angel was originally named "Bright Angel shale" by Levi Noble, but the formation contains a mix of lithologies, including a significant amount of sandstone as well as shale, and it has been recommended that the formation name be changed to Bright Angel Formation (Rose, 2011), a suggestion with which I agree and which is adopted here.

20. Quoted in McKee and Resser (1945).

21. An early geological description of the Bright Angel Formation is in Noble (1922); see also McKee and Resser (1945).

22. See chapter 5 for more on Cambrian trace fossils.

23. Lichens, familiar to anyone who has spent time hiking in the mountains, forest, or even higher deserts, are symbiotic associations of fungi and algae or cyanobacteria, and in modern settings they love growing on boulders. Slime molds, on the other hand, are amoeboid protists and are just similar to fungi in structure and life cycle.

24. For a summary of Bright Angel Formation geology, see Middleton and Elliott (2003); for reinterpretation of the environments and possible moss spores in the Bright Angel Formation see Baldwin et al. (2004) and Rose (2006, 2011); terrestrial algae appearing by the

late Proterozoic is treated in Knauth and Kennedy (2009); cryptospores similar to those in the Bright Angel have also been reported from the Pioche Formation of Nevada and from the Kaili Formation in China (Yin et al., 2012). Possible fossil lichens and slime molds in Cambrian paleosols were found in Australia and are reported in Retallack (2011); this report has inspired some debate (Jago et al., 2012; Retallack, 2012).

25. The nearest deep water to the Grand Canyon seems to have been in what is now northern Baja California; see Blakey and Ranney (2008) for nice paleogeographic maps of Cambrian and other times in the Four Corners region.

26. This interlacing of the Muav and Bright Angel is illustrated in a diagram by McKee and Resser (1945); it is also discussed and illustrated by Huntoon (1989), in an article subtitled "Mapper's Nightmare"!

27. Noble (1922).

28. For the geology of the Muav, see Middleton and Elliott (2003) and also Wood (1966).

29. Data on modern carbonate shelf environments are in Enos (1983) and Wilson and Jordan (1983).

30. I made IFPCs up. There's no official acronym that I'm aware of; I'm just saving myself the typing.

31. Elston (1989a).

32. Transgressions of the Paleozoic caused more extensive flooding of continental areas; since that time few continents have approached being totally flooded.

33. See Sloss (1963) for the identification and naming of North America's major transgressive sequences.

34. Walther's observation, as he stated it, was this: "The various deposits of the same facies areas and similarly the sum of the rocks of different facies areas are formed beside each other in space, though in cross-section we see them lying on top of each other [and] . . . only those facies and facies areas can be superimposed primarily which can be observed beside each other at the present time." Quoted in Middleton (1973).

35. As mentioned briefly earlier in this chapter, a fourth formation has recently been added to the Tonto Group: the Frenchman Mountain Dolomite, a unit that includes beds originally designated "undifferentiated dolomites" in the Grand Canyon. See Rowland and Korolev (2011).

36. Palmer (1960).

37. E. Rose, pers. comm. (2011).

38. For more on the new interpretation of the Bright Angel Formation see Rose (2006, 2011); Baldwin et al. (2004); and Gallagher (2003). For more on cryptospores of possible Cambrian land plants (mosses or moss-grade plants) in the Bright Angel, see Strother et al. (2004).

3. A Long Strange Trip

1. See Horodyski (1977), Walter et al. (1992), Knoll and Semikhatov (1998), and Grotzinger and Knoll (1999) on Precambrian stromatolites.

2. 1.1 billion is the same as 1100 million, so this is a little more than twice as long ago as the Cambrian.

3. The Precambrian is now a general term for everything before the Cambrian and is divided into the Hadean, Archean, and Proterozoic. The Hadean is technically an informal term, but the Archean and Proterozoic are formal eons. For the rest of this book, I'll use those three if I'm being specific. If I use Precambrian, it refers to the whole stretch of all three or to an unspecified time that may fall anywhere within one of the three.

4. Information on Precambrian atmospheric conditions, UV intensity, and surface water is available in Holland (1984), Kasting (1993), and Mojzsis et al. (2001).

5. Harrison et al. (2005).

6. Bowring and Housh (1995).

7. For comparison, modern oceans are about pH 8.3; fish die when levels reach about pH 3–4. Battery acid is a lot worse than the Archean ocean at pH 1 (the pH scale is logarithmic so there is a tenfold increase or decrease for each whole number change).

8. Schopf and Packer (1987); Schopf (1993); Schopf et al. (2002).

9. Brasier et al. (2002).

10. Shen et al. (2001).

11. Wacey et al. (2011).

12. Rasmussen (2000).

13. A. Knoll, pers. comm. (2011).

14. Des Marais (2000).

15. Lowe (1983).

16. Rye and Holland (2000).

17. Holland (1984); Fedonkin et al. (2007).

18. Even as oxygen built up in the surface ocean, the deep ocean may have been not just low in oxygen but possibly sulfidic (Anbar and Knoll, 2002).

19. Awramik and Barghoorn (1977).

20. One estimate is that by 2.2 billion years ago oxygen levels were already up to ~3%; cited in Knoll (2003a).

21. See Catling et al. (2001) on hydrogen escape, and Canfield (1998) on sedimentary burial of organic matter.

22. Oxygen levels have varied through time since the Precambrian but have never again been as low as the < 1% they started out as.

23. For general information on Precambrian Earth history see Holland (1984); Cloud (1988); Knoll (2003a), and Fedonkin et al. (2007).

24. White (1979).

25. For geology and paleontology summaries of Glacier National Park, see Rezak (1957), Ross (1959), Ross and Rezak (1959), Horodyski (1976, 1977, 1983), and Hunt (2006).

26. For more about the late Precambrian supercontinents see: Condie (2001), Fedonkin et al. (2007), and Scotese (2009).

27. Sagan (1967); Margulis (1981).

28. Douglas et al. (2001).

29. See also Nursall (1959), and Derry (2006).

30. Brocks et al. (1999).

31. Think about that time span though! After traveling forward through nearly 2 billion years of Earth history we now hit a "short" but critical interval of eukaryotic evolution spanning just the late Archean into the early to middle Proterozoic (ending still nearly a billion years before the Cambrian would eventually begin), and this window we are talking about spans a length of time that is as much as separates *us* from the late Proterozoic!

32. See Knoll et al. (1978) on Lake Superior fossils, and Javaux et al. (2001) on the early eukaryote diversification.

33. See Butterfield (2000) and Knoll (2003a).

34. Ratcliff et al. (2012).

35. Knoll et al. (1975).

36. For more on geology of Grand Canyon's Unkar Group, see Hendricks and Stevenson (2003).

37. For more on Chuar geology and paleontology see Horodyski and Bloeser (1983), Elston (1989b), Karlstrom et al. (2000), and Ford and Dehler (2003).

38. See Nagy and Porter (2005) for more on the Uinta Mountain Group.

39. Porter and Knoll (2000).

40. See Baldauf et al. (2000); Javaux (2007).

41. Not all autotrophs photosynthesize; there are chemosynthetic organisms. The difference is self-production of energy versus other production.

42. The ice ages were the time when so much of Utah and parts of Nevada

and Oregon were covered with lakes; the cool to downright frozen time of mammoths and mastodons and *Smilodon*, dire wolves, and North American camels. During the last glacial maximum some 18,000 years ago the Great Lakes region, the Northeast, most of Canada, and Scandinavia and the northern United Kingdom were all under ice, along with most of the Southern Ocean.

43. One could picture places like Hawaii and Iceland serving as such refuges in a Snowball Earth scenario superimposed on modern geography.

44. Strother et al. (1983).

45. For more on Snowball Earth, see Hoffman et al. (1998), Hyde et al. (2000), Knoll (2003a), Calver et al. (2004), Fanning and Link (2004), and Macdonald et al. (2010). See also Kennedy et al. (1998) and Hoffmann et al. (2004) for the view that there may have been more than two late Proterozoic glaciations and that some of these may not have been as globally extensive as originally proposed. This is the "slush ball" scenario in which the equatorial regions may not have been frozen solid but may have been seasonally free of ice.

46. Knoll et al. (2004, 2006).

47. See Knoll et al. (2004), Peterson and Butterfield (2005), and Peterson et al. (2008) on the Ediacaran emergence of animals.

48. The Ediacaran period is a time division; the Ediacaran system refers to the rocks of this age. Neither is to be confused with the informal group of organisms known as "Ediacarans" (which occur temporally during the Ediacaran period). When referring to the time or rocks instead of the animals I will make that clear.

49. J. W. Hagadorn, pers. comm. (2011). For reports and descriptions of possible Ediacaran sponges see Brasier et al. (1997), Li et al. (1998), and Fedonkin et al. (2007).

50. Love et al. (2009).

51. Antcliffe, Callow, and Brasier (2011).

52. Fedonkin and Waggoner (1997).

53. Dzik (2011).

54. Clites et al. (2012).

55. For information on rangeomorphs, see Narbonne (2004), LaFlamme et al. (2004), Antcliffe and Brasier (2008), and Bamforth et al. (2008).

56. Liu et al. (2011).

57. Glacier National Park's Precambrian sections are about 1.1 billion, too old by about 500 million; Grand

Canyon about 750 million years old, about 100 million years too old. Plenty of other areas in North America have Precambrian rocks but they are too old to have Ediacarans or are igneous or metamorphic.

58. See Hagadorn and Waggoner (2000) on Wood Canyon Ediacarans. Form of *Swartpuntia* is in Narbonne et al. (1997). *Cloudina* in Mexico is in Sour-Tovar et al. (2007).

59. The protozoan hypothesis is by Seilacher et al. (2003); subsequent studies have suggested that it is possible that one Ediacaran genus was a proto-zoan (Antcliffe, Gooday, and Brasier, 2011).

60. Xiao and Knoll (2000).

61. For more on the Ediacaran biota, see Runnegar and Fedonkin (1992), Waggoner and Collins (2004), Droser et al. (2006), and Fedonkin et al. (2007). See also Liu et al. (2010) on possible mobility in some Ediacarans.

62. Knoll and Carroll (1999).

63. See Erwin (2007) and Ohno (1996).

64. Otherwise (in low oxygen) animals would have needed to have had unexpectedly advanced circulatory systems (Runnegar, 1982); ironically, the models shown by Bruce Runnegar were being used in his paper to argue for just such advanced circulatory systems in what was then thought to be a very oxygen poor atmosphere in the late Precambrian; now, newer interpreta-tions of Proterozoic atmosphere and of *Dickinsonia* itself seem to have turned the calculations in his 1982 paper on their heads so that the numbers seem to indicate that the oxygen levels must have been nearly modern in order for a diffusion-based animal to have functioned.

65. Kennedy et al. (2006).

66. See Prave (2002) and Baldwin et al. (2004) on evidence for early ter-restrial plants.

67. We've seen several instances in this chapter of both life affecting the planet and the planet affecting life; here we see life (land plants) affecting Earth (clays and carbon burial and oxygen levels) affecting life (the growing complexity and diversity of multicellular organisms made possible by chain of events started by the first two).

68. The impact event that caused the Acraman crater in Australia (~580 mil-lion years ago) has also been suggested as having caused a bottleneck effect around this time (Grey et al., 2003). It is a significant crater (40–90 km/24–54

mi. diameter) but possibly less than half the size of the Chicxulub crater (65 million years ago), so its impact (so to speak) on the biota of the time may have been much less than the extinction at the end of the Cretaceous.

69. Chakraborty and Nei (1977); Araki and Tachida (1997).

70. For more on the rise of grazing and evolution of animals, see Stanley (1973). Types of grazing are outlined in Seilacher (1999); see also Butterfield (2007).

71. Brasier (2009).

72. We will return to the Precam-brian-Cambrian diversification of animals in chapter 9. For more on Edia-caran organisms, also see McMenamin (1998) and Fedonkin et al. (2007); on Precambrian life in general, Darwin, and the rise of animals, see Brasier (2009).

73. Landing (1994). The end of the Cambrian, by the way, is defined by the first appearance datum of the conodont chordate *Iapetognathus fluctivagus,* which is the datum for the beginning of the Ordovician system.

74. Valentine et al. (1999).

75. Gehling et al. (2001); and Jensen et al. (2000).

4. Welcome to the Boomtown

1. See Bengtson (1992), Dzik (1994), Skovsted and Peel (2007), and Skovsted et al. (2011) for more on small shelly fossils.

2. Kouchinsky et al. (1999).

3. See Mount and Signor (1985).

4. For more on the Archaeocyatha, see Hill (1964), Wood et al. (1992), Row-land (2001), and Debrenne et al. (2012).

5. There is disagreement about this last point. Some authors have suggested archaeocyathids did have algal symbi-onts (Rowland and Gangloff, 1988) but others believe this is unlikely (Surge et al., 1997).

6. We briefly mentioned this in chapter 2 with regard to dolomitization in carbonate environments.

7. Some of these cyanobacteria may have been photosynthetic in shallow water but chemosynthetic in deep water or total darkness.

8. Savarese et al. (1993); Fuller and Jenkins (2007).

9. More on Lower Cambrian corals from California and Nevada is available in Tynan (1983) and Hicks (2006).

10. For more on Cambrian reefs, see Rowland (1984), Rowland and Gangloff (1988), Wood (1993, 1995, 1998), Kruse et al. (1995), Riding and Zhuravlev

(1995), Pratt et al. (2001), and Yuan et al. (2002).

11. Negative carbon isotope excursions involve a massive input of ^{13}C-depleted carbon to the ocean or at-mosphere and are associated at various times in Earth history with significant warming episodes.

12. See Corsetti and Hagadorn (2000, 2003) for more on the Precam-brian–Cambrian boundary in the Death Valley area.

13. I refer to trilobites as the "dino-saurs of the Cambrian" only because they are big, abundant, and famous. Big and abundant as fossils in their respective formations, compared with other fossil forms from the Cambrian, and about the only fossils recognized by members of the public from most Cambrian formations, except perhaps the Burgess Shale.

14. Hollingsworth (2011a).

15. For more on these formations see Stewart (1970), Moore (1976), Ahn et al. (2011), and Hollingsworth (2011a, 2011b, 2011c).

16. For more on these earliest occur-rence of trilobites in Laurentia, see Hol-lingsworth (2005, 2006, 2011a, 2011b, 2011c).

17. Lieberman and Karim (2010).

18. See chapter 7 for more on helicoplacoids; data on Indian Springs is in Hollingsworth and Babcock (2011); stratigraphy of the *Fallotaspis* and *Nev-adella* zone formations in the Mackenzie Mountains is in Fritz (1976).

19. See McMenamin (1987) on Early Cambrian trilobites of Mexico.

20. By comparison, dinosaurs, the rock stars of the Mesozoic, are known from only about 1200 species (so far) and were around for a "mere" 160 mil-lion years.

21. St. John (2007).

22. See Kihm and St. John (2007) for more on J. Walch and a translation of his trilobite chapter.

23. Some of us still dream that a liv-ing trilobite might someday be dredged up from the deep ocean, like a modern-day coelacanth story!

24. The hard exoskeletons of trilobites lead to their relative over-representation in the fossil record. There is a similar effect in the Mesozoic with dinosaurs. Dinosaurs are big (many of them) and so are their skeletons. Because of this they fossilize easily, so they, too, probably are overrepresented to some degree. In the Late Jurassic, for example, there are as many species of mammal fossils known as those of

dinosaurs. Given their small size, there were probably more individual mammals at any given time, but you wouldn't know this from most books or news stories (or from just walking the outcrop and seeing bone fragments) . . . the big and flashy species get the ink!

25. See Fortey and Whittington (1989).

26. Details of trilobite anatomy can be found in Whittington et al. (1997) and Harrington et al. (1959); color patterns in trilobites, from Schoenemann and Clarkson (2012) and Harrington et al. (1959).

27. Whittington (2007).

28. For more on the Olenelloidea, see Whittington (1989).

29. Palmer and Gatehouse (1972).

30. For more on the Olenellina, see Whittington et al. (1997).

31. For more on agnostids and eodiscoids, see Whittington et al. (1997).

32. For more on corynexochids, see Harrington et al. (1959).

33. For more on ptychopariids, see Harrington et al. (1959).

34. For information on general trilobite systematics see Whittington et al. (1997), Fortey (1990), Jell (2003), and Gon (2009).

35. Dorsal circulation and ventral nervous systems are so much more common, indeed, that there is some thought, going back as far as the 1820s, that at some point very early in the vertebrate/chordate/deuterostome story our ancestors branched off from an arthropod- or wormlike ancestor and became inverted! Up was down and down, up. Are vertebrates "upside down" compared to most other animals? See Dawkins (2004) for more on this bizarre story of evolution.

36. For more on trilobite eye structure see Whittington et al. (1997) and Schoenemann (2007).

37. There was an extra suture that often separated both the free check and cranidium from the eyes; this suture was lost, preserving the eyes on the free cheek after the Ordovician; Olenellids frequently are just crushed. So eyes are generally poorly known except in some juveniles of the Cambrian, which suggests that the loss of the suture was a retention of a juvenile character. The sequence appears to have been: juveniles with one suture → adults with two sutures → Ordovician adults with one suture.

38. Horváth et al. (1997) and Schoenemann (2007).

39. Peng et al. (2008); see Schoenemann et al. (2010) on the eyes of a planktonic trilobite.

40. Chatterton et al. (1994).

41. The possible new view of trilobite respiration is in Suzuki and Bergström (2008).

42. For more on trilobite anatomy in general see: Levi-Setti (1993), Whittington et al. (1997), Fortey (2000), and Gon (2009); reproductive evidence of brood pouches is in Fortey and Hughes (1998), and of mating clusters in Karim and Westrop (2002), Paterson et al. (2008), and Gutiérrez-Marco et al. (2009); general arthropod anatomy is in Brusca and Brusca (2003).

43. Trilobite walking was described in Whittington et al. (1997).

44. Whittington (1989).

45. For more on mode of life in agnostids see Havlíček et al. (1993), Fatka et al. (2009), and Fatka and Szabad (2011).

46. See Fortey and Owens (1999) on trilobite feeding modes.

47. Including onychophorans, priapulans, nematodes, and tardigrades.

48. Gutiérrez-Marco et al. (2009).

49. See Whittington et al. (1997) and Hughes et al. (2006) on growth and stages; and Chatterton and Speyer (1989) on life history ecologies.

50. See Lochman-Balk and Wilson (1958) and Whittington et al. (1997).

51. Ohio has Cambrian trilobites, found in a core collected 830 m (2722 ft.) down in a drill hole.

52. For general trilobite ecology, see Hughes (2001).

53. Early work toward a zonation for North America was summarized by Howell et al. (1944).

54. Robison (1976).

55. See Budd (1997, 1998, 2001a), Conway Morris (1998), and Peel and Ineson (2009, 2011a, 2011b) on the Sirius Passet site and fauna; and Conway Morris and Peel (2008), Lagebro et al. (2009), Peel (2010a, 2010b), Daley and Peel (2010), Stein (2010), and Vinther et al. (2011) on new elements of the fauna.

56. Hazzard (1933, 1954); Hazzard and Mason (1936).

57. For more on the Wood Canyon Formation, see Fedo and Cooper (1990) and Bahde et al. (1997).

58. See Prave and Wright (1986) and Prave (1991) on the Zabriskie Quartzite.

59. For more on Latham geology see Gaines and Droser (2002) and Foster (2011a). Latham paleontology is in Mount (1973, 1974a, 1974b, 1976, 1980a) for general information;

Durham (1978) on eocrinoid; Webster et al. (2003) on trilobite stratigraphy; Foster (2011b) on trilobite taphonomy; Waggoner and Hagadorn (2004) on algae; Waggoner and Hagadorn (2005) on hyoliths and cnidarians; and Conway Morris and Peel (2010) on the palaeoscolecidan worm.

60. For more on Palaeoscolecidans, ecology, structure and classification, see Hou and Bergström (1994), Ivantsov and Wrona (2004), Lehnert and Kraft (2006), Han et al. (2007), Zhuravlev et al. (2011), and Huang et al. (2012).

61. More information on these olenellids is available in Walcott (1910), Resser (1928), Bell (1931), Harrington (1956), Fritz (1974), Whittington (1989), and Lieberman (1999a).

62. Briggs and Mount (1982).

63. Foster (2011c).

64. This estimation is based on the modern distance of about 200 miles and accounting for about 63% increase of the original east-west distance by tectonic extension of the Basin and Range province during the Cenozoic (based on data in Snow and Wernicke, 2000). Distances between the locations that are now modern fossil sites were approximately one-third less during the Cambrian. But the width of Laurentia's Cambrian shelves was still significant; 120 miles of relatively shallow water is an impressive amount of habitat for trilobites and other animals, especially considering that this is the width of the shelf band and ran *around* much of Laurentia at the time.

65. More on the geology of the Early Cambrian margin is in Nelson (1978), Stewart (1970), and Stevens and Greene (1999).

66. Perry (1871).

67. Shaw (1958).

68. Shaw (1955).

69. For the Kinzers, see Briggs (1978) on *Serracaris;* Rigby (1987) on sponges; Capdevila and Conway Morris (1999) on worms including *Kinzeria;* Skinner (2005) on Kinzer biota, taphonomy, and geology; Skovsted and Peel (2010) on small shelly fossils; and Babcock (2007) on *Olenellus* injury.

70. McKee and Resser (1945), plate 19, fig. 25.

71. For more on the Bright Angel Formation at Frenchman Mountain and its trilobites see: Pack and Gayle (1971), Palmer and Halley (1979), Webster (2003, 2011b), and Foster (2011d).

72. Fritz (1972, 1991); Randell et al. (2005).

73. See Wallin (1990) on beach dunes; Dalrymple et al. (1985) on inland dunes.

5. On Top of the World

1. In fact, these Pliocene sediments, similar in age to those that have yielded remains of our australopithecine ancestors in eastern Africa, have produced from outcrops south of Pioche fossil rodents, rabbits, dogs, cats, weasels, camels, horses, and proboscideans.

2. By relatively low, I mean 1921–2591 m (6300–8500 ft.) elevation; not much higher than the surrounding plains.

3. We do still see a few rare redlichioids later in the Cambrian in North America but the olenellids disappear at the Lower–Middle Cambrian boundary.

4. For more on latest Early Cambrian fossils at Ruin Wash see Palmer (1998b), Lieberman (2003a), Webster (2007a), Webster et al. (2008), and Hopkins and Webster (2009).

5. For more information on *Tuzoia* see Vannier, Caron, Yuan, Briggs, Collins, Zhao, and Zhu (2007) and Vannier et al. (2006); eyes possibly of *Tuzoia* in Lee et al. (2011).

6. Palmer (1998b).

7. In 2011, a meeting of the International Subcommission on Cambrian Stratigraphy suffered five flat tires on just 10 vehicles in a single day while visiting Split Mountain. I had the pleasure of having the last of those flat tires, but thanks to having witnessed and assisted in so many previous changes that day, Mark Webster and I were able to change this one in near-record time.

8. See Palmer and Halley (1979), Mount (1980b), and Fowler (1999) for more information on Carrara Formation sites and taxa.

9. Sundberg and McCollum (2003a).

10. Rowell (1980).

11. For more on Pioche Formation geology, see Merriam (1964), Sundberg and McCollum (2000), and Sundberg (2011a); on trilobite turnover at the boundary and early Middle Cambrian trilobites, see Eddy and McCollum (1998), Sundberg and McCollum (2000, 2003b), and Sundberg (2004, 2011a).

12. In some cases the animal is divided into a cephalothorax and abdomen.

13. For more information on *Canadaspis, Perspicaris,* phyllocarids, and Crustacea in general, see Walcott (1912), Robison and Richards (1981), Briggs et al. (1994), Lieberman (2003a), and Brusca and Brusca (2003).

14. Branchiopoda is not to be confused with the Brachiopoda. One additional letter, totally separate phyla. Isn't biology fun?

15. Hou (1999); Hou, Aldridge, Bergström, et al. (2004); Hou, Bergström, and Xu (2004).

16. For more on the Caborca area and its Cambrian rocks and fossils see Lochman (1948), Cooper et al. (1952), and Stewart et al. (2002).

17. *Mollisonia* also occurs in the Burgess Shale in British Columbia and in the Kaili fauna from China (Zhang et al., 2002), but it is not a well-known form from any of its occurrences.

18. See Robison and Babcock (2011) on new trilobite species from Spence Shale.

19. For more on Spence Shale geology, see Maxey (1958), Liddell et al. (1997), and Garson et al. (2011). For Spence Shale paleontology see Resser (1939) on general paleontology; Robison (1969), Willoughby and Robison (1979) on *Brooksella;* Rigby (1980) and Aase (1992) on sponges; Gunther and Gunther (1981) and Robison and Babcock (2011) on trilobites and others; Briggs and Robison (1984) and Briggs et al. (2008) on arthropods; Conway Morris and Robison (1988) on algae; Sumrall and Sprinkle (1999) on echinoderms; and mode of life of *Pagetia* in Lin and Yuan (2008) *Thoracocare* is in Robison and Campbell (1974); oryctocephalids such as those at Oneida Narrows are in Whittington (1995); Spence Shale depth approximation is based in part on data provided by D. Liddell (pers. comm., 2011).

20. Conway Morris and Robison (1988).

21. Hou, Aldridge, Bergström, Siveter, Siveter, and Feng, (2004); others have placed it with the priapulids.

22. Brusca and Brusca (2003).

23. Robison and Wiley (1995).

24. McKee and Resser (1945); trilobite list updated in Foster (2011e).

25. For more on the Bradoriida see Hou, Siveter, Williams, Walossek, and Bergström (1996), Hou, Aldridge, Bergström, Siveter, Siveter, and Feng (2004), and Duan et al. (2012).

26. For more on Bright Angel Formation geology and fauna in Grand Canyon, see McKee and Resser (1945) and Middleton and Elliott (2003); on paleoenvironments, see Baldwin et al. (2004), Gallagher (2003), and Rose (2006); on cryptospores, see Strother et al. (2004); on trilobites and other fauna, see Walcott (1889, 1898, 1916a, 1916b, 1924, 1925)–these references relate

to the Bright Angel in part but also include taxa from many other formations around North America and are good general references; on the McKee trilobite site, see Foster (2009, 2011e); the Peach Springs Canyon site is treated in Sundberg (2011b).

27. The name of these structures relates to more uncertainty about what paleontologists were finding early on. C. D. Walcott found these structures isolated as early as 1890 but did not recognize them at the time as relating to hyoliths, so he named them as a new species of animal, *Helenia,* after his daughter Helen. By 1911, Walcott saw that hyoliths had such structures, and it was in the 1970s that the structures were named helens in keeping with Walcott's original intentions–although it was recognized that the name now was for a hyolith structure and not an independent animal.

28. See Yochelson (1961a, 1961b), Babcock and Robison (1988), Malinky (1988), and Martí Mus and Bergström (2007) for more on hyoliths.

29. For information on the Cadiz Formation, see Mason (1935) on overall fauna, Foster (1994) on geology, Fuller (1980) on high Middle Cambrian locality, and Waggoner and Collins (1995) on hydrozoans.

30. Chisholm edrioasteroids are in Bell and Sprinkle (1978); for more on the *Glossopleura* zone of the Chisholm Formation, see Sundberg (2005).

31. The Rachel Limestone in the Groom Range was named by McCollum and McCollum (2011).

32. *Glossopleura* zone trilobites of Washington are in Hamilton et al. (2003); Montana occurrences in Schwimmer (1975).

33. Sundberg (1983); Droser (1991).

34. Pemberton and Frey (1982).

35. Getty and Hagadorn (2008).

36. For more on trace fossils in general, see Häntzschel (1962) and Pemberton et al. (2001); traces of the Bright Angel Formation are in Martin (1985).

37. Cambrian burrowing was deeper than it had been during the Ediacaran, but it was not as deep as it would get later on. For more on Cambrian burrowing and trace patterns, see Droser and Bottjer (1988), Droser et al. (1999), Droser et al. (2002), and Mángano (2011).

6. Magical Mystery Tour

1. We will only dedicate a chapter to the Burgess Shale here. But for more details, see Gould (1989), a classic that instantly influenced a small army of paleontologists, contemporary and

future; Briggs et al. (1994), a detailed and information-rich catalog of species with some of the most beautiful photos of these specimens ever taken; Conway Morris (1998), which includes a different perspective on some of the evolutionary significance of the Burgess Shale from that of Gould, and which also includes a fun, imaginary time-travel dive to the Burgess Shale's living animals; and Coppold and Powell (2006), an up-to-date and nicely illustrated general guide to the fossils and geology. Erwin and Valentine (2013) covers the Burgess Shale fauna, its evolutionary significance, and the Cambrian radiation in general with significant detail and data. Also see the Royal Ontario Museum's Burgess Shale fossil website.

2. The idea of the Cathedral Escarpment as a real, Cambrian bathymetric feature has not been without its controversy, but it seems to have now met a general level of acceptance. See Ludvigsen (1989), Fritz (1990), Aitken and McIlreath (1990), and Ludvigsen (1990) for some of the early debate.

3. See Fritz (1971) on the Burgess Shale as part of the Stephen Formation, Fletcher and Collins (1998, 2009) and Fletcher (2011) on Burgess Shale Formation naming and geology, Caron (2009) on burial of animals, Allison and Brett (1995) and Powell et al. (2003) on oxygenation, Johnston et al. (2009) on Burgess Shale brine seeps, Turnipseed et al. (2004) on Florida Escarpment modern brine seep, and Parsons-Hubbard et al. (2008) on modern brine seep taphonomic experiments.

4. See Rudkin (2009) on the Mount Stephen Trilobite Beds biota.

5. See Whittington (1980) on *Olenoides.*

6. Whiteaves (1892).

7. Paterson et al. (2011).

8. See my near-rant along these lines in chapter 1.

9. In addition to differences in the feeding appendages, the mouth, or "oral cone," of *Anomalocaris* seems to have been rather different from other anomalocaridid taxa (Daley and Bergström, 2012). This same study also indicated, incidentally, that *Laggania* now should be referred to as *Peytoia;* see also Daley et al. (2012). The study suggesting interspecific feeding strategy differences within the genus *Anomalocaris* is Daley, Paterson, Edgecombe, García-Bellido, and Jago (2013).

10. For more on *Anomalocaris,* see Whittington and Briggs (1985) and Collins (1996); on *Hurdia,* see Daley et al. (2009) and Daley et al. (2013);

on anomalocaridid diversity, see Daley (2010) and Daley and Budd (2009, 2010); on anomalocaridid feeding, see Nedin (1999) and Hagadorn (2009); and on swimming pattern, see Usami (2006).

11. In fact, it is hard to visit a particularly productive Cambrian fossil locality in North America that Walcott *hasn't* left his boot prints on. Many of the localities in this book were visited early on by Walcott.

12. See Yochelson (1967, 2001) for biographical details on C. D. Walcott.

13. The crew was split up between Burgess Pass and Takakkaw Falls at this time, but Walcott's wife, Helena, was present at the pass on the day (or days) the first fossils were noted; see Yochelson (2001) on discovery stories.

14. Walcott's actual entry for Tuesday, August 31, 1909, reads, "Out with Helena and Stuart collecting fossils from the Stephen formation. We found a remarkable group of Phyllopod crustaceans. . . . [draws what appear to be *Marella, Naraoia,* and a crustacean] . . . Took a large number of fine specimens to camp."

15. Thus the improbability of the discovery happening while Walcott was leading the entire pack train along the traverse below the quarry.

16. See Yochelson (2001).

17. See Collins (2009) and Bruton (2011) for histories of excavations in the Burgess Shale.

18. See Hagadorn (2002a) and Fritz (1971).

19. See Rasetti (1951) on Burgess Shale trilobites; Briggs et al. (1994) includes photos of the trilobite species along with many others from the quarry.

20. Rigby and Collins (2004) has details on the diversity of sponges from the Burgess Shale.

21. Conway Morris and Collins (1996); Hou, Aldridge, Bergström, et al. (2004); Chen et al. (2007).

22. Briggs et al. (1994).

23. Brasier (2009).

24. In fact, many annelids can similarly regenerate lost appendages or posterior (and sometime anterior) body segments if one is lost. This regenerative capacity seems to vary by species, with the most resilient ones being able to regenerate even cranial structures.

25. More on Burgess Shale polychaetes appears in Walcott (1911a), Conway Morris (1979), and Briggs et al. (1994).

26. Hou, Aldridge, Bergström, et al. (2004).

27. See Conway Morris (1977a) for more on *Dinomischus.*

28. O'Brien and Caron (2012).

29. Szaniawski (2005) and Conway Morris (2009) are the most recent salvos in this debate.

30. In fact *Hallucigenia* was described in 1977 by Simon Conway Morris (Conway Morris, 1977b) and was popularly reported in Gould's book (Gould, 1989).

31. Indeed, the lobopodians have even been included in the Onychophora in some classifications; e.g., Robison (1985) and Briggs et al. (1994); a fossil tardigrade has been found in Cambrian rocks in Siberia (Budd, 2001b; Maas and Waloszek, 2001).

32. Trilobites, as a contrasting example, are hatched with only a few or none of their adult complement of segments and grow them during ontogeny.

33. Brusca and Brusca (2003).

34. *Hallucigenia hongmeia* in Steiner et al. (2012); undescribed Burgess Shale lobopods from A. R. C. Milner (pers. comm., 2012); microstructure of *Hallucigenia* dorsal spines and identification of isolated hallucigeniid spines in other formations are in Caron et al (2013).

35. See Liu, Steiner, Dunlop, et al. (2011) for original description of *Diania;* reanalysis summarized in Ma et al. (2011).

36. For more on Cambrian lobopodians, see Walcott (1911a), Whittington (1978), Robison (1985), Budd (2001c), Hou, Aldridge, Bergström, et al. (2004); Liu et al. (2006), Schoenemann et al. (2009), Liu et al. (2008), Liu, Steiner, Dunlop, et al. (2011), and Ou et al. (2011).

37. These *Scenella* fossils have also been interpreted as parts of a cnidarian species, but they are included as molluscs here, and other very similar Cambrian forms that clearly appear to be molluscs have been found recently as well (Tortello and Sabattini, 2011).

38. This plane of symmetry is modified in oysters; it also contrasts with the plane of symmetry in brachiopods, in which the plane runs through the center of the valves perpendicular to them so that in brachiopods the shells are left-right symmetrical but each valve may have a different convexity. Bivalve shells are generally mirror images of each other.

39. Sigwart and Sutton (2007).

40. Giribet et al. (2006).

41. More on *Scenella* is in Rasetti (1954); Cambrian molluscs in general in Kouchinsky (2000, 2001), Elicki (2009), and Chaffee and Lindberg (1986);

Cambrian bivalves in Fang and Sánchez (2012).

42. For more information on *Odontogriphus* see: Conway Morris (1976), Caron et al. (2006), and Smith (2012).

43. These body sclerites are what remind me of the almond slices of a bear claw – perhaps I am just particularly hungry right now. Should have had a larger lunch.

44. Details on *Wiwaxia* available in Conway Morris (1985), Eibye-Jacobsen (2004), Smith (2012), and Fatka et al. (2011).

45. Conway Morris and Caron (2007) named and described *Orthrozanclus* and the Halwaxiida; Sigwart and Sutton (2007) supported the position of these forms relative to molluscs; Vinther and Nielsen (2005) and Vinther (2009) argued halkieriids were crown molluscs; Butterfield (2008), meanwhile, remained unconvinced that the radula of some of these halwaxiid forms was really a radula and questioned the molluscan affinities of the genera.

46. Whittington (1975).

47. The main monograph on the morphology of *Opabinia* is Whittington (1975).

48. A few of the papers debating the morphology of *Opabinia* and the significance of different interpretations for arthropod evolution are: Budd (1996), Zhang and Briggs (2007), and Budd and Daley (2012).

49. Zhang and Briggs (2007).

50. Budd (1996) and Budd and Daley (2012).

51. In fairness to the audience at that Palaeontological Association meeting, *Opabinia* is admittedly a strange-looking animal by our familiarity standards, and the scientific world had not seen anything like it previous to that time.

52. Hendricks et al. (2008).

53. For more on Cambrian non-trilobite arthropods in general see Budd (2001b).

54. See Walcott (1912) and García-Bellido and Collins (2004, 2006) for more on *Marrella;* occurrence of *Marrella* in China is in Liu (2013); the study of the possible color pattern of *Marrella* is Parker (1998).

55. Original reference, Walcott (1912); telson function in Lin (2009); redescription of *Burgessia* is in Hughes (1975).

56. The description of *Sanctacaris* is in Briggs and Collins (1988).

57. See Walcott (1912) and Bruton and Whittington (1983) for more on *Emeraldella.*

58. See Walcott (1911b), Bruton (1981), and Stein (2013) for more on *Sidneyia.*

59. For more on *Leanchoilia,* see Walcott (1912), Bruton and Whittington (1983), Butterfield (2002), and García-Bellido and Collins (2007).

60. See Walcott (1912) and Whittington (1974) for earlier descriptions of *Yohoia.* A recent study with a morphological analysis of *Yohoia's* feeding appendages is Haug et al. (2012).

61. More on *Alalcomenaeus* in Briggs and Collins (1999) and Tanaka et al. (2013).

62. Briggs et al. (1994).

63. Haug et al. (2011).

64. Strausfeld (2011) pointed out that the eyes of *Waptia,* while extending laterally from the head on short bases, are not strictly speaking on stalks.

65. Until recently little detailed work had been done on *Waptia.* See Briggs et al. (1994) for a summary, Strausfeld (2011) for detailed study of sensory structures, and Walcott (1912) for the original description.

66. See Briggs (1977) on *Perspicaris.*

67. For more on *Isoxys* see Walcott (1908) and García-Bellido et al. (2009).

68. The description of *Nereocaris* and its significance is in Legg et al. (2012).

69. See Collins and Rudkin (1981) for original description and Briggs, Sutton, Siveter, and Siveter (2005) for a summary of the oldest recorded barnacle fossils.

70. Recent analyses suggest that the pycnogonids may belong outside the chelicerates as a sister group to other euarthropods (Legg et al., 2011).

71. For more on classification of Cambrian and recent arthropods, see Wills et al. (1998), Bergström and Hou (2003), Waloszek et al. (2005), Scholtz and Edgecombe (2005, 2006), Hendricks and Lieberman (2008), Edgecombe (2010), and Budd and Legg (2011).

72. The second part of the species name honors paleontologist Desmond Collins.

73. Description of *Herpetogaster* is in Caron, Conway Morris, and Shu (2010); for more on echinoderms and *Eldonia,* see chapter 7.

74. Caron (2006).

75. See Sprinkle and Collins (1998, 2011) for the case of *Echmatocrinus* being a crinoid, and Ausich and Babcock (1998) for the octocoral case.

76. Ruedemann (1931).

77. See Ruedemann (1931) on *Chaunograptus* and Cambrian graptolites,

and Durman and Sennikov (1993) and Maletz et al. (2005) for more on Cambrian pterobranchs; both are covered by Bulman (1970).

78. This is in addition to the vertebrate conodonts; see chapter 8.

79. Chen et al. (2003).

80. More on chordates and early vertebrates is in Donoghue and Purnell (2009).

81. Chen et al. (1999).

82. Shu et al. (2003).

83. Shu et al. (1999).

84. The original description of *Pikaia* is by Walcott (1911a). The recent description and confirmation of its chordate affinities is Conway Morris and Caron (2012). See also Briggs et al. (1994) for a short summary of the genus.

85. Conway Morris (2008).

86. Sansom et al. (2010, 2013).

87. For more on the Stanley Glacier site, see Gaines (2011) and Caron, Gaines, Mángano, Streng, M., and Daley (2010); *Kootenichela deppi* is in Legg (2013).

7. Glory Days

1. See Robison (1964a), Rees (1986), Hintze and Robison (1975), Howley et al. (2006), and Brett et al. (2009) on the geology of the Wheeler and the House Range.

2. See Robison (1964b), Gunther and Gunther (1981), and Robison and Babcock (2011) on Wheeler Formation trilobites.

3. This *Elrathia,* with a healed, arcing bite mark out of it, is at the University of Kansas.

4. See Bright (1959) and Gaines and Droser (2003) on ecology of *Elrathia kingii.*

5. For more on paleontology of the Wheeler Formation, see Robison (1964b), Gunther and Gunther (1981), Ubaghs and Robison (1988); also Conway Morris and Robison (1988) and Briggs et al. (2008) on soft-bodied forms from the Wheeler Formation.

6. See Briggs et al. (2008) and Hendricks and Lieberman (2008) on *Dicranocaris.*

7. The description of *Cambropodus* is by Robison (1990).

8. More on *Pseudoarctolepis* is in Brooks and Caster (1956) and Robison and Richards (1981).

9. For more on Cambrian eocrinoids, see Robison (1965), Ubaghs (1967), Spinkle (1973), Sprinkle and Collins (2006), Zamora et al. (2009), Parsley (2012), and Parsley and Zhao

(2012); more on modern echinoderms in Brusca and Brusca (2003).

10. For more on carpoids, see Ubaghs and Robison (1985, 1988).

11. For information on helicoplacoids, see Dornbos and Bottjer (2000).

12. Odds are it would probably be like a bacterium, because such microbial life forms simply seem more likely in a wider variety of planetary environments.

13. I'm not likely the first or only person to feel this way about sea cucumbers – they are strange animals. Incidentally, paleontologist Simon Conway Morris might argue that the chances of an alien being similar to a vertebrate or arthropod are actually better than we might think. In fact, part of his argument in *Life's Solution* (2003) is approached from this angle and takes the pervasiveness of convergence beyond Earth's biosphere. I am not disputing this here (in fact, I find much of his data compelling) but am simply stating that I'd be willing to bet a few pints with fellow paleontologists that the first multicellular alien life we encounter would most likely be that which is abundant on its home planet, and that those forms would probably be small and potentially like echinoderms. Tetrapod-like forms and *Men in Black*–style arthropods and cephalopods I see as obviously less likely, at least due simply to percentage likelihood of encounter. This is assuming for the sake of argument, of course, that we find our way deep into the galaxy – and not that the aliens land on our doorstep.   I'd bet heavily against that. Far more likely, I believe we'd find something on Titan or exobacteria in Mars's soil – or possibly fossils in Martian rocks. Conway Morris also argues that while the chances of something familiar in space might be more than we think, assuming it's there in the first place, the likelihood of encountering much of anything at all may in fact be less.   My whole discussion here probably gives me away as being more of an exobiology optimist.

14. For more on *Eldonia* and modern holothuroids, see Walcott (1911c), Durham (1974), Briggs et al. (1994), Kerr and Kim (1999), and Brusca and Brusca (2003); new interpretations of *Eldonia* classification in Chen et al. (1995) and ecology in Dzik et al. (1997).

15. For additional information on Cambrian echinoderms in general, see Sprinkle (1976), Smith (1988), Guensburg and Sprinkle (2001), Shu et al. (2002), and Dornbos (2006); in addition to China, odd Cambrian stem-echinoderms have been found in France and Spain recently too (Zamora et al., 2012).

16. More on *Chancelloria* and other forms is in Briggs et al. (1994) and Janussen et al. (2002).

17. For more on *Wiwaxia,* see Conway Morris (1985).

18. See Hintze and Robison (1975), Elrick and Snider (2002), and Brett et al. (2009) for more on Marjum Formation geology.

19. Briggs et al. (2008).

20. For more on Marjum Formation paleontology, see Robison (1964b), Gunther and Gunther (1981), Robison (1984), Briggs and Robison (1984), Conway Morris and Robison (1988), Briggs et al. (2008), and Hendricks and Lieberman (2008).

21. See Maas et al. (2007), Dong et al. (2004), and Zhang, Pratt, and Shen (2011) for embryos.

22. Dornbos and Chen (2008).

23. Briggs et al. (1994); original reference on *Ottoia* is Walcott (1911a); more on *Ottoia* and priapulids appears in Conway Morris and Robison (1986) and Huang et al. (2004).

24. For more on Marjum Formation sponges, see Rigby (1983) and Rigby et al. (1997); on Cambrian sponges in general, see Rigby (1976).

25. Carrera and Botting (2008).

26. See Brusca and Brusca (2003) on sponge biology.

27. Debrenne and Reitner (2001).

28. Briggs et al. (2008); Hendricks and Lieberman (2008).

29. Hintze (1988).

30. *Skeemella* was described by Briggs, Lieberman, Halgedahl, and Jarrard (2005); Vetulicolia named by Shu et al. (2001); for more on vetulicolians, see Aldridge et al. (2007) and Shu (2005).

31. For Conasauga Formation paleontology, see Schwimmer (1989), Ciampaglio et al. (2006), Schwimmer and Montante (2007), and Resser (1938).

8. Taking Off

1. Note on figure 4.19 that although the *Cedaria* zone rocks of the Weeks Formation align with the traditional Upper Cambrian and are thus included in this chapter on the Late Cambrian, under the recently adopted four-series system, the *Cedaria* zone is mostly in Series 3.

2. See Adrain et al. (2009) and Lerosey-Aubril et al. (2012a, 2012b) on Weeks Formation trilobites and gut preservation.

3. See Robison and Babcock (2011) on these new trilobites.

4. Sutton et al. (2000).

5. We discussed this symmetry briefly with regards to the difference with bivalves in chapter 6 in the section on molluscs.

6. For more on brachiopods, see Williams et al. (1965) and Ushatinskaya (2001) on anatomy, systematics, and general ecology; Bullivant (1968), Brusca and Brusca (2003), and Kuzmina and Malakhov (2007) on morphology; Yang et al. (2013) on genetic variability in modern *Lingula;* Balthasar (2007) on shell composition; Cohen (2000) on brachiopods as protostomes; Brusca and Brusca (2003) on brachiopods as deuterostomes; Zhang et al. (2010) on brachiopod epibionts; Zhang et al. (2011) on brachiopod attachment to a trilobite; Freeman and Miller (2011) on healed predation scar; and Skovsted et al. (2009) on stem group brachiopods.

7. See Lerosey-Aubril et al. (2013) on *Tremaglaspis* in the Weeks Formation; Hesselbo (1989) on Weeks *Beckwithia;* and Størmer et al. (1955), Briggs et al. (1979), and Van Roy (2006) on general aglaspidids; original description of *Beckwithia* is Resser (1931).

8. The geology of these formations (Orr and Notch Peak) is treated in Lohmann (1976) and Brady and Rowell (1976); paleontology is treated in McBride (1976) and Rowell and Brady (1976).

9. Minnesota and Wisconsin Cambrian geology is in Clayton (1989) and Runkel et al. (1998); paleontology see Nelson (1951), Bell et al. (1952), and Westrop et al. (2005); the unusual *Pemphigaspis* is in Palmer (1951); over splitting of *Dikelocephalus* is in Labandeira and Hughes (1994).

10. See Dott (1974).

11. See Collette and Hagadorn (2010) for details on these Wisconsin arthropods, and MacNaughton et al. (2002) for more on the terrestrial arthropod trackways.

12. See Hagadorn and Seilacher (2009) for more on the Cambrian hermit arthropod trails of Wisconsin.

13. Hagadorn et al. (2002).

14. Babcock and Robison (1988).

15. Han et al. (2011).

16. For more on Cambrian jellyfish fossils, see Cartwright et al. (2007) and Hou, Aldridge, Bergström, et al. (2004).

17. Sea anemones, of course, are soft and lack the shell, however.

18. For more on conulariids, see Moore and Harrington (1956), Hughes et al. (2000), Ivantsov and Fedonkin

(2002), Waggoner and Hagadorn (2005), and Van Iten (1992), and Van Iten et al. (2013).

19. For more on Cambrian corals, see Fuller and Jenkins (2007), Savarese et al. (1993), and Reich (2009); also see chapter 4.

20. The event started out originally as the Steptoean Positive Carbon Isotope Excursion, but SPCIE wasn't much of an acronym, so it became SPICE, with the slight modification to the title to accommodate it.

21. For more on the SPICE event see Saltzmann et al. (1998), Saltzman et al. (2000), and Auerbach (2004).

22. Palmer (1965, 2003).

23. Landing et al. (2011).

24. Ahlberg et al. (2009). Is it only a matter of time until someone identifies a SLICE event?

25. "Biomere" is less commonly used now (Hopkins, 2011), in part probably because the intervals identified now correspond more or less with the refined Cambrian stage boundaries.

26. Palmer (1965); Stitt (1971).

27. Lochman-Balk and Wilson (1958); Lochman-Balk (1970).

28. Sundberg (1996).

29. Westrop and Cuggy (1999); Stitt (1971).

30. McGhee et al. (2004).

31. Myrow et al. (2004).

32. Chronic (1988).

33. For more on Wyoming trilobites of the Upper Cambrian, see Miller (1936), Lochman and Hu (1960, 1961), Kurtz (1976), and Chronic (1988); Gallatin sponge is treated in Okulitch and Bell (1955).

34. See chapter 6 for the Cambrian chordate relatives of conodonts.

35. Szaniawski (2009).

36. Even Pander in the 1850s suspected that conodonts were vertebrates.

37. Nearly complete conodonts were also being found in the Silurian of Wisconsin as it turns out; by 1995 a dozen complete specimens were known.

38. Purnell (1995a).

39. For more on conodonts, see Aldridge et al. (1993), Aldridge and Purnell (1996), Donoghue et al. (1998), and Sweet and Donoghue (2001) on general anatomy and classification; Purnell (1995b), Donoghue and Purnell (1999a, 1999b), and Donoghue (2001) on teeth and occlusion; Bergström (1990) on paleobiogeography; Clark and Miller (1969) and Miller (1980) on evolution of Cambrian lines; and Koucky et al.

(1961) on Upper Cambrian forms from Wyoming.

40. Acritarchs are also found in the Upper Cambrian in the Nolichucky Formation of Tennessee.

41. Moczydlowska et al. (2011).

42. In fact, abandonment of the formal term "acritarch" and its replacement by "green microalgae" has been recommended recently (Moczydlowska, 2011a).

43. Gros Ventre Group and Snowy Range Formation acritarchs are treated in Pedder (2010); Knoll (1994), Albani et al. (2007), Moczydlowska (2011a, 2011b), and Moczydlowska et al. (2011) discuss the Cambrian acritarch pattern of diversity increase. For more on general Cambrian acritarchs, see Strother (2005), Olcott Marshall et al. (2009), Moczydlowska (2010), and Buick (2010).

44. More on the geology of the Cambrian of Colorado is in Bassett (1939), Bass and Northrop (1953), and Myrow et al. (2003).

45. Dotsero Formation paleontology is treated in Berg and Ross (1959), Mc-Menamin and Ryan (2002), and Myrow et al. (1999); graptolite identifications are in the geology paper by Bassett (1939).

46. For more on the Lodore Formation, see Herr and Picard (1981) and Herr et al. (1982).

47. The Black Hills are also famous for Devil's Tower in the northwestern part of the uplift close to the Bear Lodge Mountains.

48. For more on Deadwood Formation geology, see Darton (1909), Kulik (1965), LeFever (1996), Stitt (1998) and DeWitt et al. (1989).

49. For more on Deadwood Formation trilobites, see Stitt and Straatmann (1997), Stitt (1998), Perfetta et al. (1999), and Stitt and Perfetta (2000); brachiopods in Robson et al. (2003); trilobite traces in Callison (1970).

50. Harvey et al. (2011).

51. For more on other Late Cambrian trilobite occurrences, see Lochman (1936) for Missouri; Resser (1938) on the southeast; Frederickson (1941, 1948), Stitt (1977), and Waskiewicz et al. (2006) for Oklahoma; Lochman (1950) on Montana; Wilson (1951) on Pennsylvania; Rasetti (1961) on Virginia and Maryland; Taylor (1976) on Nevada; Hohensee and Stitt (1989) on Arkansas, where shelf trilobites got redeposited to the basinal plain; Chatterton and Ludvigsen (1998) on British Columbia; Landing et al. (2003) on New York and

Vermont; and Westrop and Adrain (2007) on Nevada, Utah, and Oklahoma.

52. For Cambrian cephalopods, see Chen and Teichert (1983), Crick (1990), Mutvei et al. (2007), and Landing and Kröger (2009).

53. For *Nectocaris* debates, see Conway Morris (1976), Smith and Caron (2010), and Mazurek and Zatón (2011).

9. Home by the Sea

1. Conway Morris (1986). Simon Conway Morris once described the fossil record metaphorically as having been documented by an amnesic scribe. "Like the worst of chroniclers," he writes, "the scribe responsible suffers from shocking lapses of memory and much prefers the broad brush of interpretation to finicky minutia" (Conway Morris, 1989).

2. Africa has a BST deposit but it is Ordovician in age; see also Hagadorn (2002b) and Conway Morris (1989).

3. We need at least a sufficient minimum sample size (~100 specimens) from each quarry to pick up something resembling comparable faunas; these are preliminary data we are using, so one or two of the quarry samples may be a little on the small side. We are also assuming that our relative abundance proportions are accurate; this may or may not be the case, but for now the data we have are all there is to work with. Also, we are only including metazoans; no algae or cyanobacteria.

4. See Olszewski (2010) for more on use of the Shannon index. The log base is most commonly the natural logarithm. Inaccuracy due to sample size differences becomes negligible with collections of ~100 or more specimens.

5. To borrow a term from planetary astronomy and astrobiologists. Traditionally the Goldilocks Zone or Habitable Zone relates to Earth-like distance from a star in another solar system (relative to the intensity of the star); not too close and not too far, with potentially life-friendly conditions. I am using this term loosely only in noting the apparently higher diversity in parts of the Cambrian continental shelf not out on the edges of deep water nor closest to shore, where the abiotic conditions may have been less harsh.

6. See Jablonski et al. (1983) and Sepkoski (1991).

7. A good resource on general taphonomy is Martin (1999).

8. Estimates of representation of shelled taxa are in Foote and Raup (1996); relative preservation of durable

versus fragile shells is shown in Behrensmeyer et al. (2005).

9. Moore and Lieberman (2009).

10. Thick deposits of multiple, stacked, fine-grained sediment flows are known as turbidites.

11. See Caron and Jackson (2006) and Caron (2009).

12. See Plotnick (1986), Briggs (1995), and Hammarlund et al. (2011) on experimental decay studies; Gaines, Hammarlund, Hou, et al. (2012) on seawater conditions; a recent study by Garson et al. (2011) suggests that bottom-water anoxia strongly increases the chances of Burgess Shale-type preservation but is not always necessary for it.

13. The similarity of figure 9.1 to fig. 1 in Conway Morris (1986) is by happy coincidence and not, consciously, by plagiarism. Although I had a copy of that paper for some time, I drew this figure based on the cited Gaines et al. references while the Conway Morris paper sat on a shelf awaiting use later in this chapter. When I pulled it down to start that section I took one look at Dr. Conway Morris's fig. 1 and thought, "Hmm."

14. Van Roy et al. (2010)

15. Gaines, Hammarlund, Hou, et al. (2012); Gaines, Droser, Orr, et al. (2012).

16. For more on BST preservation see Babcock (2001), Gabbott et al. (2004), Gabbott et al. (2008), Gabbott and Zalasiewicz (2009), Gaines et al. (2005), Butterfield et al. (2007), and Gaines et al. (2008).

17. Wheeler Formation soft-body preservation is in Gaines et al. (2001, 2005).

18. Gaines and Droser (2005).

19. Brett and Baird (1986).

20. For more on the basics of trilobite taphonomy and biostratinomy see Speyer and Brett (1986, 1988), Brett and Baird (1986), and Speyer (1987, 1991).

21. Foster (2011b).

22. See Webster et al. (2008).

23. Foster (2011d).

24. Pratt (1998).

25. Brett-Surman is quoted in Psihoyos (1994).

26. All animals were marine except for the few examples we've discussed briefly earlier that appear to have ventured into tidal areas, possible some freshwater settings, and a few that seem to have ventured temporarily onto land.

27. For anomalocaridid discussion above, see also Daley and Budd (2010). With this discussion: Biomass is often used as a metric in paleoecological studies, but in the case of these Cambrian animals, they run such a spectrum of phyla that the densities probably vary quite a bit, and thus the calculated biomasses would be unreliable. It is much easier to quantify the estimated volume of most Cambrian animals than it would be to estimate their densities as well.

28. Not per-unit energy; that's higher in small animals.

29. Zhao et al. (2009)

30. Burzin et al. (2001)

31. Much of this section is summarized from data in Conway Morris (1986), and also from Caron (2009) and Devereux (2001); Dunne et al. (2008) discusses food webs; section also based on analysis of Burgess Shale taxa from Smithsonian Institution collections and catalog.

32. This sample is from the Smithsonian Institution's National Museum of Natural History computer catalog.

33. This group of phyla includes such household-name taxa as the Gastrotrichia and the Acanthocephala. "Oh, those!" you say. . . . I don't know much of anything about them, either.

34. Valentine et al. (1999).

35. The Cambrian radiation was fast, even at 15 million years length. These 15 million years represented just 0.37% of Earth history up to that point.

36. Erwin (2011); Shu (2008).

37. Erwin et al. (2011).

38. Chen et al. (2004).

39. Some studies have suggested that the splits between most animal groups occurred much farther back in the Precambrian, as much as 1.2 billion years ago; most studies find that the divergence times were much later.

40. Lieberman (2003b); Peterson et al. (2005, 2008).

41. Sepkoski (1988).

42. These patterns of Ordovician diversity rise, slight fall, followed by dramatic Mesozoic-Cenozoic rise again now appear to be less dramatic than it seemed twenty years ago (Alroy et al., 2008), but the post-Cambrian generic diversification is still real.

43. Lieberman (1999b); Hendricks and Lieberman (2007).

44. Valentine (2000).

45. Boucot (1983).

46. See Peters and Gaines (2012) on calcium levels and the Cambrian radiation, and Elser et al. (2006) on phosphorus levels.

47. Zhuravlev (2001).

48. Lieberman (2003b) and Lieberman and Meert (2004); also Scotese (2009).

49. E.g., Sepkoski (1998).

50. Lieberman (2001, 2003b); Briggs and Fortey (2005).

51. Foote (1988).

52. Fortey et al. (1996).

53. Webster (2007b); Hughes (2007).

54. Valentine (1995).

55. Bengtson (2002).

56. Quotation is from Peterson et al. (2005).

57. For more on herbivory in marine settings see Vermeij and Lindberg (2000).

58. By macro-predators, of course, I mean multicellular predators visible to the naked eye, as opposed to truly microscopic, unicellular organisms. "Macro-predators" sounds a bit like cable-channel dinosaur-documentary hyperbole, so I wanted to clarify that.

59. Bambach et al. (2002).

60. Butterfield (2001) and Bengtson (2002).

61. Vermeij (1989); Knoll (2003b); Dzik (2005).

62. Butterfield (2001).

63. Vannier, Steiner, Renvoisé, Hu, and Casanova (2007).

64. This would of course be in the well-oxygenated shallow to mid-depths, above the possible anoxia of the deeper-water areas.

65. Dornbos et al. (2005); Dzik (2005); Buatois and Mángano (2012).

66. Valentine et al. (1999).

67. Plotnick et al. (2010).

68. For great discussions of eye evolution, see Dawkins (1996) and Conway Morris (2003).

69. Although we still haven't found evidence of eyes in *Pikaia*, some Cambrian vertebrates did have eyes.

70. For more on the ecological hypothesis see Valentine (1995) and Erwin (1994).

71. Butterfield (2007).

72. Valentine (1969, 1995).

73. Davidson and Erwin (2006).

74. For more on GRNs, see Davidson and Erwin (2006), Erwin (2007), Erwin and Davidson (2009), Erwin (2011), and Erwin and Valentine (2013).

75. Peterson et al. (2009).

76. Dawkins (2004, p. 445).

77. Gould (1989, 1991); Gould's idea of an early maximum in disparity followed by reduction was based in part on the anatomical work of H. B. Whittington and colleagues at a time when many of the Burgess Shale's "weird wonders" were thought to

be representatives, sometimes sole representatives, of extinct phyla; most researchers now place many of these species as stem taxa related to modern phyla.

78. Briggs et al. (1992) and Wills (1998).

79. Gould (1989). This mental exercise of "seeing what the world would be like without you" and its loose parallel to a Jimmy Stewart movie was part of the inspiration for Gould's book title *Wonderful Life*—the other, implied I believe, must have been the fossils of the Burgess Shale itself.

80. Travisano et al. (1995).

81. McIver and Stonedahl (1993); also summarized in Conway Morris (2003).

82. See Conway Morris (1998, 2003).

83. Bedford Falls was Jimmy Stewart's character George Bailey's hometown in the movie *It's A Wonderful Life*. On a rerun of the world without him, George finds his hometown turned into a very unfamiliar place called Pottersville, named after the town Scrooge.

84. See Sepkoski (1984) for a summary of the Cambrian, Paleozoic, and Modern evolutionary faunas. Also see Sepkoski (1997) and Peters (2004).

85. Zhuravlev (2001).

86. See Peters and Foote (2001) on the question of a Phanerozoic overall increase in diversity; Sepkoski (1988) on the Cambrian-Ordovician increase.

87. Of possible categories in a classification system of feeding, habitat,

and locomotion modes, Cambrian forms may have occupied only half as many as do modern forms. See Zhuravlev (2001).

88. Conway Morris (1989).

89. Boucot (1983).

90. Conway Morris (2006, p. 1078).

10. On and On

1. Indeed, there will be no birds for about another 338 million years.

2. Fine and Tchernov (2007); see Stanley (2003) on the history and origin of modern corals.

3. Rudkin et al. (2008).

4. As always, a great resource on all modern invertebrate groups is Brusca and Brusca (2003).

References

Aase, A. K. 1992. New occurrence of the Cambrian sponge *Hamptonia bowerbanki* in the Spence Shale, Utah strengthens ties to the Burgess Shale. *Geological Society of America Abstracts with Programs* 24(6):1.

Adrain, J. M., Peters, S. E., and Westrop, S. R. 2009. The Marjuman trilobite *Cedarina* Lochman: Thoracic morphology, systematics, and new species from western Utah and eastern Nevada, USA. *Zootaxa* 2218:35–58.

Ahlberg, P., Axheimer, N., Babcock, L. E., Eriksson, M. E., Schmitz, B., and Terfelt, F. 2009. Cambrian high-resolution biostratigraphy and carbon isotope chemostratigraphy in Scania, Sweden: First record of the SPICE and DICE excursions in Scandinavia. *Lethaia* 42:2–16.

Ahn, S. Y., Babcock, L. E., and Hollingsworth, J. S. 2011. Revised stratigraphic nomenclature for parts of the Ediacaran-Cambrian Series 2 succession in the southern Great Basin, USA. *Memoirs of the Association of Australasian Palaeontologists* 42:105–114.

Aitken, J. D., and McIlreath, I. A. 1990. Comment. *Geoscience Canada* 17:111–116.

Albani, R., Bagnoli, G., Ribecai, C., and Raevskaya, E. 2007. Late Cambrian acritarch *Lusatia:* Taxonomy, palaeogeography, and biostratigraphic implications. *Acta Palaeontologica Polonica* 52:809–818.

Aldridge, R. J., and Purnell, M. A. 1996. The conodont controversies. *Trends in Ecology and Evolution* 11:463–468.

Aldridge, R. J., Briggs, D. E. G., Smith, M. P., Clarkson, E. N. K., and Clark, N. D. L. 1993. The anatomy of conodonts. *Philosophical Transactions of the Royal Society of London B* 340:405–421.

Aldridge, R. J., Hou, X.-G., Siveter, David J., Siveter, Derek J., and Gabbott, S. E. 2007. The systematics and phylogenetic relationships of vetulicolians. *Palaeontology* 50:131–168.

Allison, P. A., and Brett, C. E. 1995. In situ benthos and paleo-oxygenation in the Middle Cambrian Burgess Shale, British Columbia, Canada. *Geology* 23:1079–1082.

Alroy, J., Aberhan, M., Bottjer, D. J., Foote, M., Fürsich, F. T., and 30 others. 2008. Phanerozoic trends in the global diversity of marine invertebrates. *Science* 321:97–100.

Anbar, A. D., and Knoll, A. H. 2002. Proterozoic ocean chemistry and evolution: A bioinorganic bridge? *Science* 297:1137–1142.

Antcliffe, J. B., and Brasier, M. D. 2008. *Charnia* at 50: Developmental models for Ediacaran fronds. *Palaeontology* 51:11–26.

Antcliffe, J. B., Callow, R. H. T., and Brasier, M. D. 2011. The origin of sponges: Examination of Precambrian metazoan diversifications. *The Palaeontological Association 55th Annual Meeting Programme and Abstracts, The Palaeontological Association Newsletter* 78:A15–A16.

Antcliffe, J. B., Gooday, A. J., and Brasier, M. D. 2011. Testing the protozoan hypothesis for Ediacaran fossils: A developmental analysis of *Palaeopascichnus. Palaeontology* 54:1157–1175.

Araki, H., and Tachida, H. 1997. Bottleneck effect on evolutionary rate in the nearly neutral mutation model. *Genetics* 147:907–914.

Auerbach, D. J. 2004. The Steptoean Positive Isotopic Carbon Excursion (SPICE) in siliciclastic facies of the upper Mississippi valley: Implications for mass extinction and sea level change in the Upper Cambrian. BS thesis, Carleton College, Northfield, Minnesota, 35 p.

Ausich, W. I., and Babcock, L. E. 1998. The phylogenetic position of *Echmatocrinus brachiatus,* a probable octocoral from the Burgess Shale. *Palaeontology* 41:193–202.

Awramik, S. M., and Barghoorn, E. S. 1977. The Gunflint microbiota. *Precambrian Research* 5:121–142.

Babcock, L. E. 2001. Interpretation of biological and environmental changes across the Neoproterozoic–Cambrian boundary: Developing tools for predicting the occurrence of Burgess Shale–type deposits. North American Paleontological Convention, Program and Abstracts, *PaleoBios* 21(supp. 2):27.

———. 2007. Role of malformations in elucidating trilobite paleobiology: A historical synthesis. *In* Mikulic, D. G., Landing, E., and Kluessendorf, J., eds., Fabulous Fossils–300 Years of Worldwide Research on Trilobites, *New York State Museum Bulletin* 507:3–19.

Babcock, L. E., and Robison, R. A. 1988. Taxonomy and paleobiology of some Middle Cambrian *Scenella* (Cnidaria) and hyolithids (Mollusca) from western North America. *The University of Kansas Paleontological Contributions* 121:1–22.

Babcock, L. E., Peng, S., Geyer, G., and Shergold, J. H. 2005. Changing perspectives on Cambrian chronostratigraphy and progress toward subdivision of the Cambrian System. *Geosciences Journal* 9:101–106.

Bahde, J., Barretta, C., Cederstrand, L., Flaugher, M., Heller, R., Irwin, M., Swartz, C., Traub, S., Cooper, J., and Fedo, C. 1997. Neoproterozoic–Lower Cambrian sequence stratigraphy, eastern Mojave Desert, California: Implications for base of the Sauk Sequence, craton-margin hinge zone, and evolution of the Cordilleran continental margin. *In* Girty, G. H., Hanson, R. E., and Cooper, J. D., eds., *Geology of the Western Cordillera: Perspectives from Undergraduate Research,* Pacific Section SEPM, Fullerton, CA, 82:1–20.

Baldauf, S. L., Roger, A. J., Wenk-Siefert, I., and Doolittle, W. F. 2000. A kingdom-level phylogeny of eukaryotes based on combined protein data. *Science* 290:972–976.

Baldwin, C. T., Strother, P. K., Beck, J. H., and Rose, E. 2004. Palaeoecology of the Bright Angel Shale in the eastern Grand Canyon, Arizona, USA, incorporating sedimentological, ichnological and palynological data. *Geological Society London Special Publications* 228:213–236.

Balthasar, U. 2007. An Early Cambrian organophosphatic brachiopod with calcitic granules. *Palaeontology* 50:1319–1325.

Bambach, R. K., Knoll, A. H., and Sepkoski, J. J. 2002. Anatomical and ecological constraints on Phanerozoic animal diversity in the marine realm. *Proceedings of the National Academy of Sciences* 99:6854–6859.

Bamforth, E. L., Narbonne, G. M., and Anderson, M. M. 2008. Growth and ecology of a multi-branched Ediacaran rangeomorph from the Mistaken Point assemblage, Newfoundland. *Journal of Paleontology* 82:763–777.

Bass, N. W., and Northrop, S. A. 1953. Dotsero and Manitou Formations, White River Plateau, Colorado, with special reference to Clinetop Algal Limestone Member of Dotsero Formation. *Bulletin of the American Association of Petroleum Geologists* 37:889–912.

Bassett, C. F. 1939. Paleozoic section in the vicinity of Dotsero, Colorado. *Bulletin of the Geological Society of America* 50:1851–1866.

Behrensmeyer, A. K., Fürsich, F. T., Gastaldo, R. A., Kidwell, S. M., Kosnik, M. A., Kowalewski, M., Plotnik, R. E., Rogers, R. R., and Alroy, J. 2005. Are the most durable shelly taxa also the most common in the marine fossil record? *Paleobiology* 31:607–623.

Bell, B. M., and Sprinkle, J. 1978. *Totiglobus,* an unusual new edrioasteroid from the Middle Cambrian of Nevada. *Journal of Paleontology* 52:243–266.

Bell, G. K. 1931. The disputed structures of the Mesonacidae and their significance. *American Museum Novitates* 475:1–23.

Bell, W. C., Feniak, O. W., and Kurtz, V. E. 1952. Trilobites of the Franconia Formation, southeast Minnesota. *Journal of Paleontology* 26:29–38.

Bengtson, S. 1992. Proterozoic and earliest Cambrian skeletal metazoans. *In* Schopf, J. W., and Klein, C., eds., *The Proterozoic Biosphere: A Multidisciplinary Study,* Cambridge University Press, New York, p. 397–411.

———. 2002. Origins and early evolution of predation. *The Paleontological Society Papers* 8:289–317.

Berg, R. R., and Ross, R. J. 1959. Trilobites from the Peerless and Manitou Formations, Colorado. *Journal of Paleontology* 33:106–119.

Bergström, J., and Hou, X.-G. 2003. Arthropod origins. *Bulletin of Geosciences* 78:323–334.

Bergström, S. M. 1990. Relations between conodont provincialism and the changing palaeogeography during the Early Palaeozoic. *In* McKerrow, W. S., and Scotese, C. R., eds., *Palaeozoic Palaeogeography and Biogeography,* Geological Society Memoir 12, Geological Society, London, p. 105–121.

Beus, S. S. 2003. Redwall Limestone and Surprise Canyon Formation. *In* Beus, S. S., and Morales, M., eds., *Grand Canyon Geology,* Oxford University Press, New York, p. 115–135.

Blakey, R. C. 2003. Supai Group and Hermit Formation. *In* Beus, S. S., and Morales, M., eds., *Grand Canyon Geology,* Oxford University Press, New York, p. 136–162.

Blakey, R. C., and Ranney, W. 2008. *Ancient Landscapes of the Colorado Plateau.* Grand Canyon Association, Grand Canyon, AZ, 156 p.

Boggs, S. 1987. *Principles of Sedimentology and Stratigraphy.* Merrill, Columbus, OH, 784 p.

Boucot, A. J. 1983. Does evolution take place in an ecological vacuum? II. *Journal of Paleontology* 57:1–30.

Bowring, S. A., and Housh, T. 1995. The Earth's early evolution. *Science* 269:1535–1540.

Brady, M. J., and Rowell, A. J. 1976. Upper Cambrian subtidal blanket carbonate of the miogeocline, eastern Great Basin. *Brigham Young University Geology Studies* 23:153–163.

Brasier, M. 2009. *Darwin's Lost World: The Hidden History of Animal Life.* Oxford University Press, New York, 304 p.

Brasier, M., Green, O., and Shields, G. 1997. Ediacaran sponge spicule clusters from southwestern Mongolia and the origins of the Cambrian fauna. *Geology* 25:303–306.

Brasier, M. D., Green, O. R., Jephcoat, A. P., Kleppe, A. K., Van Kranendonk, M. J., Lindsay, J. F., Steele, A., and Grassineau, N. V. 2002. Questioning the evidence for Earth's oldest fossils. *Nature* 416:76–81.

Brett, C. E., and Baird, G. C. 1986. Comparative taphonomy: A key to paleoenvironmental interpretation based on fossil preservation. *Palaios* 1:207–227.

Brett, C. E., Allison, P. A., DeSantis, M. K., Liddell, W. D., and Kramer, A. 2009. Sequence stratigraphy, cyclic facies, and lagerstätten in the Middle Cambrian Wheeler and Marjum

Formations, Great Basin, Utah. *Palaeogeography, Palaeoclimatology, Palaeoecology* 277:9–33.

Briggs, D. E. G. 1977. Bivalved arthropods from the Cambrian Burgess Shale of British Columbia. *Palaeontology* 20:595–621.

———. 1978. A new trilobite-like arthropod from the Lower Cambrian Kinzers Formation, Pennsylvania. *Journal of Paleontology* 52:132–140.

———. 1995. Experimental taphonomy. *Palaios* 10:539–550.

Briggs, D. E. G., and Collins, D. 1988. A Middle Cambrian chelicerate from Mount Stephen, British Columbia. *Palaeontology* 31:779–798.

———. 1999. The arthropod *Alalcomenaeus cambricus* Simonetta, from the Middle Cambrian Burgess Shale of British Columbia. *Palaeontology* 42:953–977.

Briggs, D. E. G., and Fortey, R. A. 2005. Wonderful strife: Systematics, stem groups, and the phylogenetic signal of the Cambrian radiation. *Paleobiology* 31(supp. 2):94–112.

Briggs, D. E. G., and Mount, J. D. 1982. The occurrence of the giant arthropod *Anomalocaris* in the Lower Cambrian of southern California, and the overall distribution of the genus. *Journal of Paleontology* 56:1112–1118.

Briggs, D. E. G., and Robison, R. A. 1984. Exceptionally preserved nontrilobite arthropods and *Anomalocaris* from the Middle Cambrian of Utah. *The University of Kansas Paleontological Contributions* 111:1–23.

Briggs, D. E. G., Bruton, D. L., and Whittington, H. B. 1979. Appendages of the arthropod *Aglaspis spinifer* (Upper Cambrian, Wisconsin) and their significance. *Palaeontology* 22:167–180.

Briggs, D. E. G., Erwin, D. H., and Collier, F. J. 1994. *The Fossils of the Burgess Shale.* Smithsonian Institution Press, Washington, DC, 238 p.

Briggs, D. E. G., Fortey, R. A., and Wills, M. A. 1992. Morphological disparity in the Cambrian. *Science* 256:1670–1673.

Briggs, D. E. G., Lieberman, B. S., Halgedahl, S. L., and Jarrard, R. D. 2005. A new metazoan from the Middle Cambrian of Utah and the nature of the Vetulicolia. *Palaeontology* 48:681–686.

Briggs, D. E. G., Sutton, M. D., Siveter, David J., and Siveter, Derek, J. 2005. Metamorphosis in a Silurian barnacle. *Proceedings of the Royal Society B* 272:2365–2369.

Briggs, D. E. G., Lieberman, B. S., Hendricks, J. R., Halgedahl, S. L., and Jarrard, R. D. 2008. Middle Cambrian arthropods from Utah. *Journal of Paleontology* 82:238–254.

Bright, R. C. 1959. A paleoecologic and biometric study of the Middle Cambrian trilobite *Elrathia kingii* (Meek). *Journal of Paleontology* 33:83–98.

Brocks, J. J., Logan, G. A., Buick, R., and Summons, R. E. 1999. Archean molecular fossils and the early rise of eukaryotes. *Science* 285:1033–1036.

Brooks, H. K., and Caster, K. E. 1956. *Pseudoarctolepis sharpi,* n. gen., n. sp. (Phyllocarida), from the Wheeler Shale (Middle Cambrian) of Utah. *Journal of Paleontology* 30:9–14.

Brusca, R. C., and Brusca, G. J. 2003. *Invertebrates* (2nd ed.). Sinauer Associates, Sunderland, MA, 936 p.

Bruton, D. L. 1981. The arthropod *Sidneyia inexpectans,* Middle Cambrian, Burgess Shale, British Columbia. *Philosophical Transactions of the Royal Society of London B* 295:619–653.

———. 2011. The Cambridge University–Geological Survey of Canada excavation of the Burgess Shale in 1967. *In* Johnston, P. A., and Johnston, K. J., eds., International Conference on the Cambrian Explosion, Proceedings, *Palaeontographica Canadiana* 31:9–17.

Bruton, D. L., and Whittington, H. B. 1983. *Emeraldella* and *Leanchoilia,* two arthropods from the Burgess Shale, Middle Cambrian, British Columbia. *Philosophical Transactions of the Royal Society of London B* 300:553–582.

Buatois, L. A., and Mángano, M. G. 2012. An Early Cambrian shallow-marine ichnofauna from the Puncoviscana Formation of northwest Argentina: The interplay between sophisticated feeding behaviors, matgrounds and sea-level changes. *Journal of Paleontology* 86:7–18.

Budd, G. E. 1996. The morphology of *Opabinia regalis* and the reconstruction of the arthropod stem-group. *Lethaia* 29:1–14.

———. 1997. Stem group arthropods from the Lower Cambrian Sirius Passet fauna of North Greenland. *In* Fortey, R. A., and Thomas, R. H., eds., *Arthropod Relationships,* Systematics Association Special Volume Series, Chapman and Hall, London, p. 125–138.

———. 1998. The morphology and phylogenetic significance of *Kerygmachela kierkegaardi* Budd (Buen Formation, Lower Cambrian, N Greenland). *Transactions of the Royal Society of Edinburgh: Earth Sciences* 89:249–290.

———. 2001a. A new stem-arachnate from the Sirius Passet (Lower Cambrian of North Greenland) and the basal euarthropod problem. *North American Paleontological Convention 2001, Program and Abstracts* 21(supp. 2):37.

———. 2001b. Tardigrades as "stem-group arthropods": The evidence from the Cambrian fauna. *Zoologischer Anzeiger* 240:265–279.

———. 2001c. Ecology of nontrilobite arthropods and lobopods in the Cambrian. *In* Zhuravlev, A. Yu., and Riding, R., eds., *The Ecology of the Cambrian Radiation,* Columbia University Press, New York, p. 404–427.

Budd, G. E., and Daley, A. C. 2012. The lobes and lobopods of *Opabinia regalis* from the middle Cambrian Burgess Shale. *Lethaia* 45:83–95.

Budd, G. E., and Legg, D. 2011. Up the spout? Climbing up the chelicerate stem-group. *The Palaeontological Association 55th Annual Meeting Programme and Abstracts, The Palaeontological Association Newsletter* 78:A17.

Buick, R. 2010. Ancient acritarchs. *Nature* 463:885–886.

Bullivant, J. S. 1968. The method of feeding of lophophorates (Bryozoa, Phoronida, Brachiopoda). *New Zealand Journal of Marine and Freshwater Research* 2:135–146.

Bulman, O. M. B. 1970. Graptolithina. *In* Teichert, C., and Moore, R. C., eds., *Treatise on Invertebrate Paleontology Part V,* Geological Society of America, University of Kansas Press, Lawrence, p. V1–V163.

Burzin, M. B., Debrenne, F., and Zhuravlev, A. Yu. 2001. Evolution of shallow-water level-bottom communities. *In* Zhuravlev, A. Yu., and Riding, R., eds., *The Ecology of the Cambrian Radiation,* Columbia University Press, New York, p. 217–237.

Butterfield, N. J. 2000. *Bangiomorpha pubescens* n. gen., n. sp.: Implications for the evolution of sex, multicellularity, and the Mesoproterozoic/Neoproterozoic radiation of eukaryotes. *Paleobiology* 26:386–404.

———. 2001. Ecology and evolution of Cambrian plankton. *In* Zhuravlev, A. Yu., and Riding, R., eds., *The Ecology of the Cambrian Radiation,* Columbia University Press, New York, p. 200–216.

———. 2002. *Leanchoilia* guts and the interpretation of three-dimensional structures in Burgess Shale–type fossils. *Paleobiology* 28:155–171.

———. 2007. Macroevolution and macroecology through deep time. *Palaeontology* 50:41–55.

———. 2008. An early Cambrian radula. *Journal of Paleontology* 82:543–554.

Butterfield, N. J., Balthasar, U., and Wilson, L. 2007. Fossil diagenesis in the Burgess Shale. *Palaeontology* 50:537–543.

Callison, G. 1970. Trace fossils of trilobites from the Deadwood Formation (Upper Cambrian) of western South Dakota. *Bulletin of the Southern California Academy of Sciences* 69:20–26.

Calver, C. R., Black, L. P., Everard, J. L., and Seymour, D. B. 2004. U-Pb zircon age constraints on late Neoproterozoic glaciation in Tasmania. *Geology* 32:893–896.

Campbell, N. A. 1993. *Biology.* Benjamin/Cummings, New York, 1190 p.

Canfield, D. E. 1998. A new model for Proterozoic ocean chemistry. *Nature* 396:450–453.

Capdevila, D. G.-B., and Conway Morris, S. 1999. New fossil worms from the Lower Cambrian of the Kinzers Formation, Pennsylvania, with some comments on Burgess Shale–type preservation. *Journal of Paleontology* 73:394–402.

Caron, J.-B. 2006. *Banffia constricta,* a putative vetulicolid from the Middle Cambrian Burgess Shale. *Transactions of the Royal Society of Edinburgh: Earth Sciences* 96:95–111.

———. 2009. The greater Phyllopod Bed community, historical variations and quantitative approaches. *In* Caron, J.-B., and Rudkin, D., eds., *A Burgess Shale Primer: History, Geology, and Research Highlights.* The Burgess Shale Consortium, Toronto, p. 71–89.

Caron, J.-B., and Jackson, D. A. 2006. Taphonomy of the greater Phyllopod Bed community, Burgess Shale. *Palaios* 21:451–465.

Caron, J.-B., Conway Morris, S., and Shu, D. 2010. Tentaculate fossils from the Cambrian of Canada (British Columbia) and China (Yunnan) interpreted as primitive deuterostomes. *PLoS ONE* 5(3):e9586, doi:10.1371/journal.pone.0009586.

Caron, J.-B., Smith, M. R., and Harvey, T. H. P. 2013. Beyond the Burgess Shale: Cambrian microfossils track the rise and fall of hallucigeniid lobopodians. *Proceedings of the Royal Society B* 280:20131613.

Caron, J.-B., Scheltema, A., Schander, C., and Rudkin, D. 2006. A soft-bodied mollusc with radula from the Middle Cambrian Burgess Shale. *Nature* 442:159–163.

Caron, J.-B., Gaines, R. R., Mángano, M. G., Streng, M., and Daley, A. C. 2010. A new Burgess Shale–type assemblage from the "thin" Stephen Formation of the southern Canadian Rockies. *Geology* 38:811–814.

Carrera, M. G., and Botting, J. P. 2008. Evolutionary history of Cambrian spiculate sponges: Implications for the Cambrian Evolutionary Fauna. *Palaios* 23:124–138.

Cartwright, P., Halgedahl, S. L., Hendricks, J. R., Jarrard, R. D., Marques, A. C., Collins, A. G., and Lieberman, B. S. 2007. Exceptionally preserved jellyfishes from the Middle Cambrian. *PLoS ONE* 2(10):e1121.

Catling, D. C., Zahnle, K. J., and McKay, C. P. 2001. Biogenic methane, hydrogen escape, and the irreversible oxidation of early Earth. *Science* 293:839–843.

Chaffee, C., and Lindberg, D. R. 1986. Larval biology of Early Cambrian molluscs: The implications of small body size. *Bulletin of Marine Science* 39:536–549.

Chakraborty, R., and Nei, M. 1977. Bottleneck effects on average heterozygosity and genetic distance with the stepwise mutation model. *Evolution* 31:347–356.

Chatterton, B. D. E., and Ludvigsen, R. 1998. Upper Steptoean (Upper Cambrian) trilobites from the McKay Group of southeastern British Columbia, Canada. *The Paleontological Society Memoir* 49:1.

Chatterton, B. D. E., and Speyer, S. E. 1989. Larval ecology, life history strategies, and patterns of extinction and survivorship among Ordovician trilobites. *Paleobiology* 15:118–132.

Chatterton, B. D. E., Johanson, Z., and Sutherland, G. 1994. Form of the trilobite digestive system: Alimentary structures in *Pterocephalia*. *Journal of Paleontology* 68:294–305.

Chen J.-Y., and Teichert, C. 1983. Cambrian cephalopods. *Geology* 11:647–650.

Chen, J.-Y., Huang, D.-Y., and Li, C.-W. 1999. An Early Cambrian craniate-like chordate. *Nature* 402:518–522.

Chen, J.-Y., Zhu, M.-Y., and Zhou, G.-Q. 1995. The Early Cambrian medusiform metazoan *Eldonia* from the Chengjiang Lagerstätte. *Acta Paleontologica Polonica* 40:213–244.

Chen, J.-Y., Huang, D.-Y., Peng, Q.-Q., Chi, H.-M., Wang, X.-Q., and Feng, M. 2003. The first tunicate from the Early Cambrian of South China. *Proceedings of the National Academy of Sciences* 100:8314–8318.

Chen, J.-Y., Bottjer, D. J., Oliveri, P., Dornbos, S. Q., Gao, F., Ruffins, S., Chi, H., Li, C.-W., and Davidson, E. H. 2004. Small bilaterian fossils from 40 to 55 million years before the Cambrian. *Science* 305:218–222.

Chen, J.-Y., Schopf, J. W., Bottjer, D. J., Zhang, C.-Y., Kudryavtsev, A. B., Tripathi, A. B., Wang, X.-Q., Yang, Y.-H., Gao, X., and Yang, Y. 2007. Raman spectra of a Lower Cambrian ctenophore embryo from southwestern Shaanxi, China. *Proceedings of the National Academy of Sciences* 104:6289–6292.

Chronic, L. M. 1988. *The interrelation of fauna and lithology across a Late Cambrian biomere boundary in Wyoming.* MS thesis, University of Wyoming, Laramie.

Ciampaglio, C. N., Babcock, L. E., Wellman, C. L., York, A. R., and Brunswick, H. K. 2006. Phylogenetic affinities and taphonomy of *Brooksella* from the Cambrian of Georgia and Alabama, USA. *Palaeoworld* 15:256–265.

Clark, D. L., and Miller, J. F. 1969. Early evolution of conodonts. *Geological Society of America Bulletin* 80:125–134.

Clayton, L. 1989. Geology of Juneau County, Wisconsin. *Wisconsin Geological and Natural History Survey Information Circular* 66, 16 p.

Clites, E. C., Droser, M. L., and Gehling, J. G. 2012. The advent of hard-part structural support among the Ediacara biota: Ediacaran harbinger of a Cambrian mode of body construction. *Geology* 40, doi:10.1130/G32828.1.

Cloud, P. 1988. *Oasis in Space: Earth History from the Beginning.* W. W. Norton, New York, 508 p.

Cohen, B. L. 2000. Monophyly of brachiopods and phoronids: reconciliation of molecular evidence with Linnaean classification (the subphylum Phoroniformea nov.). *Proceedings of the Royal Society of London B* 267:225–231.

Collette, J. H., and Hagadorn, J. W. 2010. Three-dimensionally preserved arthropods from Cambrian lagerstätten of Quebec and Wisconsin. *Journal of Paleontology* 84:646–667.

Collins, D. 1996. The "evolution" of *Anomalocaris* and its classification in the arthropod Class Dinocarida (nov.) and Order Radiodonta (nov.). *Journal of Paleontology* 70:280–293.

———. 2009. A brief history of field research on the Burgess Shale. *In* Caron, J.-B., and Rudkin, D., eds., *A Burgess Shale Primer: History, Geology, and Research Highlights.* The Burgess Shale Consortium, Toronto, p. 15–31.

Collins, D., and Rudkin, D. M. 1981. *Priscansermarinus barnetti,* a probable lepadomorph barnacle from the Middle Cambrian Burgess Shale of British Columbia. *Journal of Paleontology* 55:1006–1015.

Condie, K. C. 2001. Rodinia and continental growth. *Gondwana Research* 4:154–155.

Conway Morris, S. 1976. A new Cambrian lophophorate from the Burgess Shale of British Columbia. *Palaeontology* 19:199–222.

———. 1977a. A new entoproct-like organism from the Burgess Shale of British Columbia. *Palaeontology* 20:833–845.

———. 1977b. A new metazoan from the Cambrian Burgess Shale, British Columbia. *Palaeontology* 20:623–640.

———. 1979. Middle Cambrian polychaetes from the Burgess Shale of British Columbia. *Philosophical Transactions of the Royal Society of London B* 285:227–274.

———. 1985. The Middle Cambrian metazoan *Wiwaxia corrugata* (Matthew) from the Burgess Shale and *Ogygopsis* shale, British Columbia, Canada. *Philosophical Transactions of the Royal Society of London B* 307:507–582.

———. 1986. The community structure of the Middle Cambrian Phyllopod Bed (Burgess Shale). *Palaeontology* 29:423–467.

———. 1989. Burgess Shale faunas and the Cambrian explosion. *Science* 246:339–346.

———. 1993. The fossil record and the early evolution of the Metazoa. *Nature* 361:219–225.

———. 1998. *The Crucible of Creation: The Burgess Shale and the Rise of Animals.* Oxford University Press, New York, 242 p.

———. 2003. *Life's Solution: Inevitable Humans in a Lonely Universe.* Cambridge University Press, Cambridge, 464 p.

———. 2006. Darwin's dilemma: The realities of the Cambrian "explosion." *Philosophical Transactions of the Royal Society B* 361:1069–1083.

———. 2008. A redescription of a rare chordate, *Metaspriggina walcotti* Simonetta and Insom, from the Burgess Shale (Middle Cambrian), British Columbia, Canada. *Journal of Paleontology* 82:424–430.

———. 2009. The Burgess Shale animal *Oesia* is not a chaetognath: A reply to Szaniawski (2005). *Acta Palaeontologica Polonica* 54:175–179.

Conway Morris, S., and Caron, J.-B. 2007. Halwaxiids and the early evolution of the lophotrochozoans. *Science* 315:1255–1258.

———. 2012. *Pikaia gracilens* Walcott, a stem-group chordate from the Middle Cambrian of British Columbia. *Biological Reviews* 87:480–512.

Conway Morris, S., and Collins, D. H. 1996. Middle Cambrian ctenophores from the Stephen Formation, British Columbia, Canada. *Philosophical Transactions of the Royal Society B* 351:279–308.

Conway Morris, S., and Peel, J. S. 2008. The earliest annelids: Lower Cambrian polychaetes from the Sirius Passet Lagerstätte, Peary Land, North Greenland. *Acta Palaeontologica Polonica* 53:137–148.

———. 2010. New palaeoscolecidan worms from the Lower Cambrian: Sirius Passet, Latham Shale and Kinzers Shale. *Acta Palaeontologica Polonica* 55:141–156.

Conway Morris, S., and Robison, R. A. 1986. Middle Cambrian priapulids and other soft-bodied fossils from Utah and Spain. *The University of Kansas Paleontological Contributions* 117:1–22.

———. 1988. More soft-bodied animals and algae from the Middle Cambrian of Utah and British Columbia. *The University of Kansas Paleontological Contributions* 122:1–8.

Cooper, G. A., Arellano, A. R. V., Johnson, J. H., Okulitch, V. J., Stoyanow, A., and Lochman, C. 1952. Cambrian stratigraphy and paleontology near Caborca, northwestern Sonora, Mexico. *Smithsonian Miscellaneous Collections* 119(1):1–184.

Coppold, M., and Powell, W. 2006. *A Geoscience Guide to the Burgess Shale.* The Burgess Shale Geoscience Foundation, Field, BC, 75 p.

Corsetti, F. A., and Hagadorn, J. W. 2000. Precambrian–Cambrian transition: Death Valley, United States. *Geology* 28:299–302.

———. 2003. The Precambrian–Cambrian transition in the southern Great Basin, USA. *The Sedimentary Record* May 2003:4–8.

Cox, A., and Hart, R. B. 1986. *Plate Tectonics: How It Works.* Blackwell Scientific Publications, Palo Alto, CA, 392 p.

Crick, R. E. 1990. Cambro-Devonian biogeography of nautiloid cephalopods. *In* McKerrow, W. S., and Scotese, C. R., eds., *Palaeozoic Palaeogeography and Biogeography,* Geological Society Memoir 12, Geological Society, London, p. 147–161.

Daley, A. C. 2010. The morphology and evolutionary significance of the anomalocaridids. PhD diss., Uppsala Universitet, 40 p.

Daley, A. C., and Bergström, J. 2012. The oral cone of *Anomalocaris* is not a classic "peytoia." *Naturwissenschaften* 99:501–504.

Daley, A. C., and Budd, G. E. 2009. Anomalocaridid diversity in the Middle Cambrian Burgess Shale, Canada. *International Conference on the Cambrian Explosion Abstract Volume,* The Burgess Shale Consortium, Toronto, p. 28.

———. 2010. New anomalocaridid appendages from the Burgess Shale, Canada. *Palaeontology* 53:721–738.

Daley, A. C., and Peel, J. S. 2010. A possible anomalocaridid from the Cambrian Sirius Passet Lagerstätte, North Greenland. *Journal of Paleontology* 84:352–355.

Daley, A. C., Budd, G. E., and Caron, J.-B. 2013. Morphology and systematics of the anomalocaridid arthropod *Hurdia* from the Middle Cambrian of British Columbia and Utah. *Journal of Systematic Palaeontology,* dx.doi.org/10.1080/14772019.2012.732723.

Daley, A. C., Budd, G. E., Caron, J.-B., Edgecombe, G. D., and Collins, D. 2009. The Burgess Shale anomalocaridid *Hurdia* and its significance for early euarthropod evolution. *Science* 323:1597–1600.

Daley, A. C., Edgecombe, G. D., Bergström, J., Paterson, J., and García-Bellido, D. C. 2012. The morphology of *Anomalocaris* from the Burgess Shale and Emu Bay Shale. *In* Budil, P., and Fatka, O., eds., *The 5th Conference on Trilobites and Their Relatives Abstracts,* Czech Geological Survey and Charles University, Prague, p. 19.

Daley, A. C., Paterson, J. R., Edgecombe, G. D., García-Bellido, D. C., and Jago, J. B. 2013. New anatomical information on *Anomalocaris* from the Cambrian Emu Bay Shale of South Australia and a reassessment of its inferred predatory habits. *Palaeontology* 56:971–990.

Dalrymple, R. W., Narbonne, G. M., and Smith, L. 1985. Eolian action and the distribution of Cambrian shales in North America. *Geology* 13:607–610.

Darton, N. H. 1907. Discovery of Cambrian rocks in southeastern California. *Journal of Geology* 15:470–473.

———. 1909. Geology and water resources of the northern portion of the Black Hills and adjoining regions in South Dakota and Wyoming. *United States Geological Survey Professional Paper* 65, 105 p.

Darton, N. H., et al. 1915. Guidebook of the western United States, part C: The Santa Fe Route. *United States Geological Survey Bulletin* 613, p. 81–100.

Davidson, E. H., and Erwin, D. H. 2006. Gene regulatory networks and the evolution of animal body plans. *Science* 311:796–800.

Dawkins, R. 1996. *Climbing Mount Improbable.* W. W. Norton, New York, 340 p.

———. 2004. *The Ancestor's Tale.* Houghton Mifflin, Boston, 673 p.

Debrenne, F., and Reitner, J. 2001. Sponges, cnidarians, and ctenophores. *In* Zhuravlev, A. Yu., and Riding, R., eds., *The Ecology of the Cambrian Radiation,* Columbia University Press, New York, p. 301–325.

Debrenne, F., Zhuravlev, A. Yu., and Kruse, P. D. 2012. *Treatise Online, Number 38, Part E, Revised, Volume 4, Chapter 18: General Features of the Archaeocyatha.* KU Paleontological Institute, University of Kansas, Lawrence, 102 p.

Derry, L. A. 2006. Fungi, weathering, and the emergence of animals. *Science* 311:1386–1387.

Des Marais, D. J. 2000. When did photosynthesis emerge on Earth? *Science* 289:1703–1705.

Devereux, M. G. 2001. Palaeoecology of the Middle Cambrian Raymond Quarry fauna, Burgess Shale, British Columbia. MS thesis, Department of Earth Sciences (Geology), The University of Western Ontario, 391 p.

DeWitt, E., Redden, J. A., Buscher, D., and Wilson, A. B. 1989. Geologic map of the Black Hills area, South Dakota and Wyoming. *United States Geological Survey Miscellaneous Investigations Series Map* I-1910.

Dong, X.-P., Donoghue, P. C. J., Cheng, H., and Liu, J.-B. 2004. Fossil embryos from the Middle and Late Cambrian period of Hunan, south China. *Nature* 427:237–240.

Donoghue, P. C. J. 2001. Microstructural variation in conodont enamel is a functional adaptation. *Proceedings of the Royal Society of London B* 268:1691–1698.

Donoghue, P. C. J., and Purnell, M. A. 1999a. Mammal-like occlusion in conodonts. *Paleobiology* 25:58–74.

———. 1999b. Growth, function, and the conodont fossil record. *Geology* 27:251–254.

———. 2009. The evolutionary emergence of vertebrates from among their spineless relatives. *Evolution: Education and Outreach* 2, doi:10.1007/s12052-009-0134-3.

Donoghue, P. C. J., Purnell, M. A., and Aldridge, R. J. 1998. Conodont anatomy, chordate phylogeny and vertebrate classification. *Lethaia* 31:211–219.

Dornbos, S. Q. 2006. Evolutionary palaeoecology of early epifaunal echinoderms: Response to increasing bioturbation levels during the Cambrian radiation. *Palaeogeography, Palaeoclimatology, Palaeoecology* 237:225–239.

Dornbos, S. Q., and Bottjer, D. J. 2000. Evolutionary paleoecology of the earliest echinoderms: Helicoplacoids and the Cambrian substrate revolution. *Geology* 28:839–842.

Dornbos, S. Q., and Chen, J.-Y. 2008. Community ecology of the early Cambrian Maotianshan Shale biota: Ecological dominance of priapulid worms. *Palaeogeography, Palaeoclimatology, Palaeoecology* 258:200–212.

Dornbos, S. Q., Bottjer, D. J., and Chen, J.-Y. 2005. Paleoecology of benthic metazoans in the Early Cambrian Maotianshan Shale biota and the Middle Cambrian Burgess Shale biota: Evidence for the Cambrian substrate revolution. *Palaeogeography, Palaeoclimatology, Palaeoecology* 220:47–67.

Dott, R. H., Jr. 1974. Cambrian tropical storm waves in Wisconsin. *Geology* 2:243–246.

Dott, R. H., and Batten, R. L. 1981. *Evolution of the Earth*. McGraw-Hill, New York, 573 p.

Douglas, S., Zauner, S., Fraunholz, M., Beaton, M., Penny, S., Deng, L.-T., Wu, X., Reith, M., Cavalier-Smith, T., and Maier, U.-G. 2001. The highly reduced genome of an enslaved algal nucleus. *Nature* 410:1091–1096.

Droser, M. L. 1991. Ichnofabric of the Paleozoic *Skolithos* ichnofacies and the nature and distribution of *Skolithos* piperock. *Palaios* 6:316–325.

Droser, M. L., and Bottjer, D. J. 1988. Trends in depth and extent of bioturbation in Cambrian carbonate marine environments, western United States. *Geology* 16:233–236.

Droser, M. L., Gehling, J. G., and Jensen, S. 1999. When the worm turned: Concordance of Early Cambrian ichnofabric and trace-fossil record in siliciclastic rocks of South Australia. *Geology* 27:625–628.

———. 2006. Assemblage palaeoecology of the Ediacara biota: The unabridged edition? *Palaeogeography, Palaeoclimatology, Palaeoecology* 232:131–147.

Droser, M. L., Jensen, S., Gehling, J. G., Myrow, P. M., and Narbonne, G. M. 2002. Lowermost Cambrian ichnofabrics from the Chapel Island Formation, Newfoundland: Implications for Cambrian substrates. *Palaios* 17:3–15.

Duan, Y., Han, J., Fu, D., Zhang, X., Yang, X., Komiya, T. 2012. Reproductive strategy of the bradoriid arthropod *Kunmingella douvillei* from the Lower Cambrian Chengjiang Lagerstätte, south China. *Journal of Guizhou University* 29:159.

Dunne, J. A., Williams, R. J., Martinez, N. D., Wood, R. A., and Erwin, D. H. 2008. Compilation and network analyses of Cambrian food webs. *PLoS Biology* 6:693–708.

Durham, J. W. 1974. Systematic position of *Eldonia ludwigi* Walcott. *Journal of Paleontology* 48:750–755.

———. 1978. A Lower Cambrian eocrinoid. *Journal of Paleontology* 52:195–199.

Durman, P. N., and Sennikov, N. V. 1993. A new rhabdopleurid hemichordate from the Middle Cambrian of Siberia. *Palaeontology* 36:283–296.

Dzik, J. 1994. Evolution of "small shelly fossils" assemblages of the Early Paleozoic. *Acta Palaeontologica Polonica* 39:247–313.

———. 2005. Behavioral and anatomical unity of the earliest burrowing animals and the cause of the "Cambrian explosion." *Paleobiology* 31:503–521.

———. 2011. Possible Ediacaran ancestry of the halkieriids. *In* Johnston, P. A., and Johnston, K. J., eds., International Conference on the Cambrian Explosion, Proceedings, *Palaeontographica Canadiana* 31:205–218.

Dzik, J., Zhao, Y.-L., and Zhu, M.-Y. 1997. Mode of life of the Middle Cambrian eldonioid lophophorate *Rotadiscus*. *Palaeontology* 40:385–396.

Eddy, J. D., and McCollum, L. B. 1998. Early Middle Cambrian *Albertella* biozone trilobites of the Pioche Shale, southeastern Nevada. *Journal of Paleontology* 72:864–887.

Edgecombe, G. D. 2010. Arthropod phylogeny: An overview from the perspectives of morphology, molecular data and the fossil record. *Arthropod Structure & Development* 39:74–87.

Eibye-Jacobsen, D. 2004. A reevaluation of *Wiwaxia* and the polychaetes of the Burgess Shale. *Lethaia* 37:317–335.

Elicki, O. 2009. Palaeoecological aspects of Cambrian bivalves: conclusions from Perigondwanan occurrences. *International Conference on the Cambrian Explosion Abstract Volume,* The Burgess Shale Consortium, Toronto, p. 69.

Elliott, D. K., and Martin, D. L. 1987. A new trace fossil from the Cambrian Bright Angel Shale, Grand Canyon, Arizona. *Journal of Paleontology* 61:641–648.

Elrick, M., and Snider, A. C. 2002. Deep-water stratigraphic cyclicity and carbonate mud mound development in the Middle Cambrian Marjum Formation, House Range, Utah, USA. *Sedimentology* 49:1021–1047.

Elser, J. J., Watts, J., Schampel, J. H., and Farmer, J. 2006. Early Cambrian food webs on a trophic knife edge? A hypothesis and preliminary data from a modern stromatolite-based ecosystem. *Ecology Letters* 9:295–303.

Elston, D. P. 1989a. Correlations and facies changes in Lower and Middle Cambrian Tonto Group, Grand Canyon, Arizona. *In* Elston, D. P., Billingsley, G. H., and Young, R. A., eds, *Geology of Grand Canyon, Northern Arizona (with Colorado River Guides).* American Geophysical Union, Washington, DC, p. 131–136.

———. 1989b. Middle and Late Proterozoic Grand Canyon Supergroup, Arizona. *In* Elston, D. P., Billingsley, G. H., and Young, R. A., eds., *Geology of Grand Canyon, Northern Arizona (with Colorado River Guides),* American Geophysical Union, Washington, DC, p. 94–105.

Enos, P. 1983. Shelf environment. *In* Scholle, P. A., Bebout, D. G., and Moore, C. H., eds., *Carbonate Depositional Environments,* American Association of Petroleum Geologists Memoir 33, American Association of Petroleum Geologists, Tulsa, OK, p. 267–295.

Erwin, D. H. 1994. Early introduction of major morphological innovations. *Acta Palaeontologica Polonica* 38:281–294.

———. 2007. Disparity: Morphological pattern and developmental context. *Palaeontology* 50:57–73.

———. 2011. Evolutionary uniformitarianism. *Developmental Biology* 357:27–34.

Erwin, D. H., and Davidson, E. H. 2009. The evolution of hierarchical gene regulatory networks. *Nature Reviews* 10:141–148.

Erwin, D. H., and Valentine, J. W. 2013. *The Cambrian Explosion: The Construction of Animal Biodiversity.* Roberts and Company, Greenwood Village, CO, 406 p.

Erwin, D. H., Laflamme, M., Tweedt, S. M., Sperling, E. A., Pisani, D., and Peterson, K. J. 2011. The Cambrian conundrum: Early divergence and later ecological success in the early history of animals. *Science* 334:1091–1097.

Fang, Z.-J., and Sánchez, T. M. 2012. *Treatise Online, Number 43, Part N, Revised, Volume 1, Chapter 16: Origin and Early Evolution of the Bivalvia.* KU Paleontological Institute, University of Kansas, Lawrence, 21 p.

Fanning, C. M., and Link, P. K. 2004. U-Pb SHRIMP ages of Neoproterozoic (Sturtian) glaciogenic Pocatello Formation, southeastern Idaho. *Geology* 32:881–884.

Fatka, O., and Szabad, M. 2011. Agnostids entombed under exoskeletons of paradoxidid trilobites. *Neues Jahrbuch für Geologie und Paläontologie-Abhandlungen* 259:207–215.

Fatka, O., Kraft, P., and Szabad, M. 2011. Shallow-water occurrence of *Wiwaxia* in the middle Cambrian of the Barrandian area, Czech Republic. *Acta Palaeontologica Polonica* 56:871–875.

Fatka, O., Václav, Vokác, V., Moravec, J., Sinágl, M., and Valent, M. 2009. Agnostids entombed in hyolith conchs. *Memoirs of the Association of Australasian Palaeontologists* 37:481–489.

Fedo, C. M., and Cooper, J. D. 1990. Braided fluvial to marine transition: The basal Lower Cambrian Wood Canyon Formation, southern Marble Mountains, Mojave Desert, California. *Journal of Sedimentary Petrology* 60:220–234.

Fedonkin, M. A., and Waggoner, B. M. 1997. The Late Precambrian fossil *Kimerella* is a mollusc-like bilaterian organism. *Nature* 388:868–871.

Fedonkin, M. A., Gehling, J. G., Grey, K., Narbonne, G. M., and Vickers-Rich, P. 2007. *The Rise of Animals: Evolution and Diversification of the Kingdom Animalia.* Johns Hopkins University Press, Baltimore, 326 p.

Fine, M., and Tchernov, D. 2007. Scleractinian coral species survive and recover from decalcification. *Science* 315:1811.

Fletcher, T. P. 2011. The development of mid-Cambrian lithostratigraphical nomenclature in the vicinity of the Burgess Shale, Kicking Horse Belt of Canada. *In* Johnston, P. A., and Johnston, K. J., eds., International Conference on the Cambrian Explosion, Proceedings, *Palaeontographica Canadiana* 31:39–72.

Fletcher, T. P., and Collins, D. H. 1998. The Middle Cambrian Burgess Shale and its relationship to the Stephen Formation in the southern Canadian Rocky Mountains. *Canadian Journal of Earth Sciences* 35:413–436.

———. 2009. Geology and stratigraphy of the Burgess Shale Formation on Mount Stephen and Fossil Ridge. *In* Caron, J.-B., and Rudkin, D., eds., *A Burgess Shale Primer: History, Geology, and Research Highlights.* The Burgess Shale Consortium, Toronto, p. 33–53.

Foote, M. 1988. Survivorship analysis of Cambrian and Ordovician trilobites. *Paleobiology* 14:258–271.

Foote, M., and Raup, D. M. 1996. Fossil preservation and the stratigraphic ranges of taxa. *Paleobiology* 22:121–140.

Ford, T. D., and Dehler, C. M. 2003. Grand Canyon Supergroup: Nankoweap Formation, Chuar Group, and Sixtymile Formation. *In* Beus, S. S., and Morales, M., eds., *Grand Canyon Geology.* Oxford University Press, New York, p. 53–75.

Fortey, R. 2000. *Trilobite: Eyewitness to Evolution.* Vintage, New York, 284 p.

Fortey, R. A. 1990. Ontogeny, hypostome attachment and trilobite classification. *Palaeontology* 33:529–576.

Fortey, R. A., and Hughes, N. C. 1998. Brood pouches in trilobites. *Journal of Paleontology* 72:638–649.

Fortey, R. A., and Owens, R. M. 1999. Feeding habits in trilobites. *Palaeontology* 42:429–465.

Fortey, R. A., and Whittington, H. B. 1989. The Trilobita as a natural group. *Historical Biology* 2:125–138.

Fortey, R. A., Briggs, D. E. G., and Wills, M. A. 1996. The Cambrian evolutionary "explosion": Decoupling cladogenesis from morphological disparity. *Biological Journal of the Linnean Society* 57:13–33.

Foster, J. R. 1994. A note on depositional environments of the Lower–Middle Cambrian Cadiz Formation, Marble Mountains, California. *The Mountain Geologist* 31:29–36.

———. 2009. Taphonomic characteristics of a quarry in the Bright Angel Shale (Middle Cambrian), Grand Canyon National Park, Arizona: A preliminary look. *In* Baltzer, E., ed., American Institute of Professional Geologists, Conference Proceedings, *Rocky Mountains and the Colorado Plateau: Canyons, Resources, and Hazards,* p. 77–80.

———. 2011a. A short review of the geology and paleontology of the Cambrian sedimentary rocks of the southern Marble Mountains, Mojave Desert, California. *New Mexico Museum of Natural History and Science Bulletin* 53: 38–51.

———. 2011b. Trilobite taphonomy of the Latham Shale (Lower Cambrian), Mojave Desert, California: An Inner Detrital Belt Burgess Shale–type deposit of western Laurentia. *In* Johnston, P. A., and Johnston, K. J., eds., International Conference on the Cambrian Explosion, Proceedings, *Palaeontographica Canadiana* 31:119–140.

———. 2011c. *Bonnima* sp. (Trilobita; Corynexochida) from the Chambless Limestone (Lower Cambrian) of the Marble Mountains, California: First Dorypygidae in a cratonic region of the southern Cordillera. *PaleoBios* 30:45–49.

———. 2011d. Trilobite taphonomy in the lower Pioche Formation (Dyeran; Global Stage 4) at Frenchman Mountain, Nevada. *In* Hollingsworth, J. S., Sundberg, F. A., and Foster, J. R., eds., Cambrian Stratigraphy and Paleontology of Northern Arizona and Southern Nevada, *Museum of Northern Arizona Bulletin* 67:282–283.

———. 2011e. Trilobites and other fauna from two quarries in the Bright Angel Shale (middle Cambrian, Series 3; Delamaran), Grand Canyon National Park, Arizona. *In* Hollingsworth, J. S., Sundberg, F. A., and Foster, J. R., eds., Cambrian Stratigraphy and Paleontology of Northern Arizona and Southern Nevada, *Museum of Northern Arizona Bulletin* 67:99–120.

Fowler, E. 1999. Stop 13, Biostratigraphy of upper Dyeran strata of the Carrara Formation, Emigrant Pass, Nopah Range, California. *In* Palmer,

A. R., ed., *Laurentia 99,* 5th Field Conference of the Cambrian Stage Subdivision Working Group, Institute for Cambrian Studies, Boulder, CO, p. 46–50.

Frakes, L. A. 1979. *Climates Throughout Geologic Time.* Elsevier, New York, 310 p.

Frederickson, E. A. 1941. Correlation of Cambro-Ordovician trilobites of Oklahoma. *Journal of Paleontology* 15:160–163.

———. 1948. Upper Cambrian trilobites from Oklahoma. *Journal of Paleontology* 22:798–803.

Freeman, R. L., and Miller, J. F. 2011. First report of a larval shell repair scar on a lingulate brachiopod: Evidence of durophagus predation in the Cambrian pelagic realm? *Journal of Paleontology* 85:695–702.

Fritz, W. H. 1971. Geological setting of the Burgess Shale. *Proceedings of the North American Paleontological Convention September 1969 Part I,* p. 1155–1170.

———. 1972. Lower Cambrian trilobites from the Sekwi Formation type section, Mackenzie Mountains, northwestern Canada. *Geological Survey of Canada Bulletin* 212:1–90.

———. 1974. The Early Cambrian trilobites *Olenellus, Fremontia,* and *Paedeumias. Bulletin of the Southern California Paleontological Society* 6:6–7, 10.

———. 1976. Lower Cambrian stratigraphy, Mackenzie Mountains, northwestern Canada. *Brigham Young University Geology Studies* 23:7–22.

———. 1990. Comment: In defense of the escarpment near the Burgess Shale fossil locality. *Geoscience Canada* 17:106–110.

———. 1991. Lower Cambrian trilobites from the Illtyd Formation, Wernecke Mountains, Yukon Territory. *Geological Survey of Canada Bulletin* 409:1–77.

Fuller, J. E. 1980. A Middle Cambrian fossil locality in the Cadiz Formation, Marble Mountains, California. *Southern California Paleontological Society Special Publications* 2:30–33.

Fuller, M., and Jenkins, R. 2007. Reef corals from the Lower Cambrian of the Flinders Ranges, South Australia. *Palaeontology* 50:961–980.

Gabbott, S. E., and Zalasiewicz, J. 2009. Sedimentation of the Phyllopod Bed within the Cambrian Burgess Shale Formation. *In* Caron, J.-B., and Rudkin, D., eds., *A Burgess Shale Primer: History, Geology, and Research Highlights.* The Burgess Shale Consortium, Toronto, p. 55–61.

Gabbott, S. E., Zalasiewicz, J., and Collins, D. 2008. Sedimentation of the Phyllopod Bed within the Cambrian Burgess Shale Formation of British Columbia. *Journal of the Geological Society* 165:307–318.

Gabbott, S. E., Hou, X.-G., Norry, M. J., and Siveter, D. J. 2004. Preservation of Early Cambrian animals of the Chengjiang biota. *Geology* 32:901–904.

Gaines, R. R. 2011. A new Burgess Shale-type locality in the "thin" Stephen Formation, Kootenay National Park, British Columbia: stratigraphic and paleoenvironmental setting. *In* Johnston, P. A., and Johnston, K. J., eds., International Conference on the Cambrian Explosion Proceedings, *Palaeontographica Canadiana* 31:73–88.

Gaines, R. R., and Droser, M. L. 2002. Depositional environments, ichnology, and rare soft-bodied preservation in the Lower Cambrian Latham Shale, east Mojave. *In* Corsetti, F. A., ed., Proterozoic-Cambrian of the Great Basin and Beyond, *SEPM Pacific Section Book* 93:153–164.

———. 2003. Paleoecology of the familiar trilobite *Elrathia kingii:* an early exaerobic zone inhabitant. *Geology* 31:941–944.

———. 2005. Paleoenvironmental and paleoecological dynamics of Burgess Shale-type deposits: Evidence from the three Utah lagerstätten. North American Paleontology Convention, Programme and Abstracts, *PaleoBios* 25(supp. 2):48.

Gaines, R. R., Briggs, D. E. G., and Zhao, Y.-L. 2008. Cambrian Burgess Shale-type deposits share a common mode of fossilization. *Geology* 36:755–758.

Gaines, R. R., Droser, M. L., and Kennedy, M. J. 2001. Taphonomy of soft-bodied preservation and ptychopariid lagerstätte in the Wheeler Shale (Middle Cambrian), House Range, USA: Controls and implications. North American Paleontological Convention, Program and Abstracts, *PaleoBios* 21(supp. 2):55.

Gaines, R. R., Kennedy, M. J., and Droser, M. L. 2005. A new hypothesis for organic preservation of Burgess Shale taxa in the middle Cambrian Wheeler Formation, House Range, Utah. *Palaeogeography, Palaeoclimatology, Palaeoecology* 220:193–205.

Gaines, R. R., Droser, M. L., Orr, P. J., Garson, D., Hammarlund, E., Qi, C., and Canfield, D. E. 2012. Burgess Shale–type biotas were not entirely burrowed away. *Geology* 40:283–286.

Gaines, R. R., Hammarlund, E. U., Hou, X., Qi, C., Gabbott, S. E., Zhao, Y., Peng, J., and Canfield, D. E. 2012. Mechanism for Burgess Shale–type preservation. *Proceedings of the National Academy of Sciences* 109:5180–5184.

Gallagher, A. K. 2003. Reconstructing the paleoenvironment of the Bright Angel Shale, Grand Canyon Arizona. *Geological Society of America, Northeastern Section, Abstracts with Programs:* n.p.

García-Bellido, D. C., and Collins, D. H. 2004. Moulting arthropod caught in the act. *Nature* 429:40.

———. 2006. A new study of *Marrella splendens* (Arthropoda, Marrellomorpha) from the Middle Cambrian Burgess Shale, British Columbia, Canada. *Canadian Journal of Earth Sciences* 43:721–742.

———. 2007. Reassessment of the genus *Leanchoilia* (Arthropoda, Arachnomorpha) from the Middle Cambrian Burgess Shale, British Columbia, Canada. *Palaeontology* 50:693–709.

García-Bellido, D. C., Vannier, J., and Collins, D. 2009. Soft-part preservation in two species of the arthropod *Isoxys* from the middle Cambrian Burgess Shale of British Columbia, Canada. *Acta Palaeontologica Polonica* 54:699–712.

Garson, D. E., Gaines, R. R., Droser, M. L., Liddell, W. D., and Sappenfield, A. 2011. Dynamic palaeoredox and exceptional preservation in the Cambrian Spence Shale of Utah. *Lethaia* 45:164–177.

Gehling, J. G., Jensen, S., Droser, M. L., Myrow, P. M., and Narbonne, G. M. 2001. Burrowing below the basal Cambrian GSSP, Fortune Head, Newfoundland. *Geological Magazine* 138:213–218.

Getty, P. R., and Hagadorn, J. W. 2008. Reinterpretation of *Climactichnites* Logan 1860 to include subsurface burrows, and erection of *Musculopodus* for resting traces of the trailmaker. *Journal of Paleontology* 82:1161–1172.

Geyer, G., and Shergold, J. 2000. The quest for internationally recognized divisions of Cambrian time. *Episodes* 23:188–195.

Giribet, G., Okusu, A., Lindgren, A. R., Huff, S. W., Schrödl, M., and Nishiguchi, M. K. 2006. Evidence for a clade composed of molluscs with serially repeated structures: Monoplacophorans are related to chitons. *Proceedings of the National Academy of Sciences* 103:7723–7728.

Gon, S. M. 2009. *A Pictorial Guide to the Orders of Trilobites.* Available at www.trilobites.info.

Gould, S. J. 1989. *Wonderful Life: The Burgess Shale and the Nature of History.* W. W. Norton, New York, 347 p.

———. 1991. The disparity of the Burgess Shale arthropod fauna and the limits of cladistic analysis: Why we must strive to quantify morphospace. *Paleobiology* 17:411–423.

Grey, K., Walter, M. R., and Calver, C. R. 2003. Neoproterozoic biotic diversification: Snowball Earth or aftermath of the Acraman impact? *Geology* 31:459–462.

Grotzinger, J. P., and Knoll, A. H. 1999. Stromatolites in Precambrian carbonates: Evolutionary mileposts of environmental dipsticks? *Annual Review of Earth and Planetary Sciences* 27:313–358.

Guensburg, T. E., and Sprinkle, J. 2001. Ecologic radiation of Cambro-Ordovician echinoderms. *In* Zhuravlev, A. Yu., and Riding, R., eds., *The Ecology of the Cambrian Radiation,* Columbia University Press, New York, p. 428–444.

Gunther, L. F., and Gunther, V. G. 1981. Some Middle Cambrian fossils of Utah. *Brigham Young University Geology Studies* 28:1–81.

Gutiérrez-Marco, J. C., Sá, A. A., García-Bellido, D. C., Rábano, I., and Valério, M. 2009. Giant trilobites and trilobite clusters from the Ordovician of Portugal. *Geology* 37:443–446.

Hagadorn, J. W. 2002a. Burgess Shale: Cambrian explosion in full bloom. *In* Bottjer, D. J., Etter, W., Hagadorn, J. W., and Tang, C., eds., *Exceptional Fossil Preservation: A Unique View on the Evolution of Marine Life,* Columbia University Press, New York, p. 61–89.

———. 2002b. Burgess Shale-type localities: the global picture. *In* Bottjer, D. J., Etter, W., Hagadorn, J. W., and Tang, C., eds., *Exceptional Fossil Preservation: A Unique View on the Evolution of Marine Life,* Columbia University Press, New York, p. 91–116.

———. 2009. Taking a bite out of *Anomalocaris. International Conference on the Cambrian Explosion Abstract Volume,* The Burgess Shale Consortium, Toronto, p. 28.

Hagadorn, J. W., and Seilacher, A. 2009. Hermit arthropods 500 million years ago? *Geology* 37:295–298.

Hagadorn, J. W., and Waggoner, B. 2000. Ediacaran fossils from the southwestern Great Basin, United States. *Journal of Paleontology* 74:349–359.

Hagadorn, J. W., Dott, R. H., Jr., Damrow, D. 2002. Stranded on a Late Cambrian shoreline: medusae from central Wisconsin. *Geology* 30:147–150.

Hagadorn, J. W., Kirschvink, J. L., Raub, T. D., and Rose, E. C. 2011. Above the Great Unconformity: A fresh look at the Tapeats Sandstone, Arizona-Nevada, U.S.A. *In* Hollingsworth, J. S., Sundberg, F. A., and Foster, J. R., eds., Cambrian Stratigraphy and Paleontology of Northern Arizona and Southern Nevada, *Museum of Northern Arizona Bulletin* 67:63–77.

Hamilton, M. M., Derkey, R. E., and McCollum, L. B. 2003. Early Middle Cambrian (*Glossopleura* biozone) trilobites in a steptoe surrounded by Columbia River basalt near Spokane, Washington. *Geological Society of America Abstracts with Programs* 35(6):159.

Hammarlund, E., Canfield, D. E., Bengtson, S., Leth, P. M., Schillinger, B., and Calzada, E. 2011. The influence of sulfate concentration on soft-tissue decay and preservation. *In* Johnston, P. A., and Johnston, K. J., eds., International Conference on the Cambrian Explosion, Proceedings, *Palaeontographica Canadiana* 31:141–155.

Han, J., Liu, J., Zhang, Z.-F., Zhang, X.-L., and Shu, D.-G. 2007. Trunk ornament on the palaeoscolecid worms *Cricocosmia* and *Tabelliscolex* from the Early Cambrian Chengjiang deposits of China. *Acta Palaeontologica Polonica* 52:423–431.

Han, J., Kubota, S., Uchida, H.-O., Stanley, G. D., Yao, X.-Y., Shu, D.-G., Li, Y., and Yasui, K. 2011. Tiny sea anemone with soft-tissue preservation from the lower Cambrian Kuanchuanpu Formation, South China. *In* Hollingsworth, J. S., Sundberg, F. A., and Foster, J. R., eds., Cambrian Stratigraphy and Paleontology of Northern Arizona and Southern Nevada, *Museum of Northern Arizona Bulletin* 67:26–42.

Häntzschel, W. 1962. Trace fossils and Problematica. *In* Hass, W. H., Häntzschel, W., Fisher, D. W., Howell, B. F., Rhodes, F. H. T., Müller, K. J., and Moore, R. C., eds., *Treatise On Invertebrate Paleontology Part W Miscellanea,* Geological Society of America and University of Kansas Press, p. W177–W245.

Haq, B. U., and Schutter, S. R. 2008. A chronology of Paleozoic sea-level changes. *Science* 322:64–68.

Hargraves, R. B. 1976. Precambrian geologic history. *Science* 193:363–371.

Harrington, H. J. 1956. Olenellidae with advanced cephalic spines. *Journal of Paleontology* 30:56–61.

Harrington, H. J., Henningsmoen, G., Howell, B. F., Jaanusson, V., Lochman-Balk, C., Moore, R. C., Poulsen, C., Rasetti, F., Richter, E., Richter, R., Schmidt, H., Sdzuy, K., Struve, W., Stormer, L., Stubblefield, C. J., Tripp, R., Weller, J. M., and Whittington, H. B. 1959. *Treatise On Invertebrate Paleontology, Part O, Arthropoda 1. Arthropoda–General Features. Protarthropoda. Euarthropoda–General Features. Trilobitamorpha.* Geological Society of America and University of Kansas Press, Boulder, CO, and Lawrence, KS, 560 p.

Harrison, T. M., Blicert-Toft, J., Müller, W., Albarede, F., Holden, P., and Mojzsis, S. J. 2005. Heterogeneous hadean hafnium: evidence of continental crust at 4.4 to 4.5 Ga. *Science* 310:1947–1950.

Harvey, T. H. P., Velez, M. I., and Butterfield, N. J. 2011. A cryptic radiation of crustaceans. *The Palaeontological Association 55th Annual Meeting Programme and Abstracts, The Palaeontological Association Newsletter* 78:A22.

Haug, J. T., Maas, A., Haug, C., and Waloszek, D. 2011. *Sarotrocercus oblitus:* Small arthropod with great impact on the understanding of arthropod evolution? *Bulletin of Geosciences* 86:725–736.

Haug, J. T., Waloszek, D., Maas, A., Liu, Y., and Haug, C. 2012. Functional morphology, ontogeny and evolution of mantis shrimp-like predators in the Cambrian. *Palaeontology* 55:369–399.

Havlícek, V., Vanek, J., and Fatka, O. 1993. Floating algae of the genus *Krejciella* as probable hosts of epiplanktic organisms (Dobrotivá Series, Ordovician; Prague Basin). *Journal of the Czech Geological Society* 38:79–88.

Hazzard, J. C. 1933. Notes on the Cambrian rocks of the eastern Mohave Desert, California. *University of California Publications in Geological Sciences* 23: 57–77.

———. 1954. Rocks and structure of the northern Providence Mountains,

San Bernardino County, California. *California Division of Mines Bulletin* 170:27–35.

Hazzard, J. C., and Mason, J. F. 1936. Middle Cambrian formations of the Providence and Marble Mountains, California. *Bulletin of the Geological Society of America* 47:229–240.

Hendricks, J. D., and Stevenson, G. M. 2003. Grand Canyon Supergroup: Unkar Group. *In* Beus, S. S., and Morales, M., eds., *Grand Canyon Geology*. Oxford University Press, New York, p. 39–52.

Hendricks, J. R., and Lieberman, B. S. 2007. Biogeography and the Cambrian radiation of arachnomorph arthropods. *Memoirs of the Association of Australasian Palaeontologists* 34:461–471.

———. 2008. New phylogenetic insights into the Cambrian radiation of arachnomorph arthropods. *Journal of Paleontology* 82:585–594.

Hendricks, J. R., Lieberman, B. S., and Stigall, A. L. 2008. Using GIS to study palaeobiogeographic and macroevolutionary patterns in soft-bodied Cambrian arthropods. *Palaeogeography, Palaeoclimatology, Palaeoecology* 264:163–175.

Hereford, R. 1977. Deposition of the Tapeats Sandstone (Cambrian) in central Arizona. *Geological Society of America Bulletin* 88:199–211.

Herr, R. G., and Picard, M. D. 1981. Petrography of Upper Cambrian Lodore Formation, northeast Utah and northwest Colorado. *Contributions to Geology University of Wyoming* 20:1–21.

Herr, R. G., Picard, M. D., and Evans, S. H., Jr. 1982. Age and depth of burial, Cambrian Lodore Formation, northeastern Utah and northwestern Colorado. *Contributions to Geology University of Wyoming* 21:115–121.

Hesselbo, S. P. 1989. The aglaspidid arthropod *Beckwithia* from the Cambrian of Utah and Wisconsin. *Journal of Paleontology* 63:636–642.

Hicks, M. 2006. A new genus of Early Cambrian coral in Esmeralda County, southwestern Nevada. *Journal of Paleontology* 80:609–615.

Hill, D. 1964. The phylum Archaeocyatha. *Biological Reviews* 39:232–258.

Hintze, L. F. 1988. Geologic History of Utah. *Brigham Young University Geology Studies, Special Publication* 7, 202 p.

Hintze, L. F., and Robison, R. A. 1975. Middle Cambrian stratigraphy of the House, Wah Wah, and adjacent ranges in western Utah.

Geological Society of America Bulletin 86:881–891.

Hoffmann, K.-H., Condon, D. J., Bowring, S. A., and Crowley, J. L. 2004. U-Pb zircon date from the Neoproterozoic Ghaub Formation, Namibia: constraints on Marinoan glaciation. *Geology* 32:817–820.

Hoffman, P. F., Kaufman, A. J., Halverson, G. P., and Schrag, D. P. 1998. A Neoproterozoic Snowball Earth. *Science* 281:1342–1346.

Hohensee, S. R., and Stitt, J. H. 1989. Redeposited *Elvinia* zone (Upper Cambrian) trilobites from the Collier Shale, Ouachita Mountains, west-central Arkansas. *Journal of Paleontology* 63:857–879.

Holland, H. D. 1984. *The Chemical Evolution of the Atmosphere and Oceans*. Princeton University Press, Princeton, NJ, 582 p.

Hollingsworth, J. S. 2005. The earliest occurrence of trilobites and brachiopods in the Cambrian of Laurentia. *Palaeogeography, Palaeoclimatology, Palaeoecology* 220:153–165.

———. 2006. Homiidae (Trilobita: Olenellina) of the Montezuman Stage (Early Cambrian) in western Nevada. *Journal of Paleontology* 80:309–332.

———. 2011a. Base of Cambrian Series 2, Stage 3 in the Gold Coine Mine area, Esmeralda County, Nevada. *In* Hollingsworth, J. S., Sundberg, F. A., and Foster, J. R., eds., Cambrian Stratigraphy and Paleontology of Northern Arizona and Southern Nevada, *Museum of Northern Arizona Bulletin* 67:263–266.

———. 2011b. The base of the *Fallotaspis* zone (base of the Montezuman stage, Cambrian Stage 3), Montezuma Range, Nevada. *In* Hollingsworth, J. S., Sundberg, F. A., and Foster, J. R., eds., Cambrian Stratigraphy and Paleontology of Northern Arizona and Southern Nevada, *Museum of Northern Arizona Bulletin* 67:252–256.

———. 2011c. Lithostratigraphy and biostratigraphy of Cambrian Stage 3 in western Nevada and eastern California. *In* Hollingsworth, J. S., Sundberg, F. A., and Foster, J. R., eds., Cambrian Stratigraphy and Paleontology of Northern Arizona and Southern Nevada, *Museum of Northern Arizona Bulletin* 67:26–42.

Hollingsworth, J. S., and Babcock, L. E. 2011. Base of Dyeran stage (Cambrian Stage 4) in the middle member of the Poleta Formation, Indian Springs Canyon, Montezuma Range, Nevada. *In* Hollingsworth, J.

S., Sundberg, F. A., and Foster, J. R., eds., Cambrian Stratigraphy and Paleontology of Northern Arizona and Southern Nevada, *Museum of Northern Arizona Bulletin* 67:26–42.

Hopkins, M. J. 2011. Species-level phylogenetic analysis of pterocephaliids (Trilobita, Cambrian) from the Great Basin, western USA. *Journal of Paleontology* 85:1128–1153.

Hopkins, M. J., and Webster, M. 2009. Ontogeny and geographic variation of a new species of the corynexochine trilobite *Zacanthopsis* (Dyeran, Cambrian). *Journal of Paleontology* 83:5524–547.

Horodyski, R. J. 1976. Stromatolites from the Middle Proterozoic Altyn Limestone, Belt Supergroup, Glacier National Park, Montana. *In* Walter, M. R., ed., *Stromatolites, Developments in Sedimentology* 20, Elsevier, Amsterdam, p. 585–597.

———. 1977. Environmental influences on columnar stromatolite branching patterns: examples from the Middle Proterozoic Belt Supergroup, Glacier National Park, Montana. *Journal of Paleontology* 51:661–671.

———. 1983. Sedimentary geology and stromatolites of the Middle Proterozoic Belt Supergroup, Glacier National Park, Montana. *Precambrian Research* 20:391–425.

Horodyski, R. J., and Bloeser, B. 1983. Possible eukaryotic algal filaments from the Late Proterozoic Chuar Group, Grand Canyon, Arizona. *Journal of Paleontology* 57:321–326.

Horváth, G., Clarkson, E., and Pix, W. 1997. Survey of modern counterparts of schizochroal trilobite eyes: Structural and functional similarities and differences. *Historical Biology* 12:229–263.

Hou, X.-G. 1999. New rare bivalved arthropods from the Lower Cambrian Chengjiang fauna, Yunnan, China. *Journal of Paleontology* 73:102–116.

Hou, X.-G., and Bergström, J. 1994. Palaeoscolecid worms may be nematomorphs rather than annelids. *Lethaia* 27:11–17.

Hou, X.-G., Bergström, J., and Xu, G.-H. 2004. The Lower Cambrian crustacean *Pectocaris* from the Chengjiang biota, Yunnan, China. *Journal of Paleontology* 78:700–708.

Hou, X.-G., Siveter, D. J., Williams, M., Walossek, D., and Bergström, J. 1996. Appendages of the arthropod *Kunmingella* from the early Cambrian of China: Its bearing on the systematic position of the Bradoriida and the fossil record of the

Ostracoda. *Philosophical Transactions of the Royal Society of London B* 351:1131–1145.

Hou, X.-G., Aldridge, R. J., Bergström, J., Siveter, David J., Siveter, Derek J., and Feng, X.-H. 2004. *The Cambrian Fossils of Chengjiang, China: The Flowering of Early Animal Life.* Blackwell Publishing, Oxford, 233 p.

Howell, B. F., Bridge, J., Deiss, C. F., Edwards, I., Lochman, C., Raasch, G. O., Resser, C. E., Duncan, D. C., Mason, J. F., and Denson, N. M. 1944. Correlation of the Cambrian formations of North America. *Bulletin of the Geological Society of America* 55:993–1004.

Howley, R. A., Rees, M. N., and Jiang, G. 2006. Significance of Middle Cambrian mixed carbonate-siliciclastic units for global correlation: Southern Nevada, USA. *Palaeoworld* 15:360–366.

Hua, H., Chen, Z., Yuan, X., Zhang, L., and Xiao, S. 2005. Skeletogenesis and asexual reproduction in the earliest biomineralizing animal *Cloudina*. *Geology* 33:277–280.

Huang, D., Vannier, J., and Chen, J. Y. 2004. Recent Priapulidae and their Early Cambrian ancestors: Comparisons and evolutionary significance. *Geobios* 37:217–228.

Huang, D., Zhu, M., Chen, J., and Zhao, F. 2012. The burrowing behavior of palaeoscolecidan worms: New fossil evidence from the Cambrian Chengjiang Fauna. *Journal of Guizhou University* 29:169.

Hughes, C. P. 1975. Redescription of *Burgessia bella* from the Middle Cambrian Burgess Shale, British Columbia. *Fossils and Strata* 4:415–435.

Hughes, N. C. 2001. Ecologic evolution of Cambrian trilobites. *In* Zhuravlev, A. Yu., and Riding, R., eds., *The Ecology of the Cambrian Radiation,* Columbia University Press, New York, p. 370–403.

———. 2007. Strength in numbers: High phenotypic variance in early Cambrian trilobites and its evolutionary implications. *BioEssays* 29:1081–1084.

Hughes, N. C., Gunderson, G. O., and Weedon, M. J. 2000. Late Cambrian conulariids from Wisconsin and Minnesota. *Journal of Paleontology* 74:828–838.

Hughes, N. C., Minelli, A., and Fusco, G. 2006. The ontogeny of trilobite segmentation: A comparative approach. *Paleobiology* 32:602–627.

Hunt, R. K. 2006. Middle Proterozoic paleontology of the Belt Supergroup,

Glacier National Park. *New Mexico Museum of Natural History and Science Bulletin* 34:57–62.

Huntoon, P. W. 1989. Cambrian stratigraphic nomenclature, Grand Canyon, Arizona–mappers nightmare. *In* Elston, D. P., Billingsley, G. H., and Young, R. A., eds, *Geology of Grand Canyon, Northern Arizona (with Colorado River Guides),* American Geophysical Union, Washington, DC, p. 128–130.

Huntoon, P. W., Billingsley, G. H., Sears, J. W., Ilg, B. R., Karlstrom, K. E., Williams, M. L., Hawkins, D., Breed, W. J., Ford, T. D., Clark, M. D., Babcock, R. S., and Brown, E. H. 1996. Geologic map of the eastern part of the Grand Canyon National Park, Arizona. Grand Canyon Association, AZ.

Hyde, W. T., Crowley, T. J., Baum, S. K., and Peltier, W. R. 2000. Neoproterozoic "snowball Earth" simulations with a coupled climate/ice-sheet model. *Nature* 405:425–429.

Ivantsov, A. Y., and Wrona, R. 2004. Articulated palaeoscolecid sclerite arrays from the Lower Cambrian of eastern Siberia. *Acta Geologica Polonica* 54:1–22.

Ivantsov, A. Yu., and Fedonkin, M. A. 2002. Conulariid-like fossil from the Vendian of Russia: A metazoan clade across the Proterozoic/Palaeozoic boundary. *Palaeontology* 45:1219–1229.

Jablonski, D., Sepkoski, J. J., Bottjer, D. J., and Sheehan, P. M. 1983. Onshore-offshore patterns in the evolution of Phanerozoic shelf communities. *Science* 222:1123–1125.

Jago, J. B., Gehling, J. G., Paterson, J. R., and Brock, G. A. 2012. Comments on Retallack, G. J. 2011: Problematic megafossils in Cambrian palaeosols of South Australia. *Palaeontology* 55:913–917.

Janussen, D., Steiner, M., and Maoyan, Z. 2002. New well-preserved scleritomes of Chancelloridae from the Early Cambrian Yuanshan Formation (Chengjiang, China) and the Middle Cambrian Wheeler Shale (Utah, USA) and paleobiological implications. *Journal of Paleontology* 76:596–606.

Javaux, E. J. 2007. Patterns of diversification in early eukaryotes. *In* Steemans, P., and Javaux, E., eds., *Recent Advances in Palynology,* Carnets de Géologie/Notebooks on Geology, Brest, Memoir 2007/01, p. 38–42.

Javaux, E. J., Knoll, A. H., and Walter, M. R. 2001. Morphological and

ecological complexity in early eukaryotic ecosystems. *Nature* 412:66–69.

Jell, P. A. 2003. Phylogeny of Early Cambrian trilobites. *Palaeontology* 70:45–57.

Jensen, S., Saylor, B. Z., Gehling, J. G., and Germs, G. 2000. Complex trace fossils from the terminal Proterozoic of Namibia. *Geology* 28:143–146.

Johnston, P. A., Johnston, K. J., Collom, C. J., Powell, W. G., and Pollock, R. J. 2009. Palaeontology and depositional environments of ancient brine seeps in the Middle Cambrian Burgess Shale at the Monarch, British Columbia, Canada. *Palaeogeography, Palaeoclimatology, Palaeoecology* 277:86–105.

Karim, T., and Westrop, S. R. 2002. Taphonomy and paleoecology of Ordovician trilobite clusters, Bromide Formation, south-central Oklahoma. *Palaios* 17:394–403.

Karlstrom, K. E., Bowring, S. A., Dehler, C. M., Knoll, A. H., Porter, S. M., Des Marais, D. M., Weil, A. B., Sharp, Z. D., Geissman, J. W., Elrick, M. B., Timmons, J. M., Crossey, L. J., and Davidek, K. L. 2000. Chuar Group of the Grand Canyon: Record of breakup of Rodinia, associated change in the global carbon cycle, and ecosystem expansion by 740 Ma. *Geology* 28:619–622.

Kasting, J. F. 1993. Earth's early atmosphere. *Science* 259:920–926.

Kennedy, M., Droser, M., Mayer, L. M., Pevear, D., and Mrofka, D. 2006. Late Precambrian oxygenation; inception of the clay mineral factory. *Science* 311:1446–1449.

Kennedy, M. J., Runnegar, B., Prave, A. R., Hoffmann, K.-H., and Arthur, M. A. 1998. Two or four Neoproterozoic glaciations? *Geology* 26:1059–1063.

Kerr, A. M., and Kim, J. 1999. Bi-penta-bi-decaradial symmetry: a review of evolutionary and developmental trends in Holothuroidea (Echinodermata). *Journal of Experimental Zoology* 285:93–103.

Kihm, R., and St. John, J. 2007. Walch's trilobite research: A translation of his 1771 trilobite chapter. *In* Mikulic, D. G., Landing, E., and Kluessendorf, J., eds., Fabulous Fossils–300 Years of Worldwide Research on Trilobites, *New York State Museum Bulletin* 507:115–140.

Knauth, L. P., and Kennedy, M. J. 2009. The late Precambrian greening of the Earth. *Nature* 460:728–732.

Knoll, A. H. 1994. Proterozoic and Early Cambrian protists: Evidence for accelerating evolutionary tempo. *Proceedings of the National Academy of Sciences USA* 91:6743–6750.

———. 2003a. *Life on a Young Planet.* Princeton University Press, Princeton, NJ, 277 p.

———. 2003b. Biomineralization and evolutionary history. *Reviews in Mineralogy and Geochemistry* 54:329–356.

Knoll, A. H., and Carroll, S. B. 1999. Early animal evolution: Emerging views from comparative biology and geology. *Science* 284:2129–2137.

Knoll, A. H., and Semikhatov, M. A. 1998. The genesis and time distribution of two distinctive Proterozoic stromatolite microstructures. *Palaios* 13:408–422.

Knoll, A. H., Barghoorn, E. S., and Awramik, S. M. 1978. New microorganisms from the Aphebian Gunflint Iron Formation, Ontario. *Journal of Paleontology* 52:976–992.

Knoll, A. H., Barghoorn, E. S., and Golubic, S. 1975. *Paleopleurocapsa wopfnerii* gen. et sp. nov.: A Late Precambrian alga and its modern counterpart. *Proceedings of the National Academy of Sciences USA* 72:2488–2492.

Knoll, A. H., Walter, M. R., Narbonne, G. M., and Christie-Blick, N. 2004. A new period for the geologic time scale. *Science* 305:621–622.

———. 2006. The Ediacaran Period: A new addition to the geologic time scale. *Lethaia* 39:13–30.

Kouchinsky, A. 2000. Shell microstructures in Early Cambrian molluscs. *Acta Palaeontologica Polonica* 45:119–150.

Kouchinsky, A., Bengtson, S., and Gershwin, L. 1999. Cnidarian-like embryos associated with the first shelly fossils in Siberia. *Geology* 27:609–612.

Kouchinsky, A. V. 2001. Mollusks, hyoliths, stenothecoids, and coelosclaritophorans. *In* Zhuravlev, A. Yu., and Riding, R., eds., *The Ecology of the Cambrian Radiation*, Columbia University Press, New York, p. 326–349.

Koucky, F. L., Cygan, N. E., and Rhodes, F. H. T. 1961. Conodonts from the eastern flank of the central part of the Big Horn Mountains, Wyoming. *Journal of Paleontology* 35:877–879.

Kruse, P. D., Zhuravlev, A. Y., and James, N. P. 1995. Primordial metazoan-calcimicrobial reefs: Tommotian (Early Cambrian) of the Siberian Platform. *Palaios* 10:291–321.

Kulik, J. W. 1965. *Stratigraphy of the Deadwood Formation, Black Hills, South Dakota and Wyoming.* MS thesis, South Dakota School of Mines and Technology, Rapid City.

Kurtz, V. E. 1976. Biostratigraphy of the Cambrian and lowest Ordovician, Bighorn Mountains and associated uplifts in Wyoming and Montana. *Brigham Young University Geology Studies* 23:215–227.

Kuzmina, T. V., and Malakhov, V. V. 2007. Structure of the brachiopod lophophore. *Paleontological Journal* 41:520–536.

Labandeira, C. C., and Hughes, N. C. 1994. Biometry of the Late Cambrian trilobite genus *Dikelocephalus* and its implications for trilobite systematics. *Journal of Paleontology* 68:492–517.

LaFlamme, M., Narbonne, G. M., and Anderson, M. M. 2004. Morphometric analysis of the Ediacaran frond *Charniodiscus* from the Mistaken Point Formation, Newfoundland. *Journal of Paleontology* 78:827–837.

Lagebro, L., Stein, M., and Peel, J. S. 2009. A new ?lamellipedian arthropod from the Early Cambrian Sirius Passet fauna of North Greenland. *Journal of Paleontology* 83:820–82.

Lamar, D. L., and Merifield, P. M. 1967. Cambrian fossils and origin of Earth-Moon system. *Geological Society of America Bulletin* 78:1359–1368.

Landing, E. 1994. Precambrian-Cambrian boundary global stratotype ratified and a new perspective of Cambrian time. *Geology* 22:179–182.

Landing, E., and Kröger, B. 2009. The oldest cephalopods from east Laurentia. *Journal of Paleontology* 83:123–127.

Landing, E., Westrop, S. R., and Adrain, J. M. 2011. The Lawsonian Stage: The *Eoconodontus notchpeakensis* (Miller, 1969) FAD and HERB carbon isotope excursion define a globally correlatable terminal Cambrian stage. *Bulletin of Geosciences* 86:621–640.

Landing, E., Westrop, S. R., and Van Aller Hernick, L. 2003. Uppermost Cambrian–Lower Ordovician faunas and Laurentian platform sequence stratigraphy, eastern New York and Vermont. *Journal of Paleontology* 77:78–98.

Lee, M. S. Y., Jago, J. B., García-Bellido, D. C., Edgecombe, G. D., Gehling, J. G., and Paterson, J. R. 2011. Modern optics in exceptionally preserved eyes of Early Cambrian arthropods from Australia. *Nature* 474:631–634.

LeFever, R. D. 1996. Sedimentology and stratigraphy of the Deadwood-Winnipeg Interval (Cambro-Ordovician), Williston Basin. *In* Longman, M. W., and Sonnenfeld, M. D., eds., *Paleozoic Systems of the Rocky Mountain Region,* The Rocky Mountain Section of the SEPM (Society for Sedimentary Geology), Tulsa, OK, p. 11–28.

Legg, D. 2013. Multi-segmented arthropods from the Middle Cambrian of British Columbia (Canada). *Journal of Paleontology* 87:493–501.

Legg, D. A., Sutton, M. D., Edgecombe, G. D., and Caron, J.-B. 2011. The origin and early evolution of the arthropods. *The Palaeontological Association 55th Annual Meeting Programme and Abstracts, The Palaeontological Association Newsletter* 78:A28.

———. 2012. Cambrian bivalved arthropod reveals origin of arthrodization. *Proceedings of the Royal Society B* 279:4699–4704.

Lehnert, O., and Kraft, P. 2006. *Manitouscolex,* a new palaeoscolecidan genus from the Lower Ordovician of Colorado. *Journal of Paleontology* 80:386–391.

Lerosey-Aubril, R., Ortega-Hernández, J., Kier, C., and Bonino, E. 2013. Occurrence of the Ordovician-type aglaspidid *Tremaglaspis* in the Cambrian Weeks Formation (Utah, USA). *Geological Magazine* 150:945–951.

Lerosey-Aubril, R., Hegna, T. A., Kier, C., Bonino, E., Habersetzer, J., and Carré, M. 2012a. Controls on gut phosphatisation: The trilobites from the Weeks Formation lagerstätte (Cambrian; Utah). *PLoS ONE* 7(3):e32934, doi:10.1371/journal.pone.0032934.

———. 2012b. Exploring the preservation of digestive structures of fossil arthropods: the phosphatisation of trilobite guts in the Weeks Formation Lagerstätte (Cambrian; Utah). *Journal of Guizhou University* 29:171–172.

Levinton, J., Dubb, L., and Wray, G. A. 2004. Simulations of evolutionary radiations and their application to understanding the probability of a Cambrian explosion. *Journal of Paleontology* 78:31–38.

Levi-Setti, R. 1993. *Trilobites* (2nd ed.). University of Chicago Press, Chicago, 342 p.

Li, C.-W., Chen, J.-Y., and Hua, T.-E. 1998. Precambrian sponges with cellular structures. *Science* 279:879–882.

Liddell, W. D., Wright, S. H., and Brett, C. E. 1997. Sequence stratigraphy and paleoecology of the Middle Cambrian Spence Shale in northern Utah and southern Idaho. *Brigham Young University Geology Studies* 42:59–78.

Lieberman, B. S. 1999a. Systematic revision of the Olenelloidea (Trilobita, Cambrian). *Bulletin of the Peabody Museum of Natural History* 45:1–150.

———. 1999b. Testing the Darwinian legacy of the Cambrian radiation using trilobite phylogeny and biogeography. *Journal of Paleontology* 73:176–181.

———. 2001. A test of whether rates of speciation were unusually high during the Cambrian radiation. *Proceedings of the Royal Society of London B* 268:1707–1714.

———. 2003a. A new soft-bodied fauna: The Pioche Formation of Nevada. *Journal of Paleontology* 77:674–690.

———. 2003b. Taking the pulse of the Cambrian radiation. *Integrative and Comparative Biology* 43:229–237.

Lieberman, B. S., and Karim, T. S. 2010. Tracing the trilobite tree from the root to the tips: A model marriage of fossils and phylogeny. *Arthropod Structure & Development* 39:111–123.

Lieberman, B. S., and Meert, J. G. 2004. Biogeography and the nature and timing of the Cambrian radiation. *In* Lipps, J. H., and Waggoner, B. M., eds., *Neoproterozoic-Cambrian Biological Revolutions,* The Paleontological Society Papers 10, p. 79–91.

Lin, J.-P. 2009. Function and hydrostatics in the telson of the Burgess Shale arthropod *Burgessia. Biology Letters* 5:376–379.

Lin, J.-P., and Yuan, J.-L. 2008. Reassessment of the mode of life of *Pagetia* Walcott, 1916 (Trilobita: Eodiscidae) based on a cluster of intact exuviae from the Kaili Formation (Cambrian) of Guizhou, China. *Lethaia* 42:67–73.

Liu, A. G., McIlroy, D., and Brasier, M. D. 2010. First evidence for locomotion in the Ediacara biota from the 565 Ma Mistaken Point Formation, Newfoundland. *Geology* 38:123–126.

Liu, A. G., McIlroy, D., Antcliffe, J. B., and Brasier, M. D. 2011. Effaced preservation in the Ediacara Biota and its implications for the early macrofossil record. *Palaeontology* 54:607–630.

Liu, J., Shu, D., Han, J., Zhang, Z., and Zhang, X. 2006. A large xenusiid lobopod with complex appendages from the Lower Cambrian Chengjiang Lagerstätte. *Acta Palaeontologica Polonica* 51:215–222.

———. 2008. Origin, diversification, and relationships of Cambrian lobopods. *Gondwana Research* 14:277–283.

Liu, J., Steiner, M., Dunlop, J. A., Keupp, H., Shu, D., Ou, Q., Han, J., Zhang, Z., and Zhang, Z. 2011. An armored Cambrian lobopodian from China with arthropod-like appendages. *Nature* 470:526–530.

Liu, Q. 2013. The first discovery of *Marrella* (Arthropoda, Marrellamorpha) from the Balang Formation (Cambrian Series 2) in Hunan, China. *Journal of Paleontology* 87:391–394.

Lochman, C. 1936. New trilobite genera from the Bonneterre Dolomite (Upper Cambrian) of Missouri. *Journal of Paleontology* 10:35–43.

———. 1948. New Cambrian trilobite genera from northwest Sonora, Mexico. *Journal of Paleontology* 22:451–464.

———. 1950. Upper Cambrian faunas of the Little Rocky Mountains, Montana. *Journal of Paleontology* 24:322–349.

Lochman, C., and Hu, C.-H. 1960. Upper Cambrian faunas from the northwest Wind River Mountains, Wyoming. Part I. *Journal of Paleontology* 34:793–834.

———. 1961. Upper Cambrian faunas from the northwest Wind River Mountains, Wyoming. Part II. *Journal of Paleontology* 35:125–146.

Lochman-Balk, C. 1970. Upper Cambrian faunal patterns on the craton. *Geological Society of America Bulletin* 81:3197–3224.

Lochman-Balk, C., and Wilson, J. L. 1958. Cambrian biostratigraphy in North America. *Journal of Paleontology* 32:312–350.

Lohmann, K. G. 1976. Lower Dresbachian (Upper Cambrian) platform to deep-shelf transition in eastern Nevada and western Utah: An evaluation through lithologic cycle correlation. *Brigham Young University Geology Studies* 23:111–122.

Love, G. D., Grosjean, E., Stalvies, C., Fike, D. A., Grotzinger, J. P., Bradley, A. S., Kelly, A. E., Bhatia, M., Meredith, W., Snape, C. E., Bowring, S. A., Condon, D. J., and Summons, R. E. 2009. Fossil steroids record the appearance of Demospongiae during the Cryogenian period. *Nature* 457:718–721.

Lowe, D. R. 1983. Restricted shallow-water sedimentation of Early Archean stromatolitic and evaporitic strata of the Strelley Pool Chert, Pilbara Block, Western Australia. *Precambrian Research* 19:239–283.

Ludvigsen, R. 1989. The Burgess Shale: Not in the shadow of the Cathedral escarpment. *Geoscience Canada* 16:51–59.

———. 1990. Reply to comments by Fritz and Aitken and McIlreath. *Geoscience Canada* 17:116–118.

Ma, X., Edgecombe, G. D., Legg, D. A., and Hou, X. 2011. Is *Diania cactiformis* the "missing link" between lobopodians and arthropods? *The Palaeontological Association 55th Annual Meeting Programme and Abstracts, The Palaeontological Association Newsletter* 78:A29.

Maas, A., and Waloszek, D. 2001. Cambrian derivatives of the early arthropod stem lineage, pentastomids, tardigrades, and lobopodians–an "Orsten" perspective. *Zoologischer Anzeiger* 240:451–459.

Maas, A., Huang, D.-Y., Chen, J.-Y., Waloszek, D., and Braun, A. 2007. Maotianshan-Shale nemathelminths–morphology, biology, and the phylogeny of Nemathelminthes. *Palaeogeography, Palaeoclimatology, Palaeoecology* 254:288–306.

Macdonald, F. A., Schmitz, M. D., Crowley, J. L., Roots, C. F., Jones, D. S., Maloof, A. C., Strauss, J. V., Cohen, P. A., Johnston, D. T., and Schrag, D. P. 2010. Calibrating the Cryogenian. *Science* 327:1241–1243.

MacNaughton, R. B., Cole, J. M., Dalrymple, R. W., Braddy, S. J., Briggs, D. E. G., and Lukie, T. D. 2002. First steps on land: Arthropod trackways in Cambrian-Ordovician eolian sandstone, southeastern Ontario, Canada. *Geology* 30:391–394.

Maletz, J., Steiner, M., and Fatka, O. 2005. Middle Cambrian pterobranchs and the question: What is a graptolite? *Lethaia* 38:73–85.

Malinky, J. M. 1988. Early Paleozoic Hyolitha from North America: Reexamination of Walcott's and Resser's type specimens. *Journal of Paleontology* 62:218–233.

Mángano, M. G. 2011. Trace-fossil assemblages in a Burgess Shale–type deposit from the Stephen Formation at Stanley Glacier, Canadian Rocky Mountains: unraveling ecologic and evolutionary controls. *In* Johnston, P. A., and Johnston, K. J., eds., International Conference on the Cambrian Explosion, Proceedings, *Palaeontographica Canadiana* 31:89–107.

Margulis, L. 1981. *Symbiosis in Cell Evolution: Life and Its Environment on the Early Earth.* W. H. Freeman, San Francisco, 419 p.

Martí Mus, M., and Bergström, J. 2007. Skeletal microstructure of helens, lateral spines of hyolithids. *Palaeontology* 50:1231–1243.

Martin, D. L. 1985. Depositional systems and ichnology of the Bright Angel Shale (Cambrian), eastern Grand Canyon, Arizona. MS thesis, Northern Arizona University, Flagstaff, 365 p.

Martin, R. E. 1999. *Taphonomy: A Process Approach.* Cambridge University Press, New York, 508 p.

Mason, J. F. 1935. Fauna of the Cambrian Cadiz Formation, Marble Mountains, California. *Bulletin of the Southern California Academy of Sciences* 34:97–114.

Maxey, G. B. 1958. Lower and Middle Cambrian stratigraphy in northern Utah and southeastern Idaho. *Geological Society of America Bulletin* 69:647–688.

Mazurek, D., and Zatón, M. 2011. Is *Nectocaris pteryx* a cephalopod? *Lethaia* 44:2–4.

McBride, D. J. 1976. Outer shelf communities and trophic groups in the Upper Cambrian of the Great Basin. *Brigham Young University Geology Studies* 23:139–152.

McCollum, M. B., and McCollum, L. B. 2011. Depositional sequences in the Laurentian Delamaran Stage, southern Great Basin, U.S.A. *In* Hollingsworth, J. S., Sundberg, F. A., and Foster, J. R., eds., Cambrian Stratigraphy and Paleontology of Northern Arizona and Southern Nevada, *Museum of Northern Arizona Bulletin* 67:154–173.

McCubbin, D. G. 1982. Barrier-island and strand-plain facies. *In* Scholle, P. A., and Spearing, D., eds., *Sandstone Depositional Environments,* American Association of Petroleum Geologists Memoir 31, American Association of Petroleum Geologists, Tulsa, OK, p. 247–279.

McGhee, G. R., Sheehan, P. M., Bottjer, D. J., and Droser, M. L. 2004. Ecological ranking of Phanerozoic biodiversity crises: ecological and taxonomic severities are decoupled. *Palaeogeography, Palaeoclimatology, Palaeoecology* 211:289–297.

McIver, J. D., and Stonedahl, G. 1993. Myrmecomorphy: morphological and behavioral mimicry of ants. *Annual Review of Entomology* 38:351–379.

McKee, E. D. 1982. The Supai Group of Grand Canyon. *United States Geological Survey Professional Paper* 1173, 504 p.

McKee, E. D., and Bigarella, J. J. 1979. Ancient sandstones considered to be eolian. *In* McKee, E. D., ed., A Study of Global Sand Seas, *United States Geological Survey Professional Paper* 1052, p. 187–238.

McKee, E. D., and Resser, C. E. 1945. *Cambrian History of the Grand Canyon Region.* Carnegie Institution of Washington Publication 563, Washington, DC, 232 p.

McMenamin, M. A. S. 1987. Lower Cambrian trilobites, zonation, and correlation of the Puerto Blanco Formation, Sonora, Mexico. *Journal of Paleontology* 61:738–749.

———. 1998. *The Garden of Ediacara.* Columbia University Press, New York, 295 p.

McMenamin, M. A. S., and Ryan, T. E. 2002. Cambrian echinoderms, brachiopods and silicified microfossils from the Peerless Formation, Colorado. Geological Society of America Abstracts with Programs.

Merriam, C. W. 1964. Cambrian rocks of the Pioche Mining District Nevada. *United States Geological Survey Professional Paper* 469, 59 p.

Middleton, G. V. 1973. Johannes Walther's Law of Correlation of Facies. *American Association of Petroleum Geologists Bulletin* 84:979–988.

Middleton, L. T., and Elliott, D. K. 2003. Tonto Group. *In* Beus, S. S., and Morales, M., eds., *Grand Canyon Geology,* Oxford University Press, New York, p. 90–106.

Middleton, L. T., Elliott, D. K., and Morales, M. 2003. Coconino Sandstone. *In* Beus, S. S., and Morales, M., eds., *Grand Canyon Geology,* Oxford University Press, New York, p. 163–179.

Miller, B. M. 1936. Cambrian trilobites from northwestern Wyoming. *Journal of Paleontology* 10:23–34.

Miller, J. F. 1980. Taxonomic revisions of some Upper Cambrian and Lower Ordovician conodonts with comments on their evolution. *The University of Kansas Paleontological Contributions* 99:1–39.

Moczydlowska, M. 2010. Life cycle of Early Cambrian microalgae from the *Skiagia*-plexus acritarchs. *Journal of Paleontology* 84:216–230.

———. 2011a. The Early Cambrian phytoplankton radiation: Acritarch evidence from the Lükati Formation, Estonia. *Palynology* 35:103–145.

———. 2011b. The Early Cambrian phytoplankton radiation at the base of Stage 3. *In* Hollingsworth, J. S., Sundberg, F. A., and Foster, J. R., eds., Cambrian Stratigraphy and Paleontology of Northern Arizona and Southern Nevada, *Museum of Northern Arizona Bulletin* 67:26–42.

Moczydlowska, M., Landing, E., Zang, W., and Palacios, T. 2011. Proterozoic phytoplankton and timing of chlorophyte algae origins. *Palaeontology* 54:721–733.

Mojzsis, S. J., Harrison, T. M., and Pidgeon, R. T. 2001. Oxygen-isotope evidence from ancient zircons for liquid water at the Earth's surface 4,300 Myr ago. *Nature* 409:178–181.

Moore, J. N. 1976. Depositional environments of the Lower Cambrian Poleta Formation and its stratigraphic equivalents, California and Nevada. *Brigham Young University Geology Studies* 23:23–38.

Moore, R. A., and Lieberman, B. S. 2009. Preservation of Early and Middle Cambrian soft-bodied arthropods from the Pioche Shale, Nevada, USA. *Palaeogeography, Palaeoclimatology, Palaeoecology* 277:57–62.

Moore, R. C., and Harrington, H. J. 1956. Conulata. *In* Moore, R. C., ed., *Treatise on Invertebrate Paleontology Part F, Coelenterata,* Geological Society of America, University of Kansas Press, p. F54–F66.

Mount, J. D. 1973. Early Cambrian fauna of the Latham Shale, southern California. *Southern California Academy of Sciences Annual Meeting Abstracts,* 1973, p. 4–6.

———. 1974a. Early Cambrian faunas from the Marble and Providence Mountains, San Bernardino County, California. *Bulletin of the Southern California Paleontological Society* 6:1–5.

———. 1974b. Early Cambrian articulate brachiopods from the Marble Mountains, San Bernardino County, California. *Bulletin of the Southern California Paleontological Society* 6:47–52.

———. 1976. Early Cambrian faunas from eastern San Bernardino County, California. *Bulletin of the Southern California Paleontological Society* 8:173–182.

———. 1980a. Characteristics of Early Cambrian faunas from eastern San Bernardino County, California. *Southern California Paleontological Society Special Publications* 2:19–29.

———. 1980b. An Early Cambrian fauna from the Carrara Formation, Emigrant Pass, Nopah Range, Inyo County, California: A preliminary note. Paleontological Tour of the Mojave Desert, California-Nevada, *Southern California Paleontological Society Special Publications* 2:78–80.

Mount, J. F., and Signor, P. W. 1985. Early Cambrian innovation in shallow subtidal environments: paleoenvironments of Early Cambrian shelly fossils. *Geology* 13:730–733.

Mutvei, H., Zhang, Y.-B., and Dunca, E. 2007. Late Cambrian plectronocerid nautiloids and their role in cephalopod evolution. *Palaeontology* 50:1327–1333.

Myrow, P. M., Taylor, J. F., Miller, J. F., Ethington, R. L., Ripperdan, R. L., and Allen, J. 2003. Fallen arches: Dispelling myths concerning Cambrian and Ordovician paleogeography of the Rocky Mountain region. *Geological Society of America Bulletin* 115:695–713.

Myrow, P. M., Taylor, J. F., Miller, J. F., Ethington, R. L., Ripperdan, R. L., and Brachle, C. M. 1999. Stratigraphy, sedimentology, and paleontology of the Cambrian-Ordovician of Colorado. *In* Lageson, D. R., Lester, A. P., and Trudgill, B. D., eds., *Colorado and Adjacent Areas,* Geological Society of America Field Guide 1, Boulder, CO, p. 157–176.

Myrow, P. M., Tice, L., Archuleta, B., Clark, B., Taylor, J. F., and Ripperdan, R. L. 2004. Flat-pebble conglomerate: Its multiple origins and relationship to metre-scale depositional cycles. *Sedimentology* 51:973–996.

Nagy, R. M., and Porter, S. M. 2005. Paleontology of the Neoproterozoic Uinta Mountain Group. *Utah Geological Association Publication* 33:49–62.

Narbonne, G. M. 2004. Modular construction of early Ediacaran complex life forms. *Science* 305:1141–1144.

Narbonne, G. M., Saylor, B. Z., and Grotzinger, J. P. 1997. The youngest Ediacaran fossils from southern Africa. *Journal of Paleontology* 71:953–967.

Nedin, C. 1999. *Anomalocaris* predation on nonmineralized and mineralized trilobites. *Geology* 27:987–990.

Nelson, C. A. 1951. Cambrian trilobites from the St. Croix valley. *Journal of Paleontology* 25:765–784.

Nelson, C. A. 1978. Late Precambrian-Early Cambrian stratigraphic and faunal succession of eastern California and the Precambrian-Cambrian boundary. *Geological Magazine* 115:121–126.

Noble, L. F. 1914. The Shinumo quadrangle, Grand Canyon district, Arizona. *United States Geological Survey Bulletin* 549:1–100.

———. 1922. A section of the Paleozoic formations of the Grand Canyon at the Bass Trail. *United States Geological Survey Professional Paper* 131, p. 23–73.

Nursall, J. R. 1959. Oxygen as a prerequisite to the origin of the Metazoa. *Nature* 183:1170–1172.

O'Brien, L. J., and Caron, J.-B. 2012. A new stalked filter-feeder from the Middle Cambrian Burgess Shale, British Columbia, Canada. *PLoS ONE* 7:1–21.

Ohno, S. 1996. The notion of the Cambrian pananimalia genome. *Proceedings of the National Academy of Sciences* 93:8475–8478.

Okulitch, V. J., and Bell, W. G. 1955. *Gallatinospongia,* a new siliceous sponge from the Upper Cambrian of Wyoming. *Journal of Paleontology* 29:460–461.

Olcott Marshall, A. N., Marshall, C. P., Moczydlowska, M., and Willman, S. 2009. Multiple lines of chemical evidence for pelagic Neoproterozoic acritarchs. *International Conference on the Cambrian Explosion Abstract Volume,* The Burgess Shale Consortium, Toronto, p. 47.

Olszewski, T. D. 2010. Diversity partitioning using Shannon's entropy and its relationship to rarefaction. *In* Alroy, J., and Hunt, G., eds., *Quantitative Methods in Paleobiology,* The Paleontological Society Papers 16, p. 95–116.

Ou, Q., Liu, J., Shu, D., Han, J., Zhang, Z., Wan, X., and Lei, Q. 2011. A rare onychophoran-like lobopodian from the Lower Cambrian Chengjiang lagerstätte, southwestern China, and its phylogenetic implications. *Journal of Paleontology* 85:587–594.

Pack, P. D., and Gayle, H. B. 1971. A new olenellid trilobite, *Biceratops nevadensis,* from the Lower Cambrian near Las Vegas, Nevada. *Journal of Paleontology* 45:893–898.

Palmer, A. R. 1951. *Pemphigaspis,* a unique Upper Cambrian trilobite. *Journal of Paleontology* 25:762–764.

———. 1960. Some aspects of the early Upper Cambrian stratigraphy of White Pine County, Nevada and vicinity. *Guidebook to the Geology of East Central Nevada,* Intermountain Association of Petroleum Geologists, Eleventh Annual Field Conference 1960, Intermountain Association of Petroleum Geologists, Salt Lake City, UT, p. 53–58.

———. 1965. Biomere: A new kind of biostratigraphic unit. *Journal of Paleontology* 39:149–153.

———. 1998a. A proposed nomenclature for stages and series for the Cambrian of Laurentia. *Canadian Journal of Earth Sciences* 35:323–328.

———. 1998b. Terminal Early Cambrian extinction of the Olenellina: documentation from the Pioche Formation, Nevada. *Journal of Paleontology* 72:650–672.

———. 2003. Sauk sequence divisions: International implications. *Geological Society of America Abstracts with Programs* 35(6):543.

Palmer, A. R., and Gatehouse, C. G. 1972. Early and Middle Cambrian trilobites from Antarctica. *United States Geological Survey Professional Paper* 456-D, p. D1–D37.

Palmer, A. R., and Halley, R. B. 1979. Physical stratigraphy and trilobite biostratigraphy of the Carrara Formation (Lower and Middle Cambrian) in the southern Great Basin. *United States Geological Survey Professional Paper* 1047, p. 1–131.

Parker, A. R. 1998. Colour in Burgess Shale animals and the effect of light on evolution in the Cambrian. *Proceedings of the Royal Society of London* B 265:967–972.

Parsley, R. L. 2012. Ontogeny, functional morphology, and comparative morphology of lower (Stage 4) and basal middle (Stage 5) Cambrian gogiids, Guizhou Province, China. *Journal of Paleontology* 86:569–583.

Parsley, R. L., and Zhao, Y. 2012. Gogiid eocrinoids of the Kaili Biota, Guizhou Province, China. *Journal of Guizhou University* 29:16–22.

Parsons-Hubbard, K. M., Powell, E. N., Raymond, A., Walker, S. E., Brett, C., Ashton-Alcox, K., Shepard, R. N., Krause, R., Deline, G. 2008. The taphonomic signature of a brine seep and the potential for Burgess Shale style preservation. *Journal of Shellfish Research* 27:227–239.

Paterson, J. R., Hughes, N. C., and Chatterton, B. D. E. 2008. Trilobite clusters: What do they tell us? A preliminary investigation. *In* Rábano, I., Gozalo, R., and García-Bellido, D., eds., *Advances in Trilobite Research,* Instituto Geológico y Minero de España, Madrid, p. 313–318.

Paterson, J. R., García-Bellido, D. C., Lee, M. S. Y., Brock, G. A., Jago, J. B., and Edgecombe, G. D. 2011. Acute vision in the giant Cambrian predator *Anomalocaris* and the origin of compound eyes. *Nature* 480:237–240.

Pedder, B. E. 2010. Large spinose acritarchs (LSAS) from Cambrian Laurentian sediments in the U.S.A. *Commission Internationale de Microflore du Paléozoique 2010 Warsaw Abstracts,* p. 54–55.

Peel, J. S. 2010a. A corset-like fossil from the Cambrian Sirius Passet Lagerstätte of North Greenland and its implications for cycloneuralian evolution. *Journal of Paleontology* 84:332–340.

———. 2010b. Articulated hyoliths and other fossils from the Sirium Passet Lagerstätte (early Cambrian) of North Greenland. *Bulletin of Geosciences* 85:385–394.

Peel, J. S., and Ineson, J. R. 2009. The Sirius Passet Lagerstätte (Early Cambrian) of North Greenland. *In* Smith, M., O'Brien, L., and Caron, J.-B., eds., *International Conference on the Cambrian Explosion Abstract Volume,* The Burgess Shale Consortium, Toronto, p. 50–51.

———. 2011a. The extent of the Sirius Passet Lagerstätte (early Cambrian) of North Greenland. *Bulletin of Geosciences* 86:535–543.

———. 2011b. The Sirius Passet Lagerstätte (early Cambrian) of North Greenland. *In* Johnston, P. A., and Johnston, K. J., eds., International Conference on the Cambrian Explosion, Proceedings, *Palaeontographica Canadiana* 31:109–118.

Pemberton, S. G., and Frey, R. W. 1982. Trace fossil nomenclature and the *Planolites-Palaeophycus* dilemma. *Journal of Paleontology* 56:843–881.

Pemberton, S. G., Spila, M., Pulham, A. J., Saunders, T., MacEachern, J. A., Robbins, D., and Sinclair, I. K. 2001. *Ichnology & Sedimentology of Shallow to Marginal Marine Systems: Ben Nevis & Avalon Reservoirs, Jeanne D'Arc Basin,* Geological Association of Canada Short Course 15, Geological Association of Canada, St. John's, NL, 340 p.

Peng, S., and Babcock, L. E. 2011. Continuing progress on chronostratigraphic subdivision of the Cambrian System. *Bulletin of Geosciences* 86:391–396.

Peng, S., Yang, X., and Hughes, N. C. 2008. The oldest known stalk-eyed trilobite, *Parablackwelderia* Kobayashi, 1942 (Damesellinae, Cambrian), and its occurrence in Shandong, China. *Journal of Paleontology* 82:842–850.

Perfetta, P. J., Shelton, K. L., and Stitt, J. H. 1999. Carbon isotope evidence for deep-water invasion at the Marjumiid-Pterocephaliid biomere boundary, Black Hills, USA: a common origin for biotic crises on Late Cambrian shelves. *Geology* 27:403–406.

Perry, J. B. 1871. Natural history of Chittenden, Franklin, Grand-Isle and Lamoille counties. *The Vermont Historical Gazetteer* 2:21–89 (A. M. Hemenway, ed.).

Peters, S. E. 2004. Relative abundance of Sepkoski's evolutionary faunas in Cambrian-Ordovician deep subtidal environments in North America. *Paleobiology* 30:543–560.

Peters, S. E., and Foote, M. 2001. Biodiversity in the Phanerozoic: A reinterpretation. *Paleobiology* 27:583–601.

Peters, S. E., and Gaines, R. R. 2012. Formation of the "Great Unconformity" as a trigger for the Cambrian explosion. *Nature* 484:363–366.

Peterson, K. J., and Butterfield, N. J. 2005. Origin of the Eumetazoa: Testing ecological predictions of molecular clocks against the Proterozoic fossil record. *Proceedings of the National Academy of Sciences* 102:9547–9552.

Peterson, K. J., Dietrich, M. R., and McPeek, M. A. 2009. MicroRNAs and metazoan macroevolution: insights into canalization, complexity, and the Cambrian explosion. *BioEssays* 31:736–747.

Peterson, K. J., McPeek, M. A., and Evans, D. A. D. 2005. Tempo and mode of early animal evolution: Inferences from rocks, Hox, and molecular clocks. *Paleobiology* 31(supp. 2):36–55.

Peterson, K. J., Cotton, J. A., Gehling, J. G., and Pisani, D. 2008. The Ediacaran emergence of bilaterians: Congruence between the genetic and the geological fossil records. *Philosophical Transactions of the Royal Society B* 363:1435–1443.

Plotnick, R. E. 1986. Taphonomy of a modern shrimp: Implications for the arthropod fossil record. *Palaios* 1:286–293.

Plotnick, R. E., Dornbos, S. Q., and Chen, J.-Y. 2010. Information landscapes and sensory ecology of the Cambrian Radiation. *Paleobiology* 36:303–317.

Porter, S. M., and Knoll, A. H. 2000. Testate amoebae in the Neoproterozoic Era: Evidence from vase-shaped microfossils in the Chuar Group, Grand Canyon. *Paleobiology* 26:360–385.

Powell, W. G., Johnston, P. A., and Collom, C. J. 2003. Geochemical evidence for oxygenated bottom waters during deposition of fossiliferous strata of the Burgess Shale Formation. *Palaeogeography, Palaeoclimatology, Palaeoecology* 201:249–268.

Pratt, B. R. 1998. Probable predation on Upper Cambrian trilobites and its relevance for the extinction of soft-bodied Burgess Shale-type animals. *Lethaia* 31:73–88.

Pratt, B. R., Spincer, B. R., Wood, R. A., and Zhuravlev, A. Yu. 2001. Ecology and evolution of Cambrian reefs. *In* Zhuravlev, A. Yu., and Riding, R., eds., *The Ecology of the Cambrian Radiation,* Columbia University Press, New York, p. 254–274.

Prave, A. R. 1991. Depositional and sequence stratigraphic framework of the Lower Cambrian Zabriskie Quartzite: Implications for regional correlations and the Early Cambrian paleogeography of the Death Valley region of California and Nevada. *Geological Society of America Bulletin* 104:505–515.

———. 2002. Life on land in the Proterozoic: Evidence from the Torridonian rocks of northwest Scotland. *Geology* 30:811–814.

Prave, A. R., and Wright, L. A. 1986. Isopach pattern of the Lower Cambrian Zabriskie Quartzite, Death Valley region, California-Nevada: How useful in tectonic reconstructions? *Geology* 14:251–254.

Prothero, D. R. 1990. *Interpreting the Stratigraphic Record.* W. H. Freeman, New York, 410 p.

Psihoyos, L. 1994. *Hunting Dinosaurs.* Random House, New York, 267 p.

Purnell, M. A. 1995a. Large eyes and vision in conodonts. *Lethaia* 28:187–188.

———. 1995b. Microwear on conodont elements and macrophagy in the first vertebrates. *Nature* 374:798–800.

Randell, R. D., Lieberman, B. S., Hasiotis, S. T., and Pope, M. C. 2005. New chancelloriids from the Early Cambrian Sekwi Formation with a comment on chancelloriid affinities. *Journal of Paleontology* 79:987–996.

Rasetti, F. 1951. Middle Cambrian stratigraphy and faunas of the Canadian Rocky Mountains. *Smithsonian Miscellaneous Collections* 116(5):1–270.

———. 1954. Internal shell structures in the Middle Cambrian gastropod *Scenella* and the problematic genus *Stenothecoides. Journal of Paleontology* 28:59–66.

———. 1961. Dresbachian and Franconian trilobites of the Conococheague and Frederick limestones of the central Appalachians. *Journal of Paleontology* 35:104–124.

Rasmussen, B. 2000. Filamentous microfossils in a 3,235-million-year-old volcanogenic massive sulphide deposit. *Nature* 405:676–679.

Ratcliff, W. C., Denison, R. F., Borrello, M., and Travisano, M. 2012. Experimental evolution of multicellularity. *Proceedings of the National Academy of Sciences* 109:1595–1600.

Rees, M. N. 1986. A fault-controlled trough through a carbonate platform: The Middle Cambrian House Range embayment. *Geological Society of America Bulletin* 97:1054–1069.

Reich, M. 2009. A critical review of the octocoralian fossil record (Cnidaria: Anthozoa). *International Conference on the Cambrian Explosion Abstract Volume,* The Burgess Shale Consortium, Toronto, p. 85.

Resser, C. E. 1928. Cambrian fossils from the Mohave Desert. *Smithsonian Miscellaneous Collections* 81(2):1–15.

———. 1931. A new Middle Cambrian merostome crustacean. *Proceedings of the United States National Museum* 79:1–4.

———. 1938. Cambrian system (restricted) of the southern Appalachians. *Geological Society of America Special Papers* 15:1–140.

———. 1939. The Spence Shale and its fauna. *Smithsonian Miscellaneous Collections* 97(12): 1–29.

Retallack, G. J. 2011. Problematic megafossils in Cambrian palaeosols of South Australia. *Palaeontology* 54:1223–1242.

———. 2012. Reply to comments on Retallack 2011: Problematic megafossils in Cambrian palaeosols of South Australia. *Palaeontology* 55:919–921.

Rezak, R. 1957. Stromatolites of the Belt Series in Glacier National Park and vicinity, Montana. *United States Geological Survey Professional Paper* 294-D, p. 127–151.

Riding, R. 2009. An atmospheric stimulus for cyanobacterial-bioinduced calcification ca. 350 million years ago? *Palaios* 24:685–696.

Riding, R., and Zhuravlev, A. Y. 1995. Structure and diversity of oldest sponge-microbe reefs: Lower Cambrian, Aldan River, Siberia. *Geology* 23:649–652.

Rigby, J. K. 1976. Some observations on occurrences of Cambrian porifera in western North America and their evolution. *Brigham Young University Geology Studies* 23:51–60.

———. 1980. The new Middle Cambrian sponge *Vauxia magna* from the Spence Shale of northern Utah and taxonomic position of the Vauxidae. *Journal of Paleontology* 54:234–240.

———. 1983. Sponges of the Middle Cambrian Marjum Limestone from the House Range and Drum Mountains of western Millard County, Utah. *Journal of Paleontology* 57:240–270.

———. 1987. Early Cambrian sponges from Vermont and Pennsylvania, the only ones described from North America. *Journal of Paleontology* 61:451–461.

Rigby, J. K., and Collins, D. 2004. Sponges of the Middle Cambrian Burgess Shale and Stephen formations, British Columbia. *Royal Ontario Museum Contributions in Science* 1:1–155.

Rigby, J. K., Gunther, L. F., and Gunther, F. 1997. The first occurrence of the Burgess Shale demosponge *Hazelia palmata* Walcot, 1920, in the Cambrian of Utah. *Journal of Paleontology* 71:994–997.

Robison, R. A. 1964a. Upper Middle Cambrian stratigraphy of western Utah. *Geological Society of America Bulletin* 75:995–1010.

———. 1964b. Late Middle Cambrian faunas from western Utah. *Journal of Paleontology* 38:510–566.

———. 1965. Middle Cambrian eocrinoids from western North America. *Journal of Paleontology* 39:355–364.

———. 1969. Annelids from the Middle Cambrian Spence Shale of Utah. *Journal of Paleontology* 43:1169–1173.

———. 1976. Middle Cambrian trilobite biostratigraphy of the Great Basin. *Brigham Young University Geology Studies* 23:93–109.

———. 1984. New occurrences of the unusual trilobite *Naraoia* from the Cambrian of Idaho and Utah. *University of Kansas Paleontological Contributions* 112:1–8.

———. 1985. Affinities of *Aysheaia* (Onychophora), with description of a new Cambrian species. *Journal of Paleontology* 59:226–235.

———. 1990. Earliest-known uniramous arthropod. *Nature* 343:163–164.

Robison, R. A., and Babcock, L. E. 2011. Systematics, paleobiology, and taphonomy of some exceptionally preserved trilobites from Cambrian Lagerstätten of Utah. *University of Kansas Paleontological Contributions* 5, 47 p.

Robison, R. A., and Campbell, D. P. 1974. A Cambrian corynexochid trilobite with only two thoracic segments. *Lethaia* 7:273–282.

Robison, R. A., and Richards, B. C. 1981. Larger bivalve arthropods from the Middle Cambrian of Utah. *The University of Kansas Paleontological Contributions* 106:1–19.

Robison, R. A., and Wiley, E. O. 1995. A new arthropod, *Meristosoma:* More fallout from the Cambrian explosion. *Journal of Paleontology* 69:447–459.

Robson, S. P., Nowlan, G. S., and Pratt, B. R. 2003. Middle to Upper Cambrian linguliformean brachiopods from the Deadwood Formation of subsurface Alberta and Saskatchewan, Canada. *Journal of Paleontology* 77:201–211.

Rose, E. C. 2006. Nonmarine aspects of the Cambrian Tonto Group of the Grand Canyon, USA, and broader implications. *Palaeoworld* 15:223–241.

———. 2011. Modification of the nomenclature and a revised depositional model for the Cambrian Tonto Group of the Grand Canyon, Arizona. *In* Hollingsworth, J. S., Sundberg, F. A., and Foster, J. R., eds., Cambrian Stratigraphy and Paleontology of Northern Arizona and Southern Nevada, *Museum of Northern Arizona Bulletin* 67:77–98.

Ross, C. P. 1959. Geology of Glacier National Park and the Flathead region, northwestern Montana. *United States Geological Survey Professional Paper* 296, p. 1–1125.

Ross, C. P., and Rezak, R. 1959. The rocks and fossils of Glacier National Park: The story of their origin and history. *United States Geological Survey Professional Paper* 294-K, p. 401–438.

Rowell, A. J. 1980. Inarticulate brachiopods of the Lower and Middle Cambrian Pioche Shale of the Pioche District, Nevada. *The University of Kansas Paleontological Contributions* 98:1–26.

Rowell, A. J., and Brady, M. J. 1976. Brachiopods and biomeres. *Brigham Young University Geology Studies* 23:165–180.

Rowland, S. M. 1984. Were there framework reefs in the Cambrian? *Geology* 12:181–183.

———. 2001. Archaeocyaths: A history of phylogenetic interpretation. *Journal of Paleontology* 75:1065–1078.

Rowland, S. M., and Gangloff, R. A. 1988. Structure and paleoecology of Lower Cambrian reefs. *Palaios* 3:111–135.

Rowland, S. M., and Korolev, S. S. 2011. How old is the top of the Tonto Group in the Grand Canyon? *In* Hollingsworth, J. S., Sundberg, F. A., and

Foster, J. R., eds., Cambrian Stratigraphy and Paleontology of Northern Arizona and Southern Nevada, *Museum of Northern Arizona Bulletin* 67:26–42.

Rudkin, D. 2009. The Mount Stephen Trilobite Beds. *In* Caron, J.-B., and Rudkin, D., eds., *A Burgess Shale Primer: History, Geology, and Research Highlights*. The Burgess Shale Consortium, Toronto, p. 91–102.

Rudkin, D. M., Young, G. A., and Nowlan, G. S. 2008. The oldest horseshoe crab: A new xiphosurid from Late Ordovician konservat-lagerstätten deposits, Manitoba, Canada. *Palaeontology* 51:1–9.

Ruedemann, R. 1931. Some new Middle Cambrian fossils from British Columbia. *Proceedings of the United States National Museum* 79:1–18.

Runkel, A. C., McKay, R. M., and Palmer, A. R. 1998. Origin of a classic cratonic sheet sandstone: Stratigraphy across the Sauk II–Sauk III boundary in the Upper Mississippi Valley. *Geological Society of America Bulletin* 110:188–210.

Runnegar, B. 1982. Oxygen requirements, biology and phylogenetic significance of the late Precambrian worm *Dickinsonia*, and the evolution of the burrowing habit. *Alcheringa* 6:223–229.

Runnegar, B. N. 1992. The tree of life. *In* Schopf, J. W., and Klein, C., eds., *The Proterozoic Biosphere: A Multidisciplinary Study*, Cambridge University Press, New York, p. 471–475.

Runnegar, B. N., and Fedonkin, M. A. 1992. Proterozoic metazoan body fossils. *In* Schopf, J. W., and Klein, C., eds., *The Proterozoic Biosphere: A Multidisciplinary Study*, Cambridge University Press, New York, p. 369–388.

Rye, R., and Holland, H. D. 2000. Life associated with a 2.76 Ga ephemeral pond? Evidence from Mount Roe #2 paleosol. *Geology* 28:483–486.

Sagan, L. 1967. On the origin of mitosing cells. *Journal of Theoretical Biology* 14:225–274.

Saltzman, M. R., Runnegar, B., and Lohmann, K. C. 1998. Carbon isotope stratigraphy of Upper Cambrian (Steptoean Stage) sequences of the eastern Great Basin: Record of a global oceanographic event. *Geological Society of America Bulletin* 110:285–297.

Saltzman, M. R., Ripperdan, R. L., Brasier, M. D., Lohmann, K. C., Robison, R. A., Chang, W. T., Peng, S., Ergaliev, E. K., and Runnegar, B.

2000. A global carbon isotope excursion (SPICE) during the Late Cambrian: relation to trilobite extinctions, organic-matter burial and sea level. *Palaeogeography, Palaeoclimatology, Palaeoecology* 162:211–223.

Sansom, R. S., Gabbott, S. E., and Purnell, M. A. 2010. Non-random decay of chordate characters causes bias in fossil interpretation. *Nature* 463:797–800.

———. 2013. Atlas of vertebrate decay: A visual and taphonomic guide to fossil interpretation. *Palaeontology* 56:457–474.

Savarese, M., Mount, J. F., Sorauf, J. E., and Bucklin, L. 1993. Paleobiologic and paleoenvironmental context of coral-bearing Early Cambrian reefs: Implications for Phanerozoic reef development. *Geology* 21:917–920.

Schoenemann, B. 2007. Trilobite eyes and a new type of neural superposition eye in an ancient system. *Palaeontographica Abteilung A* 281:63–9.

Schoenemann, B., and Clarkson, E. N. K. 2012. Colour patterns of Devonian trilobites. *In* Budil, P., and Fatka, O., eds., *The 5th Conference on Trilobites and Their Relatives Abstracts*, Czech Geological Survey and Charles University, Prague, p. 54.

Schoenemann, B., Clarkson, E. N. K., Ahlberg, P., and Álvarez, M. E. D. 2010. A tiny eye indicating a planktonic trilobite. *Palaeontology* 53:695–701.

Schoenemann, B., Liu, J.-N., Shu, D.-G., Han, J., and Zhang, Z.-F. 2009. A miniscule optimized visual system in the Lower Cambrian. *Lethaia* 42:265–273.

Scholtz, G., and Edgecombe, G. D. 2005. Heads, Hox and the phylogenetic position of trilobites. *In* Koenemann, S., and Jenner, R., eds., *Crustacea and Arthropod Relationships*, Crustacean Issues 16, CRC Press, Taylor & Francis, Boca Raton, FL, p. 139–165.

———. 2006. The evolution of arthropod heads: Reconciling morphological, developmental and palaeontological evidence. *Development Genes and Evolution* 216:395–415.

Schopf, J. W. 1993. Microfossils of the Early Archean Apex Chert: New evidence of the antiquity of life. *Science* 260:640–646.

Schopf, J. W., and Packer, B. M. 1987. Early Archean (3.3-billion to 3.5-billion-year-old) microfossils from Warrawoona Group, Australia. *Science* 237:70–73.

Schopf, J. W., Kudryavtsev, A. B., Agresti, D. G., Wdowiak, T. J., and Czaja, A. D. 2002. Laser-Raman imagery of Earth's earliest fossils. *Nature* 416:73–76.

Schwimmer, D. R. 1975. Quantitative taxonomy and biostratigraphy of Middle Cambrian trilobites from Montana and Wyoming. *Mathematical Geology* 7:149.

———. 1989. Taxonomy and biostratigraphic significance of some Middle Cambrian trilobites from the Conasauga Formation in western Georgia. *Journal of Paleontology* 63:484–494.

Schwimmer, D. R., and Montante, W. M. 2007. Exceptional fossil preservation in the Conasauga Formation, Cambrian, northwestern Georgia, USA. *Palaios* 22:360–372.

Scotese, C. R. 2009. Late Proterozoic plate tectonics and palaeogeography: A tale of two supercontinents, Rodinia and Pannotia. *Geological Society, London, Special Publications* 326:67–83.

Scotese, C. R., and McKerrow, W. S. 1990. Revised world maps and introduction. *In* McKerrow, W. S., and Scotese, C. R., eds., *Palaeozoic Palaeogeography and Biogeography*, Geological Society Memoir 12, Geological Society, London, p. 1–21.

Seilacher, A. 1999. Biomat-related lifestyles in the Precambrian. *Palaios* 14:86–93.

Seilacher, A., Grazhdankin, D., and Legouta, A. 2003. Ediacaran biota: The dawn of animal life in the shadow of giant protists. *Paleontological Research* 7:43–54.

Sepkoski, J. J. 1984. A kinetic model of Phanerozoic taxonomic diversity, III: Post-Paleozoic families and mass extinctions. *Paleobiology* 10:246–267.

———. 1988. Alpha, beta, or gamma: Where does all the diversity go? *Paleobiology* 14:221–234.

———. 1991. A model of onshore-offshore change in faunal diversity. *Paleobiology* 17:58–77.

———. 1997. Biodiversity: Past, present, and future. *Journal of Paleontology* 7:533–539.

———. 1998. Rates of speciation in the fossil record. *Philosophical Transactions of the Royal Society of London B* 353:315–326.

Shaw, A. B. 1955. Paleontology of northwestern Vermont, V: The Lower Cambrian fauna. *Journal of Paleontology* 29:775–805.

———. 1958. Stratigraphy and structure of the St. Albans area, northwestern

Vermont. *Bulletin of the Geological Society of America* 69:519–568.

Shen, Y., Buick, R., and Canfield, D. E. 2001. Isotopic evidence for microbial sulphate reduction in the early Archaean era. *Nature* 410:77–81.

Shu, D. 2005. On the Phylum Vetulicolia. *Chinese Science Bulletin* 50:2342–2354.

———. 2008. Cambrian explosion: Birth of tree of animals. *Gondwana Research* 14:219–240.

Shu, D.-G., Conway Morris, S., Han, J., Zhang, Z.-F., and Liu, J.-N. 2002. Ancestral echinoderms from the Chengjiang deposits of China. *Nature* 430:422–428.

Shu, D.-G., Conway Morris, S., Han, J., Chen, L., Zhang, X.-L., Zhang, Z.-F., Liu, H.-Q., Li, Y., and Liu, J.-N. 2001. Primitive deuterostomes from the Chengjiang Lagerstätte (Lower Cambrian, China). *Nature* 414:419–424.

Shu, D.-G., Luo, H.-L., Conway Morris, S., Zhang, X.-L., Hu, S.-X., Chen, L., Han, J., Zhu, M., Li, Y., and Chen, L.-Z. 1999. Lower Cambrian vertebrates from South China. *Nature* 402:42–46.

Shu, D.-G., Conway Morris, S., Han, J., Zhang, Z.-F., Yasui, K., Janvier, P., Chen, L. Zhang, X.-L., Liu, J.-N., Li, Y., and Liu, H.-Q. 2003. Head and backbone of the Early Cambrian vertebrate *Haikouichthys*. *Nature* 421:526–529.

Sigwart, J. D., and Sutton, M. D. 2007. Deep molluscan phylogeny: Synthesis of palaeontological and neontological data. *Proceedings of the Royal Society B* 274:2413–2419.

Skinner, E. S. 2005. Taphonomy and depostional circumstances of exceptionally preserved fossils from the Kinzers Formation (Cambrian), southeastern Pennsylvania. *Palaeogeography, Palaeoclimatology, Palaeoecology* 220:167–192.

Skovsted, C. B., and Peel, J. S. 2007. Small shelly fossils from the argillaceous facies of the Lower Cambrian Forteau Formation of western Newfoundland. *Acta Palaeontologica Polonica* 52:729–748.

———. 2010. Early Cambrian brachiopods and other shelly fossils from the basal Kinzers Formation of Pennsylvania. *Journal of Paleontology* 84:754–762.

Skovsted, C. B., Brock, G. A., Topper, T. P., Paterson, J. R., and Holmer, L. E. 2011. Scleritome construction, biofacies, biostratigraphy and systematics of the tommotiid *Eccentrotheca helenia* sp. nov. from the early Cambrian

of South Australia. *Palaeontology* 54:253–286.

Skovsted, C. B., Holmer, L. E., Larsson, C. M., Högström, A. E. S., Brock, G. A., Topper, T. P., Balthasar, U., Petterson Stolk, S., and Paterson, J. R. 2009. The scleritome of *Paterimitra:* An Early Cambrian stem group brachiopod from South Australia. *Proceedings of the Royal Society B* 276:1651–1656.

Sloss, L. L. 1963. Sequences in the cratonic interior of North America. *Geological Society of America Bulletin* 74:93–114.

Smith, A. B. 1988. Patterns of diversification and extinction in early Palaeozoic echinoderms. *Palaeontology* 31:799–828.

Smith, L. H., and Lieberman, B. S. 1999. Disparity and constraint in olenelloid trilobites and the Cambrian Radiation. *Paleobiology* 25:459–470.

Smith, M. R. 2012. Mouthparts of the Burgess Shale fossils *Odontogriphus* and *Wiwaxia:* Implications for the ancestral molluscan radula. *Proceedings of the Royal Society B* 279:4287–4295.

Smith, M. R., and Caron, J.-B. 2010. Primitive soft-bodied cephalopods from the Cambrian. *Nature* 465:469–472.

Snoke, A. W. 2003. Nelson Horatio Darton: The quintessential reconnaissance geologist of the Rocky Mountains and Great Plains. *Rocky Mountain Geology* 38:283–287.

Snow, J. K., and Wernicke, B. P. 2000. Cenozoic tectonism in the central Basin and Range: Magnitude, rate, and distribution of upper crustal strain. *American Journal of Science* 300:659–719.

Sonett, C. P., Kvale, E. P., Zakharian, A., Chan, M. A., and Demko, T. M. 1996. Late Proterozoic and Paleozoic tides, retreat of the moon, and rotation of the Earth. *Science* 273:100–104.

Sour-Tovar, F., Hagadorn, J. W., and Huitrón-Rubio, T. 2007. Ediacaran and Cambrian index fossils from Sonora, Mexico. *Palaeontology* 50:169–175.

Speyer, S. E. 1987. Comparative taphonomy and palaeoecology of trilobite lagerstätten. *Alcheringa* 11:205–232.

———. 1991. Trilobite taphonomy: A basis for comparative studies of arthropod preservation, functional anatomy, and behaviour. *In* Donovan, S. K., ed., *The Processes of Fossilization,* Columbia University Press, New York, p. 194–219.

Speyer, S. E., and Brett, C. E. 1986. Trilobite taphonomy and Middle Devonian taphofacies. *Palaios* 1:312–327.

———. 1988. Taphofacies models for epeiric sea environments: middle Paleozoic examples. *Palaeogeography, Palaeoclimatology, Palaeoecology* 63:225–262.

Sprinkle, J. 1973. Morphology and evolution of blastozoan echinoderms. *Harvard University Museum of Comparative Zoology Special Publication,* 237 p.

———. 1976. Biostratigraphy and paleoecology of Cambrian echinoderms from the Rocky Mountains. *Brigham Young University Geology Studies* 23:61–73.

Sprinkle, J., and Collins, D. 1998. Revision of *Echmatocrinus* from the Middle Cambrian Burgess Shale of British Columbia. *Lethaia* 31:269–282.

———. 2006. New eocrinoids from the Burgess Shale, southern British Columbia, Canada, and the Spence Shale, northern Utah, USA. *Canadian Journal of Earth Sciences* 43:303–322.

———. 2011. *Echmatocrinus* from the Middle Cambrian Burgess Shale: A crinoid echinoderm or an octocoral cnidarian? *In* Johnston, P. A., and Johnston, K. J., eds., International Conference on the Cambrian Explosion, Proceedings, *Palaeontographica Canadiana* 31:169–176.

St. John, J. 2007. The earliest trilobite research (antiquity to the 1820s). *In* Mikulic, D. G., Landing, E., and Kluessendorf, J., eds., Fabulous Fossils–300 Years of Worldwide Research on Trilobites, *New York State Museum Bulletin* 507:201–211.

Stanley, G. D. 2003. The evolution of modern corals and their early history. *Earth-Science Reviews* 60:195–225.

Stanley, S. M. 1973. An ecological theory for the sudden origin of multicellular life in the Late Precambrian. *Proceedings of the National Academy of Sciences* 70:1486–1489.

Stein, M. 2010. A new arthropod from the Early Cambrian of North Greenland, with a "great appendage"-like antennula. *Zoological Journal of the Linnean Society* 158:477–500.

Stein, M. 2013. Cephalic and appendage morphology of the Cambrian arthropod *Sidneyia inexpectans* Walcott, 1911. *Zoologischer Anzeiger,* dx.doi.org/10.1016/j.jcz.2013.05.001.

Steiner, M., Hu, S., Liu, J., and Keupp, H. 2012. A new species of *Hallucigenia* from the Cambrian Stage 4 Wulongqing Formation of Yunnan (South

China) and the structure of sclerites in lobopodians. *Bulletin of Geosciences* 87, doi:10.3140/bull.geosci.1280.

Stevens, C. H., and Greene, D. C. 1999. Stratigraphy, depositional history, and tectonic evolution of Paleozoic continental-margin rocks in roof pendants of the eastern Sierra Nevada, California. *Geological Society of America Bulletin* 111:919–933.

Stewart, J. H. 1970. *Upper Precambrian and Lower Cambrian Strata in the Southern Great Basin, California and Nevada.* United States Geological Survey Professional Paper 620, U.S. Government Printing Office, Washington, DC, 203 p.

Stewart, J. H., Amaya-Martinez, R., and Palmer, A. R. 2002. Neoproterozoic and Cambrian strata of Sonora, Mexico: Rodinian supercontinent to Laurentian Cordilleran margin. *Geological Society of America Special Paper* 365:5–48.

Stitt, J. H. 1971. Repeating evolutionary pattern in Late Cambrian trilobite biomeres. *Journal of Paleontology* 45:178–181.

———. 1977. Late Cambrian and earliest Ordovician trilobites, Wichita Mountains area, Oklahoma. *Oklahoma Geological Survey Bulletin* 124:1–79.

———. 1998. Trilobites from the *Cedarina dakotaensis* zone, lowermost part of the Deadwood Formation (Marjuman stage, Upper Cambrian), Black Hills, South Dakota. *Journal of Paleontology* 72:1030–1046.

Stitt, J. H., and Perfetta, P. J. 2000. Trilobites, biostratigraphy, and lithostratigraphy of the *Crepicephalus* and *Aphelaspis* zones, lower Deadwood Formation (Marjuman and Steptoean stages, Upper Cambrian), Black Hills, South Dakota. *Journal of Paleontology* 74:199–223.

Stitt, J. H., and Straatmann, W. M. 1997. Trilobites from the upper part of the Deadwood Formation (upper Franconian and Trempealeauan stages, Upper Cambrian), Black Hills, South Dakota. *Journal of Paleontology* 71:86–102.

Størmer, L., Petrunkevitch, A., and Hedgpeth, J. W. 1955. *Treatise on Invertebrate Paleontology Part P, Arthropoda 2, Chelicerata with sections on Pycnogonida and* Palaeoisopus, Geological Society of America and University of Kansas Press, p. P1–P181.

Strausfeld, N. J. 2011. Some observations on the sensory organization of the crustaceomorph *Waptia fieldensis*

Walcott. *In* Johnston, P. A., and Johnston, K. J., eds., International Conference on the Cambrian Explosion, Proceedings, *Palaeontographica Canadiana* 31:157–168.

Strother, P. K. 2005. Middle and Late Cambrian acritarchs from the Inner Clastic Belt of Laurentia. *Geological Society of America Northeastern Section Abstracts with Programs* 37(1):19.

Strother, P. K., Knoll, A. H., and Barghoorn, E. S. 1983. Microorganisms from the Late Precambrian Narssârssuk Formation, northwestern Greenland. *Palaeontology* 26:1–32.

Strother, P. K., Wood, G. D., Taylor, W. A., and Beck, J. H. 2004. Middle Cambrian cryptospores and the origin of land plants. *Memoirs of the Association of Australasian Palaeontologists* 29:99–113.

Sumrall, C. D., and Sprinkle, J. 1999. *Ponticulocarpus,* a new cornute-grade stylophoran from the Middle Cambrian Spence Shale of Utah. *Journal of Paleontology* 73:886–891.

Sundberg, F. A. 1983. *Skolithos linearis* Haldeman from the Carrara Formation (Cambrian) of California. *Journal of Paleontology* 57:145–149.

———. 1996. Morphological diversification of Ptychopariida (Trilobita) from the Marjumiid biomere (Middle and Upper Cambrian). *Paleobiology* 22:49–65.

———. 2004. Cladistic analysis of Early–Middle Cambrian kochaspid trilobites (Ptychopariida). *Journal of Paleontology* 78:920–940.

———. 2005. The Topazan Stage, a new Laurentian stage (Lincolnian Series–"Middle" Cambrian). *Journal of Paleontology* 79:73–71.

———. 2011a. Delamaran biostratigraphy and lithostratigraphy of southern Nevada. *In* Hollingsworth, J. S., Sundberg, F. A., and Foster, J. R., eds., Cambrian Stratigraphy and Paleontology of Northern Arizona and Southern Nevada, *Museum of Northern Arizona Bulletin* 67:174–185.

———. 2011b. Cambrian of Peach Springs Canyon, Hualapai Indian Reservation, Arizona. *In* Hollingsworth, J. S., Sundberg, F. A., and Foster, J. R., eds., Cambrian Stratigraphy and Paleontology of Northern Arizona and Southern Nevada, *Museum of Northern Arizona Bulletin* 67:186–191.

Sundberg, F. A., and McCollum, L. B. 2000. Ptychopariid trilobites of the Lower–Middle Cambrian boundary

interval, Pioche Shale, southeastern Nevada. *Journal of Paleontology* 74:604–630.

———. 2003a. Early and mid Cambrian trilobites from the outer-shelf deposits of Nevada and California, USA. *Palaeontology* 46:945–986.

———. 2003b. Trilobites of the lower Middle Cambrian *Poliella denticulata* biozone (new) of southeastern Nevada. *Journal of Paleontology* 77:331–359.

Surge, D. M., Savarese, M., Dodd, J. R., and Lohmann, K. C. 1997. Carbon isotopic evidence for photosynthesis in Early Cambrian oceans. *Geology* 25:503–506.

Sutton, M. D., Bassett, M. G., and Cherns, L. 2000. The type species of *Lingulella* (Cambrian Brachiopoda). *Journal of Paleontology* 74:426–438.

Suzuki, Y., and Bergström, J. 2008. Respiration in trilobites: A reevaluation. *GFF* 130:211–229.

Sweet, W. C., and Donoghue, P. C. J. 2001. Conodonts: Past, present, future. *Journal of Paleontology* 75:1174–1184.

Szaniawski, H. 2005. Cambrian chaetognaths recognized in Burgess Shale fossils. *Acta Palaeontologica Polonica* 50:1–8.

———. 2009. The earliest known venomous animals recognized among conodonts. *Acta Palaeontologica Polonica* 54:669–676.

Tanaka, G., Hou, X., Ma, X., Edgecombe, G. D., and Strausfeld, N. J. 2013. Chelicerate neural ground pattern in a Cambrian great appendage arthropod. *Nature* 502:364–367.

Taylor, M. E. 1976. Indigenous and redeposited trilobites from Late Cambrian basinal environments of central Nevada. *Journal of Paleontology* 50:668–700.

Terfelt, F., Bagnoli, G., and Stouge, S. 2011. The Cambrian–Ordovician Boundary: An anachronistic point in time. *In* Hollingsworth, J. S., Sundberg, F. A., and Foster, J. R., eds., Cambrian Stratigraphy and Paleontology of Northern Arizona and Southern Nevada, *Museum of Northern Arizona Bulletin* 67:308–309.

Tortello, M. F., and Sabattini, N. M. 2011. *Totoralia,* a new conical-shaped mollusk from the middle Cambrian of western Argentina. *Geologica Acta* 9:175–185.

Travisano, M., Mongold, J. A., Bennett, A. F., and Lenski, R. E. 1995. Experimental tests of the roles of adaptation, chance, and history in evolution. *Science* 267:87–90.

Turner, C. E. 2003. Toroweap Formation. *In* Beus, S. S., and Morales, M., eds., *Grand Canyon Geology,* Oxford University Press, New York, p. 180–195.

Turnipseed, M., Jenkins, C. D., Van Dover, C. L. 2004. Community structure in Florida Escarpment seep and Snake Pit (mid-Atlantic Ridge) vent mussel beds. *Marine Biology* 145:121–132.

Tynan, M. C. 1983. Coral-like microfossils from the Lower Cambrian of California. *Journal of Paleontology* 57:1188–1211.

Ubaghs, G. 1967. Eocrinoidea. *In* Moore, R. C., ed., *Treatise on Invertebrate Paleontology Part S, Echinodermata 1,* Geological Society of America, University of Kansas Press, Lawrence, p. S455–S495.

Ubaghs, G., and Robison, R. A. 1985. A new homoiostelean and a new eocrinoid from the Middle Cambrian of Utah. *The University of Kansas Paleontological Contributions* 115:1–24.

———. 1988. Homalozoan echinoderms of the Wheeler Formation (Middle Cambrian) of western Utah. *The University of Kansas Paleontological Contributions* 120:1–17.

Usami, Y. 2006. Theoretical study on the body form and swimming pattern of *Anomalocaris* based on hydrodynamic simulation. *Journal of Theoretical Biology* 238:11–17.

Ushatinskaya, G. T. 2001. Brachiopods. *In* Zhuravlev, A. Yu., and Riding, R., eds., *The Ecology of the Cambrian Radiation,* Columbia University Press, New York, p. 350–369.

Valentine, J. W. 1969. Patterns of taxonomic and ecological structure of the shelf benthos during Phanerozoic time. *Palaeontology* 12:684–709.

———. 1995. Why no new phyla after the Cambrian? Genome and ecospace hypotheses revisited. *Palaios* 10:190–194.

———. 2000. Two genomic paths to the evolution of complexity in bodyplans. *Paleobiology* 26:513–519.

Valentine, J. W., Jablonski, D., and Erwin, D. H. 1999. Fossils, molecules and embryos: New perspectives on the Cambrian explosion. *Development* 126:851–859.

Van Iten, H. 1992. Microstructure and growth of the conulariid test: Implications for conulariid affinities. *Palaeontology* 35:359–372.

Van Iten, H., Tollerton, V. P., Jr., Ver Straeten, C. A., De Moraes Leme, J., Guimaraes Simões, M., and Coelho Rodrigues, S. 2013. Life mode of in situ *Conularia* in a Middle Devonian epibole. *Palaeontology* 56:29–48.

Vannier, J., Chen, J.-Y., Huang, D.-Y., Charbonnier, S., and Wang, X.-Q. 2006. The Early Cambrian origin of thylacocephalan arthropods. *Acta Palaeontologica Polonica* 51:201–214.

Vannier, J., Steiner, M., Renvoisé, E., Hu, S.-X., and Casanova, J.-P. 2007. Early Cambrian origin of modern food webs: Evidence from predator arrow worms. *Proceedings of the Royal Society B* 274:627–633.

Vannier, J., Caron, J.-B., Yuan, J.-L., Briggs, D. E. G., Collins, D., Zhao, Y.-L., and Zhu, M.-Y. 2007. *Tuzoia:* Morphology and lifestyle of a large bivalved arthropod of the Cambrian seas. *Journal of Paleontology* 81:445–471.

Van Roy, P. 2006. An aglaspidid arthropod from the Upper Ordovician of Morocco with remarks on the affinities and limitations of Aglaspidida. *Transactions of the Royal Society of Edinburgh: Earth Sciences* 96:327–350.

Van Roy, P., Orr, P. J., Botting, J. P., Muir, L. A., Vinther, J., Lefebvre, B., el Hariri, K., and Briggs, D. E. G. 2010. Ordovician faunas of Burgess Shale type. *Nature* 465:215–218.

Vermeij, G. J. 1989. The origin of skeletons. *Palaios* 4:585–589.

Vermeij, G. J., and Lindberg, D. R. 2000. Delayed herbivory and the assembly of marine benthic ecosystems. *Paleobiology* 26:419–430.

Vinther, J. 2009. The canal system in sclerites of Lower Cambrian *Sinosachites* (Halkieriidae: Sachitida): Significance for the molluscan affinities of the sachitids. *Palaeontology* 52:689–712.

Vinther, J., and Nielsen, C. 2005. The Early Cambrian *Halkieria* is a mollusc. *Zoologica Scripta* 34:81–89.

Vinther, J., Smith, M. P., and Harper, D. A. T. 2011. Vetulicolians from the Lower Cambrian Sirius Passet Lagerstätte, North Greenland, and the polarity of morphological characters in basal deuterostomes. *Palaeontology* 54:711–719.

Wacey, D., Kilburn, M. R., Saunders, M., Cliff, J., and Brasier, M. D. 2011. Microfossils of sulphur-metabolizing cells in 3.4-billion-year-old rocks of Western Australia. *Nature Geoscience* 4, doi:10.1038/ngeo1238.

Waggoner, B. M., and Collins, A. G. 1995. A new chondrophorine (Cnidaria, Hydrozoa) from the Cadiz Formation (Middle Cambrian) of California. *Paläontologische Zeitschrift* 69:7–17.

———. 2004. *Reductio ad absurdum:* Testing the evolutionary relationships of Ediacaran and Paleozoic problematic fossils using molecular divergence dates. *Journal of Paleontology* 78:51–61.

Waggoner, B., and Hagadorn, J. W. 2004. An unmineralized alga from the Lower Cambrian of California, USA. *Neues Jahrbuch fur Geologie und Paläontologie Abhandlungen* 231:67–83.

———. 2005. Conical fossils from the Lower Cambrian of eastern California. *PaleoBios* 25:1–10.

Walcott, C. D. 1889. Description of new genera and species of fossils from the Middle Cambrian. *United States National Museum Proceedings* 11:441–446.

———. 1898. Cambrian Brachiopoda: *Obolus* and *Lingulella,* with description of new species. *Proceedings of the United States National Museum* 21:385–420.

———. 1908. Mount Stephen rocks and fossils. *Canadian Alpine Journal* 1(2):232–248.

———. 1910. *Olenellus* and other genera of the Mesonacidae. Cambrian Geology and Paleontology I, no. 6, *Smithsonian Miscellaneous Collections* 53(6), 442 p.

———. 1911a. Middle Cambrian annelids. Cambrian Geology and Paleontology II, *Smithsonian Miscellaneous Collections* 57:109–144.

———. 1911b. Middle Cambrian Merostomata. Cambrian Geology and Paleontology II, *Smithsonian Miscellaneous Collections* 57:17–40.

———. 1911c. Middle Cambrian holothurians and medusae. Cambrian Geology and Paleontology II, *Smithsonian Miscellaneous Collections* 57:41–68.

———. 1912. Middle Cambrian Branchiopoda, Malacostraca, Trilobita and Merostomata. Cambrian Geology and Paleontology II, *Smithsonian Miscellaneous Collections* 57:145–228.

———. 1916a. Cambrian trilobites. Cambrian Geology and Paleontology III, no. 3. *Smithsonian Miscellaneous Collections* 64(3):157–258.

———. 1916b. Cambrian trilobites. Cambrian Geology and Paleontology III, no. 5. *Smithsonian Miscellaneous Collections* 64(5):303–456.

———. 1924. Cambrian and lower Ozarkian trilobites. Cambrian Geology and Paleontology V, no. 2. *Smithsonian Miscellaneous Collections* 75(2):53–60.

———. 1925. Cambrian and Ozarkian trilobites. Cambrian Geology and Paleontology V, no. 3. *Smithsonian Miscellaneous Collections* 75(3):61–146.

Walker, J. D., and Geissman, J. W. (comps.). 2009. Geologic Time Scale. *Geological Society of America*, doi:10.1130/2009CTS004R2C.

Wallin, B. 1990. Early Cambrian coastal dunes at Vassbo, Sweden. *Geological Society of America Bulletin* 102:1535–1543.

Waloszek, D., Chen, J., Maas, A., and Wang, X. 2005. Early Cambrian arthropods: New insights into arthropod head and structural evolution. *Arthropod Structure & Development* 34:189–205.

Walter, M. R., Grotzinger, J. P., and Schopf, J. W. 1992. Proterozoic stromatolites. *In* Schopf, J. W., and Klein, C., eds., *The Proterozoic Biosphere: A Multidisciplinary Study,* Cambridge University Press, New York, p. 253–260.

Waskiewicz, R. A., Westrop, S. R., and Adrain, J. M. 2006. Cambrian (Steptoean-basal Sunwaptan) trilobite biostratigraphy of the Honey Creek Formation, Wichita Mountains, Oklahoma. *Geological Society of America Abstracts with Programs* 38(1):12.

Webster, M. 2003. Ontogeny and phylogeny of Early Cambrian olenelloid trilobites, with emphasis on the late Dyeran Biceratopsidae. PhD diss., University of California Riverside, 538 p.

———. 2007a. Ontogeny and evolution of the Early Cambrian trilobite genus *Nephrolenellus* (Olenelloidea). *Journal of Paleontology* 81:1168–1193.

———. 2007b. A Cambrian peak in morphological variation within trilobite species. *Science* 317:499–502.

———. 2011a. Trilobite biostratigraphy and sequence stratigraphy of the upper Dyeran (traditional Laurentian "Lower Cambrian") in the southern Great Basin, U.S.A. *In* Hollingsworth, J. S., Sundberg, F. A., and Foster, J. R., eds., Cambrian Stratigraphy and Paleontology of Northern Arizona and Southern Nevada, *Museum of Northern Arizona Bulletin* 67:121–154.

———. 2011b. Litho- and biostratigraphy of the Dyeran-Delamaran boundary interval at Frenchman Mountain, Nevada. *In* Hollingsworth, J. S., Sundberg, F. A., and Foster, J. R., eds., Cambrian Stratigraphy and Paleontology of Northern Arizona and Southern Nevada, *Museum of Northern Arizona Bulletin* 67:195–203.

Webster, M., Gaines, R. R., and Hughes, N. C. 2008. Microstratigraphy, trilobite biostratinomy and depositional environment of the "Lower Cambrian" Ruin Wash lagerstätte, Pioche Formation, Nevada. *Palaeogeography, Palaeoclimatology, Palaeoecology* 264:100–122.

Webster, M., Sadler, P. M., Kooser, M. A., and Fowler, E. 2003. Combining stratigraphic sections and museum collections to increase biostratigraphic resolution. *In* Harries, P. J., ed., Approaches In High-Resolution Stratigraphic Paleontology, *Topics In Geobiology* 1:95–128.

Weimer, R. J., Howard, J. D., and Lindsay, D. R. 1982. Tidal flats and associated tidal channels. *In* Scholle, P. A., and Spearing, D., eds., *Sandstone Depositional Environments,* American Association of Petroleum Geologists Memoir 31, Tulsa, OK, p. 191–245.

Westrop, S. R., and Adrain, J. M. 2007. *Bartonaspis* new genus, a trilobite species complex from the base of the Upper Cambrian Sunwaptan stage in North America. *Canadian Journal of Earth Sciences* 44:987–1003.

Westrop, S. R., and Cuggy, M. B. 1999. Comparative paleoecology of Cambrian trilobite extinctions. *Journal of Paleontology* 73:337–354.

Westrop, S. R., Palmer, A. R., and Runkel, A. 2005. A new Sunwaptan (Late Cambrian) trilobite fauna from the upper Mississippi valley. *Journal of Paleontology* 79:72–88.

White, B. 1979. Stratigraphy and microfossils of the Precambrian Altyn Formation of Glacier National Park, Montana. *In* Linn, R. M., ed., *Proceedings of the First Conference of Scientific Research in the National Parks* 2:727–735.

Whiteaves, J. F. 1892. Description of a new genus and species of phyllocarid crustacea from the Middle Cambrian of Mount Stephen, B. C. *Canadian Record of Science* 5:205–208.

Whittington, H. B. 1974. *Yohoia* Walcott and *Plenocaris* n. gen., arthropods from the Burgess Shale, Middle Cambrian, British Columbia. *Geological Survey of Canada Bulletin* 231:1–21.

———. 1975. The enigmatic animal *Opabinia regalis,* Middle Cambrian, Burgess Shale, British Columbia. *Philosophical Transactions of the Royal Society of London B* 271:1–43.

———. 1978. The lobopod animal *Aysheaia pedunculata* Walcott, Middle Cambrian, Burgess Shale, British Columbia. *Philosophical Transactions of the Royal Society of London B* 284:165–197.

———. 1980. Exoskeleton, moult stage, appendage morphology, and habits of the Middle Cambrian trilobite *Olenoides serratus. Palaeontology* 23:171–204.

———. 1989. Olenelloid trilobites: type species, functional morphology and higher classification. *Philosophical Transactions of the Royal Society of London B* 324:111–147.

———. 1995. Oryctocephalid trilobites from the Cambrian of North America. *Palaeontology* 38:543–562.

———. 2007. Reflections on the classification of the Trilobita. *In* Mikulic, D. G., Landing, E., and Kluessendorf, J., eds., Fabulous Fossils–300 Years of Worldwide Research on Trilobites, *New York State Museum Bulletin* 507:225–230.

Whittington, H. B., and Briggs, D. E. G. 1985. The largest Cambrian animal, *Anomalocaris,* Burgess Shale, British Columbia. *Philosophical Transactions of the Royal Society of London B* 309:569–609.

Whittington, H. B., Chatterton, B. D. E., Speyer, S. E., Fortey, R.A., Owens, R. M., Chang, W. T., Dean, W. T., Jell, P. A., Laurie, J. R., Palmer, A. R., Repina, L. N., Rushton, W. A., Shergold, J. H., Clarkson, E. N. K., Wilmot, N. V., and Kelly, S. R. A. 1997. *Treatise on Invertebrate Paleontology Part O, Arthropoda 1, Trilobita, Revised (Volume 1: Introduction, Order Agnostida, Order Redlichiida).* Geological Society of America and University of Kansas, Boulder, CO, and Lawrence, KS, 530 p.

Wilhelms, D. E. 1984. Moon. *In* Carr, M. H., ed., *The Geology of the Terrestrial Planets.* National Aeronautics and Space Administration SP-469, p. 107–205.

Williams, A., Rowell, A. J., Muir-Wood, H. M., and 16 others. 1965. *Treatise on Invertebrate Paleontology Part H, Brachiopoda,* Geological Society of America and University of Kansas Press, p. H1–H927.

Williams, K. E. 1997. Early Paleozoic paleogeography of Laurentia and western Gondwana: Evidence from tectonic subsidence analysis. *Geology* 25:747–750.

Willoughby, R. G., and Robison, R. A. 1979. Medusoids from the Middle Cambrian of Utah. *Journal of Paleontology* 53:494–500.

Wills, M. A. 1998. Cambrian and Recent disparity: The picture from priapulids. *Paleobiology* 24:177–199.

Wills, M. A., Briggs, D. E. G., Fortey, R. A., Wilkinson, M., and Sneath, P. H. A. 1998. An arthropod phylogeny based on fossil and recent taxa. *In* Edgecombe, G. D., ed., *Arthropod Fossils and Phylogeny,* Columbia University Press, New York, p. 33–105.

Wilson, J. L. 1951. Franconian trilobites of the central Appalachians. *Journal of Paleontology* 25:617–654.

Wilson, J. L., and Jordan, C. 1983. Middle shelf environment. *In* Scholle, P. A., Bebout, D. G., and Moore, C. H., eds., *Carbonate Depositional Environments,* American Association of Petroleum Geologists Memoir 33, American Association of Petroleum Geologists, Tulsa, OK, p. 297–343.

Witzke, B. J. 1990. Palaeoclimatic constraints for Palaeozoic palaeolatitudes of Laurentia and Euramerica. *In* McKerrow, W. S., and Scotese, C. R., eds., *Palaeozoic Palaeogeography and Biogeography,* Geological Society Memoir 12, Geological Society, London, p. 57–73.

Wood, R. 1993. Nutrients, predation, and the history of reef-building. *Palaios* 8:526–543.

———. 1995. The changing biology of reef-building. *Palaios* 10:517–529.

———. 1998. The ecological evolution of reefs. *Annual Review of Ecology and Systematics* 29:179–206.

Wood, R., Zhuravlev, A. Y., and Debrenne, F. 1992. Functional biology and ecology of Archaeocyatha. *Palaios* 7:131–156.

Wood, W. H. 1966. Facies changes in the Cambrian Muav Limestone, Arizona. *Geological Society of America Bulletin* 77:1235–1246.

Xiao, S., and Knoll, A. H. 2000. Phosphatized animal embryos from the Neoproterozoic Doushantuo Formation at Weng'An, Guizou,

south China. *Journal of Paleontology* 74:767–788.

Yang, S., Lai, X., Sheng, G., and Wang, S. 2013. Deep genetic divergence within a "living fossil" brachiopod *Lingula anatine. Journal of Paleontology* 87:902–908.

Yin, L., Zhao, Y., Peng, J., Yang, X., and Li, X. 2012. Cryptospore-like microfossils from the Cambrian Kaili Formation of eastern Guizhou Province, China. *Journal of Guizhou University* 29:23–27.

Yochelson, E. L. 1961a. The operculum and mode of life of *Hyolithes. Journal of Paleontology* 35: 152–161.

———. 1961b. Notes on the class Coniconchia. *Journal of Paleontology* 35:162–167.

———. 1967. Charles Doolittle Walcott 1850–1927. *National Academy of Sciences, Biographical Memoirs* 39:471–540.

———. 2001. *Smithsonian Institution Secretary, Charles Doolittle Walcott.* The Kent State University Press, Kent, OH, 589 p.

Yuan, X., Xiao, S., Parsley, R. L., Zhou, C., Chen, Z., and Hu, J. 2002. Towering sponges in an Early Cambrian Lagerstätte: Disparity between nonbilaterian and bilaterian epifaunal tiers at the Neoproterozoic-Cambrian transition. *Geology* 30:363–366.

Zamora, S., Gozalo, R., and Liñán, E. 2009. Middle Cambrian gogiid echinoderms from Northeast Spain: Taxonomy, palaeoecology, and palaeogeographic implications. *Acta Palaeontologica Polonica* 54:253–265.

Zamora, S., Rahman, I. A., and Smith, A. B. 2012. Plated Cambrian bilaterians reveal the earliest stages of echinoderm evolution. *PLoS ONE* 7(6):1–11, e38296.

Zhang, X., and Briggs, D. E. G. 2007. The nature and significance of the appendages of *Opabinia* from the

Middle Cambrian Burgess Shale. *Lethaia* 40:161–173.

Zhang, X., Zhao, Y., Yang, R., and Shu, D. 2002. The Burgess Shale arthropod *Mollisonia* (*M. sinica* new species): New occurrence from the Middle Cambrian Kaili fauna of southwest China. *Journal of Paleontology* 76:1106–1108.

Zhang, X.-G., Pratt, B. R., and Shen, C. 2011. Embryonic development of a middle Cambrian (500 myr old) scalidophoran worm. *Journal of Paleontology* 85:898–903.

Zhang, Z., Han, J., Wang, Y., Emig, C. C., and Shu, D. 2010. Epibionts on the lingulate brachiopod *Diandongia* from the Early Cambrian Chengjiang lagerstätte, south China. *Proceedings of the Royal Society B* 277:175–181.

Zhang, Z., Holmer, L. E., Popov, L., and Shu, D. 2011. An obolellate brachiopod with soft-part preservation from the early Cambrian Chengjiang fauna of China. *Journal of Paleontology* 85:460–463.

Zhao, F.-C., Caron, J.-B., Hu, S.-X., and Zhu, M.-Y. 2009. Quantitative analysis of taphofacies and paleocommunities in the early Cambrian Chengjiang lagerstätte. *Palaios* 24:826–839.

Zhuravlev, A. Yu. 2001. Biotic diversity and structure during the Neoproterozoic-Ordovician transition. *In* Zhuravlev, A. Yu., and Riding, R., eds., *The Ecology of the Cambrian Radiation,* Columbia University Press, New York, p. 173–199.

Zhuravlev, A. Yu., Gámez Vintaned, J. A., and Liñán, E. 2011. The Palaeoscolecida and the evolution of the Ecdysozoa. *In* Johnston, P. A., and Johnston, K. J., eds., International Conference on the Cambrian Explosion, Proceedings, *Palaeontographica Canadiana* 31:177–204.

Index

JOHN FOSTER is Director of the Museum of Moab in Moab, Utah. He worked for thirteen years as Curator of Paleontology at the Museum of Western Colorado in Grand Junction, Colorado. He has worked in Cambrian deposits in several areas of the western states, including California, Nevada, Arizona, Utah, Colorado, Idaho, and South Dakota. He is also author of *Jurassic West: The Dinosaurs of the Morrison Formation and Their World* (Indiana University Press, 2007).